Probability for Engineering, Mathematics, and Science

Probability for Engineering, Mathematics, and Science

Chris P. Tsokos

BROOKS/COLE
CENGAGE Learning™

Australia • Brazil • Japan • Korea • Mexico • Singapore • Spain • United Kingdom • United States

Probability for Engineering, Mathematics, and Science
Chris P. Tsokos

Publisher: Richard Stratton
Senior Sponsoring Editor: Molly Taylor
Assistant Editor: Shaylin Walsh
Editorial Assistant: Alexander Gontar
Associate Media Editor: Andrew Coppola
Marketing Manager: Ashley Pickering
Marketing Coordinator: Michael Ledesma
Marketing Communications Manager: Mary Anne Payumo
Content Project Managers: Susan Miscio and Alison Eigel Zade
Senior Art Director: Linda Helcher
Print Buyer: Diane Gibbons
Permissions Editor: Shalice Shah-Caldwell
Production Service and Compositor: Graphic World Inc.
Interior Designer: Pier 1 Design
Cover Designer: Rokusek Design
Cover Image Credits: Sergej Khakimullin/©Shutterstock
David Brimm/©Shutterstock
stee/©Shutterstock

© 2012 Brooks/Cole, Cengage Learning

ALL RIGHTS RESERVED. No part of this work covered by the copyright herein may be reproduced, transmitted, stored, or used in any form or by any means, graphic, electronic, or mechanical, including but not limited to photocopying, recording, scanning, digitizing, taping, Web distribution, information networks, or information storage and retrieval systems, except as permitted under Section 107 or 108 of the 1976 United States Copyright Act, without the prior written permission of the publisher.

> For product information and technology assistance, contact us at
> **Cengage Learning Customer & Sales Support, 1-800-354-9706**
> For permission to use material from this text or product, submit all requests online at **www.cengage.com/permissions**
> Further permissions questions can be emailed to
> **permissionrequest@cengage.com**

Library of Congress Control Number: 2011926935

ISBN-13: 978-1-111-43027-6

ISBN-10: 1-111-43027-6

Brooks/Cole
20 Channel Center Street
Boston, MA 02210
USA

Cengage Learning is a leading provider of customized learning solutions with office locations around the globe, including Singapore, the United Kingdom, Australia, Mexico, Brazil, and Japan. Locate your local office at **international.cengage.com/region**

Cengage Learning products are represented in Canada by Nelson Education, Ltd.

For your course and learning solutions, visit **www.cengage.com**

Purchase any of our products at your local college store or at our preferred online store **www.cengagebrain.com**

Printed in the United States of America
1 2 3 4 5 6 7 15 14 13 12 11

CONTENTS

Preface viii

1 Probability 1

Introduction 2

1.1 Definitions of Probability 3
1.2 Axiomatic Definition of Probability 5
1.3 Conditional Probability 13
1.4 Marginal Probabilities 18
1.5 Bayes's Theorem 24
1.6 Independent Events 27
1.7 Combinatorial Probability 30

Summary 38
Theoretical Exercises 39
Applied Problems 43
References 46

2 Discrete Probability Distributions 47

Introduction 48

2.1 Discrete Probability Density Function 51
2.2 Cumulative Distribution Function 56
2.3 Point Binomial Distribution 60
2.4 Binomial Probability Distribution 61
2.5 Poisson Probability Distribution 65
2.6 Hypergeometric Probability Distribution 73
2.7 Geometric Probability Distribution 77
2.8 Negative Binomial Probability Distribution 79

Summary 81
Theoretical Exercises 81
Applied Problems 84
References 88

3 Probability Distributions of Continuous Random Variables 89

Introduction 90

3.1 Continuous Random Variable and Probability Density Function 90

3.2 Cumulative Distribution Function of a Continuous Random Variable 96
3.3 Continuous Probability Distributions 100

Summary: Important Concepts 128
Theoretical Exercises 129
Applied Problems 134
References 139

4 Functions of a Random Variable 141

Introduction 142

4.1 Distribution of a Continuous Function of a Discrete Random Variable 143
4.2 Distribution of a Continuous Function of a Continuous Random Variable 148
4.3 Other Types of Derived Distribution 160

Summary 162
Theoretical Exercises 162
Applied Problems 165

5 Expected Values, Moments, and the Moment-Generating Function 167

Introduction 168

5.1 Mathematical Expectation 168
5.2 Properties of Expectation 174
5.3 Moments 182
5.4 Moment-Generating Function 191

Summary 200
Theoretical Exercises 202
Applied Problems 206

6 Two Random Variables 221

Introduction 222

6.1 Joint Probability Density Function 222
6.2 Bivariate Cumulative Distribution Function 229
6.3 Marginal Probability Distributions 237
6.4 Conditional Probability Density and Cumulative Distribution Functions 244
6.5 Independent Random Variables 251
6.6 Function of Two Random Variables 256
6.7 Expected Value and Moments 271
6.8 Conditional Expectations 293
6.9 Bivariate Normal Distribution 302

Summary 314
Theoretical Exercises 318
Applied Problems 325

7 Sequence of Random Variables 331

Introduction *332*

7.1 Multivariate Probability Density Functions 332
7.2 Multivariate Cumulative Distribution Functions 337
7.3 Marginal Probability Distributions 339
7.4 Conditional Probability Density and Cumulative Distribution Functions 342
7.5 Sequence of Independent Random Variables 345
7.6 Functions of Random Variables 348
7.7 Expected Value and Moments 354
7.8 Conditional Expectations 357

Summary *360*
Applied Problems *362*
References *368*

8 Limit Theorems 369

Introduction *369*

8.1 Chebyshev's Inequality 370
8.2 Bernoulli's Law of Large Numbers 372
8.3 Weak and Strong Laws of Large Numbers 375
8.4 Central Limit Theorem 386
8.5 DeMoivre-Laplace Theorem 389
8.6 Normal Approximation to the Poisson Distribution 399
8.7 Normal Approximation to the Gamma Distribution 401

Summary *403*
Applied Problems *404*

9 Finite Markov Chains 411

Introduction *411*

9.1 Basic Concepts in Markov Chains 412
9.2 n-Step Transition Probabilities 424
9.3 Evaluation of P^n 429
9.4 Classification of States 449

Applied Problems *467*
References *473*

List of Tables 475

Answers to Selected Exercises and Problems 539

Appendix 553

Index 583

PREFACE

We live in a global probabilistic society. Probability plays a major role in everything in our daily lives. Thus, it is fairly important that we have some basic understanding of the definitions and methods of probability. The aim of this textbook is, through the teaching of the basic concepts of probability, to enhance our understanding of this subject. In the study of probability, the most important concept—beyond the definitions of probability—is that of a random variable that characterizes the behavior of phenomena we are interested in understanding.

Probability has been defined, successively, from its earliest approach as a *measure of belief,* to the *classical* or *a priori,* to the *relative frequency* or *a posteriori,* to the most modern axiomatic definitions of probability. Several examples and real-world problems are used to enhance its understanding and usefulness. However, the primary goal of this textbook is the study of the *probability distribution* that characterizes the probabilistic behavior of a random variable and its functional forms. Probability distributions play a major role in model building in every aspect of our society. Knowledge of the probability distribution that characterizes the behavior of a random variable automatically enhances our knowledge of the phenomenon that we are investigating. Parametrically, we cannot probabilistically characterize the behavior of a random variable without knowledge of its probability distribution. This aspect is essential in performing parametric statistical analysis. The importance of this last statement has led to the probability distribution theme of this book.

Distinguishing features of the book include the following:

- Each new concept, theorem, definition, and corollary is followed by one or more examples as aids to thorough understanding.
- Important concepts and results, along with tabular display of the probability distributions, are highlighted by boxes.
- New concepts are illustrated and interpreted both graphically and analytically.
- A summary at the end of each chapter reviews the main objectives and highlights of the chapter.
- Numerous exercises appear at the end of each chapter. These are divided into two sections: theoretical and applied.

In Chapter 1, probability is defined and developed from a historical point of view, with emphasis on its classical definition as a device for evaluating probabilities and the axiomatic approach as a manipulate method.

The definition of a random variable is introduced early in the book. One-dimensional discrete and continuous random variables are then addressed. Emphasis is placed on explaining the *why,* as well as the *how,* of the probabilistic characterization of the random variable with respect to its probability distribution. Next, the concept of derived distributions is considered. Here, emphasis is placed on continuous functions of one-dimensional discrete random variables and continuous functions of continuous one-dimensional random variables. We then turn to n-dimensional continuous variates. Consideration is devoted to probability densities, cumulative distribution functions, and many of their properties. Strong emphasis is placed upon the expected value of a random variable, moments, and moment-generating functions.

Finite Markov chains play an important part in probabilistic modeling of real-world problems in health sciences, engineering, and social sciences, among others. In Chapter 9, I introduce the basic aspects of Markov chains and illustrate their usefulness with several real examples. To understand this class of stochastic processes, you are required to have some knowledge of matrices. A review of this requirement can be found in the Appendix.

Prerequisites and Use of This Book

This book is written with emphasis on the basic needs of probability for undergraduate mathematics, engineering, physical science, and biology majors. No matter what your professional goals are, the knowledge of probability enhances these goals.

This book requires knowledge of some basic concepts of calculus. It should be easily accessible if you have had the service course of one-semester undergraduate calculus. Furthermore, the book serves the needs of undergraduate students with other educational interests. This book can be used to fulfill the following needs:

- To learn the basic concepts and usefulness of probability
- To use this background for more advanced courses in probability
- To prepare for a subsequent study of statistics
- To provide the required knowledge of classical probability distributions for engineering and physical science students

This book is intended for use in a one-semester or two-quarter course on the basic aspects of probability and its applications. Chapters 1 to 6 could be covered in a one-semester course, with additional material from remaining chapters included at the instructor's discretion. It is suggested, however, that in a two-quarter course the entire book should be covered.

The examples and exercises have been selected from a variety of interests to illustrate the usefulness of probability and probability distributions. The problems at the end of each chapter are split into applied and theoretical exercises to enhance understanding of both concepts. The necessary tables appear in the Tables section.

Acknowledgments

I am grateful to my colleagues and friends for many stimulating and helpful suggestions during the writing of this book. To my undergraduate and graduate students, too many to mention, I express my thanks for reading numerous drafts of this manuscript and for their useful comments and assistance. I also thank Taysseer Sharaf and Beverly Devine-Hoffmeyer for their excellent assistance in typing and proofreading this book. Finally, I am grateful to my wife, Debbie, for her assistance in the preparation of the final form of this manuscript and for providing unbounded patience and encouragement.

Chris P. Tsokos
Tampa, Florida

SPECIAL FEATURES

- A clear and illustrative development of the fundamental aspects of the meaning and usefulness of probability.
- A focus on the importance of probability throughout the book, with real-world problems from engineering, sciences, ecology, and operations research, among other areas.
- A step-by-step approach to the use of mathematics that develops the subject, with graphical illustrations throughout the book.
- Emphasis on the development of the classical discrete and continuous probability distributions, their properties, and interdisciplinary relevance to real-life problems in our global society.
- Numerous exercises that blend theory with applications from several areas.
- A self-contained chapter on the theory and applications of discrete finite Markov chains. A step-by-step development of the theory is clearly illustrated by state and tree diagrams. The relevance of Markov chains is illustrated with several applications.
- An extensive review of mathematics that may not be covered in a basic course of calculus, such as *set theory, computational methods, binomial and multinomial theorems, matrices, Jacobians,* and *gamma and beta functions.* This review is presented in the Appendix for convenience.

Very Special Acknowledgment

"Probability is the very guide to life."

—*Cicero*

I wish to express my appreciation to the following academic educators and distinguished professors for having reviewed my probability book and expressed their opinions:

"This book is a masterful treatment of calculus-based probability for undergraduates. Its thorough coverage of families of discrete and continuous distributions with numerous applications and exercises makes it especially valuable for students in engineering, science, and mathematics."

—**Professor James J. Higgins, Department of Statistics, Kansas State University**

"An excellent book on the basic aspects of probability for undergraduate majors in mathematics, engineering, and sciences. It will certainly inspire the students to learn probability."

—**Dr. V. Lakshmikantham, Distinguished Professor, Mathematical Sciences, Florida Institute of Technology**

"The undergraduate student with basic knowledge of calculus will find this book enlightening, motivating, and [helpful in] understanding the importance of the subject of probability."

—**Dr. G. Aryal, Department of Mathematics, Computer Science, and Stats, Purdue University Calumet**

"A well-written book on calculus-based probability theory, which emphasizes the importance of probability distributions through numerous realistic exercises and applications."

—**Dr. William J. Padgett, Distinguished Professor Emeritus, Statistics, University of South Carolina, Columbia**

"A very constructive and motivating treatment of basic probability theory and its applications, with a special emphasis on the importance of discrete and continuous probability distributions for undergraduates."

—**Professor M. Sambandham, Chief Editor, Journal of Dynamic Systems**

"An excellent calculus-based introduction to the probability theory, with an abundant mix of interdisciplinary applications."

—**Professor D. Kannan, Department of Mathematics, University of Georgia**

"A very nice feature of the book is that many interesting problems in both theory and applications illustrate the usefulness, relevance, and importance of probability, especially in engineering."
—Dr. G. Okogbaa, College of Engineering,
University of Southern Florida

"I look forward to teaching from this excellent book. It is complete, from the basic theory of probability to its applicability in real-world problems."
—Dr. A. Hoare, Department of Environmental Science,
Policy, and Geography, University of South Florida, St. Petersburg

"The student will acquire a good understanding of the meaning of 'probability' and its importance in every aspect of our society—excellent book."
—Dr. L. Camara, Department of Mathematics,
Middle Georgia College

"This excellent book offers the undergraduate student a justifiable, useful, and motivating approach to learn probability and relate it to their professional interests."
—Dr. Constantin Corduneanu, Emeritus Professor, Mathematics,
University of Texas at Arlington

"This book provides an excellent treatment of probability for undergraduate students with unique emphasis on its applicability to real-world problems. Students in math, engineering, and sciences will find it very inspiring to learn the subject."
—Dr. B. Miladinovic, Center for Evidence-Based Medicine
& Health Outcomes Research, College of Medicine,
University of Southern Florida

"The scope of undergraduate probability theory is presented, emphasizing methods and unifying principles. Numerous examples are given with detailed development and solutions. Students will find this a most useful textbook and reference."
—Dr. Robert Jernigan, Professor, Mathematics and Statistics,
American University, Washington, D.C.

"This is a well-written, well-organized book for the introduction of probability for scientists and engineers, with inspiring examples. It is a valuable textbook and a reference book for undergraduate students of all majors."
—Dr. A. Vatsala, Professor, Mathematics,
University of Louisiana at Lafayette

ABOUT THE AUTHOR

Chris P. Tsokos is Distinguished University Professor of Mathematics and Statistics at the University of South Florida (USF). Dr. Tsokos received his B.S. in engineering sciences/mathematics, his M.A. in mathematics from the University of Rhode Island (URI), and his Ph.D. in statistics and probability from the University of Connecticut. Dr. Tsokos has also served on the faculty at Virginia Polytechnic Institute and State University and URI.

Dr. Tsokos's research has extended into a variety of areas, including stochastic systems, statistical models, reliability analysis, ecological systems, operations research, time series, Bayesian analysis, and mathematical and statistical modeling of global warming, with both parametric and nonparametric survival analysis. He is the author of more than 250 research publications in these areas.

For the past 4 years, Dr. Tsokos's research efforts have focused on developing probabilistic models and parametric and nonparametric statistical models for cancer data. His research aims are data-driven and are oriented specifically toward understanding the behavior of breast, lung, brain, and colon cancers. Information on the subject can be found on his website.

Dr. Tsokos has published more than 300 articles in his current area of interest. He is the author of several research monographs and books, including *Random Integral Equations with Applications to Life Sciences and Engineering, Probability Distribution: An Introduction to Probability Theory with Applications, Mainstreams of Finite Mathematics with Applications, Probability with the Essential Analysis, Applied Probability Bayesian Statistical Methods with Applications to Reliability,* and *Mathematical Statistics with Applications.*

Dr. Tsokos has been the recipient of many distinguished awards and honors, including Fellow of the American Statistical Association, USF Distinguished Scholar Award, Sigma Xi Outstanding Research Award, USF Outstanding Undergraduate Teaching Award, USF Professional Excellence Award, URI Alumni Excellence Award in Science and Technology, Pi Mu Epsilon, election to the International Statistical Institute, Sigma Pi Sigma, USF Teaching Incentive Program, and several humanitarian and philanthropic recognitions and awards.

In addition, Dr. Tsokos is a member of several academic and professional societies. He is currently serving as honorary editor, chief editor, editor, or associate editor of more than 12 academic professional journals.

Dedicated to   Debbie
Matthew
Jonathan
Maria

"The probable is what usually happens."
— Aristotle

CHAPTER 1

Probability

OBJECTIVES:

1.1 Definitions of Probability
1.2 Axiomatic Definition of Probability
1.3 Conditional Probability
1.4 Marginal Probabilities
1.5 Bayes's Theorem
1.6 Independent Events
1.7 Combinatorial Probability

Blaise Pascal (1623 – 1662)

"The excitement that a gambler feels when making a bet is equal to the amount he might win times the probability of winning it."
—B. Pascal

Blaise Pascal was born in the French province of Auvergne in 1623 and showed phenomenal ability in mathematics at an early age. At 14, he was admitted to the weekly meetings of Roberval, Mersenne, Mydorge, and other French geometricians from which the French Academy ultimately sprung. At 16, Pascal wrote an essay on conic sections; in 1641, at the age of 18, he constructed the first arithmetical machine, an instrument that he further improved 8 years later.

In 1650, suffering from ill health, Pascal decided to abandon mathematics and science and devote himself to religious contemplation. Three years later, however, he returned briefly to mathematical research. At this time, he wrote *Traite Du Triangle Arithmetique*[1], conducted several experiments on fluid pressure, and in correspondence with Fermat assisted in laying the foundation of probability theory.

Only once, in 1658, did Pascal return to mathematics. While suffering with a toothache some geometric ideas occurred to him, and his teeth suddenly ceased to ache. Regarding this

continued

as a sign of divine will, he obediently applied himself toward developing these ideas, producing in 8 days a fairly full account of the geometry of the cycloid curve and solving some problems that subsequently, when issued as challenge problems, baffled other mathematicians.

Pascal's famous *Provinciales* and *Pensées,* which are read today as models of early French literature, were written toward the close of his brief life. He died in Paris in 1662 at the age of 39. His father, Etienne Pascal (1588–1640), was also an able mathematician; it is for the father that the limaçon of Pascal is named.

---

[1]*Treatise on the Arithmetic Triangle.*

INTRODUCTION

Globally we live in a probabilistic society.
Everything in our daily lives is probabilistically driven.
—Chris Tsokos

Probability theory is concerned with the construction of mathematical models that enable us to make predictions about certain mass phenomena from the necessarily incomplete information derived from sampling techniques.

Probability theory had its beginnings with games of chance in the seventeenth century. The earliest mathematical thought regarding probability arose out of the collaboration of the eminent mathematicians Blaise Pascal and Pierre Fermat and a gambler, Chevalier de Méré. They were interested in what seemed to be contradictions between mathematical calculations and actual games of chance, such as throwing dice, tossing a coin, or spinning a roulette wheel. For example, in repeated throws of a die, it was observed that each number, 1 to 6, appeared approximately equally often; that is, each number appeared with a frequency of approximately one-sixth. However, if two dice were rolled, the sum of numbers showing on the two dice, that is, 2 to 12 did not appear equally often. It was then recognized that as the number of throws increased, the frequency of these possible results could be predicted by following some simple rules. Similar basic experiments were conducted using other games of chance, which resulted in the establishment of various basic rules of probability. Probability theory was developed solely to be applied to games of chance until the eighteenth century, when Pierre Laplace and Karl F. Gauss applied the basic probabilistic rules to other physical problems. A complete treatment of the classical approach to probability theory is given by Isaac Todhunter.[1] Some classical concepts and definitions are discussed in Section 1.2.

The modern theory of probability was initiated by the Russian mathematician A.N. Kolmogorov in 1933. He developed the subject from an axiomatic point of view using advanced mathematics (measure theory); however, it is quite possible to develop and apply the basic concepts of probability using the less difficult techniques of set theory. It is the objective of this book, using the set theory techniques, to develop probability theory by the axiomatic approach, which is the clearest and most useful way to handle both theory and applications.

In recent years, probability theory, always one of the most interesting of studies, has emerged as one of the most important mathematical disciplines. Some areas in which probability has been successfully applied are statistics, engineering, operations research, physics, medicine, business, economics, accounting, education, sociology, psychology, agriculture, meteorology, linguistics, and political science.

1.1 | Definitions of Probability

The development of the concept of probability from its earliest formulations to the modern approach can be studied by analyzing the various definitions of the term. It is the objective of this section to introduce these definitions and to illustrate both their uses and their limitations. We discuss the following four **definitions of probability**:

Definition 1.1.1

1. Probability as a measure of belief
2. Classical or a priori probability
3. Relative-frequency or a posteriori probability
4. Axiomatic definition of probability

1. **Probability as a measure of belief.** The value of the probability of an event may sometimes be assessed by one's own judgment. For example, one might say, concerning a woman's pregnancy, "I'm almost certain that she will have a boy." This statement may be interpreted as meaning that the probability of the woman having a son is being assessed at a high value or that the probability of the woman having a boy is greater than the probability of her having a girl. As a second example, if one says, "It is probable that we shall land on Mars," he or she is saying that the probability of humans landing on Mars is high. In each of these examples, the high probability assigned to the events involved was not the result of an exact method of determination but was the exercising of a belief.

    This approach to measuring probability is certainly not sufficient to solve the various physical problems of today, although it is often a useful approach for the layman. However, one must remember that the solutions of various open theoretical and experimental problems may owe their success to the mathematician's or scientist's *belief* that a solution is attainable.

2. **Classical or a priori probability.** The probability of an event S_1, written $Pr(S_1)$, may be expressed as the ratio n_{S_1}/n, where n_{S_1} is the number of ways in which S_1 may occur in the particular situation considered and n is the total number of possible outcomes in the given situation. In each case, these outcomes must be equally likely.

    As the name implies, n, the total number of outcomes of a given situation, and n_{S_1}, the number of outcomes favorable to S_1 of the same situation, are found *a priori*, that is, without actually conducting the experiment. The following examples illustrate the meaning of the classical or a priori definition of probability.

Example 1.1.1

In tossing a fair die, there are six possible outcomes of the experiment, each of which is equally likely and three of which result in an odd number. Therefore, the probability of obtaining an odd number is given by

$$\frac{n_{S1}}{n} = \frac{3}{6} = \frac{1}{2}.$$

◀

Example 1.1.2

Consider the experiment of drawing a card from an ordinary deck of 52 playing cards. There are 52 possible outcomes in the experiment, and we assume that each card has the same chance of being chosen. Because 13 of the 52 cards are diamonds, the probability that a card drawn at random will be a diamond is

$$\frac{n_{S1}}{n} = \frac{13}{52} = \frac{1}{4},$$

the probability that it will be a picture card is

$$\frac{n_{S1}}{n} = \frac{12}{52} = \frac{3}{13},$$

and the probability that it will be an ace is

$$\frac{n_{S1}}{n} = \frac{4}{52} = \frac{1}{13}.$$

It is important to note that all possible outcomes in a given situation must be equally likely: If one tossed a coin twice, desiring to obtain "tails" on both occasions, one could reason, according to the classical definitions of probability, that there are three possible outcomes to the experiment; that is, two heads (HH), one head and one tail (HT), and two tails (TT). Thus, the required probability would be

$$\frac{n_{S1}}{n} = \frac{1}{3}.$$

◀

However, this reasoning is not correct because the preceding outcomes are not equally likely. There are *four* equally likely outcomes: HH, HT, TH, and TT. The correct probability is therefore $1/4$.

If in Example 1.1.1 the die is not fair or, in other words, the appearance of each of the six faces is not equally likely, the classical definition of probability is not applicable. Moreover, the classical definition of probability cannot measure the probability of events when the total number of possible outcomes in a physical phenomenon is infinite. For this reason, we turn to a third definition of probability.

3. **Relative-frequency or a posteriori probability.** This definition was developed as the result of the work of R. Von Mises[2] and is stated as

$$\lim_{n \to \infty} \frac{n_{S1}}{n} = p, \tag{1.1.1}$$

where n is the total number of identical trials in a given problem, n_S is the number of occurrences of the event S_1, and p is the probability of the event.[1] (*Identical* simply means that each trial is conducted under identical conditions.)

The concept of the limit in Equation (1.1.1) suggests that its applicability to various scientific problems cannot be extensive because, in such problems, n is usually finite and the limit of the ratio n_{S1}/n cannot be taken. Despite this theoretical limitation, the relative-frequency definition of probability is perhaps one of the most popular probability definitions among scientists: From a practical point of view, the limitation means that the accuracy of measuring the probability of a "true state of nature," p, increases as the number of trials becomes very large. For example, in tossing a coin that need not be fair, we are interested in obtaining the probability p of getting a head, which is given by

$$\lim_{n \to \infty} \frac{n_H}{n}.$$

This does not offer us a specific answer, but it does give us an approximate value of p for a particular number of trials, and this approximation improves as n increases. If we were to use the a priori definition of probability, we would have estimated p to be equal to 1/2, but this may be only a lucky guess because we have no evidence that the coin is fair. In tossing a symmetrical coin, the probability of obtaining a head approaches that of obtaining a tail as $n \to 2000$ trials.

As the name *a posteriori* implies, p is calculated *after* the experiment has been performed, whereas in the classical definition of probability, it is determined *before* conducting the experiment.

4. **Axiomatic definition of probability.** This definition results from the 1933 studies of A. N. Kolmogorov[3] and is the most important definition of probability because it eliminates most of the difficulties that are encountered in using the other definitions and provides a solid basis for further study of probability theory. Although the original development of the axiomatic approach depended on advanced mathematics (measure theory), we introduce it here by means of finite set operations. Section 1.2 is devoted to the development of the axiomatic definition of probability.

1.2 Axiomatic Definition of Probability

We begin the formulation of the axiomatic definition of probability by defining certain essential basic concepts.

By an **experiment**, we mean a specific procedure that we follow and at the completion of which we observe certain results. Each possible outcome of the experiment is represented by a **sample point**. The set of all possible outcomes of an experiment is represented by a **sample space** denoted by S. For example, if our experiment is to toss a fair coin once, the sample space consists of two possible outcomes, a head and a tail, so that $S = \{H, T\}$. If we were to roll a die, the total number of possible outcomes of this experiment would be six; thus,

$$S = \{x : x = 1, 2, 3, 4, 5, 6\}$$

If the experiment were to flip a coin twice, the sample space would consist of four sample points:

$$S = \{(H,H), (H,T), (T,H), (T,T)\}$$

We define an **event** as a set of sample points with some specified property. Thus, in the example of tossing a coin twice, the event S_1 of getting heads on both the first and the second tosses is a subset of S, and we write $S_1 = \{(H,H)\} \subset S$. An event that consists of only one sample point is called a **simple event** or an **elementary event**; thus, $S_1 = \{(H,H)\}$ is a simple event. An event that consists of more than one sample point is called a **compound event**. In the experiment of rolling a die, the event S_2 of getting an odd number consists of outcomes 1, 3, and 5. Therefore, $S_2 = \{x : x = 1, 3, 5\}$. We refer to the sample space S as a universal set; all other sets are considered subsets (events) of S.

For our purposes, consider the sample space S from either of two points of view: *discrete* or *continuous*.

■ Definition 1.2.1

A sample space S is **discrete** if it contains a finite number of points or if it contains an infinite number of points that can be put into a one-to-one correspondence with the positive integers.

Example 1.2.1

The sample space of positive integers $1, 2, 3, \ldots, 100$ —that is, $S = \{x : x = 1, 2, 3, \ldots, 100\}$— contains a finite number of points, so it is a discrete sample space.

■ Definition 1.2.2

A sample space S is **continuous** if it contains a continuum of points.

◀

Example 1.2.2

Consider an experiment in which we wish to observe the lifetime of a certain type of lightbulb. The outcome of this experiment forms a continuous sample, that is,

$$S = \{t : 0 \le t < \infty\}.$$

We now state some basic definitions of those sets that are subsets of S.

◀

■ Definition 1.2.3

Two events, S_1 and S_2, of the sample space S are said to be **equal** if every sample point of S_1 is also a sample point of S_2 and every sample point of S_2 is also a sample point of S_1. We denote this by $S_1 = S_2$.

■ Definition 1.2.4

If a subset S_1 of S contains no sample points, it is called the **impossible event** and is denoted by $\varnothing$, which designates the empty set.

■ Definition 1.2.5

The **complement** of an event S_1 of the sample space S is the set of sample points that are in S but not in S_1; we denote this by $S - S_1$ or $\overline{S_1}$.

Definition 1.2.6

If S_1 and S_2 are two events of the sample space S, then the event that consists of all sample points in S_1, S_2, or both is called the **union** of S_1 and S_2 and is denoted by $S_1 \cup S_2$.

Definition 1.2.7

If S_1 and S_2 are two events of the sample space S, then the event that consists of all points in both S_1 and S_2 is called the **intersection** of S_1 and S_2 and is denoted by $S_1 \cap S_2$ or $S_1 S_2$.

Definition 1.2.8

Two events, S_1 and S_2, of the sample space S are said to be **mutually exclusive events** or **disjoint events** if $S_1 \cap S_2 = \emptyset$.

Example 1.2.3

Let the sample space S be defined as $S = \{t : 0 \leq t < \infty\}$, and let two events of S be given as

$$S_1 = \{t : 25 < t \leq 100\}$$

and

$$S_2 = \{t : 60 < t \leq 140\}.$$

Then,

$$S_1 \cup S_2 = \{t : 25 < t \leq 140\},$$

$$S_1 \cap S_2 = \{t : 60 < t \leq 100\},$$

$$\overline{S_1 \cap S_2} = \{t : 0 \leq t \leq 60 \text{ or } 100 < t < \infty\},$$

$$\overline{S_1 \cup S_2} = \{t : 0 < t \leq 25 \text{ or } 140 < t < \infty\},$$

$$\overline{S_1} \cup \overline{S_2} = \{t : 0 < t \leq 60 \text{ or } 100 < t < \infty\},$$

and

$$\overline{S_1} \cup \overline{S_2} = \{t : 0 \leq t \leq 25 \text{ or } 140 < t < \infty\}.$$

◄

Note that $\overline{S_1 \cap S_2} = \overline{S_1} \cup \overline{S_2}$ and $\overline{S_1 \cup S_2} = \overline{S_1} \cap \overline{S_2}$. These two relationships are called **De Morgan's laws**.

If a sample space is finite or countably infinite, then every subset of the sample space S is an event. If the sample space consists of n sample points, then there are 2^n possible events in S. On the other hand, if the set is uncountable, then certain subsets of S cannot be events. The reason for this is beyond the scope of this book.

We now state the axiomatic definition of probability: For every event S_i, $i = 1, 2, 3, \ldots, n, \ldots$ of the sample space S, there is a number, denoted by $Pr(S_i)$, that satisfies the following axioms:

Axiom 1.2.1

$$0 \leq Pr(S_i); \ 0 \leq Pr(S_i) \leq 1$$

Axiom 1.2.2:

$$Pr(S) = 1$$

Axiom 1.2.3

If $S_1, S_2, \ldots, S_n, \ldots$ is a sequence of mutually exclusive events, that is,

$$S_i \cap S_j = \emptyset \text{ for } i \neq j = 1, 2, 3, \ldots n, \ldots,$$

then

$$Pr(S_1 \cup S_2 \cup S_3 \cup \ldots \cup S_n \cup \ldots) = Pr(S_1) + Pr(S_2) + \ldots + Pr(S_n) + \ldots$$

or

$$Pr\left(\bigcup_{i=1}^{\infty} S_i\right) = \sum_{i=1}^{\infty} Pr(S_i).$$

$Pr(S_i)$ is called the **probability of the event** S_i. If we can determine the probability of every event of the sample space S, then we say that S is a **probability space**. It should be emphasized that the axiomatic approach to probability in no way assigns a numerical value to the probability of any event. It is used primarily to manipulate probabilities of compound events so that they may be expressed in terms of probabilities of simple events, which in turn may be evaluated by one of the other definitions of probability already mentioned, most usually by the classical definition. For example, suppose that two balanced dice are tossed once and we are interested in the following two probabilities:

1. The probability that the sum showing on the two dice is less than six.

2. The probability that the absolute difference of the values on the dice is greater than or equal to four.

For the first probability, we have

$$Pr(\text{Sum} < 6) = Pr\left[(\text{Sum} = 2) \cup (\text{Sum} = 3) \cup (\text{Sum} = 4) \cup (\text{Sum} = 5)\right]$$
$$= Pr(\text{Sum} = 2) + Pr(\text{Sum} = 3) + Pr(\text{Sum} = 4) + Pr(\text{Sum} = 5),$$

using Axiom 1.2.3.

Up to this point, we have used the axiomatic approach to probability, which does not suffice to evaluate the probabilities $Pr(\text{Sum} = 2)$, $Pr(\text{Sum} = 3)$, and so on. Therefore, we must now use the classical definition of probability. The event $(\text{Sum} = 2)$ may occur in only one way out of 36 possible outcomes; thus, $Pr(\text{Sum} = 2) = 1/36$. The event $(\text{Sum} = 3)$ may occur in either of two ways; thus, $Pr(\text{Sum} = 3) = 2/36$. Similarly, we obtain $Pr(\text{Sum} = 4) = 3/36$ and $Pr(\text{Sum} = 5) = 4/36$. Therefore,

$$Pr(\text{Sum} < 6) = \frac{1}{36} + \frac{2}{36} + \frac{3}{36} + \frac{4}{36} = \frac{5}{18}.$$

In the second probability, we let E represent the absolute difference showing on the dice. Thus, we want to determine $Pr(E \geq 4)$, which may be written as

$$Pr[(E=4) \cup (E=5)]$$

or (using Axiom 1.2.3),

$$Pr(E=4) + Pr(E=5).$$

Again, we can go no further using the axiomatic approach; we must shift to the classical approach to evaluate $Pr(E=4)$ and $Pr(E=5)$. The event $E=4$ may occur in one of four ways, as shown in the following diagram:

Die 1   1   2   5   6
Die 2   5   6   1   2

Therefore, $Pr(E=4) = 4/36$, because the total number of possible outcomes is 36. In a similar way, it can be determined that $Pr(E=5) = 2/36$.

Thus,

$$Pr(E \geq 4) = \frac{4}{36} + \frac{2}{36} = \frac{1}{6}.$$

Now we establish some elementary theorems of probability.

Theorem 1.2.1

If S_1 and S_2 are any two events of the sample space S, such that $S_1 \subset S_2$, then $Pr(S_1) \leq Pr(S_2)$.

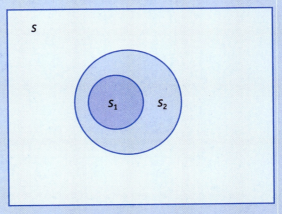

Figure 1.2.1 Venn Diagram

Proof From Figure 1.2.1, we can see that $S_2 = S_1 \cup (S_2 \cap \overline{S_1})$. Because S_1 and $S_2 \cap \overline{S_1}$ are mutually exclusive events, we apply Axiom 1.2.3 and have

$$Pr(S_2) = Pr(S_1 \cup (S_2 \cap \overline{S_1}))$$
$$= Pr(S_1) + Pr(S_2 \cap \overline{S_1}).$$

However, from Axiom 1.2.1,

$$Pr(S_1) \leq Pr(S_1) + Pr(S_2 \cap \overline{S_1}),$$

which implies that

$$Pr(S_1) \leq Pr(S_2).$$

Theorem 1.2.2

If S_k is any event of the sample space S, then $Pr(S_k) \leq 1$.

Proof Because $S_k \subset S$ and from Theorem 1.2.1, we can write

$$Pr(S_k) \leq Pr(S).$$

Applying Axiom 1.2.2, we have

$$Pr(S_k) \leq 1.$$

Theorem 1.2.3

If S_k is any event of the sample space S, then $Pr(\overline{S_k}) = 1 - Pr(S_k)$.

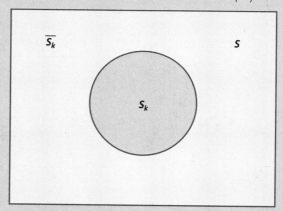

Figure 1.2.2 Venn Diagram

Proof From Figure 1.2.2,

$$S = S_k \cup \overline{S_k},$$

because S_k and $\overline{S_k}$ are mutually exclusive. Applying Axiom 1.2.3, we have

$$Pr(S) = Pr(S_k) + Pr(\overline{S_k}).$$

Hence, from Axiom 1.2.2,

$$1 = Pr(S_k) + Pr(\overline{S_k}).$$

$$Pr(\overline{S_k}) = 1 - Pr(S_k).$$

Theorem 1.2.4

If $\emptyset$ is the impossible event of the sample space S, then $Pr(\emptyset) = 0$.

Proof Because we know that $\emptyset \cup S = S$, where $\emptyset$ and S are mutually exclusive, we have, by applying Axioms 1.2.2 and 1.2.3,

$$Pr(\emptyset \cup S) = Pr(S)$$

or

$$Pr(\emptyset) + Pr(S) = Pr(S),$$

which implies

$$Pr(\emptyset) = 0.$$

Theorem 1.2.5

If S_1 and S_2 are any two events of the sample space S, then

$$Pr(S_1 \cup S_2) = Pr(S_1) + Pr(S_2) - Pr(S_1 \cap S_2).$$

Proof From Figure 1.2.3, we see that $S_1 \cup S_2 = (S_1 \cap \overline{S_2}) \cup S_2$ because $S_1 \cap \overline{S_2}$ and S_2 are mutually exclusive events. Applying Axiom 1.2.3, we have

$$Pr(S_1 \cup S_2) = Pr(S_1 \cap \overline{S_2}) + Pr(S_2) \tag{1.2.1}$$

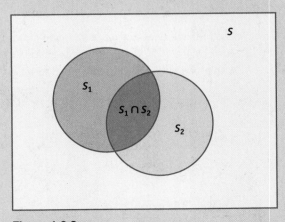

Figure 1.2.3

We can also write the event S_1 as a pair of mutually exclusive events, that is,

$$S_1 = (S_1 \cap S_2) \cup (S_1 \cap \overline{S_2})$$

or

$$Pr(S_1) = Pr(S_1 \cap S_2) + Pr(S_1 \cap \overline{S_2}). \qquad (1.2.2)$$

Solving Equation (1.2.2) for $Pr(S_1 \cap \overline{S_2})$ and substituting it into Equation (1.2.1), we have

$$Pr(S_1 \cup S_2) = Pr(S_1) + Pr(S_2) - Pr(S_1 \cap S_2).$$

If the events S_1 and S_2 are mutually exclusive, the, according to Axiom 1.2.3, the last expression becomes

$$Pr(S_1 \cup S_2) = Pr(S_1) + Pr(S_2). \qquad ◄$$

In certain cases in which the sample space is infinite, it is sometimes useful to choose the sample points, say, n of them, in such a way that the probability of each point is equal to $1/n$. This technique often facilitates the process of evaluating the probabilities of compound events, as shown in the following example.

Example 1.2.4

Consider the experiment of drawing cards from a well-shuffled deck. The space consists of 52 sample points corresponding to the 52 possible cards that might be drawn. We would like to find the probability of the following events:

a. S_1 = the occurrence of a diamond in a single draw

b. S_2 = the occurrence of a picture card in a single draw

c. The event of a diamond or a picture card, or a *picture diamond*, that is, $S_1 \cup S_2$

d. The event of not obtaining event $S_1 \cup S_2$, that is, $\overline{S_1 \cup S_2}$

Solution:

a. The probability of any one card being drawn is the same as that of any other card, that is, $Pr(E_1) = Pr(E_2) = \ldots = Pr(E_{52}) = 1/52$. The event S_1 consists of 13 sample points, each with probability of occurrence equal to $1/52$; therefore,

$$Pr(S_1) = \sum_{i=1}^{13} Pr(E_i) = 13 \frac{1}{52} = \frac{1}{4}.$$

b. Similarly,

$$Pr(S_2) = \sum_{i=1}^{12} Pr(S_i) = \frac{3}{13}.$$

c. Applying Theorem 1.2.5,

$$Pr(S_1 \cup S_2) = Pr(S_1) + Pr(S_2) - Pr(S_1 \cap S_2),$$

where

$$Pr(S_1 \cap S_2) = Pr(\text{picture diamond}) = \frac{3}{52}$$

(the number of sample points with the picture diamond attribute over the total number of sample points in the sample space). Hence,

$$Pr(S_1 \cup S_2) = \frac{13}{52} + \frac{12}{52} - \frac{3}{52} = \frac{11}{26}.$$

d. We know from Theorem 1.2.3 that

$$Pr(\overline{S_1 \cup S_2}) = 1 - Pr(S_1 \cup S_2).$$

Therefore,

$$Pr(\overline{S_1 \cup S_2}) = 1 - \frac{22}{52} = \frac{15}{26}.$$

◀

Theorem 1.2.1 is verified from Example 1.2.4(a) and (b), that is, $Pr(S_2) Pr(S_1)$.

A generalization of Theorem 1.2.5, known as the general law of total probability, is attributed to Poincaré[4] and is given by the following theorem.

Theorem 1.2.6

If $S_1, S_2, \ldots, S_n$ is a sequence of events defined on the sample space S, then

$$Pr\left(\bigcup_{i=1}^{n} S_i\right) = \sum_{i=1}^{n} Pr(S_i) - \sum_{\substack{i,j=1 \\ i<j}}^{n} Pr(S_i \cap S_j) + \sum_{\substack{i,j,k \\ i<j<k}}^{n} Pr(S_i \cap S_j \cap S_k)$$

$$+ \ldots + (-1)^{n+1} Pr(S_1 \cap S_2 \cap \ldots \cap S_n).$$

The proof of this theorem is by mathematical induction.

The general law of total probability for mutually exclusive events reduces to

$$Pr\left(\bigcup_{i=1}^{n} S_i\right) = \sum_{i=1}^{n} Pr(S_i).$$

1.3 Conditional Probability

Let S be the sample space; let S_1 and S_2 be events of S. We define the **conditional probability** of the event S_i —given that event S_2 has occurred, as denoted by $Pr(S_1|S_2)$ —as the probability of $S_1 \cap S_2$ divided by the probability of S_2. That is,

$$Pr(S_1|S_2) = \frac{Pr(S_1 \cap S_2)}{Pr(S_2)}, \text{ where } Pr(S_2) > 0. \quad (1.3.1)$$

Similarly,

$$Pr(S_2|S_1) = \frac{Pr(S_1 \cap S_2)}{Pr(S_1)}, \text{ where } Pr(S_1) > 0. \tag{1.3.2}$$

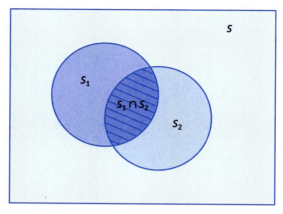

Figure 1.3.1 Venn Diagram

One can illustrate the event $S_1 \cap S_2$, as shown by the Venn diagram in Figure 1.3.1. If, for example, the sample space S consists of n elementary events, n_1 of which have the attribute associated with S_1, n_2 of which have the attribute associated with S_2, and n_{12} of which have the attribute associated with S_1 and S_2, then

$$Pr(S_1) = \frac{n_1}{n}, \quad Pr(S_2) = \frac{n_2}{n}, \quad Pr(S_1 \cap S_2) = \frac{n_{12}}{n}$$

and

$$Pr(S_1|S_2) = \frac{\frac{n_{12}}{n}}{\frac{n_2}{n}} = \frac{n_{12}}{n_2}$$

or

$$Pr(S_2|S_1) = \frac{\frac{n_{12}}{n}}{\frac{n_1}{n}} = \frac{n_{12}}{n_1}.$$

Example 1.3.1

Suppose that in a deck of cards we have removed the 3 picture diamonds (jack, queen, and king). From the remaining 49 cards, we select a card at random. Let the event S_1 denote that the chosen card is a picture card, and let the event S_2 denote that the chosen card is a spade. Event S_1 consists of 9 sample points, which implies that $Pr(S_1) = 9/49$. The event $S_1 \cap S_2$ represents the appearance of a picture spade. The number of sample points in $S_1 \cap S_2$ is therefore 3, so $Pr(S_1 \cap S_2) = 3/49$. The event S_2 consists of 13 sample points, each with the probability of $1/49$ of occurring. Thus, $Pr(S_2) = 13/49$. Now suppose that we have been told that the card drawn is a spade and we are interested in finding the probability that it is also a picture card.

Solution: We write

$$Pr(S_1|S_2) = \frac{Pr(S_1 \cap S_2)}{Pr(S_2)} = \frac{\frac{3}{49}}{\frac{13}{49}} = \frac{3}{13}.$$

Furthermore, the probability of a spade, given that the chosen card is a picture card, is shown by

$$Pr(S_2|S_1) = \frac{Pr(S_1 \cap S_2)}{Pr(S_1)} = \frac{\frac{3}{49}}{\frac{9}{49}} = \frac{1}{3}.$$

◀

Example 1.3.2

Consider the experiment of rolling a pair of fair dice. Suppose that we observed that the sum of the two dice is seven, and we want to find the probability that at least one of the dice shows a three.

Solution: The sample space in the experiment consists of 36 sample points. We must determine the number of sample points in the sample space that constitute the event S_1 (the sum of seven),

$$S_1 = \{(1,6),(2,5),(3,4),(4,3),(5,2),(6,1)\},$$

and the event S_2 (a three appears on at least one die),

$$S_2 = \{(3,1),(3,2),(3,3),(3,4),(3,5),(3,6),(6,3),(5,3),(4,3),(2,3),(1,3)\}.$$

That is, there are 6 sample points in S_1 and 11 sample points in S_2. Hence,

$$Pr(S_1) = \frac{6}{36}, \quad Pr(S_1 \cap S_2) = \frac{2}{36}$$

and

$$Pr(S_2|S_1) = \frac{\frac{2}{36}}{\frac{6}{36}} = \frac{1}{3}.$$

Equations (1.3.1) and (1.3.2) can be written as

$$Pr(S_1 \cap S_2) = Pr(S_1) Pr(S_2|S_1) = Pr(S_2) Pr(S_1|S_2); \qquad (1.3.3)$$

that is, the probability of the intersection of S_1 and S_2 equals the product of the probability of S_1 and the conditional probability of S_2, given that S_1 has occurred, or the probability of event S_2 times the conditional probability of S_1, given that S_2 has occurred. Furthermore,

$$Pr(S_2|S_1) = \frac{\frac{2}{36}}{\frac{11}{36}} = \frac{2}{11},$$

the probability of obtaining a sum of seven, given that a three has appeared on at least one die.

◀

Example 1.3.3

In Example 1.3.1, we were interested in obtaining the probability that a picture spade had occurred. If we are given enough information so that the conditional probability $Pr(S_2|S_1)$ can be calculated, then, using Equation (1.3.3), we have the desired probability:

$$Pr(S_1 \cap S_2) = Pr(S_1)Pr(S_2|S_1) = \frac{9}{49} \times \frac{1}{3} = \frac{3}{49}.$$

◀

The concept of conditional probability can be extended to more than two events: if S_1, S_2, and S_3 are events of the sample space S, then

$$Pr(S_3|S_1 \cap S_2) = \frac{Pr(S_1 \cap S_2 \cap S_3)}{Pr(S_1 \cap S_2)}, \; Pr(S_1 \cap S_2) > 0,$$

or

$$Pr(S_1 \cap S_2 \cap S_3) = Pr(S_1 \cap S_2)Pr(S_3|S_1 \cap S_2). \tag{1.3.4}$$

But

$$Pr(S_1 \cap S_2) = Pr(S_1)Pr(S_2|S_1)$$

and Equation (1.3.4) becomes

$$Pr(S_1 \cap S_2 \cap S_3) = Pr(S_1)Pr(S_2|S_1)Pr(S_3|S_1 \cap S_2). \tag{1.3.5}$$

That is, the probability of the intersection of the three events equals the probability of S_1 times the conditional probability of S_2, given that S_1 has occurred, times the probability of event S_3, given the joint occurrence of events S_1 and S_2.

A generalization of Equation (1.3.5) to n events, which is known as the **general law of compound probability**, is given by Theorem 1.3.1.

Theorem 1.3.1

Let $S_1, S_2, S_3, \ldots, S_n$ be events of the sample space S such that

$$Pr(S_1 \cap S_2 \cap S_3 \cap \ldots \cap S_j) > 0 \text{ for } 1 \leq j \leq n-1.$$

Then,

$$Pr(S_1 \cap S_2 \cap \ldots \cap S_n) = Pr(S_1)Pr(S_2|S_1)Pr(S_3|S_1 \cap S_2) \ldots Pr(S_n|S_1 \cap S_2 \cap \ldots \cap S_{n-1}).$$

The proof is left as an exercise.

Example 1.3.4

A lot in a certain warehouse contains 40 electrical generators, of which 6 are known to be defective. On a particular day, 5 generators will be used. We are interested in the probability that all 5 will be operable.

Solution: Let S_1, S_2, S_3, S_4, and S_5 be the events that the first, second, third, fourth, and fifth generators, respectively, are operable. Then, according to Theorem 1.3.1, we have

$$Pr(S_1 \cap S_2 \cap S_3 \cap S_4 \cap S_5) = Pr(S_1) Pr(S_2|S_1) Pr(S_3|S_1 \cap S_2) \cdots$$
$$Pr(S_5|S_1 \cap S_2 \cap S_3 \cap S_4)$$
$$= \left(\frac{34}{40}\right)\left(\frac{33}{39}\right)\left(\frac{32}{38}\right)\left(\frac{31}{37}\right)\left(\frac{30}{36}\right).$$

Here, $Pr(S_1) = 34/40$ because 34 of the 40 generators are operable. $Pr(S_2|S_1)$, the probability that the second generator is operable given that the first one was, is $33/39$, because there are only 33 operable generators of the remaining 39. Similarly, $Pr(S_3|S_1 \cap S_2) = 32/38$, because there are 32 operable generators of the remaining 38, and so on. ◂

We now prove that the definition of conditional probability satisfies Axioms 1.2.1 to 1.2.3. Let S_1, S_2, and S_3 be events of the sample space S, such that $Pr(S_i) > 0$, $i = 1, 2, 3$. Then, we must show:

1. $Pr(S_1|S_2) \geq 0$ if $(S_1|S_2)$ is the sure event,
2. $Pr(S_1|S_2) = 0$ if $S_1 \cap S_2 = \emptyset$
3. $Pr(S_1 \cup S_2|S_3) = Pr(S_1|S_3) + Pr(S_2|S_3)$.

Proof of Part 1: $0 \leq Pr(S_1 \cap S_2)$ implies that $0 \leq Pr(S_2) Pr(S_1|S_2)$. Dividing both sides of the last expression by $Pr(S_2) > 0$ gives $0 \leq Pr(S_1|S_2)$, which is the condition expressed by Axiom 1.2.1.

Proof of Part 2: If S_1, given S_2, is the sure event, then $S_1 \subset S_2$ and

$$Pr(S_1|S_2) = \frac{Pr(S_1 \cap S_2)}{Pr(S_2)} = \frac{Pr(S_2)}{Pr(S_2)} = 1.$$

This condition is expressed by Axiom 1.2.2.

Proof of Part 3:

$$Pr(S_1 \cup S_2|S_3) = \frac{Pr[(S_1 \cup S_2) \cap S_3]}{Pr(S_3)}$$
$$= \frac{Pr[(S_1 \cap S_3) \cup (S_2 \cap S_3)]}{Pr(S_3)}.$$

But because S_1 and S_2 are mutually exclusive, so are the events $S_1 \cap S_3$ and $S_2 \cap S_3$. Thus, we obtain:

$$Pr(S_1 \cup S_2|S_3) = \frac{Pr(S_1 \cap S_3) + Pr(S_2 \cap S_3)}{Pr(S_3)}$$
$$= \frac{Pr(S_1 \cap S_3)}{Pr(S_3)} + \frac{Pr(S_2 \cap S_3)}{Pr(S_3)}$$
$$= Pr(S_1|S_3) + Pr(S_2|S_3).$$

By mathematical induction, this proof can be extended to $S_1, S_2, \ldots S_n$ events of S. ◂

From these remarks, we see that the axioms of the axiomatic definition hold for the concept of conditional probability; therefore, theorems similar to Theorems 1.2.1, 1.2.2, 1.2.4, and 1.2.5 can be stated for conditional probability.

1.4 Marginal Probabilities

Let S be a sample space that consists of n observations; let $G_1, G_2, \ldots, G_k$ be a sequence of pairwise mutually exclusive events such that

$$\bigcup_{i=1}^{k} G_i = S;$$

and let $E_1, E_2, \ldots, E_t$ be another sequence of pairwise mutually exclusive events defined on S such that

$$\bigcup_{i=1}^{t} E_i = S.$$

Thus, the sample space has been partitioned into two sequences of events, $\{G\}, i = 1, 2, \ldots, k$ and $\{E\}, i = 1, 2, \ldots, t$, both of which are pairwise mutually exclusive within themselves but are not necessarily pairwise mutually exclusive between the two sequences. Such a partition of S into $kt = n$ disjoint subsets of S is shown in Table 1.4.1, where n_{11} is the number of observations that have the characteristics G_1 and E_1, n_{12} is the number of observations that have the characteristics G_i and E_j, and

$$n = \sum_{i=1}^{k} \sum_{j=1}^{t} n_{ij}.$$

This condition is expressed by Axiom 1.2.2.

Table 1.4.1

				E				
G	E_1	E_2	E_3	...	...	E_t		
G_1	n_{11}	n_{12}	n_{13}	.	.	n_{1t}	$\sum_{j=1}^{t} n_{1j}$	
G_2	n_{21}	n_{22}	n_{23}	.	.	n_{2t}	$\sum_{j=1}^{t} n_{2j}$	
G_3	n_{31}	n_{32}	n_{33}	.	.	n_{3t}	$\sum_{j=1}^{t} n_{3j}$	
⋮	⋮	⋮	⋮	...	...	⋮	⋮	
G_k	n_{k1}	n_{k2}	n_{k3}	.	.	n_{kt}	$\sum_{j=1}^{t} n_{kj}$	
	$\sum_{i=1}^{k} n_{i1}$	$\sum_{i=1}^{k} n_{i2}$	$\sum_{i=1}^{k} n_{i3}$	...	...	$\sum_{i=1}^{k} n_{it}$	n	

Table 1.4.2

G	E						
	E_1	E_2	E_3	...	...	E_t	
G_1	P_{11}	P_{12}	P_{13}	.	.	P_{1t}	$\sum_{j=1}^{t} P_{1j} = P_1$
G_2	P_{21}	P_{22}	P_{23}	.	.	P_{2t}	$\sum_{j=1}^{t} P_{2j} = P_2$
G_3	P_{31}	P_{32}	P_{33}	.	.	P_{3t}	$\sum_{j=1}^{t} P_{3j} = P_3$
⋮	⋮	⋮	⋮	...	...	⋮	⋮
G_k	P_{k1}	P_{k2}	P_{k3}	.	.	P_{kt}	$\sum_{j=1}^{t} P_{kj} = P_k$
	$\sum_{i=1}^{k} P_{i1}$	$\sum_{i=1}^{k} P_{i2}$	$\sum_{i=1}^{k} P_{i3}$	...	...	$\sum_{i=1}^{k} P_{it} =$	1
	$P_{.1}$	$P_{.2}$	$P_{.3}$	.	.	$P_{.t}$	

The probability of occurrence of event $G_2 \cap E_3$ (**joint probability**) is equal to the number of observations that are common to both G_2 and E_3, n_{23}, divided by the total number of observations in the sample space n:

$$Pr(G_2 \cap E_3) = \frac{n_{23}}{n} = p_{23}.$$

And in general,

$$Pr(G_i \cap E_j) = \frac{n_{ij}}{n} = p_{ij}. \tag{1.4.1}$$

Therefore, by calculating the probability for every cell of Table 1.4.1, we construct a **probability sample space**, as shown in Table 1.4.2, where

$$\sum_{i=1}^{k}\sum_{j=1}^{t} p_{ij} = \sum_{i=1}^{k} p_{i.} = \sum_{i=1}^{t} p_{.j} = p_{..} = 1.$$

Often in physical problems, when the joint attribute of two happenings is recorded, for example, $G_i \cap E_j$, we are interested in finding the probability of the event G_i alone. From the manner in which we have partitioned the sample space, we can write G_i as

$$G_i = (G_i \cap E_1) \cup (G_i \cap E_2) \cup ... \cup (G_i \cap E_t). \tag{1.4.2}$$

Because $(G_i \cap E_j) \cap (G_i \cap E_{j^*}) = \emptyset$ for $j \neq j^*$, by applying Axiom 1.2.3 to Equation (1.4.2) we have

$$Pr(G_i) = Pr(G_i \cap E_1) + Pr(G_i \cap E_2) + ... + Pr(G_i \cap E_t)$$

$$= \sum_{j=1}^{t} Pr(G_i \cap E_j) = \frac{1}{n}\sum_{j=1}^{t} n_{ij} = p_{i..} \tag{1.4.3}$$

The sum of the probabilities of the ith row, as given in Equation (1.4.3), is called the **marginal probability** of event G_i. The last column of Table 1.4.2 gives the marginal probabilities of the events G_i, $i = 1, 2, \ldots, k$. For example, the marginal probability of G_2 is given by

$$Pr(G_2) = \sum_{j=1}^{t} Pr(G_2 \cap E_j) = \frac{1}{n}\sum_{j=1}^{t} n_{2j} = p_{2.}.$$

Similarly, the marginal probability of event E_j is given by summing all probabilities in column j, that is,

$$Pr(E_j) = \sum_{i=1}^{k} Pr(G_i \cap E_j) = \frac{1}{n}\sum_{i=1}^{k} n_{ij} = p_{.j}. \qquad (1.4.4)$$

For example, the marginal probability E_6 is given by

$$Pr(E_6) = \sum_{i=1}^{k} Pr(G_i \cap E_6) = \frac{1}{n}\sum_{i=1}^{k} n_{i6} = p_{.6}.$$

Furthermore, the conditional probability of event G_2, given that event E_1 has occurred, is given by

$$Pr(G_2 | E_1) = \frac{Pr(G_2 \cap E_1)}{Pr(E_1)} = \frac{\dfrac{n_{21}}{n}}{\dfrac{1}{n}\sum_{i=1}^{k} n_{i1}}$$

$$= \frac{p_{21}}{p_{.1}} = p_{2|1}.$$

And in general,

$$Pr(G_i | E_j) = \frac{Pr(G_i \cap E_j)}{Pr(E_j)} = \frac{\dfrac{n_{21}}{n}}{\dfrac{1}{n}\sum_{i=1}^{k} n_{ij}} = \frac{p_{ij}}{p_{.j}} = p_{i|j}. \qquad (1.4.5)$$

Similarly,

$$Pr(E_3 | G_2) = \frac{Pr(G_2 \cap E_3)}{Pr(G_2)} = \frac{\dfrac{n_{23}}{n}}{\dfrac{1}{n}\sum_{j=1}^{t} n_{2j}}$$

$$= \frac{p_{23}}{p_{2.}} = p_{3|2}$$

$$= \frac{p_{ij}}{p_{i.}} = p_{j|i}. \qquad (1.4.6)$$

Example 1.4.1

An industrial firm is trying a certain electrical component that is used in one of their systems and provided by two companies, G_1 and G_2. The components have been installed into three systems, E_1, E_2, and E_3, each operating under different environmental conditions.

1.4 Marginal Probabilities

Table 1.4.3

	E def.			
G	E_1	E_2	E_3	Total
G_1	30	20	10	60
G_2	10	15	35	60
Total	40	35	45	120

Table 1.4.4

	E			
G	E_1	E_2	E_3	Total
G_1	$\frac{1}{4}$	$\frac{1}{6}$	$\frac{1}{12}$	$\frac{1}{2}$
G_2	$\frac{1}{12}$	$\frac{1}{8}$	$\frac{7}{24}$	$\frac{1}{2}$
Total	$\frac{1}{3}$	$\frac{7}{24}$	$\frac{3}{8}$	1

After the systems have functioned for t hours, they are inspected. The number of defective systems is given in Table 1.4.3.

The joint probabilities of the events $G_i \cap E_j$, $i = 1, 2$ and $j = 1, 2, 3$, are given by Equation (1.4.1) as

$$Pr(G_i \cap E_j) = \frac{n_{ij}}{n} = p_{ij}$$

and have been calculated from Table 1.4.3 and are shown in Table 1.4.4. For example, the probability that a component obtained from company G_1 and used in system E_3 is defective is $1/12$.

a. If the system is chosen at random and the component is inspected and found to be defective, then what is the probability that it came from company G_1?

b. A system was chosen at random, and the component was found to be defective; what is the probability that it is installed in system E_3?

Solution:

a. The answer to this question is given by the marginal probability of event G_1; that is,

$$Pr(G_1) = \sum_{j=1}^{3} Pr(G_1 \cap E_j) = \frac{1}{n} \sum_{j=1}^{3} n_{1j}$$

$$= \frac{30 + 20 + 10}{120} = \frac{1}{2}.$$

Therefore, the probabilities given in the column under Total are the marginal probabilities of the events G_1 and G_2.

b. The answer to this question is given by the marginal probability of E_3:

$$Pr(E_3) = \sum_{i=1}^{2} Pr(G_i \cap E_3) = \frac{1}{n} \sum_{i=1}^{2} n_{i3}$$

$$= \frac{10 + 35}{120} = \frac{3}{8}.$$

Therefore, the probabilities given in the last row of Table 1.4.4 are the marginal probabilities of the events E_1, E_2, and E_3.

Furthermore, we might be interested in the following probabilities:

$$Pr(G_1|E_1) = \frac{Pr(G_1 \cap E_1)}{Pr(E_1)} = \frac{\frac{1}{4}}{\frac{1}{3}} = \frac{3}{4}$$

is the probability that a defective component chosen at random from a system of the type E_1 was purchased from company G_1.

$$Pr(E_1|G_2) = \frac{Pr(G_2 \cap E_1)}{Pr(G_2)} = \frac{\frac{1}{12}}{\frac{1}{2}} = \frac{1}{6}$$

is the probability that the defective component came from system E_1, given that it was purchased from company G_2. ◀

The partition of the sample space can be extended into three sequences of events, $\{G_i\}$, $i = 1, 2, \ldots, k$, $\{E_i\}$, $i = 1, 2, \ldots, t$, and $\{F_i\}$, $i = 1, 2, \ldots, r$. In this case, the sample space is partitioned into cells, where each cell would consist of a number, that is, n_{ijl}, representing the number of observations that have the attributes G_i, E_j, and F_l. The probability that such an event occurs is equal to n_{ijl} divided by n, the total number of observations in the sample space S. Here,

$$\sum_{i=1}^{r} \sum_{j=1}^{t} \sum_{i=1}^{k} n_{ijl} = n.$$

Also,

$$. Pr(G_i \cap E_j \cap F_l) = \frac{n_{ijl}}{n}$$

The marginal probabilities of the events E_m, $m = 1, 2, 3, \ldots, t$, and $E_m F_p$, $p = 1, 2, \ldots, r$ are given by

$$Pr(E_m) = \sum_{i=1}^{k} \sum_{l=1}^{r} Pr(G_i \cap E_m \cap F_l)$$

$$= \frac{1}{n} \sum_{i=1}^{k} \sum_{l=1}^{r} n_{iml}$$

$$= p_{.m.}$$

and

$$Pr(E_m \cap F_p) = \sum_{i=1}^{k} Pr(G_i \cap E_m \cap F_p)$$

$$= \frac{1}{n} \sum_{i=1}^{k} n_{imp}$$

$$= p_{.mp},$$

respectively.

The conditional probability of the event G_i, given that event $E_j \cap F_l$ has occurred, and $G_i \cap F_l$, given that the event E_j has occurred, are given by

$$Pr(G_i | E_j \cap F_l) = \frac{Pr(G_i \cap E_j \cap F_l)}{Pr(E_j \cap F_l)}$$

$$= \frac{\frac{n_{ijl}}{n}}{\frac{1}{n}\sum_{i=1}^{k} n_{ijl}} = \frac{p_{ijl}}{p_{.jl}}$$

$$= p_{i|jl}$$

and

$$Pr(G_i \cap F_l | E_j) = \frac{Pr(G_i \cap E_j \cap F_l)}{Pr(E_j)}$$

$$= \frac{\frac{n_{ijl}}{n}}{\frac{1}{n}\sum_{i=1}^{k}\sum_{l=1}^{r} n_{ijl}} = \frac{p_{ijl}}{p_{.j.}}$$

$$= p_{il|j},$$

respectively. For example,

$$Pr(E_6 | G_1 \cap F_3) = \frac{Pr(G_1 \cap E_6 \cap F_3)}{Pr(G_1 \cap F_3)}$$

$$= \frac{\frac{n_{163}}{n}}{\frac{1}{n}\sum_{j=1}^{t} n_{1j3}}$$

$$= \frac{p_{163}}{p_{1.3}} = p_{6|13}$$

and

$$Pr(E_3 \cap F_5 | G_3) = \frac{Pr(G_3 \cap E_3 \cap F_5)}{Pr(G_3)}$$

$$= \frac{\frac{n_{335}}{n}}{\frac{1}{n}\sum_{j=1}^{t}\sum_{l=1}^{r} n_{3jl}}$$

$$= \frac{p_{335}}{p_{3..}} = p_{35|3}.$$

This concept of partitioning the sample space can be easily generalized to m different partitions.

◀

1.5 Bayes's Theorem

Let the sample space S be partitioned into a finite number of mutually exclusive events, $S_1, S_2, \ldots, S_n$. Also, let S^* be any other event of the sample space S. That is,

$$S^* = S \cap S^* = (S_1 \cup S_2 \cup \ldots \cup S_n) \cap S^*$$
$$= (S_1 \cap S^*) \cup (S_2 \cap S^*) \cup \ldots \cup (S_n \cap S^*). \quad (1.5.1)$$

Because $S_i \cap S^*$, $i = 1, 2, \ldots, n$, are mutually exclusive, we can apply Axiom 1.2.3 to Equation (1.5.1); that is,

$$Pr(S^*) = Pr(S_1 \cap S^*) + Pr(S_2 \cap S^*) + \ldots + Pr(S_n \cap S^*). \quad (1.5.2)$$

But according to Equation (1.3.3), we obtain for every i

$$Pr(S_i \cap S^*) = Pr(S^* | S_i) Pr(S_i),$$

and Equation (1.5.2) can be written as

$$Pr(S_i) = Pr(S_1) Pr(S^* | S_1) + Pr(S_2) Pr(S^* | S_2) + \ldots + Pr(S_n) Pr(S^* | S_n)$$
$$= \sum_{i=1}^{n} Pr(S_i) Pr(S^* | S_i). \quad (1.5.3)$$

Equation (1.5.3) is known as the theorem of **absolute probability** of event S^*. Furthermore, the conditional probability of S_i, given S^*, is given for every i by

$$Pr(S_i | S^*) = \frac{Pr(S_i \cap S^*)}{Pr(S^*)}, \; Pr(S^*) > 0. \quad (1.5.4)$$

Substituting Equation (1.5.3) for $Pr(S^*)$ and $Pr(S_i) Pr(S^* | S_i)$ for $Pr(S_i \cap S^*)$ in Equation (1.5.4), we obtain Theorem 1.5.1, which results from the work of Thomas Bayes.

Theorem 1.5.1 (Bayes's theorem)

If $S_1, S_2, \ldots, S_n$ and S^* satisfy the preceding assumptions—that is, the theorem of absolute probability and $Pr(S^*) > 0$—then for every $i = 1, 2, \ldots, n$ we have

$$Pr(S_i | S^*) = \frac{Pr(S_i) Pr(S^* | S_i)}{Pr(S_1) Pr(S^* | S_1) + Pr(S_2) Pr(S^* | S_2) + \ldots + Pr(S_n) Pr(S^* | S_n)}$$

or

$$Pr(S_i | S^*) = \frac{Pr(S_i) Pr(S^* | S_i)}{\sum_{j=1}^{n} Pr(S_j) Pr(S^* | S_j)}$$

$$= \frac{Pr(S_i) Pr(S^* | S_i)}{Pr(S^*)}. \quad (1.5.5)$$

Equation (1.5.5) is sometimes called the **formula for posteriori probability** rather than Bayes's formula because the formula gives us the probability of S_i after S^* has been observed. Also, the probabilities $Pr(S_i)$ in Bayes's formula are called **a priori probabilities**.

The following examples illustrate various applications of Bayes's theorem.

Example 1.5.1

Consider that we have two large boxes, each of which contains 30 electrical components. It is known that the first box contains 26 operable and 4 nonoperable components and that the second box contains 28 operable and 2 nonoperable components.

a. We are interested in obtaining the probability that a component selected at random will be operable, assuming that the probability of making a selection from each of the boxes is the same.

b. Furthermore, suppose that the component chosen at random is operable and we wish to find the probability that the component was chosen from box S_1, $Pr(S_1|S^*)$.

Solution:

a. Let S_1 and S_2 be the events of selecting the first and the second boxes, respectively, and let S^* be the event that an operable component is chosen. We note that the event of selecting an operable component, S^*, can occur with either S_1 or S_2; that is,

$$S^* = (S_1 \cap S^*) \cup (S_2 \cap S^*).$$

Because $S_1 \cap S^*$ and $S_2 \cap S^*$ are mutually exclusive events, we have

$$Pr(S^*) = Pr(S_1 \cap S^*) + Pr(S_2 \cap S^*)$$

or

$$Pr(S^*) = Pr(S_1)Pr(S^*|S_1) + Pr(S_2)Pr(S^*|S_2).$$

Also,

$$Pr(S_1) = Pr(S_2) = \frac{1}{2}, \quad Pr(S^*|S_1) = \frac{13}{15}$$

and

$$Pr(S^*|S_2) = \frac{14}{15}.$$

Therefore, the probability that the component is operable is given by

$$Pr(S^*) = \left(\frac{1}{2}\right)\left(\frac{26}{30}\right) + \left(\frac{1}{2}\right)\left(\frac{28}{30}\right) = .90.$$

b. By Bayes's formula,

$$Pr(S_1|S^*) = \frac{Pr(S_1)Pr(S^*|S_1)}{Pr(S_1)Pr(S^*|S_1) + Pr(S_2)Pr(S^*|S_2)}$$

$$= \frac{\left(\frac{1}{2}\right)\left(\frac{26}{30}\right)}{\left(\frac{1}{2}\right)\left(\frac{26}{30}\right) + \left(\frac{1}{2}\right)\left(\frac{28}{30}\right)} = \frac{26}{54} \approx .48.$$

◀

Example 1.5.2

Three weapon systems are shooting at the same target. From a design standpoint, each weapon has an equally likely chance of hitting the target. However, in practice, it has been observed that the precision of these weapon systems is not the same; that is, the first weapon usually hits the target 10 out of 12 shots, the second hits it 9 out of 12 shots, and the third hits it 8 out of 12 shots. We have observed that the target has been hit, and we are interested in finding the probability that the shot was fired by the third weapon system.

Solution: Let S_1, S_2, and S_3 be the events that the target has been hit by the first, second, and third weapon system, respectively. Also, let S^* be the event that the target has been hit. We wish to find $Pr(S_3|S^*)$. Thus, according to Bayes's theorem, we have

$$Pr(S_3|S^*) = \frac{Pr(S_3)Pr(S^*|S_3)}{Pr(S_1)Pr(S^*|S_1) + Pr(S_2)Pr(S^*|S_2) + Pr(S_3)Pr(S^*|S_3)}$$

$$= \frac{\left(\frac{1}{3}\right)\left(\frac{8}{12}\right)}{\left(\frac{1}{3}\right)\left(\frac{10}{12}\right) + \left(\frac{1}{3}\right)\left(\frac{9}{12}\right) + \left(\frac{1}{3}\right)\left(\frac{8}{12}\right)} = \frac{8}{27},$$

meaning that the probability that the shot was fired from the third weapon system is $8/27$, assuming that $Pr(S_1) = Pr(S_2) = Pr(S_3) = 1/3$, as was specified in the design of the system.

Furthermore,

$$Pr(S_1|S^*) = \frac{\left(\frac{1}{3}\right)\left(\frac{10}{12}\right)}{\left(\frac{1}{3}\right)\left(\frac{10}{12}\right) + \left(\frac{1}{3}\right)\left(\frac{9}{12}\right) + \left(\frac{1}{3}\right)\left(\frac{8}{12}\right)} = \frac{10}{27}$$

and

$$Pr(S_2|S^*) = \frac{\left(\frac{1}{3}\right)\left(\frac{9}{12}\right)}{\left(\frac{1}{3}\right)\left(\frac{10}{12}\right) + \left(\frac{1}{3}\right)\left(\frac{9}{12}\right) + \left(\frac{1}{3}\right)\left(\frac{8}{12}\right)} = \frac{9}{27}$$

◀

1.6 Independent Events

The notion of independent events is quite basic and plays a central role in probability theory. Let S_1 and S_2 be two events of the sample space S that have positive probabilities. The event S_2 is said to be **independent** of event S_1 if

$$Pr(S_1 \cap S_2) = Pr(S_1) Pr(S_2). \tag{1.6.1}$$

Here, the probability of the intersection of the events $S_1 \cap S_2$ is equal to the product of the probabilities of S_1 and S_2. Therefore, from the definition of conditional probability (Section 1.3.1), if S_1 and S_2 are independent events, then

$$Pr(S_2|S_1) = \frac{Pr(S_1 \cap S_2)}{Pr(S_1)} = \frac{Pr(S_1) Pr(S_2)}{Pr(S_1)} = Pr(S_2). \tag{1.6.2}$$

That is, the probability that S_2 occurs is not affected by the fact that S_1 has occurred.

If S_2 is independent of S_1, then S_1 is independent of S_2. To show this independence, we write

$$Pr(S_1 \cap S_2) = Pr(S_1) Pr(S_2|S_1) = Pr(S_2) Pr(S_1|S_2) \tag{1.6.3}$$

and

$$Pr(S_2|S_1) = Pr(S_2).$$

Example 1.6.1

Consider a problem in which two guns are shooting at the same target. Let S_1 and S_2 be the events that the target is hit by guns 1 and 2, respectively. The probability that S_1 occurs is unaffected by the occurrence or nonoccurrence of S_2. It has been observed that $Pr(S_1) = 1/3$ and $Pr(S_2) = 1/4$. Thus, the probability of both guns hitting the target is given by

$$Pr(S_1 \cap S_2) = Pr(S_1) Pr(S_2)$$
$$= \left(\frac{1}{3}\right)\left(\frac{1}{4}\right) = \frac{1}{12}.$$

Furthermore, if we are interested in the probability that S_1 or S_2 occurs, we use Theorem 1.2.5 and write

$$Pr(S_1 \cup S_2) = Pr(S_1) + Pr(S_2) - Pr(S_1 \cap S_2)$$
$$= Pr(S_1) + Pr(S_2) - Pr(S_1) Pr(S_2)$$
$$= \left(\frac{1}{3}\right) + \left(\frac{1}{4}\right) - \left(\frac{1}{12}\right) = \frac{1}{2}.$$

If events S_1 and S_2 do not satisfy Equation (1.6.1), they are said to be **dependent** or **nonindependent**. Thus, the events S_1 and S_2 are dependent if and only if

$$Pr(S_1 \cap S_2) \neq Pr(S_1) Pr(S_2).$$

Example 1.6.2

Let S_1 and S_2 denote the events that a particular sports coat and a pair of trousers, respectively, are purchased on a specified day. It has been found that $Pr(S_1) = Pr(S_2) = 0.46$ and $Pr(S_1 \cap S_2) = 0.23$. Determine the conditional probabilities $Pr(S_1|S_2)$ and $Pr(S_2|S_1)$ and the total probability $Pr(S_1 \cup S_2)$. Are the events S_1 and S_2 independent?

Solution: The conditional probabilities are obtained by

$$Pr(S_1|S_2) = \frac{Pr(S_1 \cap S_2)}{Pr(S_2)} = \frac{0.23}{0.46} = 0.5$$

and

$$Pr(S_2|S_1) = \frac{Pr(S_1 \cap S_2)}{Pr(S_1)} = \frac{0.23}{0.46} = 0.5.$$

The total probability is given by

$$Pr(S_1 \cup S_2) = Pr(S_1) + Pr(S_2) - Pr(S_1 \cap S_2)$$
$$= 0.46 + 0.46 - 0.23 = 0.69.$$

The two events are dependent, not independent, because $Pr(S_1|S_2) \neq Pr(S_1)$ or, equivalently, $Pr(S_1 \cap S_2) \neq Pr(S_1)Pr(S_2)$. ◀

The independence of two events can be generalized to a finite number of events defined on the sample space. Let $S_1, S_2, \ldots, S_n$ be a finite sequence of events of S. These events are said to be **mutually independent** or independent if

$$Pr(S_{j1} \cap S_{j2} \cap \ldots \cap S_{jt}) = Pr(S_{j1})Pr(S_{j2})\ldots Pr(S_{jt})$$

for all integer indices $j_1, j_2, \ldots, j_t$, such that $1 \leq j_1 < j_2 \ldots < j_t \leq n$. That is, the probability of the intersection of every combination of events equals the product of their probabilities. For example, the events S_1, S_2, and S_3 are independent if

1. $Pr(S_1 \cap S_2) = Pr(S_1)Pr(S_2)$
2. $Pr(S_1 \cap S_3) = Pr(S_1)Pr(S_3)$
3. $Pr(S_2 \cap S_3) = Pr(S_2)Pr(S_3)$
4. $Pr(S_1 \cap S_2 \cap S_3) = Pr(S_1)Pr(S_2)Pr(S_3).$

Furthermore, it follows that

$$Pr(S_1|S_2 \cap S_3) = Pr(S_1|S_2) = Pr(S_1|S_3) = Pr(S_1),$$

$$Pr(S_2|S_1 \cap S_3) = Pr(S_2|S_1) = Pr(S_2|S_3) = Pr(S_2),$$

and

$$Pr(S_3|S_1 \cap S_2) = Pr(S_3|S_1) = Pr(S_3|S_2) = Pr(S_3),$$

assuming that the events $S_1, S_2, S_3, S_1 \cap S_2$ and $S_1 \cap S_3$ have nonzero probabilities. The total number of equations of the form in parts 1 to 4 that are required to establish the independence of n events equals $2^n = (n+1)$.

It is possible for the events in the sequence $S_1, S_2, \ldots, S_n$ to be independent in twos, in threes, in fours, and so on, yet $S_1, S_2, \ldots, S_n$ may not be independent. The following example illustrates this.

◀

Example 1.6.3

A sample poll of 100 voters revealed the following information concerning three candidates, $A, B,$ and $C,$ who were running for three offices:

10 voted in favor of both A and B.
    35 voted in favor of A or B but not C.
    25 voted in favor of B but not A or C.
    65 voted in favor of B or C but not A.
    25 voted in favor of C but not A or B.
    0 voted in favor of A and C but not B.
15 turned in blank ballots.

Here, $A, B,$ and C are pairwise independent but are not three-way independent:

$$Pr(A \cap B \cap C) = \frac{10}{100} = \frac{1}{10}$$

and

$$Pr(A)Pr(B)Pr(C) = \left(\frac{20}{100}\right)\left(\frac{50}{100}\right)\left(\frac{50}{100}\right) = \frac{1}{20},$$

$$Pr(A \cap B) = \frac{1}{10} \quad \text{and} \quad Pr(A)Pr(B) = \frac{1}{10},$$

$$Pr(A \cap C) = \frac{1}{10} \quad \text{and} \quad Pr(A)Pr(C) = \frac{1}{10}, \text{ and}$$

$$Pr(B \cap C) = \frac{1}{4} \quad \text{and} \quad Pr(B)Pr(C) = \frac{1}{4}.$$

We now state some obvious probabilities of independent events:

1. If S_1 and S_2 are two mutually exclusive events, then these events are independent if and only if $Pr(S_1)Pr(S_2) = 0$, and this is so if and only if either $Pr(S_1)$ or $Pr(S_2)$ or both are equal to zero.

2. If S_1 and S_2 are independent, then Theorem 1.2.5 can be written as

$$Pr(S_1 \cup S_2) = Pr(S_1) + Pr(S_2) - Pr(S_1)Pr(S_2).$$

3. If S_1 and S_2 are independent events, then their complements $\overline{S_1}$ and $\overline{S_2}$ are also independent and

$$Pr\left(\overline{S_1} \cap \overline{S_2}\right) = Pr\left(\overline{S_1}\right) Pr\left(\overline{S_2}\right).$$

4. If S_1, S_2, and S_3 are independent events, then S_1 is independent of $S_2 \cap S_3$ and $S_2 \cup S_3$.

1.7 Combinatorial Probability

As we have seen throughout our discussion, the number of observations in a sample space S is the basis for obtaining probabilities of various events defined on S. That is, if we know that in a certain problem the space consists of n equally likely observations and n_{s1} of them are favorable to event S_1, then the probability that event S_1 will occur is given by

$$Pr(S_1) = \frac{n_{s1}}{n}.$$

In this section, we discuss the manner in which we can obtain the total number of observations of the sample space under four basic sampling procedures.

The initial step in many physical problems involves the concept of a **random sample**: when we say that a random sample of k objects was drawn from a population of N objects, we mean that each of the N objects had equal probability of being selected. When a random sample of size k is taken, the sample space consists of a certain number of k-tuples, or the total number of ways in which the random sample of size k can be selected. We consider the following four sampling schemes:

1. Sampling with replacement and the objects being ordered.

2. Sampling without replacement and the objects being ordered.

3. Sampling without replacement and the objects being unordered.

4. Sampling with replacement and the objects being unordered.

For the purpose of illustrating the preceding sampling methods, consider a population of N objects numbered $1, 2, 3, \ldots, N$, from which we take a random sample of size k according to one of the four sampling schemes.

> **Sampling under Scheme 1** There are N possible ways in which the first observation in the sample can be drawn. There are also N possible ways in which the second observation in the sample can be drawn. Therefore, there are N^2 possible outcomes for the first two drawings of the sample. Continuing in this manner, it is obvious that the sample space S consists of
>
> $$N^k \quad k\text{-tuples.} \tag{1.7.1}$$

Example 1.7.1

An urn contains nine balls numbered 1 to 9. If a random sample with replacement of size $k = 6$ is taken, then the sample space S, according to Equation (1.7.1), will consist of 9^6 6-tuples.

Example 1.7.2

If a die is rolled five times, then the sample space S will consist of 6^5 5-tuples. That is, this experiment is equivalent to sampling with replacement, where the number of times that we roll the die is equivalent to the size of the random sample and N is equivalent to the number of faces on the die.

> **Sampling under Scheme 2** There are N possible ways in which the first observation of the sample can be drawn. There are $N-1$ possible ways in which the second object in the sample can be selected. Finally, there are $N-(k-1)=N-k+1$ possible ways in which the kth observation may be drawn. Therefore, the total number of k-tuples that the sample space consists of is given by
>
> $$P_k^N = N(N-1)(N-2)\ldots(N-k+1)$$
>
> $$= \frac{N!}{(N-k)!}. \quad (1.7.2)$$
>
> If in Equation (1.7.2) $k=N$, then $P_N^N = N(N-1)(N-2)\ldots(1) = N!$.

Example 1.7.3

An urn holds 11 balls that are numbered 1 to 11. A random ordered sample of size $k=3$ is drawn. Thus, according to Equation (1.7.2), the sample space S consists of $11!/(11-3)! = 990$ 3-tuples.

Example 1.7.4

If 13 cards are selected from an ordinary deck of playing cards and the order in which the cards are drawn is important, then the sample space S will consist of $52!/(52-13)!$ 13-tuples.

Example 1.7.5

An urn contains nine balls that are numbered 1 to 9. If a random sample of size $k=9$ is drawn and order is of importance, then there are $N! = k! = 9!$ possible ways in which this sample can be drawn.

> **Sampling under Scheme 3** To obtain the total number of ways in which a random sample can be drawn under this scheme, we simply need to eliminate the order in Scheme 2. This is done by dividing $N!/(N-k)!$ by the number of different sample orderings possible. If the population is of size N and the sample is of size k, then the total number of samples is
>
> $$\frac{P_k^N}{k!} = \frac{N!}{k!(N-k)!} = \binom{N}{k}, \quad k = 0, 1, 2, \ldots, N. \quad (1.7.3)$$

Example 1.7.6

If a football squad consists of 72 players, how many possible selections of 11-man teams are there?

Solution: Applying Equation (1.7.3), we have

$$\binom{72}{11} = \frac{72!}{11!(72-11)!}$$

That is, the sample space consists of $\binom{72}{11}$ 11-tuples. ◀

Example 1.7.7

What is the total number of ways in which a 13-card hand can be drawn from an ordinary deck of playing cards?

Solution: Because there is no replacement and order is irrelevant, the number of ways is

$$\binom{52}{13} = \frac{52!}{13!(52-13)!}$$

◀

> **Sampling under Scheme 4** In obtaining an unordered sample of size k, with replacement, from a population of size N, $k-1$ replacements are made before sampling ceases. Thus, N is increased by $k-1$ so that sampling in this manner may be thought of as drawing an unordered sample of size k from a population of size $N+k-1$. That is, the number of samples is
>
> $$\binom{N+k-1}{k} = \frac{(N+k-1)!}{k!(N-1)!}, \quad k = 0, 1, 2, \text{K}, N. \tag{1.7.4}$$

Example 1.7.8

An urn contains 15 balls numbered 1 to 15. If a random sample of size $k=4$ is drawn with replacement but without regard to order from the urn, then, according to Equation (1.7.4), the sample space consists of

$$\binom{15+4-1}{4} = \frac{18!}{4!14!} \text{ 4-tuples.}$$

◀

We now derive a useful formula that concerns itself with the number of ways in which a sample of size k may be partitioned into r ordered groups of unordered elements such that not only will the ith group contain k_i elements for $k = 1, 2, \ldots, r$ but also

$$\sum_{i=1}^{r} k_i = k.$$

Here, the elements of the first group may be selected in $\binom{k}{k_1}$ ways, those of the second group in $\binom{k-k_1}{k_2}$ ways, those of the third group $\binom{k-k_1-k_2}{k_3}$ ways, and so on, until those of the rth group are selected in $\binom{k-k_1-k_2-\ldots-k_{r-1}}{k_r}$ ways. Taking the product of these combinations gives

$$\frac{k!}{k_1!k_2!k_3!\ldots k_r!},$$

which is derived from

$$\frac{k!}{k_1!(k-k_1)!} \cdot \frac{(k-k_1)!}{k_2!(k-k_1-k_2)!} \cdot \frac{(k-k_1-k_2)!}{k_3!(k-k_1-k_2-k_3)!} \cdots \frac{(k-k_1-k_2-\ldots-k_{r-1})!}{k_r!(k-k_1-k_2-\ldots-k_r)!},$$

where the second component of the denominators of each factor cancels with the numerator of each succeeding factor. This last expression reduces to

$$\frac{k!}{k_1!k_2!k_3!\ldots k_r!} = \binom{k}{k_1!k_2!k_3!\ldots k_r}. \tag{1.7.5}$$

Example 1.7.9

A die is tossed 10 times. Find the total number of ways in which we can obtain the following: three ones, no twos, two threes, no fours, three fives, and two sixes.

Solution: This number is given by applying Equation (1.7.5); that is,

$$\binom{10}{3,0,2,0,3,2} = \frac{10!}{3!0!2!0!3!2!}.$$

Example 1.7.10

Four players in a game of bridge are dealt 13 cards each from an ordinary deck of 52 cards. The total number of ways in which the 13 cards can be dealt to the four players is given by Equation (1.7.5); that is

$$\binom{52}{13,13,13,13} = \frac{52!}{13!13!13!13!}.$$

We now illustrate some of the uses of the preceding counting techniques in solving various probability problems.

Example 1.7.11

If a five-volume set of books is placed on a bookshelf at random, what is the probability that the books will be in the correct order?

Solution: Let A be the event that the five-volume set of books is in the correct order. The sample space S consists of $P^5 = 5!/(5-5)! = 5 \cdot 4 \cdot 3 \cdot 2 \cdot 1$ 5-tuples, and there is only one 5-tuple favorable to event A. Therefore,

$$Pr(A) = \frac{1}{5 \cdot 4 \cdot 3 \cdot 2 \cdot 1}.$$

◀

Example 1.7.12

A certain lot of 26 mechanical components contains 6 defective items. A random sample of 4 components is drawn from the lot.

a. What is the probability that all four components drawn are operable?

b. What is the probability that two are operable and two defective?

Solution:

a. Let A be the event that all four components are operable. The sample space S consists of $\binom{26}{4}$ 4-tuples because order is not relevant. The number of 4-tuples in the sample space favorable to event A is given by $\binom{20}{4}$, because there are 20 operable components in the lot. Therefore,

$$Pr(A) = \frac{\binom{20}{4}}{\binom{26}{4}}.$$

b. Let B be the event of drawing 2 operable and 2 defective components. The sample space in this part remains the same as in Example 1.7.12(a). The number of ways in which 2 operable components out of 20 can be drawn is given by $\binom{20}{2}$; the number of ways in which 2 defective components out of 6 can be drawn is given by $\binom{6}{2}$. The total number of ways of achieving event B is given by $\binom{20}{2}\binom{6}{2}$. Therefore,

$$Pr(B) = \frac{\binom{20}{2}\binom{6}{2}}{\binom{26}{4}}.$$

◀

Example 1.7.13

Suppose that 7 cards are selected at random from an ordinary deck of 52 cards. What is the probability that 2 of the cards are aces, 1 is a nine, 1 is a ten, 1 is a jack, 1 is a queen, and 1 is a king?

Solution: Let A be the event that seven cards are drawn—two are aces, and there is one nine, one ten, one jack, one queen, and one king. The sample space S consists of $\binom{52}{7}$ 7-tuples, because order is not relevant. The number of ways of obtaining two aces is $\binom{4}{2}$; the tnumber of ways of obtaining one nine, one ten, one jack, one queen, or one king is $\binom{4}{1}$; and the total number of ways of achieving event A, that is, the total number of 7-tuples in the sample space favorable to event A, is given by

$$\binom{4}{2}\binom{4}{1}\binom{4}{1}\binom{4}{1}\binom{4}{1}\binom{4}{1}.$$

Therefore,

$$\Pr(A) = \frac{\binom{4}{2}\binom{4}{1}^5}{\binom{52}{7}}.$$

◂

Example 1.7.14

A lot of 45 electrical components numbered 1 to 45 is drawn at random, one by one, and is divided among five customers.

a. Suppose that it is known that components, 7, 11, 16, 21, and 36 are defective. What is the probability that each customer will receive one defective component?

b. What is the probability that one customer will have drawn five defective components?

c. What is the probability that two customers will receive two defective components each, two none, and the other one?

Solution:

a. Let A represent the event desired. The sample space S according to Equation (1.7.4) consists of

$$\binom{45}{9,9,9,9,9}$$

points. The total number of sample points favorable to event A is given by

$$\binom{5}{1,1,1,1,1}\binom{40}{8,8,8,8,8}.$$

That is, the 5 defective components can be distributed to the five customers in 5! ways and the 40 remaining components, according to Equation (1.7.4), can be divided among the customers in $\binom{40}{8,8,8,8,8}$ ways. We obtain

$$Pr(A) = \frac{\binom{40}{8,8,8,8,8}\binom{5}{1,1,1,1,1}}{\binom{45}{9,9,9,9,9}}.$$

b. Let B be the desired event. The totality of sample points in S remains the same as that in Example 1.7.14(a). The total number of sample points favorable to event B is given by

$$\binom{5}{1}\binom{5}{5}\binom{40}{4,9,9,9,9}$$

and

$$Pr(B) = \frac{\binom{5}{1}\binom{5}{5}\binom{40}{4,9,9,9,9}}{\binom{45}{9,9,9,9,9}}.$$

c. Let C be the desired event; by reasoning similar to that presented earlier, we have

$$Pr(C) = \frac{\binom{5}{2,2,1,0,0}\binom{40}{7,7,8,9,9}\binom{5}{2,2,1}}{\binom{45}{9,9,9,9,9}}.$$

◀

Example 1.7.15

In choosing an unordered sample of size k with replacement from a population of size N, find the probability that r specified elements are included in the sample.

Solution: Here, let $r < k < N$.

a. The sample can be obtained in $\binom{N+k-1}{k}$ ways.

b. The r elements may be selected in $\binom{r}{r} = 1$ way.

c. After the r elements are included in the sample, there are $N-r$ other elements in the population from which $k-r$ will be selected to be included in the sample. The number of ways this can be done is equal to

$$\binom{(N-r)+(k-r)-1}{k-r} = \binom{N+k-2r-1}{k-r}.$$

Therefore, the probability desired is

$$\frac{\binom{N+k-2r-1}{k-r}}{\binom{N+k-1}{k}}.$$

Example 1.7.16

From a population of size 10, an unordered sample of size 6 is selected, with replacement. Find the probability that four specified elements are included in the sample.

Solution: The probability asked for is

$$\frac{\binom{10+6-(2\cdot 4)-1}{6-4}}{\binom{10+6-1}{6}} = \frac{\binom{7}{2}}{\binom{15}{6}} = \frac{3}{715}.$$

Until now, we have been dealing with probability problems by determining the number of points in the sample space and then dividing this number into the number of points favorable to the event whose probability we wish to find. In many instances, this procedure can become exceedingly tedious, especially if the number of sample points is large, difficult to determine, or both. To escape these difficulties, we set up a real valued function on the sample space, which helps us find the probabilities we want. This function is called a **random variable** and is discussed in detail in the next chapter. But its **probability density**, which characterizes the behavior of the random variable, is what helps us in determining probabilities.

For example, suppose a coin is tossed four times and we are interested in the number of heads that occur. In this case, we let the random variable X be equal to the number of heads that occur; consequently, its probability density function is

$$f(x) = Pr(X = x) = \begin{cases} \dfrac{\binom{4}{x}}{16}, & x = 0,1,2,3,4, \\ 0, & \text{elsewhere} \end{cases}$$

Thus, we can find, without lengthy calculations, $Pr(X = x_0)$, where x_0 is any real number from $-\infty$ to $+\infty$. That is,

$$Pr(X = 3) = \frac{\binom{4}{3}}{16} = \frac{4}{16}, \quad Pr(X = -2) = 0, \quad Pr(X = 7) = 0,$$

$$Pr(X = 0) = \frac{\binom{4}{0}}{16} = \frac{1}{16},$$

and so forth.

In addition to considering in detail the concept of a random variable, we also study in detail in the next chapter the methods of determining the density of a random variable, the uses of the random variable, and some important probability densities.

Summary

Four definitions of probability have been discussed:

1. Probability as a measure belief
2. Classical or a priori probability
3. Relative-frequency or a posteriori probability
4. Axiomatic definition of probability

Let $S_i, i = 1, 2, \ldots, n, \ldots$ represent events of the sample space S. The number $Pr(S_i)$, which satisfies $0 \leq Pr(S_i)$, $Pr(S) = 1$, and for $S_i \cap S_j = \emptyset$, $i \neq j = 1, 2, 3, \ldots, n, \ldots$, $Pr\left(\sum_{i=1}^{\infty} S_i\right) = Pr \sum_{i=1}^{\infty}(S_i)$, is called the *probability* of the event S_i.

The following elementary properties of probability were studied:

1. If S_1 and S_2 are any two events of the sample space S, such that $S_1 \subset S_2$, then

$$Pr(S_1) \leq Pr(S_2).$$

2. If S_k is any event of the sample space S, then $Pr(S_k) \leq 1$ and $Pr(\overline{S_k}) = 1 - Pr(S_k)$.

3. If $\emptyset$ is the empty set, then $Pr\emptyset = 0$.

4. If S_1 and S_2 are any two events of the sample space S, then

$$Pr(S_1 \cup S_2) = Pr(S_1) + Pr(S_2) - Pr(S_1 \cap S_2).$$

Let $S_1, S_2, \ldots, S_n$ be a sequence of events defined on S. The *general law of total probability* is given by

$$Pr\left(\bigcup_{i=1}^{n} S_i\right) = \sum_{i=1}^{n} Pr(S_i) - \sum_{\substack{i,j=1 \\ i<j}}^{n} Pr(S_i \cap S_j) + \sum_{\substack{i,j=k \\ i<j<k}}^{n} Pr(S_i \cap S_j \cap S_k)$$

$$+ \ldots + (-1)^{n+1} Pr(S_1 \cap S_2 \cap \ldots \cap S_n).$$

The *conditional probability* of the event S_1, given that event S_2 has occurred, $S_1 | S_2$, is given by

$$Pr(S_1 | S_2) = \frac{Pr(S_1 \cap S_2)}{Pr(S_2)}, \quad Pr(S_2) > 0.$$

Let $S_1, S_2, S_3, \ldots, S_n$ be events defined on S. The *general law of compound probability* is given by

$$Pr(S_1 \cap S_2 \cap \ldots \cap S_n)$$

$$= Pr(S_1) Pr(S_2 | S_1) Pr(S_3 | S_1 \cap S_2) \ldots Pr(S_n | S_1 \cap S_2 \cap \cdots \cap S_{n-1}),$$

where

$$Pr(S_1 \cap S_2 \cap \ldots \cap S_j) > 0, \quad 1 \le j \le n-1.$$

The concept of *marginal probability* was studied.

Partition S into a finite number of mutually exclusive events $S_1, S_2, \ldots, S_n$ and let $S^* = (S_1 \cap S^*) \cup (S_2 \cap S^*) \cup \ldots \cup (S_n \cap S^*)$. The *absolute probability* of event S^* is

$$Pr(S^*) = \sum_{i=1}^{n} Pr(S_i) Pr(S^* | S_i).$$

If $Pr(S^*) > 0$, then for every $i = 1, 2, \ldots, n$ we have *Bayes's theorem*:

$$Pr(S_i | S^*) = \frac{Pr(S_i) Pr(S^* | S_i)}{\sum_{j=1}^{n} Pr(S_j) Pr(S^* | S_j)}.$$

Let $S_1, S_2, \ldots, S_n$ be a finite sequence of events of S. These events are said to be *mutually independent* or *independent* if

$$Pr(S_{j_1} \cap S_{j_2} \cap \ldots \cap S_{j_t}) = Pr(S_{j_1}) Pr(S_{j_2}) \ldots Pr(S_{j_t})$$

for all integer indices $j_1, j_2, \ldots, j_t$, such that $1 \le j_1 < j_2 < \ldots < j_t \le n$.

Combinatorial probability was discussed by considering four sampling schemes:

1. Sampling with replacement and the objects being ordered

2. Sampling without replacement and the objects being ordered

3. Sampling without replacement and the objects being unordered

4. Sampling with replacement and the objects being unordered

Theoretical Exercises

1.1 The following six postulates are true for all sets contained in the sample space S:

P_1: $A, B \subset S$ implies $A \cup B, A \cap B \subset S$ (existence postulate).

P_2: $A \cup B = B \cup A$ and $A \cap B = B \cap A$ (commutative postulate).

P_3: $(A \cup B) \cup C = A \cup (B \cup C)$ and $(A \cap B) \cap C = A \cap (B \cap C)$ (associative postulate).

P_4: $A \cap (B \cup C) = (A \cap B) \cup (A \cap C)$ and $A \cup (B \cap C) = (A \cup B) \cap (A \cup C)$ (distributive postulate).

P_5: $A \cap S = A \cup \emptyset = A$ (identity postulate).

P_6: $A \cap \overline{A} = \emptyset$, and $A \cup \overline{A} = S$ (complementation postulate).

Using only postulates 1 through 6 and any others you might derive, prove:
   a. $A \cup S = S$ and $A \cap \emptyset = \emptyset$
   b. $A \cup A = A$ and $A \cap A = A$
   c. $\overline{S} = \emptyset$ and $\overline{\emptyset} = S$
   d. $\overline{(\overline{A})} = A$
   e. $\overline{A \cup B} = \overline{A} \cap \overline{B}$ and $\overline{A \cap B} = \overline{A} \cup \overline{B}$
   f. $A \cup (A \cap B) = A$ and $A \cap (A \cup B) = A$.

1.2 If $S_1, S_2, \ldots, S_n \subset S$, then, for $k \leq n-1$, show that
$$Pr\left[\left(\bigcup_{i=1}^{k} S_i\right) \cup S_{k+1}\right] = Pr\left(\bigcup_{i=1}^{k+1} S_i\right).$$

1.3 If $Pr(B) > 0$, then show:
   a. $Pr(A|B) = Pr(A)/Pr(B)$ for $A \subset B$
   b. $Pr(A|B) = 1$ for $A \supset B$

1.4 If $Pr(B) > 0$ and $Pr(\overline{B}) > 0$, show that $Pr(A|B) + Pr(\overline{A}|B) + Pr(\overline{A}|B) = 1$.

1.5 Show that in general the following two statements are false:
   a. $Pr(A|B) + Pr(A|\overline{B}) = 1$
   b. $Pr(A|B) + Pr(\overline{A}|\overline{B}) = 1$

1.6 If $Pr(B) = p$, $Pr(\overline{A}|B) = q$, and $Pr(\overline{A} \cap \overline{B}) = r$, find:
   a. $Pr(A \cap \overline{B})$
   b. $Pr(A)$
   c. $Pr(B|A)$

1.7 If A and B are independent events, prove:
   a. A and $\overline{B}$ are independent.
   b. $\overline{A}$ and B are independent.
   c. $\overline{A}$ and $\overline{B}$ are independent.

1.8 Show that if two sets are mutually exclusive, they are not independent but that the converse does not hold.

1.9 If A and B each play the game of each tossing alternately a fair coin, the winner being the one who tosses a head first, find the probability that A wins, assuming A tosses first.

1.10 Find the same probability as in Exercise 1.9, but in this case let the probability that the coin comes up heads on any one toss be equal to p.

1.11 State when two independent events are also mutually exclusive.

1.12 State when two mutually exclusive events are also independent.

Theoretical Exercises

For Exercises 1.13 to 1.17, assume that n, m, and k are integers, with $n > k$ and $m > k$.

1.13 Show:

a. $(1+t)^n = \sum_{x=0}^{n} \binom{n}{x} t^x$

b. $(1-t)^n = \sum_{x=0}^{n} (-1)^x \binom{n}{x} t^x$

c. $(1+t)^{-n} = \sum_{x=0}^{\infty} (-1)^x \binom{n-1+x}{x} t^x$

d. $(1-t)^{-n} = \sum_{x=0}^{\infty} \binom{n-1+x}{x} t^x$

1.14 Show:

a. $\sum_{x=0}^{r} \binom{n}{x} \binom{m}{r-x} = \binom{n+m}{r}$

b. $\sum_{x=0}^{n} \binom{n}{x}^2 = \binom{2n}{n}$

1.15 Show:

a. $\binom{n}{k} = \binom{n-1}{k-1} + \binom{n-1}{k}$

b. $\sum_{x=0}^{n-k} \binom{n-1-x}{k-1} = \binom{n}{k}$

c. $\sum_{x=0}^{k} \binom{n-1-x}{k-x} = \binom{n}{k}$

1.16 Show:

a. $\binom{-n}{k} = (-1)^k \binom{n+k-1}{k}$

b. $\binom{-n}{-k} = (-1)^{n+k} \binom{k-1}{n-1}$, $k \geq n$

1.17 Show:

a. $\sum_{x=0}^{n,2k} (-1)^x \binom{n}{x} \binom{n}{2k-x} = (-1)^k \binom{n}{k}$,

where $n, 2k = \min(n, 2k)$ and $\frac{n}{2} > k$

b. $\sum_{x=0}^{2k} (-1)^x \binom{n+x}{x} \binom{n+2k-x}{2k-x} = \binom{n+k}{k}$

c. $\sum_{x=0}^{k} (-1)^x \binom{n+k}{x} = (-1)^k \binom{n-1+k}{k}$

d. $\sum_{x=0}^{k} (-1)^x \binom{m-1+x}{x} \binom{n}{k-x} = \binom{n-m}{k}$

e. $\sum_{x=0}^{n,k} (-1)^x \binom{n}{x} \binom{m-1+k-x}{k-x} = (-1)^k \binom{n-m}{k}$

f. $\sum_{x=k}^{n}\binom{x}{k}=\binom{n+1}{k+1}$

g. $\sum_{x=0}^{k}(-1)^{x}\binom{n}{x}=(-1)^{k}\binom{n-1}{k}$

1.18 The Fibonacci sequence of integers is 1, 1, 2, 3, 5, 8, 13,..., where each element in the sequence is the sum of the previous two terms. Show that the ith term may be expressed as

$$\sum_{x=0}^{\left[\frac{i-1}{2}\right]}\binom{i-1-x}{x},$$

where $\left[(i-1)/2\right]$ is the largest integer not exceeding $(i-1)/2$.

1.19 If n balls numbered 1 to n are put at random into n cells numbered 1 to n, where each cell is occupied by one ball, the probability that each will have the same number is given by

$$Pr(m)=\frac{1}{m!}\sum_{k=m}^{n}(-1)^{k-m}\frac{1}{(k-m)!}.$$

Verify this formula for $n=4$ and $m=0, 1, 2, 3, 4$. Derive the formula for obtaining at least $m\,(m\geq 1)$ matches and get

$$\frac{1}{(m-1)!}\sum_{k=m}^{n}(-1)^{k-m}\frac{1}{k(k-m)!}.$$

A point divides a line segment into two unequal parts. Two other points, one on either side of the first point, are chosen at random. We want to find the probability that the three segments formed by the second two points make a triangle.

Let the line segment be denoted as shown in Figure 1.E.1, where lengths a and b are fixed by the location of the first point and x and y can each vary in length from 0 to a and from 0 to b, respectively.

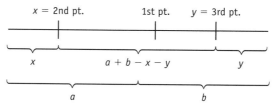

Figure 1.E.1

We want the lengths x, y, and $a=b-x-y$ to form a triangle, which means that any one of the lengths cannot exceed the sum of the other two lengths. Thus, we have that

$$x<\frac{a+b}{2},\ y<\frac{a+b}{2},\ \text{and}\ x+y>\frac{a+b}{2}.$$

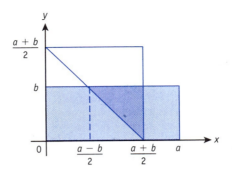

Figure 1.E.2

In Figure 1.E.2, the doubly shaded area represents these three inequalities and the singly shaded area represents all possible values that x and y may take on. Therefore, the doubly shaded area divided by the singly shaded area represents the probability that we want. This reasoning is analogous to representing a probability as the number of favorable outcomes divided by the total number of outcomes.

Our answer comes out to be $b/2a$, which depends on the assumption that a is greater than b. Otherwise, the result would be $a/2b$. Thus, the answer is (smaller length)/2(larger length).

Exercises 1.20 to 1.23 should be solved in a similar manner.

1.20 A point is chosen at random on a line segment length l. Find the probability that the two line segments thus formed, along with the distance $l/2$, make a triangle.

1.21 Three points, X_1, X_2, and X_3, are chosen at random on a line segment of length l. Find the probability that X_3 falls between X_1 and X_2.

1.22 Two points are chosen at random on a line segment of length l. Find the probability that the three line segments thus formed make a triangle.

1.23 In a different type of problem, three points are chosen at random on the circumference of a circle. Find the probability that they lie on the same semicircle.

Applied Problems

1.1 A pair of balanced dice is tossed. If X equals the total number of spots showing on the dice, then, for $k = 2, 3, \ldots, 12$, find:
 a. $Pr(X = k)$ b. $Pr(X \le k)$ c. $Pr(X \ge k)$

1.2 In Problem 1.1, find:
 a. $Pr(4 \le X \le 9)$ b. $Pr(4 < X \le 9)$
 c. $Pr(4 \le X < 9)$ d. $Pr(4 < X < 9)$

1.3 In a freshman high school class of 400 students, 110 take math, 130 take English, 150 take history, 30 take math and English, 40 take math and history, 50 take English and history, and 280 take math or English or history. If a freshman student is picked at random from the class, then, for $k = 0, 1, 2, 3$, find the probability that:

 a. He takes exactly k of the named subjects.
 b. He takes at least k of the named subjects.
 c. He takes no more than k of the named subjects.

1.4 A man tosses a fair coin twice. Find the conditional probability that he gets two heads, given that he got at least one head.

1.5 Three balanced dice are tossed. Find the probability of obtaining a nine, given:
  a. The sum is odd.
  b. The sum is less than or equal to nine.
  c. None of the dice are odd.
  d. At least one of the dice is odd.
  e. At least two of the dice are odd.
  f. All dice are odd.
  g. All dice are different.
  h. Two of the dice are the same.
  i. All dice are the same.

1.6 Three men, A, B, and C, fire at a target. A fires twice as often as B, who fires three times as often as C, who in turn fires once every 6 seconds. B's accuracy is twice that of A, which is one third that of C, which in turn is equal to 0.8. After 1 minute of shooting, firing is halted. Find the probability that the target is hit. Would this probability be any different if the firing had been halted at the end of, say, 6 seconds?

1.7 In Problem 1.6, assume that the target was hit at least once. Find the probability that any one hit was the result of:
  a. A firing   b. B firing   c. C firing

1.8 Each of 12 ordered boxes contains 12 coins, consisting of pennies and dimes. The number of dimes in each box is equal to its order among the boxes, that is, box 1 contains one dime and 11 pennies, box 2 contains two dimes and 10 pennies, and so on. A pair of fair dice is tossed, and the total showing indicates which box is chosen to have a coin selected at random from it. Find the probability that the coin selected is a dime.

1.9 In Problem 1.8, assume that the coin selected is a dime. Then, for $k=1,2,\ldots,12$, find the probability that it came from box k.

1.10 If a balanced coin is tossed n times, find the probability of getting at least one head.

1.11 In Problem 1.10, find the least number of times the coin needs to be tossed to ensure that the probability of getting at least one head is greater than three-quarters.

1.12 If a population consists of five different objects, find the probability that the third object is included in a sample of size three under:
  a. Scheme 1   b. Scheme 2   c. Scheme 3   d. Scheme 4

1.13 Suppose six balanced dice are tossed. Find the probability:
  a. All show a different value.
  b. Exactly two show the same value.
  c. At least k show the same value, where $k = 2, 3, 4, 5, 6$.

1.14 Tossing a pair of fair dice, find the probability that a total of k results before a total of 7 shows, where $k = 4, 5, 6, 7, 8, 9, 10$. *Hint:* Consider the infinite geometric series

$$Pr(k) + Pr(\bar{k} \cap \bar{7}) Pr(k) + \left[Pr(\bar{k} \cap \bar{7}) \right]^2 Pr(k) + \ldots = \frac{Pr(k)}{Pr(k) + Pr(7)},$$

the sum of which should be derived and not just accepted.

1.15 In the game of craps, the thrower wins on the first throw if he gets a 7 or an 11. He loses on the first throw if he gets a 2, a 3, or a 12. To win on any subsequent throw, he must roll the same total he rolled on the first throw before he rolls a 7, provided that the first throw was not a 2, a 3, or 12. Otherwise, he loses. Find the probability that the thrower wins in the game of craps.

1.16 In a game of poker, find the probability of being dealt:
  a. a royal flush (A, K, Q, J, 10 of the same suit)
  b. a straight flush but not a royal flush (five consecutive cards of the same suit except A, K, Q, J, 10)
  c. four of a kind (four cards of the same face value and another card)
  d. a full house (three cards of the same face value and two other cards of the same face value)
  e. a flush but not a straight flush or a royal flush (five cards of the same suit but not in consecutive order)
  f. a straight but not a flush (five cards in consecutive order that are not all from the same suit)
  g. three of a kind (three cards of the same face value and two others each of different face value with respect to one another and to the first three cards)
  h. two pairs (two cards of the same face value, two more cards of the same face value that differs from the value of the first two, and another card different in face value from the first four cards)
  i. a pair (two cards of the same face value and three others each different in face value with respect to one another and the first two cards)

1.17 (*Chevalier de Méré's problem*) Determine which outcome in the following two situations is more likely to occur: obtaining a total of 6 at least once when tossing a die four times, or obtaining a total of 12 at least once when tossing a pair of dice 24 times.

1.18 (*Banach's problem*) A man has two boxes of matches, each of which contains n matches. He selects a box at random and uses one match from it each time he wishes to light a cigarette. Find the probability that, when he selects a box to extract a match from, the box is empty and the other box has k matches in it, where $k = 0, 1, \ldots, n$.

1.19 A coin is tossed until the same result appears twice in a row. Find the probability:
  a. This event occurs on the nth toss.
  b. Six or fewer tosses are required for this event.
  c. This event occurs on an even-numbered toss.

1.20 There are s pairs of shoes in a closet. A man selects $2n$ shoes at random. Find the probability that he will have at least one pair among the $2n$ shoes selected.

1.21 An urn contains w_1 white and b_1 black balls. A second urn contains w_2 white and b_2 black balls. A ball is selected at random from the first urn and placed in the second urn. Then a ball is selected at random from the second urn. Find the probability that it is white.

1.22 In Problem 1.21, assume that the ball selected from the second urn is white. Find the probability that it came from the first urn.

1.23 If n people each flip a balanced coin, find the probability that exactly one coin is at odds with the rest.

References

[1] Todhunter, I. *A History of the Theory of Probability from the Time of Pascal to Laplace.* New York: Chelsea Publishing Company, 1949.

[2] Von Mises, R. *Warscheinlichkeit, Statistik und Wahrheit.* Vienna: Springer-Verlag, 1936.

[3] Kolmogorov, A. N. "Grundbegriffe der Wahrscheinlichkeitsrechnung." *Ergeb. Mat. und ihrer Grenzg.,* 2(3), 1933. (Translated to: *Foundations of the Theory of Probability.* New York: Chelsea Publishing Company, 1936.)

[4] Poincaré, H. *Calcul des Probabilités.* Paris: Gauthier-Villero, 1912.

> "But to us, probability is the very guide to life."
> —"The Analogs of Religion," 1736

CHAPTER 2

Discrete Probability Distributions

OBJECTIVES:

2.1 Discrete Probability Density Function
2.2 Cumulative Distribution Function
2.3 Point Binomial Distribution
2.4 Binomial Probability Distribution
2.5 Poisson Probability Distribution
2.6 Hypergeometric Probability Distribution
2.7 Geometric Probability Distribution
2.8 Negative Binomial Probability Distribution

Jakob Bernoulli (1654–1705)

> "In the middle of every difficulty lies opportunity."
> —Albert Einstein

The merchant family Bernoulli, of Basel, Switzerland, has produced scientists in each generation since the end of the seventeenth century. One of the distinguished members of this family was Jakob (also called James or Jacques), who pioneered work in differential and integral calculus and the field of probability theory. His work on the binomial distribution produced, among other achievements, Bernoulli's law and the Bernoulli trials.

Bernoulli's initial studies were in theology, primarily at his father's insistence, but he refused a church appointment and began lecturing on experimental physics. He became a mathematics professor at Basel University and later served as rector there. His earliest papers concerned astronomy (especially the motions of comets); later papers explored infinite series and their summation (the Bernoulli numbers are named for him).

continued

> Both Jakob Bernoulli and his brother Johann (also called John or Jean) constantly exchanged correspondence with Gottfried Wilhelm von Leibniz[1] (often with bitter rivalry), producing many of the fundamentals of calculus and the integration of ordinary differential equations. Jakob Bernoulli examined many special curves, including the catenary (the curve of a hanging chain), the isochron (the curve along which a body falls with uniform velocity), and the logarithmic spiral (the spiral that reproduces itself under various transformations). This last was such a delight to him that he stipulated that it be engraved on his tombstone with the inscription *Eadem Mutata Resurgo*. ("I arise the same through change.")
>
> ---
>
> [1] Wilhelm von Leibniz was a world-famous mathematician known for developing calculus independently of Sir Isaac Newton.

■ INTRODUCTION

A random variable is a function (rule) that assigns a number to every outcome of an experiment. That is, if, after an experiment, $s_1, s_2, \ldots, s_n$ are all the possible outcomes that constitute the sample space S and we wish to assign a number to each of these outcomes by using a rule $X(s_i), i = 1, 2, \ldots, n$, then we call this rule a *random variable*.

Before we proceed to define precisely the meaning of and explain the role of a random variable in probability theory and applications, we briefly review the concept of a function, which plays a central role in scientific investigation.

In mathematics, and in many physical sciences, we encounter basic formulas. For example, if r is the radius of a circle, then its area is given by $A = \pi r^2$. If we wish to determine the area of a circle, we must specify its radius. Thus, the area is a *dependent variable* whose value depends on the *independent variable* r. This means that for each value of r there is a corresponding value of area. The collection of these pairs of numbers is a *function* (or *mapping*); that is, a function is a rule or relationship between the value r and the value A. Therefore, we can say that the area of a circle is a function of its radius, or $A = f(r)$. The notion of a function involves three things: (1) a set D called the *domain* of the function, (2) a set R called the *range* (or *image*) of the function, and (3) a rule that assigns to each element $x \in D$ an element $y \in R$. Thus, the set of pairs

$$\{(x, y): x \in D, y \in R; y = f(x)\}$$

is a function if $y = f(x)$ does not assign two values of y to a single value of x. In the preceding example, let the domain be the set $D = \{r : r = 1, 2, 3, 4\}$. Applying the rule $A = f(r) = \pi r^2$ for $r \in D$, we get $f(1) = \pi$ (the value of f at $r = 1$), $f(2) = 4\pi$ (the value of f at $r = 2$), and so on. Thus, we obtain the range of the function $R = \{A : A = \pi, 4\pi, 9\pi, 16\pi\}$. Therefore, the function itself is the set

$$\{(1, \pi), (2, 4\pi), (3, 9\pi), (4, 16\pi)\}.$$

Definition 2.0.1

A **function** is a collection of ordered pairs of numbers (x, y), where $x \in D$ and $y \in R$, with no two different pairs having the same first number.

Suppose that S is the sample space of a given experiment. Let $X(s)$ be a real valued function, defined on S, that transforms points of S to the set R of real numbers. Thus, S is the domain of the function, and the range is the set of real numbers. For example, if we consider an experiment that involves flipping a pair of coins, the sample space consists of four elements:

$$S = \{(H,H), (H,T), (T,H), (T,T)\}.$$

If we let $X(s)$ be a real valued function that can assume values equal to the number of heads in s_i ($s_1 = HH$, $s_2 = HT$, $s_3 = TH$, and $s_4 = TT$), then this function may be represented by the set of ordered pairs

$$\{[s_1, X(s_1)], [s_2, X(s_2)], [s_3, X(s_3)], [s_4, X(s_4)]\}$$

or

$$\{(s_1, 2), (s_2, 1), (s_3, 1), (s_4, 0)\}.$$

As another example, consider an experiment that involves rolling a pair of dice once. The sample space consists of 36 ordered pairs of numbers:

$$S = \{(1,1), (1,2), \ldots, (6,6)\}.$$

If we let $X(e_i, e_j) = e_i + e_j$, $i, j = 1, 2, \ldots, 6$ be a real valued function that assigns to each point (e_i, e_j) in S the sum of its members, then this function can be represented by the set of ordered pairs

$$\{[(e_1, e_1), X(e_1, e_1)], [(e_1, e_2), X(e_1, e_2)], \ldots, [(e_6, e_6), X(e_6, e_6)]\}$$

or

$$\{[(1,1), 2], [(1,2), 3], \ldots, [(6,6), 12]\}.$$

We now define the concept of a random variable.

Definition 2.0.2

Let S be a sample space and $X(s)$ a real valued function (defined on S) that transforms the points of S into a subset of the real line, R_X (range space). Then, X is called a *random variable*, provided that the probabilities of $\{X(s) = x\}$ and $\{X(s) \leq x\}$ can be calculated, where s is any element in the sample space and x is any number in R_X.

For convenience, we denote $X(s)$ as X and the values of the range space R_X as $x_1, x_2, \ldots, x_n$. Using only the two events listed in the definition, the following events can be specified:

1. $\{X < x_1\}$, $\{X \geq x_1\}$, $\{X > x_1\}$
2. $\{x_1 \leq X \leq x_2\}$, $\{x_1 < X \leq x_2\}$, $\{x_1 \leq X < x_2\}$, $\{x_1 < X < x_2\}$

For example, the events

$$\{x_1 \leq X < x_2\} = \{X \leq x_2\} - \{X = x_2\} - \{X \leq x_1\} + \{X = x_1\}$$

and

$$\{x_1 \leq X \leq x_2\} = \{X \leq x_2\} - \{X \leq x_1\} + \{X = x_1\}.$$

We can illustrate the meaning of the preceding definition with an example: Consider the previous experiment, in which a pair of fair dice was rolled once. The sample space consists of 36 ordered pairs of numbers, namely,

$$S = \{(1,1), (1,2), \ldots, (6,6)\}.$$

Let the random variable X represent the total number of spots showing on the dice. Thus, X assigns to each ordered pair in S its sum; that is, $X(1,1) = 2$, $X(1,2) = 3, \ldots$, $X(6,6) = 12$, which are in R_X. So far, we have established that X is a real valued function, defined on S, that has a range that is a subset of R_X. Next, consider the events $\{X = x\}$ and $\{X \leq x\}$, where x may be any real number. It is clear that the probabilities of these events are defined.

For example,

$$Pr(X = -38) = 0, \quad Pr(X = 5.5) = 0, \quad Pr(X = 8) = \frac{5}{36},$$

$$Pr(X \leq 0) = 0, \quad Pr(X \leq 6.4) = Pr(X \leq 6) = \frac{15}{36},$$

$$Pr(X \leq 45) = 1, \quad \text{and} \quad Pr(X \leq \infty) = 1.$$

We have further shown that the probabilities of the events $\{X = x\}$, $\{X \leq x\}$, $x \in R_x$, are defined. Therefore, X is a random variable.

Synonyms for *random variable* are *chance variable*, *variate*, and *stochastic variable*. We denote random variables by the capital letters X, Y, Z, and so on; $X(s)$ indicates the value assigned to an observation of the experiment.

Random variables in this book are divided into two classes: **discrete random variables** and **continuous random variables**.

In this chapter, we are concerned with the one-dimensional discrete random variable. In Chapter 3, we discuss the one-dimensional continuous variable.

2.1 Discrete Probability Density Function

We begin by defining a one-dimensional discrete random variable.

Definition 2.1.1

Let X be a random variable. If the number of elements in the range space R_X is finite or countably infinite, then X is called a **one-dimensional discrete random variable**.

One of the primary objectives in developing the concept of a random variable is the calculation of the probability so that it assumes certain values. For example, we may be interested in finding the probabilities of some of the events given in the introduction to this chapter. We could have $Pr(X = x_1)$, the probability that the random variable is equal to x_1, and $Pr(x_1 < X < x_2)$, the probability that the random variable falls between x_1 and x_2. More precisely, if $E_1 = \{x : x_1 < x < x_2\}$, then $Pr(x_1 < X < x_2) = Pr(X \text{ in } E_1)$, the probability that the random variable assumes a value contained in a certain set $E_1 \subset E$. Similarly, we would be interested in $Pr(X < x_1)$, $Pr(X \leq x_2)$, $Pr(X > x_2)$, $Pr(X \geq x_2)$, and $Pr(x_1 \leq X \leq x_2)$.

The mathematical function used to determine these probabilities is called the **probability density function of a random variable**. For example, if the discrete random variable X may assume the values $x_1, x_2, \ldots, x_n$, we might be interested in determining a function $f(x)$ such that $f(x_1), f(x_2), \ldots, f(x_n)$ are the probabilities that $X = x_1, X = x_2, \ldots, X = x_n$. Having such a function helps us find the probability of the event $\{x_1 \leq X \leq x_2\}$; that is,

$$Pr(x_1 \leq X \leq x_2) = \sum_{x_1 \leq X \leq x_2} f(x).$$

We formulate such a mathematical function in Definition 2.1.2.

Definition 2.1.2

Let X be a one-dimensional discrete random variable. A function $f(x_i)$ is called a **discrete probability density function** or **density function** of the random variable X if the following conditions are satisfied:

1. $f(x_i) \geq 0$, for all $x_i \in R_X$

2. $\sum_{x_i \in R_X} f(x_i) = 1$

Here, the summation is taken over all x_i in the range space. Thus, we call $f(x_i)$ a discrete probability density function of the random variable X if both of the preceding conditions hold.

Other common ways in which we can phrase Definition 2.1.2 are "$f(x_i)$ is the distribution of the random variable X" and "the random variable X is distributed as $f(x_i)$." The distribution $f(x_i)$ is also referred to as the *mass* or *frequency* density function. The following examples illustrate the manner in which the distribution of a random variable can be determined and applied.

Example 2.1.1

Let an experiment consist of tossing five fair coins, and let the random variable X represent the number of heads that show. Thus, X may take on the value 0, 1, 2, 3, 4, or 5; we wish to construct its probability density function.

Solution: Because each toss of a coin has two possible outcomes, the sample space of this experiment contains 2^5 5-tuples, and the event of getting x heads can occur in $\binom{5}{x}$ ways. Therefore,

$$f(x) = \begin{cases} \dfrac{\binom{5}{x}}{2^5}, & x = 0, 1, \ldots, 5, \\ 0, & \text{elsewhere.} \end{cases}$$

We can verify that $f(x)$ is the probability distribution function of the random variable X by showing that both conditions from Definition 2.1.2 are satisfied:

$$\sum_{\text{all } x} f(x) = \sum_{x=0}^{5} \frac{\binom{5}{x}}{2^5} = 1$$

and

$$f(x_i) \geq 0,$$

for $x = 0, 1, \ldots, 5$.

We can use the discrete probability density function $f(x)$ to calculate the following probabilities, for example, concerning the random variable X:

a. The probability of obtaining *exactly* three heads in the tossing of five coins is

$$Pr(X = 3) = f(3) = \frac{\binom{5}{3}}{2^5} = \frac{5}{16}.$$

b. The probability that the random variable X is *at most* equal to four heads is

$$Pr(X \leq 4) = \sum_{x=0}^{4} \frac{\binom{5}{x}}{2^5} = 1 - Pr(X > 4) = 1 - Pr(X = 5) = 1 - \frac{\binom{5}{5}}{2^5} = \frac{31}{32}.$$

c. The probability that the random variable X is *at least* equal to three heads is

$$Pr(X \geq 3) = \sum_{x=3}^{5} \frac{\binom{5}{x}}{2^5} = 1 - Pr(X < 3) = 1 - \sum_{x=0}^{2} \frac{\binom{5}{x}}{2^5} = \frac{1}{2}.$$

d. The probability that the random variable X (the number of heads) assumes a value in the closed interval $[2, 4]$ is

$$Pr(2 \leq X \leq 4) = \sum_{x=2}^{4} \frac{\binom{5}{x}}{2^5} = \frac{25}{32}.$$

e. The probability that the random variable X equals two, given that the number of heads is less than or equal to three is

$$Pr(X = 2 | X \leq 3) = \frac{\frac{\binom{5}{2}}{2^5}}{\sum_{x=0}^{3} \frac{\binom{5}{x}}{2^5}} = \frac{\binom{5}{2}}{\sum_{x=0}^{3} \binom{5}{x}} = \frac{5}{13}.$$

f. The probability that the number of heads is less than or equal to two when fewer than four heads occurred is

$$Pr(X \leq 2 | X < 4) = \frac{\sum_{x=0}^{2} \frac{\binom{5}{x}}{2^5}}{\sum_{x=0}^{3} \frac{\binom{5}{x}}{2^5}} = \frac{\sum_{x=0}^{2} \binom{5}{x}}{\sum_{x=0}^{3} \binom{5}{x}} = \frac{8}{13}.$$

Figure 2.1.1 shows the graph of the density function $f(x)$ as vertical lines, the lengths of which indicate the probabilities for all possible values that the random variable can assume. The function is symmetrical; that is, the probability that the random variable $X = 0$, no heads, is equal to the probability that $X = 5$, all heads. Also, $Pr(X = 1) = Pr(X = 4) = 5/32$, and so on. Furthermore, the graph indicates that the probability of getting two heads is the same probability as that of getting three heads, because the probability of obtaining a head on a single toss is the same as that of obtaining a tail.

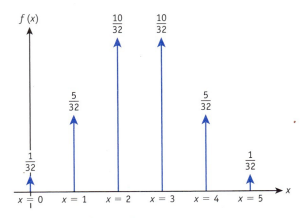

Figure 2.1.1 Discrete Probability Distribution

Example 2.1.2

Consider an experiment that involves throwing a pair of fair dice once. Let the random variable X be equal to the sum of the spots resulting. We wish to find its density, $f(x)$, where X may take on the values $2, 3, \ldots, 12$.

Solution: Here $f(2) = f(12) = 1/36$, $f(3) = f(11) = 2/36$, $f(4) = f(10) = 3/36$, $f(9) = 4/36$, $f(6) = f(8) = 5/36$, and $f(7) = 6/36$.

With a little algebraic manipulation, we obtain

$$f(x) = \begin{cases} \dfrac{6 - |7 - x|}{36}, & x = 2, 3, \ldots, 12, \\ 0, & \text{elsewhere.} \end{cases}$$

Using this density function, we can obtain, for example, the following probabilities:

a. $Pr(5 < X \leq 7) = \sum_{x=6}^{7} (6 - |7 - x|/36) = 11/36$, the probability that the sum is greater than five but less than eight.

b. $Pr(X > 9) = \sum_{x=10}^{12} (6 - |7 - x|/36) = 6/36$, the probability that the observed sum is greater than nine.

Figure 2.1.2 shows the graph of $f(x)$; the vertical lines indicate the probabilities for the values of x that the random variable assumes. Clearly, $f(x) \geq 0$ for all x, and we can verify that

$$\sum_{x=2}^{12} \frac{6 - |7 - x|}{36} = 1.$$

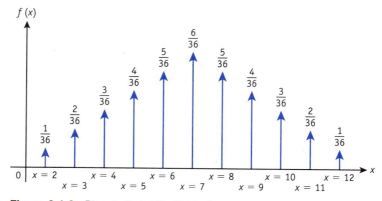

Figure 2.1.2 Discrete Probability Distribution

Example 2.1.3

In the experiment in Example 2.1.2, suppose that the random variable X is equal to the difference of the spots resulting. X could take on the values $-5, -4, \ldots, 0, \ldots, 4, 5$ and $f(-5) = f(5) = 1/36$, $f(-4) = f(4) = 2/36$, $f(-3) = f(3) = 3/36$, $f(-2) = f(2) = 4/36$, $f(-1) = f(1) = 5/36$, and $f(0) = 6/36$.

Combining these results gives

$$f(x) = \begin{cases} \dfrac{6-|x|}{36}, & x = -5, -4, \ldots, 5, \\ 0, & \text{elsewhere.} \end{cases}$$

a. Find the probability that the random variable assumes a value between -3 and 4.

b. Find the probability that the observed difference of the spots is either -4 or 0.

Solution:

a. $Pr(-3 < X < 4) = \sum_{x=-2}^{3} \dfrac{6-|x|}{36} = \dfrac{27}{36}$

b. $Pr(X = -4) + Pr(X = 0) = f(-4) + f(0) = \dfrac{2}{36} + \dfrac{6}{36} = \dfrac{2}{9}$

Example 2.1.4

Find c so that the function given by

$$f(x) = c2^{-|x|}, \quad x = 0, \pm 1, \pm 2, \ldots,$$

is a probability density function.

Solution: Here, we must find c such that

$$\sum_{x=-\infty}^{\infty} c2^{-|x|} = 1.$$

Thus,

$$\sum_{x=-\infty}^{\infty} c2^{-|x|} = c\sum_{x=-\infty}^{-1} 2^{x} + c2^{0} + c\sum_{x=1}^{\infty} 2^{-x}$$

$$= c\sum_{x=1}^{\infty} 2^{-x} + c + c\sum_{x=1}^{\infty} 2^{-x}$$

$$= c + c + c = 1,$$

$$3c = 1 \quad \text{or} \quad c = \dfrac{1}{3}.$$

Therefore,

$$f(x) = \frac{1}{3} 2^{-|x|}, \quad x = 0, \pm 1, \pm 2, \ldots$$

is a probability density function.

Another way of graphing a discrete probability density function is by a **probability histogram**, as shown in Figure 2.1.3 for Example 2.1.3. In the probability histogram, the rectangles are typically made of equal width so that the areas of the rectangles are proportional to their heights. Furthermore, the base of each rectangle extends from half a unit below to half a unit above the value that the random variable can assume.

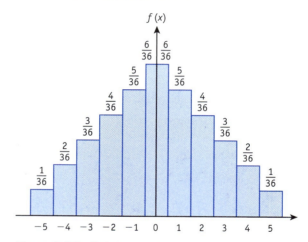

Figure 2.1.3 Probability Histogram

2.2 Cumulative Distribution Function

We have seen that, given a discrete random variable X, it is often necessary to determine various probabilities concerning the behavior of X. For example, we might be interested in calculating the probability that the random variable X takes on a value contained in the set $E_1 = \{x : x_1 \leq x \leq x_2\}$ or in the set $E_j = \{x : x \leq x_j\}$. The probability that X assumes a value in E_j can be obtained by writing

$$Pr(X \leq x_j) = \sum_{x_i \leq x_j} f(x_i), \qquad (2.2.1)$$

where $f(x_i)$ is the probability density function of the random variable X. We denote such a function by $F(x)$ and define it in Definition 2.2.1.

Definition 2.2.1

The function $F(x)$, defined by

$$F(x) = Pr(X \leq x) = \sum_{x_i \leq x} f(x_i), \qquad (2.2.2)$$

where the summation is extended over all points x_i for which $x_i \leq x$, is called the **cumulative distribution function** of the discrete random variable X, which has the probability density function $f(x_i)$.

As the name implies, the cumulative distribution function $F(x)$ accumulates the values of $f(x_i)$ as x_i goes from $-\infty$ to x. Also, its graph is a "step function" that is continuous from the right.

Example 2.2.1

The cumulative distribution function of the random variable X, whose density function was given in Example 2.1.1, is obtained by

$$F(x) = Pr(X \leq x) = \begin{cases} 0, & x < 0, \\ \sum_{x_i \leq [x]^*} \dfrac{\binom{5}{x_i}}{2^5}, & 0 \leq x \leq 5, \\ 1, & x > 5. \end{cases}$$

Table 2.2.1 gives the cumulative probabilities for all values that the random variable can assume and provides data for the graphical representation of $F(x)$, as shown in Figure 2.2.1.

Table 2.2.1 Cumulative Probability Distribution

X	0	1	2	3	4	5
$F(x)$	$\dfrac{1}{32}$	$\dfrac{6}{32}$	$\dfrac{16}{32}$	$\dfrac{26}{32}$	$\dfrac{31}{32}$	$\dfrac{32}{32}$

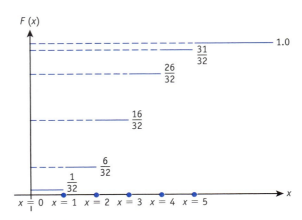

Figure 2.2.1 Cumulative Distribution Function

Theorem 2.2.1 states the probabilities of the cumulative distribution function of a discrete random variable.

---
*[x] denotes the greatest integer not exceeding x.

Theorem 2.2.1

If $F(x)$ is the cumulative distribution function of the discrete random variable X, then $F(x)$ has the following properties:

1. $F(\infty) = 1$.
2. $F(-\infty) = 0$.
3. $F(x)$ is a nondecreasing function of x.
4. $F(x)$ is continuous from the right.

Proof Properties 1 and 2 follow from the definition of a probability density function of the random variable X. To show property 3—that is, $F(x_2) \geq F(x_1)$ for x_1 and x_2, two points on the real axis, such that $x_1 < x_2$—we let the set $E_1 = \{x : x \leq x_1\}$ be a subset of the set $E_2 = \{x : x \leq x_2\}$. Thus,

$$E_1 \subset E_2 \quad \text{or} \quad \{x : x \leq x_1\} \subset \{x : x \leq x_2\} \tag{2.2.3}$$

Applying Theorem 1.2.1 to Equation (2.2.3), we have $Pr(E_1) \leq Pr(E_2)$, which is equivalent to

$$Pr(X \leq x_1) \leq Pr(X \leq x_2)$$

or

$$F(x_1) \leq F(x_2).$$

Thus, $F(x)$ is nondecreasing. To show that the distribution function is continuous from the right, we need Definition 2.2.2.

Definition 2.2.2

A sequence of events $\{E_i\}, i = 1, 2, 3, \ldots$, is a **nonincreasing sequence** if

$$E_i \supset E_{i+1}$$

for all i. The intersection of the events of the nonincreasing sequence is called the **limit** of this sequence; that is,

$$\lim_{i \to \infty} E_i = \bigcap_{i=1}^{\infty} E_i = E^*.$$

Let $x_1 > x_2 > \ldots > x^*$ be a decreasing sequence of points that converge to x^* from the right, and let $E_i, i = 1, 2, 3, \ldots$, be the event that the random variable X assumes a value from the interval $(x^*, x_i]$. That is,

$$E_i = \{x : x \in (x^*, x_i]\}.$$

If $i_1 < i_2$, the occurrence of event E_{i_2} implies the occurrence of the event E_{i_1}. Thus, the sequence of events $\{E_i\}$ is a nonincreasing sequence. The point x^* is not included in any of the intervals considered, which implies that the intersection of the events $E_i, i = 1, 2, 3, \ldots$, is the impossible event; that is, $Pr(E^*) = 0$, where $E^* = \underset{i=1}{\overset{\infty}{I}} E_i$. Therefore,

$$\lim_{i \to \infty} Pr(E^*) = \lim_{i \to \infty} Pr\left[x^* < X \leq x_i\right]$$

$$= \lim_{i \to \infty}\left\{Pr\left[X \leq x_i\right] - Pr\left[X \leq x^*\right]\right\}$$

$$= \lim_{i \to \infty}\left[F(x_i) - F(x^*)\right]$$

$$= \lim_{i \to \infty} F(x_i) - F(x^*)$$

$$= 0$$

or

$$F(x^*) = \lim_{i \to \infty} F(x_i) = \lim_{x_i \to x_i^*} F(x_i).$$

Thus, we have continuity at the point x^* from the right.

Example 2.2.2

The cumulative distribution function of the probability density of the experiment, as stated in Example 2.1.2, is given by

$$F(x) = Pr(X \leq x) = \begin{cases} 0, & x < 2, \\ \sum_{x_i \leq [x]} \dfrac{6 - |7 - x_i|}{36}, & 2 \leq x \leq 12, \\ 1, & x > 12. \end{cases}$$

Figure 2.2.2 shows that $F(x)$ is a step function and continuous from the right.

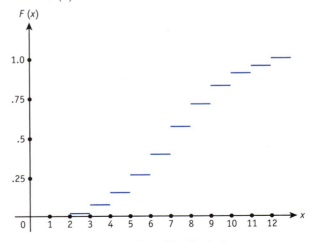

Figure 2.2.2 Cumulative Probability Distribution

In the following sections, we study some special discrete probability distributions that are applicable to various physical problems.

2.3 Point Binomial Distribution

Consider an experiment entailing a single observation, the result of which is one of two possible outcomes, for example, success or failure, operable or nonoperable, or yes or no. Whatever the event, if the probability of success is p, the probability of a failure is $q=1-p$, where $p+q=1$. We denote a failure using 0 and a success using 1, and we formulate the distribution of the experiment as explained in Definition 2.3.1.

Definition 2.3.1

A random variable X is said to be distributed as a **point binomial** if its probability density function is

$$f(x;p) = \begin{cases} p^x q^{1-x}, & x=0,1, \quad 0<p<1, \\ 0, & \text{elsewhere.} \end{cases} \quad (2.3.1)$$

It is easy to show that the conditions of a probability density function are satisfied. Equation (2.3.1) is a one-parameter distribution, where the parameter is p. In general, the word **parameter** means a constant that characterizes a probability density function. A **statistic** is a parameter-free function of a random variable, which is based on the observed values of a given experiment. Usually, a parameter is estimated by a statistic. The graph of $f(x;p)$ and $F(x)$ is shown in Figure 2.3.1.

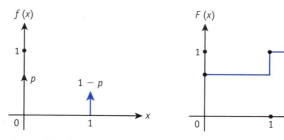

Figure 2.3.1 $f(x; p)$ and $F(x)$

Example 2.3.1

Consider an experiment that involves firing a missile that has been observed to fire successfully with a probability of $p=0.88$
The density of the random variable X is

$$f(x;0.88) = \begin{cases} (0.88)^x (0.12)^{1-x}, & x=0,1, \\ 0, & \text{elsewhere.} \end{cases}$$

The probability that the missile will fire successfully is

$$Pr(X=1) = f(1) = 0.88,$$

and the probability of a failure is

$$Pr(X=0) = f(0) = 0.12.$$

◀

2.4 Binomial Probability Distribution

The binomial, or Bernoulli, probability distribution, named after Jakob Bernoulli, who was introduced at the opening of this chapter, is one of the more widely used discrete probability distributions. It applies to problems that involve independent repeated trials, the outcomes of which have been classified into two categories, for example, success and failure, where the probability of a success is the same for each trial. The binomial distribution is thus a generalization of the point binomial. The random variable of interest in such a problem is the number of times in which the experiment results in a success.

More precisely, consider an experiment consisting of n independent trials; the outcome of each trial is either a success or a failure, with probabilities p and $1-p$, respectively. Let 0 denote a failure and 1 denote a success. If X denotes the number of successes among the n independent trials, the value that it takes on is equal to the number of ones that appear. For example, in a sequence such as $\{0, 1, 1, 0, 1, 1, 1, 0, 0, 0\}$, we have a failure on the first trial, a success on the second and third trials, a failure on the fourth trial, and so on. Therefore, the possible values that X can assume are $0, 1, \ldots, n$. To obtain the probability that $X = r$ successes, we proceed as follows: the number of r_1 successes and number of $n-r_0$ failures can occur in any of

$$\binom{n}{r} = \frac{n!}{r!(n-r)!}$$

distinct orderings of the n trials. The probability of each such ordering is equal to the product of rp results and $n = r(1-p)$ results, because all trials are independent. Adding these $\binom{n}{r}$ probabilities, we obtain

$$Pr(X=r) = \binom{n}{r} p^r (1-p)^{n-r}.$$

Thus, we state Definition 2.4.1.

Definition 2.4.1

A random variable X is said to have a **binomial distribution**, or **Bernoulli distribution**, if its probability density function is given by

$$f(x; n, p) = \begin{cases} \binom{n}{x} p^x (1-p)^{n-x}, & x = 0, 1, \ldots, n, \quad 0 < p < 1, \\ 0, & \text{elsewhere.} \end{cases} \quad (2.4.1)$$

The binomial distribution has two parameters, n and p. The parameter n is a discrete parameter because it can only have the values $1, 2, 3, \ldots$, and p is a continuous parameter because it can assume any value between 0 and 1. In any particular problem, n and p must have specific numerical values. The parameter p is usually estimated from given data in various physical problems. The maximum value of the binomial density function is obtained if we use the integral part of $(n+1)p$ for x. For example, if an experiment consists of 10 trials and $p=0.75$, then $f(x;10,0.75)$ attains a maximum value when x is equal to 8, which is the integral part of $(10+1)(0.75)$. This value of x, which maximizes $f(x;n,p)$, is called the **mode** or **modal value**.

It is clear, then, that $f(x;n,p)$ is a density function, because $f(x;n,p) \geq 0$ for $x = 0, 1, \ldots, n$ and

$$\sum_{x=0}^{n} f(x;n,p) = \sum_{x=0}^{n} \binom{n}{x} p^x (1-p)^{n-x} = [p + (1-p)]^n = 1.$$

The cumulative distribution function is given by

$$F(x) = Pr(X \leq x) = \begin{cases} 0, & x < 0, \\ \sum_{i=0}^{[x]} \binom{n}{i} p^i (1-p)^{n-i}, & 0 \leq x \leq n, \\ 1, & x > n. \end{cases}$$

The binomial distribution has been tabulated for various values of x, n, and p. Some of the more useful tables are those from Kitagawa,[1] Lieberman and Owen,[2] and Molina.[3] Tables of individual terms, binomial distributions,[4,5] and cumulative binomial distributions[6] in the Appendix contain probabilities for selected values of x, n, and p.

Example 2.4.1

A fair coin is tossed 10 times. We consider heads a success and tails a failure. It is clear that $p = 1/2$, $n = 10$, and the assumptions that underlie the binomial distribution are satisfied. Find the probability of (a) exactly seven successes, (b) at least seven successes, (c) at most seven successes, and (d) no successes.

Solution:

a. The probability of exactly seven successes is

$$Pr(X = 7) = f\left(7;10,\frac{1}{2}\right) = \binom{10}{7}\left(\frac{1}{2}\right)^7 \left(\frac{1}{2}\right)^3 = \frac{15}{128}.$$

b. The probability of at least seven successes is

$$Pr(X \geq 7) = \sum_{x=7}^{10} \binom{10}{x} \left(\frac{1}{2}\right)^x \left(\frac{1}{2}\right)^{10-x} = \frac{11}{64}.$$

c. The probability of at most seven successes is

$$Pr(X \leq 7) = 7\sum_{x=0}^{10}\binom{10}{x}\left(\frac{1}{2}\right)^x\left(\frac{1}{2}\right)^{10-x}$$

$$= 1 - Pr(X > 7) = \frac{53}{64}.$$

d. The probability of no successes is

$$Pr(X = 0) = f\left(0; 10, \frac{1}{2}\right) = \frac{1}{2}^{10} = \frac{1}{1024},$$

which is the same as the probability of exactly 10 successes.

The cumulative distribution of Example 2.4.1 is given by

$$F(x) = \begin{cases} 0, & x < 0, \\ \sum_{i=0}^{[x]}\binom{10}{i}\left(\frac{1}{2}\right)^i\left(\frac{1}{2}\right)^{10-i}, & 0 \leq x \leq 10, \\ 1, & x > 10. \end{cases}$$

Example 2.4.2

A quarterback on a football team has a pass completion average of 0.62. If, in a given game, he attempts 16 passes, what is the probability that he will complete (a) 12 passes and (b) more than half of his passes?

Solution:

a. $Pr(X = 12) = \binom{16}{12}(0.62)^{12}(0.38)^4 = 0.1224$

b. $Pr(X > 8) = \sum_{x=9}^{16}\binom{16}{x}(0.62)^x(0.38)^{16-x} = 0.7701$

Example 2.4.3

A woman fires at a target six times; the probability of her hitting it is equal to 0.40.

a. What is the probability that she will hit the target at least once?

b. How many times must she fire at the target so that the probability of hitting it at least once is greater than 0.77?

Solution:

a. $Pr(X \geq 1) = Pr(\overline{X < 1}) = 1 - Pr(X = 0) = 1 - \binom{6}{0}(0.4)^0(0.6)^6$

$= 1 - (0.6)^6.$

b. We must find n such that

$$1-(0.6)^n > 0.77. \quad (2.4.2)$$

The inequality in Equation (2.4.2) may be written as $(0.6)^n < 0.23$, which implies that

$$n \log 0.6 < \log 0.23$$

or

$$n > \frac{\log 0.23}{\log 0.6} = \frac{-0.6383}{-0.2219} = 2.9.$$

Therefore, the woman must fire at the target three or more times to maintain a probability greater than 0.77 of hitting it at least once.

If the probability of a success in the binomial distribution varies from trial to trial, we have what is known as a **generalized binomial distribution**. Consider an experiment that consists of n independent trials. On the rth trial, $r = 1, 2, \ldots, n$, a success may occur with probability p_r or a failure may occur with probability $1 - p_r = q_r$. The random variable X, the number of successes in n trials, can assume the values $1, 2, \ldots, n$, and it thus has a generalized binomial distribution. X may be represented as the sum

$$X = X_1 + X_2 + \ldots + X_n,$$

where the random variables X_i, $i = 1, 2, \ldots, n$ are independent and distributed as follows:

$$p_r = Pr(X_r = 1)$$

and

$$1 - p_r = Pr(X_r = 0).$$

The probability that $X = r$ is obtained by summing the probabilities of each possible combination of r successes and $(n-r)$ failures. Example 2.4.4 illustrates the preceding distribution.

◀

Example 2.4.4

An electronics store has received two shipments of an electronic component. The proportion of defective components in the first shipment is $q_1 = 1 - p_1 = 0.01$; in the second, the proportion is $q_2 = 1 - p_2 = 0.02$. We draw a component at random from each shipment and assign the value of 1 to the random variable X_i, $i = 1, 2$, if the component is operable and the value 0 if it is defective. Hence, the random variable $X = X_1 + X_2$ can assume the value 0, 1, or 2; that is, that none, one, or both of the components are operable, where

$$p_1 = Pr(X_1 = 1) = 0.99$$

and

$$p_2 = Pr(X_2 = 1) = 0.98.$$

To find the density of X, we proceed as follows:

$$Pr(X=0) = Pr(X_1=0)Pr(X_2=0) = q_1 \cdot q_2 = (0.01)(0.02) = 0.0002$$

$$Pr(X=1) = Pr(X_1=1)Pr(X_2=0) + Pr(X_1=0)Pr(X_2=1)$$

$$= p_1 \cdot q_2 + q_1 \cdot p_2 = (0.99)(0.02) + (0.98)(0.01) = 0.0296$$

$$Pr(X=2) = Pr(X_1=1)Pr(X_2=1) = p_1 \cdot p_2 = (0.99)(0.98) = 0.9702$$

The conditions of a density function are satisfied; that is, $Pr(X=x) \geq 0$ for $x = 0, 1, 2$, and

$$Pr(X=0) + Pr(X=1) + Pr(X=2) = 1.$$

◀

2.5 | Poisson Probability Distribution

The Poisson distribution is a limiting form of the binomial in which $p \to 0$ and $n \to \infty$ in such a way that np remains constant. It is used in many cases to approximate the binomial distribution when the number of trials n is comparatively large, p is small, and the product $\lambda = np$ is of moderate magnitude.

We derive the Poisson distribution by proving the following theorem, which is credited to the French mathematician Siméon Poisson, who announced it in 1837.

■ **Theorem 2.5.1**

Let the random variable X be distributed according to the binomial density:

$$f(x;n,p) = \begin{cases} \binom{n}{x} p^x (1-p)^{n-x}, & x = 0, 1, \ldots, n, \\ 0, & \text{elsewhere.} \end{cases}$$

If for $n = 1, 2, 3, \ldots$ the relation $p = \lambda/n$, where λ is a constant greater than zero, holds, then

$$\lim_{n \to \infty} f(x;n,p) = \begin{cases} \dfrac{e^{-\lambda} \lambda^x}{x!}, & x = 0, 1, 2, \ldots, \\ 0, & \text{elsewhere.} \end{cases}$$

continued

Proof Substituting $p = \lambda/n$ into the binomial density and taking the limit as $n \to \infty$ gives us

$$\lim_{n \to \infty} \left\{ \binom{n}{x} \left(\frac{\lambda}{n} \right)^x \left(1 - \frac{\lambda}{n} \right)^{n-x} \right\}$$

$$= \lim_{n \to \infty} \left\{ \frac{n!}{x!(n-x)!} \frac{\lambda^x}{n^x} \cdot \left(1 - \frac{\lambda}{n} \right)^n \cdot \frac{1}{\left(1 - \frac{\lambda}{n} \right)^x} \right\}$$

$$= \frac{\lambda^x}{x!} \lim_{n \to \infty} \left\{ \frac{n(n-1)(n-2)\ldots(n-x+1)}{n^x} \right\}$$

$$\cdot \lim_{n \to \infty} \left\{ \left(1 - \frac{\lambda}{n} \right)^n \lim_{n \to \infty} \frac{1}{\left(1 - \frac{\lambda}{n} \right)^x} \right\}.$$

Dividing n into each factor of $n(n+1)(n-2)\ldots(n-x+1)$, we have

$$\lim_{n \to \infty} \left\{ 1 \left(1 - \frac{1}{n} \right) \left(1 - \frac{2}{n} \right) \ldots \left(1 - \frac{x-1}{n} \right) \right\} = 1.$$

Using the facts that

$$\lim_{n \to \infty} \left(1 - \frac{\lambda}{n} \right)^n = e^{-\lambda}$$

and

$$\lim_{n \to \infty} \frac{1}{\left(1 - \frac{\lambda}{n} \right)} = 1,$$

we obtain the desired result:

$$\lim_{n \to \infty} \binom{n}{x} p^x (1-p)^{n-x} = \frac{e^{-\lambda} \lambda^x}{x!}, \quad x = 0, 1, 2, \ldots.$$

Hence, we have Definition 2.5.1.

2.5 Poisson Probability Distribution

Definition 2.5.1

A random variable X is said to have a **Poisson distribution** if its probability density function is given by

$$f(x;\lambda) = \begin{cases} \dfrac{e^{-\lambda}\lambda^x}{x!}, & x = 0, 1, 2, \ldots, \quad \lambda = np > 0 \\ 0, & \text{elsewhere.} \end{cases} \quad (2.5.1)$$

It is easy to see that the definition of a density function is satisfied: $f(x;\lambda) \geq 0$ and

$$\sum_{x=0}^{\infty} \frac{e^{-\lambda}\lambda^x}{x!} = e^{-\lambda} \sum_{x=0}^{\infty} \frac{\lambda^x}{x!} = e^{-\lambda}e^{\lambda} = 1$$

because

$$1 + \frac{\lambda^2}{2!} + \frac{\lambda^3}{3!} + \ldots + \frac{\lambda^k}{k!} + \ldots$$

converges for all real λ to e^{λ}.

The cumulative distribution of the Poisson density is given by

$$F(x) = Pr(X \leq x) = \begin{cases} 0, & x < 0, \\ \displaystyle\sum_{x_i \leq [x]} \frac{e^{-\lambda}\lambda^{x_i}}{x_i!}, & x \geq 0. \end{cases}$$

Many interesting random phenomena in the sciences satisfy the conditions of the Poisson density function. Typical problems in which the probability p that the event occurs is comparatively small and the number of observations n is large are concerned with rare occurrences of events in a fixed time interval. The following are some problems of this type:

1. The frequency of certain "peaks" per minute at a telephone switchboard
2. The number of misprints per page in a dictionary
3. The number of traffic accidents that occur per day on a certain turnpike
4. The number of α-particles emitted per hour by a radioactive source
5. The number of babies born with a heart defect in a large city during a 1-year period
6. The number of no-hitters pitched by a Hall of Fame player during his baseball career
7. The number of live viruses remaining after the production process of a certain vaccine

The Poisson distribution is characterized by the single parameter λ, which represents the average number of occurrences of an event in a given number of trials. Furthermore, it is easier to compute probabilities using the Poisson density than using the binomial probability function. The following recursive formulas are useful in calculating successive probabilities of Equation (2.5.1):

$$f(x;\lambda)\frac{\lambda}{x+1} = \frac{e^{-\lambda}\lambda^{x+1}}{(x+1)!} = f(x+1;\lambda) \qquad (2.5.2)$$

$$f(x;\lambda)\frac{x}{\lambda} = \frac{e^{-\lambda}\lambda^{x-1}}{(x-1)!} = f(x-1;\lambda) \qquad (2.5.3)$$

For example, if $\lambda = 2.5$, we evaluate $f(0;2.5)$ using Equation (2.5.1) as

$$f(0;2.5) = e^{-2.5} = 0.08208.$$

Applying the recursion formula (2.5.2), we have

$$f(1;2.5) = f(0;2.5)\frac{2.5}{1} = (0.08208)(2.5) = 0.2052$$

and

$$f(2;2.5) = f(1;2.5)\frac{2.5}{2} = (0.2052)\frac{2.5}{2} = 0.2565.$$

Edward Molina[3] has prepared excellent tables for the Poisson function for values of

$$\lambda = 0.001(0.001)0.01(0.01)0.3(0.1)15(1)100.$$

Toshio Kitagawa[1] also has tabulated the Poisson distribution for

$$\lambda = 0.001(0.001)1.00, 1.01(0.01)5$$

and

$$5.01(0.01)10.00.$$

Figure 2.5.1 illustrates the behavior of the Poisson density as we vary the size of λ. For small values of $\lambda (\lambda \leq 1)$, the line diagram of the density is skewed to the right; as λ increases, the function becomes symmetrical.

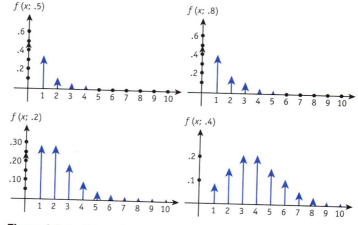

Figure 2.5.1 Poisson Probability Distribution: $\lambda = 0.2, 0.4, 0.5, 0.8$

2.5 Poisson Probability Distribution

Table 2.5.1 Comparison of Binomial and Poisson Distributions

	$n = 45$,		$p = 0.02$,		$\lambda = 0$.			
x	0	1	2	3	4	5	6	7
Binomial	0.4028	0.3699	0.1661	0.0485	0.0104	0.0017	0.0002	0.0000
Poisson	0.4065	0.3659	0.1646	0.0493	0.0111	0.0020	0.0003	0.0000

Table 2.5.1 shows that, for comparatively small x and n, the Poisson distribution gives a good approximation of the binomial distribution.

There is no set rule for the size of p and n in approximating the binomial with the Poisson distribution. In practice, if $p \leq 0.1$ and n is fairly large—say, $n \geq 40$—the approximation will be quite good.

The following examples illustrate some problems to which the Poisson distribution may be applied.

Example 2.5.1

A certain electronics company produces a particular type of vacuum tube. On the average, 3 tubes out of 100 are defective. The company packs the tubes in boxes of 400. Find the probability that a box of 400 tubes will contain (a) r defective tubes, (b) at least k defective tubes, and (c) at most 1 defective tube.

Solution: Because n is large and p is small, these probabilities can be approximated by using the Poisson density with $\lambda = np = (400)(0.03) = 12$ as follows:

a. $Pr(X = r) = \dfrac{e^{-12}(12)^r}{r!}, \quad r \leq 400$

b. $Pr(X \geq k) = \sum_{x=k}^{400} \dfrac{e^{-12}(12)^x}{x!}$

c. $Pr(X \leq 1) = \sum_{x=0}^{1} \dfrac{e^{-12}(12)^x}{x!} = 13e^{-12}$

Example 2.5.2

Suppose a mathematics textbook of 400 pages contains 200 misprints, which are randomly distributed throughout the text. We are interested in the probability that a given page will contain (a) no misprints and (b) 3 or more misprints.

Solution: Because the probability that a misprint will appear on a given page is $p = 0.0025$ and $n = 200$, it is justifiable to use the Poisson distribution and to obtain the necessary probabilities with $\lambda = 0.5$. Thus, we obtain the following results:

a. $Pr(X = 0) = e^{-0.5} = 0.6065$

b. $$Pr(X \geq 3) = \sum_{x=3}^{200} \frac{e^{-0.5}(0.5)^x}{x!}$$

$$= 1 - \sum_{x=0}^{2} \frac{e^{-0.5}(0.5)^x}{x!} = 1 - 1.625e^{-5}$$ ◀

Example 2.5.3 (Rutherford and Geiger)

In Table 2.5.2, we give the results of the famous physics experiment, conducted by Ernest Rutherford and Hans Geiger, regarding the α-particles emitted by a radioactive substance in 2608 periods of 7.5 seconds each.

In this experiment, n equals 2068, p is quite small, and the distribution of the random variable X can be approximated by the Poisson density. The average number of λ of α-particles emitted during a period of 7.5 seconds is

$$\lambda = \frac{1}{2608} \sum_{x=0}^{10} x n_x = 3.87$$

and

$$f(x; 3.87) = \frac{e^{-3.87}(3.87)^x}{x!}, \quad x = 0, 1, \ldots, 10.$$

The third row of Table 2.5.2 shows that the Poisson density gives a good approximation of the problem. Now we can calculate the following probabilities:

a. What is the probability that in a given period of 7.5 seconds we will observe five particles?

b. What is the probability that we will observe at most three particles in the same period?

Solution:

a. $$Pr(X = 5) = \frac{e^{-3.87}(3.87)^5}{5!}$$

b. $$Pr(X \leq 3) = \sum_{x=0}^{3} \frac{e^{-3.87}(3.87)^x}{x!}$$ ◀

We see that the Poisson distribution was developed as a limiting form of the binomial density; however, the Poisson function can be derived independently of the binomial distribu-

Table 2.5.2 Rutherford and Geiger Experiment

x	0	1	2	3	4	5	6	7	8	9	10
n_x	57	203	383	525	532	408	273	139	45	27	16
$f(x; 3.87)$	0.0209	0.0872	0.1562	0.2015	0.1949	0.1509	0.0973	0.0538	0.026	0.0112	0.00933

tion. We develop it by considering the following problem. Let $f(x;t)$ be the probability of getting x successes in a time interval of length t under the following conditions:

1. The probability of a success during a very small time interval Δt is $\alpha \Delta t$, $\alpha > 0$.

2. The probability that more than one success occurs during such a time interval Δt is negligible.

3. The probability of having a certain number of successes in a time interval of length t depends only on t and not on when the time interval begins or ends.

We can write

$$f(x;t+\Delta t) = Pr\{[(x \text{ successes in } t) \cap (0 \text{ successes in } \Delta t)]$$

$$\cup [(x-1 \text{ successes in } t) \cap (1 \text{ successes in } \Delta t)]\}$$

$$= Pr[(x \text{ successes in } t) \cap (0 \text{ successes in } \Delta t)]$$

$$+ Pr[(x-1 \text{ successes in } t) \cap (1 \text{ successes in } \Delta t)]$$

$$= Pr(x \text{ successes in } t) Pr(0 \text{ successes in } \Delta t)$$

$$+ Pr(x-1 \text{ successes in } t) Pr(1 \text{ successes in } \Delta t)$$

$$= f(x;t)[1-\alpha \Delta t] + f(x-1;t)[\alpha \Delta t]. \quad (2.5.4)$$

Subtracting $f(x;t)$ from both sides of Equation (2.5.4) and dividing by Δt gives us

$$\frac{f(x;t+\Delta t) - f(x;t)}{\Delta t} = \frac{\{f(x;t)[1-\alpha \Delta t] + f(x-1;t)\}[\alpha \Delta t] - f(x,t)}{\Delta t}. \quad (2.5.5)$$

Taking the limit of both sides of Equation (2.5.5) as $\Delta t \to 0$ and simplifying, the result is

$$\lim_{\Delta t \to 0} \frac{f(x;t+\Delta t) - f(x;t)}{\Delta t} = \lim_{\Delta t \to 0} \frac{\alpha \Delta t [f(x-1;t)] - f(x;t)}{\Delta t}$$

$$= \alpha [f(x-1;t) - f(x;t)].$$

Therefore,

$$\frac{df(x;t)}{dt} = \alpha [f(x-1;t) - f(x;t)]. \quad (2.5.6)$$

For $x = 0$, $f(0-1,t) = 0$, and Equation (2.5.6) becomes

$$\frac{df(0;t)}{dt} = -\alpha f(0;t)$$

or

$$\frac{df(0;t)}{f(0;t)} = -\alpha dt. \quad (2.5.7)$$

Integrating both sides of Equation (2.5.7),

$$\int \frac{df(0;t)}{f(0;t)} = -\alpha \int dt$$

or

$$\ln f(0;t) = -\alpha t + c. \tag{2.5.8}$$

For $t=0$, $f(0;0)=1$, which implies $c=0$. Thus,

$$\ln f(0;t) = -\alpha t$$

or

$$f(0;t) = e^{-\alpha t}.$$

Also, for $x=1$, Equation (2.5.6) becomes

$$\frac{df(1;t)}{dt} + \alpha f(1;t) = \alpha e^{-\alpha t}. \tag{2.5.9}$$

Solving this nonhomogeneous differential equation with the initial condition $f(1;0)=0$, we have

$$f(1;t) = \frac{e^{-\alpha t}(\alpha t)^1}{1!}. \tag{2.5.10}$$

Similarly, for $x=2$ and $f(2,0)=0$, Equation (2.5.6) becomes

$$\frac{df(2;t)}{dt} + \alpha f(2;t) = \alpha f(1;t)$$

or

$$\frac{df(2;t)}{dt} + \alpha f(2;t) = \alpha^2 t e^{-\alpha t}. \tag{2.5.11}$$

The solution of Equation (2.5.11) is given by

$$f(2;t) = \frac{(\alpha t)^2 e^{-\alpha t}}{2!}.$$

Continuing with this recursive approach, we conclude that, for any x, as long as $f(x;0)=0$, we have

$$f(x,t) = \frac{e^{-\alpha t}(\alpha t)^x}{x!}, \quad x=0, 1, \ldots, n. \tag{2.5.12}$$

Therefore, if $\lambda = \alpha t$, the average number of occurrences of successes in the period t, Equation (2.5.12) is identical to the Poisson density function. A rigorous proof of the preceding development can be achieved by induction.

Example 2.5.4

a. Find the probability that during a 3-minute interval there will be exactly five telephone calls if the probability of an incoming call to the switchboard during Δt seconds is $1/45 \Delta t$.

b. What is the probability that during a 2-minute interval there will be more than three calls but fewer than six?

Solution:

a. $\lambda = \alpha \Delta t = 1/45(180 \text{ seconds}) = 4$, and

$$Pr(X=5) = \frac{e^{-4}(4)^5}{5!} = 0.1563.$$

b. $\Delta t = 120$, so $\lambda = 2.67$ and

$$Pr(3 < X < 6) = \sum_{x=4}^{5} \frac{e^{-2.67}(2.67)^x}{x!}.$$

2.6 Hypergeometric Probability Distribution

One of the discrete distributions that is useful in industrial engineering problems (quality inspection) and in other physical problems of the combinatorial type is the hypergeometric distribution. Consider the following experiment: We are given a population of n elements of which n_1 are successes and $n_2 = n - n_1$ are failures. A random sample of size r is drawn from n, and we seek the probability that the group so chosen will contain x successes. Thus, the random variable X can assume values from 0 up to and including r or n_1, whichever is smaller. To find the probability that there are exactly x successes in the random sample, we proceed as follows: The total number of ways in which r elements can be selected out of n (disregarding order) is $\binom{n}{r}$; thus, each element in the sample space has a probability $1/\binom{n}{r}$ of being chosen. The total number of ways in which x successes can be selected from n_1 is $\binom{n_1}{x}$, and the remaining $(r-x)$ elements can be selected in $\binom{n_2}{r-x}$ ways. Therefore, the total number of ways in which we can achieve x successes and $r-x$ failures is $\binom{n_1}{x}\binom{n_2}{r-x}$, and the probability of this selection is given by

$$\binom{n_1}{x}\binom{n_2}{r-x} \Big/ \binom{n}{r}.$$

Definition 2.6.1

A random variable X has the **hypergeometric distribution** if it has the following function for its probability density:

$$f(x;r,n_1,n_2) = \begin{cases} \dfrac{\binom{n_1}{x}\binom{n_2}{r-x}}{\binom{n_1+n_2}{r}}, & \begin{array}{l} x = 0, 1, \ldots, r \leq n_1 \\ \text{or} \\ x = 0, 1, \ldots, n_1 \leq r, \end{array} \\ 0, & \text{elsewhere.} \end{cases} \quad (2.6.1)$$

The hypergeometric distribution depends on the three parameters r, n_1, and n_2. It is easy to show that it satisfied the conditions of a density function $f(x;r,n_1,n_2) \geq 0$ for $x = 0, 1, \ldots, r \leq n_1$ or $n_1 \leq r$, and

$$\sum_{x=0}^{r} \frac{\binom{n_1}{x}\binom{n_2}{r-x}}{\binom{n_1+n_2}{r}} = \frac{1}{\binom{n_1+n_2}{r}} \sum_{x=0}^{r} \binom{n_1}{x}\binom{n_2}{r-x}. \quad (2.6.2)$$

From Exercise 1.14 we saw that

$$\sum_{x=0}^{r} \binom{n_1}{x}\binom{n_2}{r-x} = \binom{n_1+n_2}{r}.$$

Equation (2.6.2) thus becomes

$$\frac{\binom{n_1+n_2}{r}}{\binom{n_1+n_2}{r}} = 1.$$

We show some applications of the hypergeometric distribution in the problems that follow.

Example 2.6.1

A lot of 60 electrical components were subjected to a quality inspection. It was found that 48 of the components were not defective and the remaining components were defective. If a random sample of 15 components is chosen from this lot, what is the probability that (a) exactly 11 of them will be operable and (b) at most 9 will be operable?

Solution: Applying the hypergeometric distribution, we have the following:

a. $\Pr(X = 11) = \dfrac{\binom{48}{11}\binom{12}{4}}{\binom{60}{15}}$

b. $\Pr(X \le 9) = \sum_{x=0}^{9} \dfrac{\binom{48}{x}\binom{12}{12-x}}{\binom{60}{15}} = \sum_{x=3}^{9} \dfrac{\binom{48}{x}\binom{12}{12-x}}{\binom{60}{15}}$

◀

Example 2.6.2

The faculty senate of a certain college consists of 66 senators, 38 of whom are from the sciences and 28 of whom are from the arts. A committee of 16 senators was chosen at random. What is the probability that the committee will have no more than 6 senators from the arts?

Solution: From Equation (2.6.2),

$$\Pr(X \le 6) = \sum_{x=0}^{6} \dfrac{\binom{28}{x}\binom{38}{16-x}}{\binom{66}{16}}.$$

Calculating probabilities using the hypergeometric distribution becomes laborious when n is large. The work, however, is simplified by first calculating $f(0;r,n_1,n_2)$ and then applying the following recursion formulas, which are derived directly from the hypergeometric density:

$$f(x-1;r,n_1,n_2) = \dfrac{(r-x)(n_1-x)}{(x+1)(n-n_1-r+x+1)} f(x;r,n_1,n_2) \qquad (2.6.3)$$

and

$$f(x+1;r,n_1,n_2) = \dfrac{x(n-n_1-r+x)}{(r-x+1)(n_1-x+1)} f(x;r,n_1,n_2). \qquad (2.6.4)$$

The recursion formula in Equation (2.6.4) was used to obtain the results in Example 2.6.3, which are given in Table 2.6.1. In Kitagawa,[1] tables have been calculated for the hypergeometric distribution for $n = n_1 + n_2 (1)100$ and $r = 1(1)50$; $r = 1(1)50$ means that the values of r begin with the value 1 and increase by jumps of 1 until the value 50 is reached.

◀

Example 2.6.3

A wholesaler who specializes in rebuilding automobile engines has 30 engines of a specific make in stock. Only 20 of the engines were overhauled; the others were only partially rebuilt. A local mechanic purchased 7 of the engines. Find the probability that (a) all 7 engines were overhauled, (b) at least 4 were overhauled, and (c) at most r were overhauled, where $r \le 7$.

Solution:

a. $\Pr(X = 7) = \dfrac{\binom{20}{7}\binom{10}{0}}{\binom{30}{7}}$

Chapter 2 Discrete Probability Distributions

Table 2.6.1 Hypergeometric Probability Distribution

x	0	1	2	3	4	5	6	7
f(x)	0.0001	0.0021	0.0235	0.1176	0.2855	0.3427	0.1904	0.03821
F(x)	0.0001	0.0022	0.02564	0.1432	0.4288	0.7714	0.9618	1.0000

b. $Pr(X \geq 4) = \sum_{x=4}^{7} \dfrac{\binom{20}{x}\binom{10}{7-x}}{\binom{30}{7}}$

c. $Pr(X \leq r) = \sum_{x=0}^{r} \dfrac{\binom{20}{x}\binom{10}{7-x}}{\binom{30}{7}}$

The graph of $f(x)$ and $F(x)$ for Example 2.6.3 is given in Figure 2.6.1.

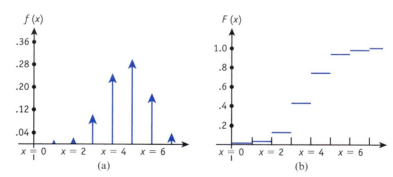

Figure 2.6.1 $f(x)$ and $F(x)$

It has been left as an exercise to show that if we were to purchase an engine from the local mechanic, the probability of the engine's having been overhauled is the same as if we had purchased the engine from the wholesaler.

The hypergeometric distribution approaches that of the binomial as n becomes large. That is, if p is the original proportion of successes in the population, the probability of obtaining x successes and $r-x$ failures in r trials, without replacement, is given according to Equation (2.6.1) by

$$f(x;r,p,n) = \begin{cases} \dfrac{\binom{np}{x}\binom{n-np}{r-x}}{\binom{n}{r}}, & \begin{array}{l} x=0,1,\ldots,r \leq n_1 \\ \text{or} \\ x=0,1,\ldots,n_1 \leq r, \end{array} \\ 0, & \text{elsewhere,} \end{cases} \quad (2.6.5)$$

where $n = n_1 + n_2$ is the total number of successes and failures, as before. Equation (2.6.5) can be written as

$$f(x;r,p,n) = \frac{(np)!(n-np)!r!(n-r)!}{x!(np-x)!(r-x)!(n-np-r+x)!n!}. \tag{2.6.6}$$

Approximating $n!$ in Equation (2.6.6) using Stirling's formula and taking the limit as $n \to \infty$, we get

$$\lim_{n \to \infty} f(x;r,p,n) = \binom{r}{x} p^x (1-p)^{r-x}, \quad x = 0, 1, \ldots, r.$$

The details of the proof are left as an exercise (see Exercise 2.15).

2.7 Geometric Probability Distribution

Consider an experiment that meets the assumptions underlying the binomial situation, except that the number of trials is not fixed. The trials are independent and identical, and each of them can result in one of two possible outcomes, a success or a failure, with probability p or $1-p$, respectively. In the binomial distribution, the random variable X, the number of successes, can assume the values $0, 1, \ldots, n$. However, in this case, we are interested in the number of trials required for the first success. Thus, the discrete random variable X can assume any one of an infinite number of positive integral values: $1, 2, 3, \ldots$. Trial x can yield the first success only if we have observed an unbroken run of $x-1$ failures during the first $x-1$ trials. The probability that this run of $x-1$ failures occurs is $(1-p)^{x-1}$. Thus, the probability of $x-1$ failures followed by a success is $(1-p)^{x-1} p$. We state the distribution of such a random variable in Definition 2.7.1.

Definition 2.7.1

A random variable X has the **geometric distribution** if it has the following function for its probability density:

$$f(x;p) = \begin{cases} p(1-p)^{x-1}, & x = 1, 2, 3\ldots, \quad 0 < p < 1, \\ 0, & \text{elsewhere.} \end{cases} \tag{2.7.1}$$

This one-parameter distribution is called the geometric distribution because its successive terms constitute a geometric progression. The conditions of a density function are satisfied, that is, $f(x) \geq 0$ for all x, and from the formula for the sum of an infinite geometric series, we have

$$\sum_{x=1}^{\infty} p(1-p)^{x-1} = 1.$$

The cumulative distribution of the geometric density is given by

$$F(x) = Pr(X \leq x) = \begin{cases} 0, & x < 1, \\ \sum_{x_i \leq [x]} p(1-p)^{x_i-1}, & x \geq 1. \end{cases}$$

The following examples illustrate some uses of the geometric distribution.

Example 2.7.1

In a game of billiards, a player continues to play until she misses a shot. If a particular player misses any of her shots with the probability $p=1/4$, what is the probability that this player's turn will last (a) exactly six shots, (b) at most five shots, and (c) at least four shots?

Solution: Assuming that the shots are independent and $p=1/4$ remains the same for every shot, we can apply the geometric distribution:

a. $Pr(X=6) = \left(\dfrac{1}{4}\right)\left(\dfrac{3}{4}\right)^5$

b. $Pr(X \leq 5) = \sum_{x=1}^{5} \left(\dfrac{1}{4}\right)\left(\dfrac{3}{4}\right)^{x-1}$

c. $Pr(X \geq 4) = \sum_{x=4}^{\infty} \left(\dfrac{1}{4}\right)\left(\dfrac{3}{4}\right)^{x-1} = 1 - \sum_{x=4}^{3}\left(\dfrac{1}{4}\right)\left(\dfrac{3}{4}\right)^{x-1}$

Example 2.7.2

What is the probability that we need six or fewer tosses of a fair pair of dice to throw a sum of five?

Solution: The probability of throwing a five on any throw is $1/9$, and the desired probability is

$$Pr(X \leq 6) = \sum_{x=1}^{6} \left(\dfrac{1}{9}\right)\left(\dfrac{8}{9}\right)^{x-1}.$$

The probability that it will take more than four throws of the pair of dice to obtain a sum of five is

$$Pr(X > 4) = 1 - Pr(X \leq 4) = 1 - \sum_{x=1}^{4}\left(\dfrac{1}{9}\right)\left(\dfrac{8}{9}\right)^{x-1}.$$

Example 2.7.3

Suppose that a student can take a certain standard examination in French as many times as he wishes. The number of times a student takes this examination is a random variable closely approximated by the geometric distribution. If the probability of his passing the examination at any time is $p=0.66$, what is the probability that (a) he passes the examination on the third try and (b) he fails to pass the examination on the first five tries?

Solution: Applying Equation (2.7.1), we have the following:

a. $Pr(X=3) = (0.66)(0.34)^2$

b. $1 - F(5) = 1 - Pr(X \leq 5) = 1 - \sum_{x=1}^{5}(0.66)(0.34)^{x-1}$

2.8 Negative Binomial Probability Distribution

Consider an experiment that meets the assumptions underlying the binomial density function. Whereas in the binomial we are interested in the number of successes achieved in n trials, in the negative binomial we are interested in the number of trials required to achieve r successes. Thus, if we let the random variable X be the number of trials required to achieve r successes, X may take on the values $r, r+1, r+2, \ldots$. If we achieve r successes in x trials, then we have obtained $x-r$ failures, and the probability that this event occurs in a particular order is $p^r(1-p)^{x-r}$. But because the last trial resulted in a success, there are $r-1$ other successes that occurred in $x-1$ trials, and this can happen in $\binom{x-1}{r-1}$ ways. The probability of achieving r successes in x trials is thus

$$\binom{x-1}{r-1} p^r (1-p)^{x-r}, \quad x = r, r+1, r+2, \ldots.$$

We define the distribution of the random variable in Definition 2.8.1.

Definition 2.8.1

A random variable X has the **negative binomial distribution** if it has the following function for its probability density:

$$f(x; p, r) = \begin{cases} \binom{x-1}{r-1} p^r (1-p)^{x-r}, & x = r, r+1, r+2, \ldots, \\ & 0 < p < 1, \\ 0, & \text{elsewhere.} \end{cases} \quad (2.8.1)$$

It is clear that this two-parameter distribution is greater than or equal to zero for all values of x, where $r > 0$. We must also show that

$$\sum_{x=r}^{\infty} f(x; p, r) = 1$$

and

$$\sum_{x=r}^{\infty} \binom{x-1}{r-1} p^r (1-p)^{x-r} = p^r \sum_{x=r}^{\infty} \binom{x-1}{x-r} (1-p)^{x-r}. \quad (2.8.2)$$

Letting $y = x - r$ in Equation (2.8.2), we have

$$p^r \sum_{x=r}^{\infty} \binom{x-1}{x-r} (1-p)^{x-r} = p^r \sum_{y=0}^{\infty} (-1)^y \binom{-r}{y} (1-p)^y$$

$$= p^r \left[1 - (1-p)\right]^{-r} \quad (2.8.3)$$

and

$$p^r p^{-r} = 1.$$

Therefore, $f(x;p,r)$ satisfies the condition of a probability density function. The negative binomial density gets its name from $\binom{-r}{y}$ in Equation (2.8.3).

We now consider the following applications of the negative binomial distribution.

Example 2.8.1

To pass a certain marksmanship test, an individual is required to shoot at a target until he scores six bull's-eyes. He is judged on the number of trials that are necessary to achieve this score. If the probability of his hitting a bull's-eye on any trial is 0.25, what is the probability that (a) he requires 9 shots and (b) he requires more than 9 but fewer than 12 shots?

Solution:

a. $p = \dfrac{1}{4}, \quad r = 6$

and
$$Pr(X=9) = \binom{9-1}{6-1}\left(\dfrac{1}{4}\right)^{6}\left(\dfrac{3}{4}\right)^{9-6} = \binom{8}{5}\left(\dfrac{1}{4}\right)^{6}\left(\dfrac{3}{4}\right)^{3}.$$

b. In this case, p and r are the same as in Example 2.8.1(a), but we want

$$Pr(9<X<12) = \sum_{x=10}^{11}\binom{x-1}{5}\left(\dfrac{1}{4}\right)^{6}\left(\dfrac{3}{4}\right)^{x-6}$$

$$= \binom{9}{5}\left(\dfrac{1}{4}\right)^{6}\left(\dfrac{3}{4}\right)^{4} + \binom{10}{5}\left(\dfrac{1}{4}\right)^{6}\left(\dfrac{3}{4}\right)^{5}.$$

◂

Example 2.8.2

Find the probability that a person flipping a fair coin will get the sixth head on the 11th trial.

Solution:

$$d = \dfrac{1}{2}$$

and

$$r = 6;$$

therefore,

$$Pr(X=11) = \binom{11-1}{6-1}\left(\dfrac{1}{2}\right)^{6}\left(\dfrac{1}{2}\right)^{11-6} = \binom{10}{5}\left(\dfrac{1}{2}\right)^{11}.$$

◂

Example 2.8.3

What is the probability that seven draws are required to get three aces from an ordinary deck of 52 cards, with replacement?

Solution:

$$r = 3$$

and

$$p = \frac{4}{52};$$

therefore,

$$Pr(X=7) = \binom{7-1}{3-1}\left(\frac{1}{13}\right)^3\left(\frac{12}{13}\right)^{7-3} = \binom{6}{2}\left(\frac{1}{13}\right)^3\left(\frac{12}{13}\right)^4.$$

In this chapter, we have studied the concept of a one-dimensional discrete random variable, developed a number of discrete probability distributions, and illustrated some of their uses. However, many physical situations must be characterized by a continuous random variable. That is, a continuous random variable may assume a nondenumerable (noncountable) number of values. Chapter 3 is devoted to studying the one-dimensional continuous random variable and to examining some useful continuous probability distributions.

Summary

Random variable
Discrete random variable
Discrete probability density function
Cumulative probability distribution function
Point binomial probability distribution
Binomial probability distribution
Poisson probability distribution
Hypergeometric probability distribution
Geometric probability distribution
Negative binomial probability distribution

Theoretical Exercises

2.1 Seven cards are dealt from an ordinary deck of playing cards.
  a. Determine the probability density function for the number of hearts dealt.
  b. Write the cumulative distribution function. Be sure to include the admissible values of the random variable.

2.2 An experiment consists of tossing six fair coins. Determine the probability density and cumulative distribution function for the number of heads that will appear in the experiment.

2.3 A fair coin is tossed until a tail appears.
  a. Determine the probability density for the number of tosses required.
  b. Obtain the cumulative distribution function.

2.4 A company received a shipment of a certain type of electronic component. The lot consists of r_1 operable and r_2 defective components. A sample of m components is drawn from the lot without replacement.
  a. What is the probability density function of the number of operable components?
  b. What is the probability that at least k of the m components to be inspected will be operable?

2.5 A fair die is rolled until a six appears. Let X be the number of rolls required in the experiment. Find $f(x)$ and $F(x)$. Sketch the graphs of $f(x)$ and $F(x)$.

2.6 An urn contains eight white balls and 12 black ones. We draw 10 balls at random from the urn. Find the probability density function for the number of white balls drawn in the following two situations:
  a. if there is replacement
  b. if there is no replacement

2.7 Find k so that the function given by

$$f(x) = \frac{k}{x+1}, \qquad x = 1, 2, 3, 4,$$

is a probability density function. Graph the density and cumulative distribution functions. Show that the properties of the cumulative distribution function are satisfied.

2.8 Find c so that the discrete function given by

$$f(x) = \frac{1}{c} 4^{-|x|}, \qquad x = 0, \pm 1, \pm 2, \ldots,$$

is a probability density function. What is the cumulative distribution function of the random variable X?

2.9 Let

$$f(x) = \begin{cases} \dfrac{k}{x^2}, & x = \pm 1, \pm 2, \pm 3, \ldots, \\ 0, & \text{elsewhere.} \end{cases}$$

  a. Find k so that $f(x)$ is a probability density function.
  b. Determine the cumulative distribution function of the random variable X.

2.10 Show that the function given by

$$f(x) = \begin{cases} \dfrac{1}{2^{|x|} + |x|}, & x = \pm 1, \pm 2, \pm 3, \ldots, \\ 0, & \text{elsewhere,} \end{cases}$$

is a probability density function of the discrete random variable X.

2.11 Show that the binomial probability density function $f(x;n,p)$ given by Equation (2.7.1) can be written as

$$f(n-x;n,1-p) = \begin{cases} \binom{n}{n-x}(1-p)^{n-x}p^{n-(n-x)}, & x=0,1,\ldots,n, \\ & 0<p<1, \\ 0, & \text{elsewhere.} \end{cases}$$

2.12 A simple way to calculate binomial probabilities is as follows:
For a given n and p, evaluate $f(0;n,p)$ and then apply the recursive relationship

$$f(x+1;n,p) = f(x;n,p)\frac{p(n-x)}{(1-p)(x+1)}$$

to obtain other binomial probabilities. Derive this recursion formula.

2.13 Show that for $x = 0, 1, 2, \ldots, 4 \leq n_1$,

$$\sum_{x=0}^{n_1,r} \binom{n_1}{x}\binom{n_2}{r-x} = \binom{n_1+n_2}{r},$$

where n_1, r means $\min(n_1, r)$.

2.14 A simple recursive formula for calculating hypergeometric probabilities is

$$f(x+1;r,n_1,n_2) = \left\{\frac{(r-x)(n_1-x)}{(x+1)(n_2-r+x+1)}\right\}f(x;r,n_1,n_2).$$

Derive this recursive relationship, and then apply it to calculate the hypergeometric probabilities for $r=6$, $n_1 = 7$, and $n_2 = 8$.

2.15 Show that the hypergeometric distribution is asymptotically a binomial distribution; that is,

$$\lim_{n\to\infty}\frac{\binom{pn}{x}\binom{n-np}{r-x}}{\binom{n}{r}} = \binom{n}{x}p^x(1-p)^{n-x},$$

where $n = n_1 + n_2$, $pn = n_1$, and $x = 0, 1, \ldots, r \leq np$, or $x = 0, 1, \ldots, np < r$.

2.16 Show that the geometric distribution function given by Equation (2.4.1) satisfies the conditions of a probability density function.

2.17 Show that

$$\sum_{y=0}^{\infty}\binom{y+r-1}{y}(1-p)^y = \sum_{y=0}^{\infty}(-1)^y\binom{-r}{y}(1-p)^y,$$

where $0 \leq p \leq 1$.

2.18 Suppose that n cards are drawn from an ordinary deck of cards with or without replacement, where $n \leq 52$. Exactly k spades are contained in the n drawn cards. Show that the probability that the jth card was a spade is k/n.

Applied Problems

2.1 In Exercise 2.1, find the probability:
  a. Exactly five hearts are dealt.
  b. The number of hearts dealt is from two to five.

2.2 Use Exercise 2.2.
  a. What is the probability of at most four heads appearing?
  b. What is the probability that the number of heads will be from two to five?

2.3 Use Exercise 2.3.
  a. What is the probability of obtaining the first tail at the fourth trial?
  b. What is the probability that the first tail will appear in at most three trials?

2.4 In Exercise 2.6, find the probability:
  a. At least five white balls are drawn.
  b. At most five white balls are drawn.

2.5 In Exercise 2.7, find:
  a. $Pr(X > 2)$
  b. $Pr(2.5 \leq X \leq 7.3)$

2.6 (Discrete uniform distribution) Let X be a discrete random variable that may assume the values $x_1, x_2, \ldots, x_n$ with equal probability. The probability density function of X is given by

$$(x) = \begin{cases} \dfrac{1}{n}, & x = x_1, x_2, \ldots, x_n, \\ 0, & \text{otherwise.} \end{cases}$$

Thus, if the range of the random variable is $x = 2, 3, \ldots 9$, then the discrete uniform distribution of X is

$$f(x) = \begin{cases} \dfrac{1}{8}, & x = 2, 3, \ldots, 9, \\ 0, & \text{otherwise.} \end{cases}$$

What is the probability?
  a. The value that the random variable assumes is greater than 6.
  b. The value that the random variable assumes is greater than 4 but less than or equal to 7.
  c. The value that the random variable assumes is greater than or equal to 8.

2.7 In the general population, the probability that an infant is male is 0.52. Assuming that the sexes of fraternal twins are independent, what is the probability that fraternal twins will have the same sex?

2.8 A die is rolled 120 times. Find the probability that 35 or more sixes will appear.

2.9 Two fair dice are rolled many times. Find the probability:
  a. In the first seven rolls, four sevens appear.
  b. The first seven appears on the seventh roll.
  c. The fourth seven appears on the seventh roll.

2.10 Consider the experiment of rolling a die. Assume that a success occurs whenever a one or a five appears and a failure occurs if another number appears. The die is rolled five times.
   a. Find the probability that three successes will occur.
   b. Find the probability that at most two successes occur.
   c. Write and graph the cumulative distribution function of the random variable that characterizes the experiment.

2.11 Suppose that a radio tube inserted into a certain type of a system has a probability of 0.2 of functioning more than 500 hours. We test 20 such tubes.
   a. Find the probability that exactly k of these tubes will function more than 500 hours.
   b. Find the probability that the number of tubes that will be operable more than 500 hours falls between 12 and 17.
   c. Sketch the cumulative distribution function of the random variable that describes the random phenomenon.

2.12 A basketball team is playing an opposing team, and both teams have the same ability.
   a. Which of the following two events is more likely to occur: one team winning two basketball games out of five or the same team winning three games out of four?
   b. Which of the following two events is more likely to occur: a team winning at least two games out of five or the same team winning at most three games out of four?

2.13 From a standard deck of 52 playing cards, 7 cards are dealt at random without replacement. What is the probability of the following?
   a. Exactly two of them will be aces.
   b. At least three will be aces.
   c. At most one will be an ace.

2.14 A distributor of a certain high-precision component determines that 5% of her product will not meet the required standards of the consumer. She sells the components in packages of 300 and guarantees that 90% of them will pass the required standard. Find the probability that a package violates the distributor's guarantee.

2.15 White mice are used to evaluate the effectiveness of a certain drug. If only 50% of the mice react to the drug, how many mice should be used to ensure that at least one of the mice will give an observable reaction with a probability of at least 95%?

2.16 Suppose that, during the course of a basketball game, a certain player is fouled three times and that each time he is allowed to make one penalty shot. Because it is his first time at the foul line, the probability that he makes his first shot is 0.3. However, he is not a bad shooter; the probability that he makes his second shot is 0.8. The third shot is taken during the last few seconds of the game and may win the game for his team. The player is extremely nervous, and the opposing team has already called time out to rattle him. The probability of his making the third shot is reduced to 0.1. What is the probability that he makes exactly two shots out of the three?

2.17 The proportion of mice that convulse when exercised after receiving a standard dose of insulin increases as the time since the last feeding increases. The first group has been without food for 4 hours, and the probability of a mouse convulsing is 0.5. A second group has been without food for 5 hours, and the probability that a mouse in this group will convulse is 0.6. A third group has been without food for 5.5 hours, and the probability that a mouse in this group will convulse is 0.65. If one mouse is selected from each group and this sample size of three is tested, what is the probability?

a. Fewer than two mice will convulse.
b. At most two mice will convulse.

2.18 A store receives three shipments of flashlight batteries. The proportion of defective batteries in the first, second, and third shipments is 0.015, 0.01, and 0.05, respectively. A battery is selected at random from each of the three shipments. Find the probability of the following:

a. At most two of the batteries are defective.
b. None of the batteries are defective.

2.19 A pharmaceutical company produces a certain type of vaccine. It has been determined that, on the average, 2 out of 100 viruses remain alive after application of the vaccine. What is the probability that a batch of 500 viruses, after application of the vaccine, will contain?

a. Exactly 10 live viruses
b. At most 6 live viruses
c. No live viruses

2.20 The probability of the birth of a "phantom" poodle (black with white feet) is 0.01. If 50 poodles are born in the kennel of a certain poodle breeder in a given year, what is the probability?

a. None of them will be of the phantom variety.
b. At least three of them will be of the phantom variety.

2.21 A drug manufacturer receives a shipment of 20,000 calibrated eyedroppers for administering the Salvin poliovirus vaccine. The calibration marks are missing on 400 of the droppers, and these defective droppers are scattered randomly throughout the shipment. Find the probability of the following:

a. At least 1 defective dropper will be detected in a random sample of 30.
b. At most 5 defective droppers will be detected in a random sample of 50.

2.22 Suppose that, in a certain production process, two defective items are produced for every 100 items. These items are sold in lots of 100. Company A has ordered two lots with the agreement that, if there are more than four defectives, the remaining items will be returned for a refund of the entire purchase price. What is the probability that the purchase price will be refunded?

2.23 The probability that a swimmer can break the world record in the butterfly event is 0.05. If she competes in 12 meets during the course of a year, what is the probability that she will break the record at least once? Construct a table comparing the binomial probabilities with those of the Poisson distribution for all possible values that the random variable can assume.

2.24 In geophysics, the age of a zeicon may be determined by counting the number of uranium fission tracks on a polished surface that has been etched with hydrofluoric acid both before and after neutron bombardment. Each track represents the fission of one atom. The probability that any atom decays by fission is very small (the half-life is equal to 8×10^{15} years), but there is a very large number of atoms in a sample. If, for a given sample, there should be five tracks on average in the area of the surface examined, what is the probability of finding fewer than three tracks in the area and thus greatly underestimating the age of the material?

2.25 Five cards are dealt from an ordinary deck of playing cards. What is the probability that the hand will contain the following?
  a. No aces
  b. At least two aces
  c. At most one ace

2.26 A men's store purchased a lot of 100 shirts from a certain manufacturer. Although 88 of the shirts were not damaged, the remaining shirts were not suitable to sell. A random sample of 40 shirts is chosen from this lot. Find the probability:
  a. Exactly 36 shirts have not been damaged.
  b. At most 22 shirts were not damaged.

2.27 Five cards are drawn, without replacement, from a deck of cards containing only four spades and six clubs. What is the probability that the spades outnumber the clubs?

2.28 Solve Problem 2.21 using the hypergeometric and binomial probability density functions, and compare your answers. Is n large enough to accept the answers given by the binomial distribution?

2.29 A lazy fisherman decides to cast until he has caught a single fish. If the stream is so well stocked that on average a person catches a fish every 10th try, what is the probability that the fisherman will catch a fish after exactly five tries?

2.30 A volleyball team has a 70% chance of winning each point while it is the serving team. As soon as the serving team fails to score, it must give up service. What is the probability that the score will be five to nothing before the serve is lost by this team for the first time?

2.31 A bridge player has a 0.25 probability of being dealt a biddable hand on each deal. What is the probability that she must be dealt three hands to get a biddable hand?

2.32 A chemist is working on a new theory, but he needs to confirm it with laboratory experiments. Each trial is considered a success if a certain electronic device that is used to trigger the reaction functions successfully. The chemist has only enough chemicals to perform the experiment five times, but four of these trials bust be "successes" to give him sufficient information about his new theory. The electronic device is built in such a way that if it breaks down it cannot be repaired, and the chemist has neither the time nor the money to order a new one. What should the device's probability of failure be on each trial if the chemist wishes to be 93% certain that the first failure occurs after the fourth trial?

2.33 A carnival worker operates a "throw-until-you-win" game; that is, each person is allowed all the throws he needs to be successful. To increase the number of people playing the game (and consequently the profit), the worker wants to design the game so that 95% of the patrons win in seven or fewer throws. What should the probability of success be on a single throw so that the carnival worker can achieve her objective?

2.34 A golf tournament is arranged in such a way that, to qualify, a golfer must win two other tournaments of a similar nature during the year. Assume that the probability of winning each tournament remains constant for each golfer. At the beginning of the golfing season, a man wishes to bet that a particular golfer either will not qualify for the tournament or will win the tournament. He wants to be 90% certain that one of these two events will occur before he places his bet. What must the golfer's probability of winning be before the man is willing to bet?

2.35 The probability of winning the local gas station's "Tigerama" game is 0.001. A woman is determined to win. What is her probability of winning if she goes to the gas station exactly 100 times?

2.36 A man is playing Russian roulette. Before each pull of the trigger, he spins the cylinder, which contains a bullet in one of the six chambers. What is the probability that he will shoot himself on the third try?

2.37 A certain type of vaccine is packaged with five vials to the box. To package the vaccine in this way, the packer picks a vial off a moving belt, checks it for defects, and puts it in the box if it is satisfactory or sets it aside if it is not. On average, 98% of the vials are satisfactory.

    a. What percentage of boxes will be filled after inspecting only six vials?
    b. For what percentage of boxes will it be necessary to inspect seven vials?

2.38 A card player draws n cards from an ordinary deck and lays them aside, unnoticed. Then she draws another. Find the probability that this last card is a spade.

References

[1] Kitagawa, T. *Tables of Poisson Distribution.* Tokyo: Baifukan Press, 1952.

[2] Lieberman, G. H. and D. B. Owen. *Tables of the Hypergeometric Probability Distribution.* Stanford, California: Stanford University Press, 1961.

[3] Molina, E. C. *Poisson's Exponential Binomial Limit.* Princeton, New Jersey: D. Van Nostrand Company, Inc., 1942.

[4] Roming, H. G. *50–100 Binomial Tables.* New York: John Wiley & Sons, Inc., 1953.

[5] *The Tables of the Binomial Probability Distribution.* Applied Mathematics Series 6. Washington, D.C.: National Bureau of Standards, 1950.

[6] *Tables of the Cumulative Binomial Probability Distribution.* Cambridge, Massachusetts: Harvard University Press, 1955.

> *"Excellence is a process, not just an outcome."*
> —Aristotle

CHAPTER 3

Probability Distributions of Continuous Random Variables

OBJECTIVES:

3.1 Continuous Random Variable and Probability Density Function
3.2 Cumulative Distribution Function of a Continuous Random Variable
3.3 Continuous Probability Distributions

Carl Friedrich Gauss (1777–1855)

> *"The probability that we may fail in the struggle ought not deter us from the support of a course we believe to be just."*
> —Abraham Lincoln

Considered by some to be the top mathematician of all time, Carl Friedrich Gauss was a child prodigy who matured to become an adult prodigy. He had an agile mind, was easily able to do incredible mental calculations, and had a photographic memory and amazing mathematical inventiveness. In addition to the discoveries that he published, he kept a scientific journal in which he noted ideas and calculations that would have made several great reputations had they been published promptly.

His *Disquisitiones Arithmeticae* (1801) is regarded as the basic work in the theory of numbers, but this was only one of his many accomplishments. In addition to his law of normal distributions (the Gaussian bell-shaped curve) used by statisticians, Gauss made contributions to astronomy, geodesy (the measurement of Earth), geometry (including non-Euclidean geometry), theoretical physics (especially electromagnetism), complex numbers, and functions. Along with his masterful theoretical research, he was a well-known inventor. In 1833, together with Wilhelm Weber, he devised an electromagnetic telegraph.

Gauss was of Nether-Saxon peasant origin. His genius became evident early in his life, and he was encouraged in his intellectual endeavors by his mother. His father was an honest, uncouth bricklayer who would have preferred the youngster to learn a trade rather than follow a learned profession. Countless anecdotes point to young Gauss's prodigious precocity. He described himself as being able to count before he could talk. When he turned 14, his abilities

continued

were brought to the attention of the Duke of Brunswick, who became Gauss's patron. At 15, Gauss entered the 3-year Collegium Carolinum in Brunswick, where he invented the method of least squares (which he later used to calculate an orbit for the first discovered minor planet, Ceres), mastered the classic languages, and studied the important works of earlier mathematicians (most notably those of Sir Isaac Newton). He resolved to follow the examples of Archimedes and Newton and leave behind him only finished works of art, with no trace remaining of the labor by which they were achieved.

Gauss later attended the University of Göttingen and the University of Helmstedt. His doctoral thesis gave the first proof of the fundamental theorem of algebra: that every algebraic equation has at least one root among the complex numbers. After the death of his sponsor, Gauss was appointed director of the Göttingen Observatory. In this position, he was able to continue his research and cultivate his interests. His hobbies included extensive reading in European literature, world politics, the study of foreign languages and new sciences (including botany and mineralogy), and music.

In his later years, Gauss received honors from scientific bodies and governments from around the world; however, his full stature was realized only in the twentieth century, because many of his important contributions came to light long after his death.

INTRODUCTION

In Chapter 2, we introduced the concepts of a random variable and discussed in detail the behavior of a discrete random variable. The aim in this chapter is to study the continuous random variable to derive most of the important continuous probability density functions and to illustrate some of their uses with physical applications.

3.1 Continuous Random Variable and Probability Density Function

A continuous random variable may assume a nondenumerable (noncountable) number of values, whereas a discrete random variable is restricted to taking on isolated values. Typical examples of a continuous random variable are the operational life of a certain electrical system, $0 \leq x < \infty$; the temperature fluctuation in the Sea of Tranquility on the moon, $-273°C \leq x < \infty$; the length of time of a telephone conversation, $t_1 \leq x \leq t_2$; and the power output of a certain machine operating in a changing environment, $0 \leq x < \infty$. Many other examples have similar characteristics. The random variable may assume values from a certain interval or from a collection of intervals from the range of the random variable.

We study the continuous random variable so that we may calculate the probabilities that the variate will assume values from a certain interval on the real axis. To obtain such probabilities, one must be able to derive the probability density function of the continuous random variable. One of the prime tasks in probability is the formulation of the distribution of a random variable. Once we know its distribution, we know the laws governing it.

We begin the development of the probability density function of a continuous random variable with a discussion of a physical phenomenon. Consider an experiment that involves

3.1 Continuous Random Variable and Probability Density Function

putting an electrical system into operation at time $t=0$, thinking of the time that elapses until its failure as a random variable X, which assumes values from the interval $(0,T]$. Here, the random variable can assume a noncountable number of values. Let us assume that we have observed n such systems and have recorded the elapsed times until failure. One could interpret X as a discrete random variable and calculate probabilities that the random variable X can assume certain individual values. However, realistically, one must deal with intervals rather than with individual points. Let us divide the time interval $(0,T]$ into r subintervals, each of equal length—$(t_0,t_1],(t_1,t_2],\ldots,(t_{r-1},t_r]$—such that

$$0 < t \le t_1 \le t_2 \le \ldots \le T.$$

In addition, let $x_i, i=1, 2, \ldots, r$ represent the number of systems that failed in the ith subinterval such that

$$\sum_{i=1}^{r} x_i = n.$$

The number of systems that have failed in the ith subinterval, divided by n, will be the relative frequency that a system will fail between t_{i-1} and t_i.

We can construct a probability histogram, as shown in Figure 3.1.1, with p as the empirical probability, such that $\sum_{i=1}^{r} p_i = 1$, where $p_i = x_i/n, i=1, 2,\ldots,r$. The rectangles of the histogram are of equal width; thus, their areas are proportional to their heights. Furthermore, the sum of all rectangles is equal to one. Therefore, we can estimate the probability that a system will fall in the interval $(t_0,t_3]$ by adding the areas of the three rectangles over this interval: If X is the random variable representing the number of failures, then

$$Pr(x_0 \le X \le x_3) = Pr\left[(X=x_1) \cup (X=x_2) \cup (X=x_3)\right]$$
$$= Pr(X=x_1) + (X=x_2) + (X=x_3)$$
$$= p_1 + p_2 + p_3.$$

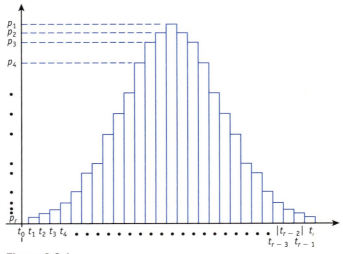

Figure 3.3.1

Using this approach, one can obtain estimates of the probabilities that a continuous random variable is contained in an interval in terms of area. These estimates can be improved by decreasing the widths of the intervals, which increases the number of such intervals so that we can approximate the area under the rectangles by a continuous function, as shown in Figure 3.1.2.

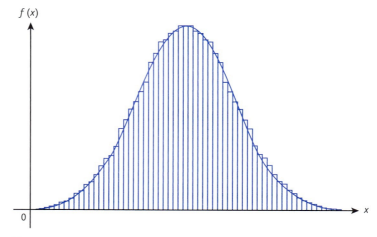

Figure 3.1.2 Continuous Function

What we described earlier is the definition of a definite (Riemann) integral. Suppose the interval $(0, T]$ has been subdivided into m intervals. If the kth subinterval is of width $(\Delta x)_k$ and height $f(x_k)$ and x_k is any point in this subinterval, then

$$\lim_{m \to \infty} \sum_{k=0}^{m} f(x_k)(\Delta x)_k = \int_0^T f(x) dx. \tag{3.1.1}$$

Accordingly, there is a function, $f(x)$, that gives the correct probability for any interval in the range of the random variable in terms of the area under the smooth curve. Such a function becomes meaningless when we speak of its ith value; the probabilities are given by areas under the curve, not by evaluating the function at a particular value. Thus, the probability that a continuous random variable is equal to a specific value is zero. That is,

$$Pr(X = x_0) = 0.$$

We can formulate the concepts developed in the preceding discussion.

Definition 3.1.1

Let X be a random variable, If the range space R_X is an interval or the union of two or more nonoverlapping intervals, then X is called a **one-dimensional continuous random variable**.

3.1 Continuous Random Variable and Probability Density Function

Definition 3.1.2

Let X be a one-dimensional continuous random variable. A function $f(x)$ is called a **continuous probability density function** of the random variable X if the following conditions are satisfied:

1. $f(x) \geq 0$ for all $x \in R_X$
2. $\int_{R_X} f(x) dx = 1$

Here, the integration is preformed over those x's that are contained only in the intervals of the range space. Furthermore, it must be true that for any real numbers α and β such that $\alpha < \beta$,

$$Pr(\alpha \leq X \leq \beta) = \int_\alpha^\beta f(x) dx.$$

We call $f(x)$ a *probability density function* or *density function* of the continuous random variable X if the preceding conditions hold. As in the discrete case, other common ways in which we can phrase the definition are "the random variable X is distributed as $f(x)$" and "$f(x)$ is the distribution of the random variable X."

For the integral in Equation (3.1.1) to exist in a given interval, it is necessary that the function $f(x)$ be continuous almost everywhere on the interval. Thus, we assume that the density function of a continuous variate is continuous except, at most, for a finite number of points.

Given the probability density function $f(x)$ of a continuous random variable X and letting $E_1 = \{x : a \leq x \leq b\}$, we can define $Pr(E_1) = Pr(a \leq X \leq b) = \int_{E_1} f(x) dx = 1 \int_a^b f(x) dx$. The probability is represented by the shaded area under the graph in Figure 3.1.3.

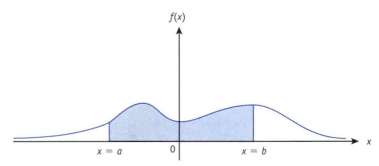

Figure 3.1.3 $Pr(E_1)$

The probability density function of a continuous variate may be obtained by some mathematical reasoning of the problem or by fitting the function to the data at hand and forcing it to satisfy the conditions of a probability density function. Thus, if we are given a continuous function, say, $g(x), -\infty < x < \infty$, we can evaluate $\int_{-\infty}^{\infty} g(x) dx = c$, where $c > 0$ is some real number not necessarily equal to one. Next, we can obtain a new function,

$$f(x) = \frac{g(x)}{c}, -\infty < x < \infty,$$

so that, if $f(x) \geq 0$ for all x, then $f(x)$ is a density function.

Finally, we must emphasize that $f(x)$ does not give the probability of anything when evaluated at a particular value of x. However, when this function is integrated between two limits, it yields a probability.

Example 3.1.1

The length of time in minutes that an individual talks on a long-distance telephone call has been found to be random. Let x be the length of the talk; assume it to be a continuous random variable with a probability density function given by

$$f(x) = \begin{cases} \alpha e^{-(1/4)x}, & x > 0, \\ 0, & \text{elsewhere.} \end{cases}$$

Find:

a. The value of α that makes $f(x)$ a probability density function

b. The probability that this individual will talk (i) between 7 and 12 minutes, (ii) less than 7 minutes, and (iii) more than 12 minutes

c. $Pr(X \leq 7 | 5 \leq X \leq 10)$

Solution:

a. Applying the definition of a probability density function, we can integrate

$$\int_0^\infty \alpha e^{-(1/4)x} dx = 1$$

and obtain $\alpha = \dfrac{1}{4}$.

b. i. $Pr(7 \leq X \leq 12) = \int_7^{12} \dfrac{1}{4} e^{-(1/4)x} dx = e^{-7/4} - e^{-3}$

ii. $Pr(X \leq 7) = \int_0^7 \dfrac{1}{4} e^{-(1/4)x} dx = 1 - e^{-7/4}$

iii. $Pr(X \geq 12) = \int_{12}^\infty \dfrac{1}{4} e^{-(1/4)x} dx = e^{-3}$

c. The concept of conditional probability discussed in Section 1.3 can be meaningfully applied here:

$$Pr(X \leq 7 | 5 \leq X \leq 10) = \dfrac{Pr(5 \leq X \leq 7)}{Pr(5 \leq X \leq 10)}$$

$$= \dfrac{\int_5^7 \dfrac{1}{4} e^{-(1/4)x} dx}{\int_5^{10} \dfrac{1}{4} e^{-(1/4)x} dx} = \dfrac{1 - e^{-1/2}}{1 - e^{-5/4}}$$

is the probability that the telephone conversation lasted less than 7 minutes, given that it lasted between 5 and 10 minutes.

Figure 3.1.4 shows the graph of the density function and the probability of Example 3.1.1(b) as areas under $f(x)$.

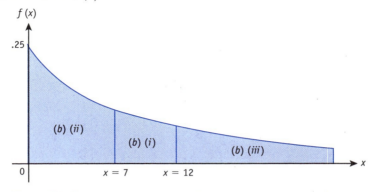

Figure 3.1.4 Graph of the Solution of Example 3.1.1(b)(i) through (iii)

Example 3.1.2

The number of automobiles that a certain dealer sells per day can be considered a random variable of the continuous type, with the following function for its probability density:

$$f(x) = \begin{cases} \beta(x+1), & 0 \leq x \leq 12, \\ \dfrac{-10(6\beta-1)}{x^2}, & 12 < x \leq 20, \\ 0, & \text{elsewhere}. \end{cases}$$

The continuity of such a physical phenomenon can be justified by a manner similar to that used in the discussion earlier in this section. Find the value of β that makes $f(x)$ a density function and the probability that the dealer will sell between 8 and 14 cars in a given day.

Solution: To determine the value of β, we must integrate $f(x)$ over all admissible values of x,

$$\int_{-\infty}^{\infty} f(x)dx = \int_{0}^{12} \beta(x+1)dx + \int_{12}^{20} \frac{-10(6\beta-1)}{x^2} dx = 1,$$

and we obtain $\beta = 1/123$. Now,

$$Pr(8 \leq X \leq 14) = \int_{8}^{14} f(x)dx$$

$$= \int_{8}^{12} \frac{1}{123}(x+1)dx + \int_{12}^{14} \frac{1170}{123x^2} dx$$

$$= \frac{811}{1722} = 0.47.$$

Figure 3.1.5 shows the graph of the preceding probability density function. It also shows the desired probability in terms of the shaded area under the curve.

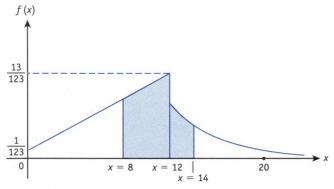

Figure 3.1.5 Probability Density Function

3.2 Cumulative Distribution Function of a Continuous Random Variable

Once the probability density function of a continuous random variable is known, we can determine various probabilities concerning its behavior. For example, as we have already seen, we might be interested in the probability that the value of the random variable X lies in a certain set $S_1 = \{x : x \leq x_1\}$; that is,

$$Pr(X \text{ in } S_1) = Pr(X \leq x_1) = \int_{-\infty}^{x_1} f(x)\,dx. \qquad (3.2.1)$$

We denote a function such as the one in Equation (3.2.1) by $F(x)$; it is defined in Definition 3.2.1.

Definition 3.2.1

The function defined by

$$F(x) = Pr(X \leq x) = \int_{-\infty}^{x} f(s)\,ds \qquad (3.2.2)$$

is called the **cumulative distribution function of the continuous random variable** X, which has the probability density function $f(x)$.

Example 3.2.1

Find the cumulative distribution function of the continuous random variable X that has for its probability density the function

$$f(x) = \begin{cases} \dfrac{3}{8}(x^2 - x + 1), & 0 \leq x \leq 2, \\ 0, & \text{elsewhere.} \end{cases}$$

Solution: Applying Equation (3.2.2), for $x < 0$ we have

$$F(x) = P(X \leq x) = \int_{-\infty}^{x} f(x)\,dx = \int_{-\infty}^{x} 0 \cdot dx = 0,$$

for $0 \leq x \leq 2$:

$$F(x) = P(X \leq x) = \int_{-\infty}^{0} f(x)\,dx + \int_{0}^{x} f(x)\,dx$$

$$= F(0) + \int_{0}^{x} \frac{3}{8}(x^2 - x + 1)\,dx = 0 + \frac{3}{8}\left[\frac{x^3}{3} - \frac{x^2}{2} + x\right]_{0}^{x}$$

$$= \frac{3(2x^3 - 3x^2 + 6x)}{8 \cdot 6} = \frac{x(2x^2 - 3x + 6)}{16},$$

and for $x > 2$:

$$F(x) = P(X \leq x) = \int_{-\infty}^{2} f(x)\,dx + \int_{2}^{x} f(x)\,dx$$

$$= F(2) + \int_{2}^{x} 0 \cdot dx = 1.$$

Therefore,

$$F(x) = \begin{cases} 0, & x < 0, \\ \dfrac{x(2x^2 - 3x + 6)}{16}, & 0 \leq x \leq 2, \\ 1, & x > 2. \end{cases} \quad (3.2.3)$$

Thus, if we wish to find the probability that the random variable can assume a value between 0 and some value of x less than 2, we only need to evaluate Equation (3.2.3) at the point. A graphical representation of Equation (3.2.3) is shown in Figure 3.2.1.

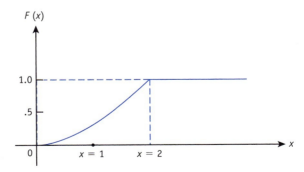

Figure 3.2.1 Cumulative Probability Distribution $F(x)$

◀

The cumulative distribution function of the continuous type has some interesting properties that are similar to those of the discrete case. We state these properties in the following theorem.

> **Theorem 3.2.1**
>
> If $F(x)$ is the cumulative distribution function of the continuous random variable X, then $F(x)$ has the following properties:
>
> 1. $F(\infty) = 1.$
> 2. $F(-\infty) = 0.$
> 3. $F(x)$ is a nondecreasing function.
> 4. $F(x)$ is continuous from the right.
> 5. $\dfrac{dF(x)}{dx} = f(x)$ for all x, at which $F(x)$ is differentiable.
>
> **Proof** Properties 1 through 4 are identical to those of Theorem 2.2.1, so the proof is omitted here. By definition, we can write
>
> $$F(x) = Pr(X \leq x) = \int_{-\infty}^{x} f(t)\,dt.$$
>
> Applying the fundamental theorem of calculus, we obtain property 5; that is, $F'(x) = f(x).$

◀

Example 3.2.2

If a continuous random variable X has for its probability density the function

$$f(x) = \begin{cases} \dfrac{1}{4} x e^{-x/2}, & x \geq 0, \\ 0, & \text{elsewhere}, \end{cases}$$

then, for $x \geq 0$:

$$F(x) = Pr(X \leq x) = \frac{1}{4} \int_0^x t e^{-t/2}\,dt = 1 - e^{-x/2}\left(1 + \frac{x}{2}\right).$$

Therefore,

$$F(x) = \begin{cases} 0, & x < 0, \\ 1 - e^{-x/2}\left(1 + \dfrac{x}{2}\right), & x \geq 0. \end{cases}$$

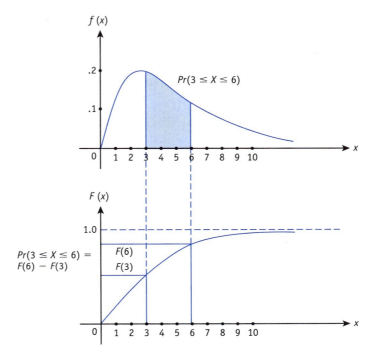

Figure 3.2.2 Graph of f(x) and F(x)

Example 3.2.3

The life length T of a mechanical system can be interpreted as behaving like a random variable of the continuous type, with probability density function $f(t)$. If the cumulative distribution of such a system is

$$F(t) = \begin{cases} 0, & t < 0, \\ 1 - \exp\left(-\frac{(t-\gamma)^\beta}{\alpha}\right), & \alpha > 0, \beta, \gamma \geq 0, t \geq 0, \end{cases}$$

find the probability density function that describes the failure behavior of such a system.

Solution: Applying property 5 of Theorem 3.2.1, we have, for $t \geq 0$,

$$\frac{dF(t)}{dt} = \frac{\beta}{\alpha}(t-\gamma)^{\beta-1} \exp\left(-\frac{(t-\gamma)^\beta}{\alpha}\right) = f(t).$$

It is easy to show that $f(t)$ satisfies the conditions of a density function.

The terminology used in this book is quite standard; however, you should keep in mind that when we speak of a **probability distribution** of a random variable, we refer to its *probability density function*. When we speak of the **distribution function**, we always mean its *cumulative distribution function*.

3.3 Continuous Probability Distributions

In this section, we study a number of continuous probability distributions that are applicable to a number of physical phenomena.

Uniform Distribution The simplest probability density function of a random variable X of the continuous type is the uniform distribution.

Definition 3.3.1

A random variable X has a **uniform distribution** or **rectangular distribution** if its probability density function is given by

$$f(x) = \begin{cases} \dfrac{1}{\beta - \alpha}, & \alpha < x < \beta, \ 0 < \alpha < \beta < \infty, \\ 0, & \text{elsewhere}. \end{cases} \quad (3.3.1)$$

Clearly, the conditions of a density function hold.
The cumulative distribution function of the random variable X is given by

$$F(x) = Pr(X \leq x) = \int_{-\infty}^{x} \frac{1}{\beta - \alpha} dx.$$

Hence,

$$F(x) = \begin{cases} 0, & x \leq \alpha, \\ \dfrac{x - \alpha}{\beta - \alpha}, & \alpha < x < \beta, \\ 1, & x \geq \beta. \end{cases}$$

Figure 3.3.1 shows the graphs of $f(x)$ and $F(x)$.

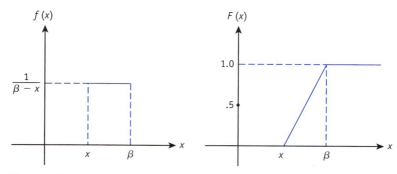

Figure 3.3.1 Graphs of $f(x)$ and $F(x)$

Applying to Equation (3.3.1) the linear transformation

$$y = \frac{x-\alpha}{\beta-\alpha},$$

we obtain the distribution of the random variable Y, which is uniform over the interval $[0,1]$; that is,

$$f(y) = \begin{cases} 1, & 0 < y < 1, \\ 0, & \text{elsewhere}. \end{cases}$$

Example 3.3.1

The melting point X of a certain specimen may be assumed to be a continuous random variable that is uniformly distributed over the interval $[110, 120]$. Then,

$$f(x) = \begin{cases} \dfrac{1}{10}, & 110 \le x \le 120, \\ 0, & \text{elsewhere}. \end{cases}$$

The probability that such a specimen will melt between 112°C and 115°C is

$$Pr(112 \le X \le 115) = \int_{112}^{115} \frac{1}{10} dx = \frac{3}{10}.$$

Example 3.3.2

The efficiency X of a certain electrical component may be assumed to be a random variable that is distributed uniformly between 0 and 100 units. What are the probabilities that X is (a) between 60 and 80 units and (b) greater than 90 units?

Solution: From Equation (3.3.1), we have:

a. $Pr(60 \le X \le 80) = \int_{60}^{80} f(x) dx = \frac{1}{100} \int_{60}^{80} dx = 0.20$

b. $Pr(X > 90) = \frac{1}{100} \int_{90}^{100} dx = 1 - \frac{1}{100} \int_{0}^{90} dx = 1 - 0.9 = 0.10$

Thus, 20% of the electrical components will have an efficiency between 60 and 80 units and only 10% will have an efficiency that is more than 90 units.

Gaussian or Normal Distribution The normal density function is the most important probability function because it has unique mathematical properties and because it can be applied to practically any physical problem. It constitutes the basis for the development of many methods of statistical theory.

The normal density function was discovered by Abraham De Moivre in 1733 as a limiting form of the binomial distribution. In 1774, Pierre Laplace studied the mathematical properties of the normal density. Through a historical error, the discovery of the normal distribution was attributed to Gauss, who, in a paper in 1809, first referred to it. The function was studied in the nineteenth century by scientists who noted that errors of measurement followed a pattern that was closely approximated by a function they called the "normal curve of error." The density function is stated in Definition 3.3.2.

Definition 3.3.2

A random variable X is said to be **Gaussian** or **normally distributed** if it has the following as its probability density function:

$$f(x,\mu,\sigma^2) = \frac{1}{\sqrt{2\pi}\sigma} \exp\left(-\frac{1}{2}\left(\frac{x-\mu}{\sigma}\right)^2\right), \qquad (3.3.2)$$

$-\infty < x < \infty, -\infty < \mu < \infty, \sigma > 0.$

The two parameters μ and σ, which completely describe the normal density function, are of significant importance and are discussed in Chapter 4. The density function $f(x;\mu,\sigma^2)$ has one mode at $x = \mu$ and two inflection points located $\pm\sigma$ from the modal point. The function is symmetrical about μ. Figures 3.3.2 and 3.3.3 show a graphical representation of the normal distribution for a fixed value of σ, with μ varying, and for a fixed value of μ, with σ varying, respectively. Thus, the parameter μ is a "location parameter"; that is, it shifts the mode of the function on the x-axis; σ is a "shape parameter"; that is, it alters the shape of the density with respect to a fixed scale. For small values of σ, the distribution is closely compact to the mode; as σ increases, the density deviates farther from μ.

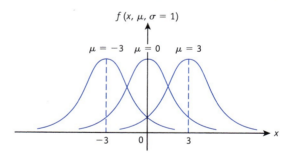

Figure 3.3.2 Normal Density Function with $\sigma = 1$ and $\mu = -3, 0, 3$.

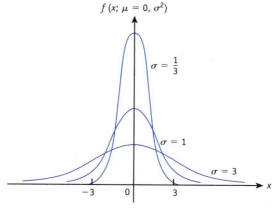

Figure 3.3.3 Normal Distribution Function with $\mu = 0$ and $\sigma^2 = \frac{1}{9}, 1, 9$.

We proceed to show that Equation (3.3.2) is a probability density function. It is clear that $f(x;\mu,\sigma^2)$ is greater than zero for $-\infty < x < \infty$, $-\infty < \mu < \infty$, and $\sigma > 0$. To show that

$$\int_{-\infty}^{\infty} f(x;\mu,\sigma^2)\,dx = 1,$$

we proceed as follows: Let I represent the value of the integral

$$I = \frac{1}{\sqrt{2\pi}\sigma}\int_{-\infty}^{\infty} \exp\left[-\frac{1}{2}\left(\frac{x-\mu}{\sigma}\right)^2\right]dx. \tag{3.3.3}$$

Applying the linear transformation $y = (x-\mu)/\sigma$, which implies $\sigma\,dy = dx$, to Equation (3.3.3), we have

$$I = \frac{1}{\sqrt{2\pi}}\int_{-\infty}^{\infty} e^{-(1/2)y^2}\,dy,$$

which, for symmetry, may be written as

$$I = \frac{2}{\sqrt{2\pi}}\int_0^{\infty} e^{-y^2/2}\,dy. \tag{3.3.4}$$

We show that $I^2 = 1$; because $f(x;\mu,\sigma^2) \geq 0$, it follows that $I = 1$. Consider

$$I^2 = \frac{2}{\sqrt{2\pi}}\int_0^{\infty} e^{-(1/2)y^2}\,dy \cdot \frac{1}{\sqrt{2\pi}}\int_0^{\infty} e^{-(1/2)z^2}\,dz$$

$$= \frac{4}{2\pi}\int_0^{\infty}\int_0^{\infty} e^{-(1/2)(y^2+z^2)}\,dy\,dz. \tag{3.3.5}$$

Changing the double integral in Equation (3.3.5) from rectangular to polar coordinates by letting $y = r\sin\theta$ and $z = r\cos\theta$ and changing the limits accordingly, we have

$$I^2 = \frac{4}{2\pi}\int_0^{\infty}\int_0^{\pi/2} |J| e^{-(1/2)r^2}\,d\theta\,dr. \tag{3.3.6}$$

The absolute value of the Jacobian is r, and Equation (3.3.6) becomes

$$I^2 = \frac{4}{2\pi}\int_0^{\infty}\int_0^{\pi/2} re^{-(1/2)r^2}\,d\theta\,dr$$

$$= \int_0^{\infty} re^{-(1/2)r^2}\,dr = 1.$$

Therefore, $I = 1$, and it has been verified that Equation (3.3.2) is a probability density function.

Applying the linear transformation $y = (x-\mu)/\sigma$ to Equation (3.3.2), as seen earlier, results in

$$f(y;\mu=0,\sigma^2=1) = \frac{1}{\sqrt{2\pi}} e^{-(1/2)y^2},\ -\infty < y < \infty. \tag{3.3.7}$$

Equation (3.3.7) is known as the **unit** or **standard normal distribution** of the random variable Y. This transformation simply locates the mean of the variate Y at $\mu = 0$, and the parameter σ becomes unity.

The cumulative distribution function of the normal density function is given by

$$F(x) = Pr(X \le x) = \frac{1}{\sqrt{2\pi}\sigma} \int_{-\infty}^{x} \exp\left(-\frac{1}{2}\left(\frac{t-\mu}{\sigma}\right)^2\right) dt, \qquad (3.3.8)$$

$-\infty < x < \infty, -\infty < \mu < \infty, \sigma > 0.$

A graph of $F(x)$ is shown in Figure 3.3.4.

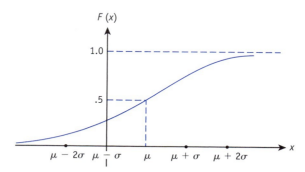

Figure 3.3.4 Cumulative Probability of the Normal Distribution

If we let $y = (t-\mu)/\sigma$ in Equation (3.3.8), we have

$$F\left(\frac{x-\mu}{\sigma}\right) = \frac{1}{\sqrt{2\pi}} \int_{-\infty}^{\frac{x-\mu}{\sigma}} e^{-(1/2)y^2} dy. \qquad (3.3.9)$$

The preceding function cannot be integrated explicitly, and the probabilities are calculated numerically. The normal density and its cumulative form, Equation (3.3.9), have been tabulated extensively. The table of normal distribution has the normal density function $f(x; \mu=0, \sigma^2=1)$ for various values of x. Table 5 has values of $F(x)$ for various values of $(x-\mu/\sigma)$. In using the tables, keep in mind that $f(x; \mu=0, \sigma^2=1) = f(-x; \mu=0, \sigma^2=1)$ and $F(x) = 1 - F(-x)$. That is, because of the symmetrical property of Equation (3.3.2), we only need to have probabilities for positive values of x. For example,

$$Pr(X \le -x) = \frac{1}{\sqrt{2\pi}\sigma} \int_{-\infty}^{-x} \exp\left(-\frac{1}{2}\left(\frac{t-\mu}{\sigma}\right)^2\right) dt$$

$$= \frac{1}{\sqrt{2\pi}} \int_{-\infty}^{\frac{-x-\mu}{\sigma}} e^{-(1/2)y^2} dy$$

$$= F\left(\frac{-x-\mu}{\sigma}\right) = 1 - F\left(\frac{x+\mu}{\sigma}\right)$$

and

$$Pr(\alpha \le X \le \beta) = \frac{1}{\sqrt{2\pi}\sigma}\int_\alpha^\beta \exp\left(-\frac{1}{2}\left(\frac{x-\mu}{\sigma}\right)^2\right)dx$$

$$= \frac{1}{\sqrt{2\pi}}\int_{\frac{\alpha-\mu}{\sigma}}^{\frac{\beta-\mu}{\sigma}} e^{-(1/2)y^2}\, dy$$

$$= Pr\left(Y \le \frac{\beta-\mu}{\sigma}\right) - Pr\left(Y \le \frac{\alpha-\mu}{\sigma}\right)$$

$$= F\left(\frac{\beta-\mu}{\sigma}\right) - F\left(\frac{\alpha-\mu}{\sigma}\right).$$

Also,

$$Pr(X \ge \gamma) = \frac{1}{\sqrt{2\pi}\sigma}\int_\gamma^\infty \exp\left(-\frac{1}{2}\left(\frac{x-\mu}{\sigma}\right)^2\right)dx$$

$$= \frac{1}{\sqrt{2\pi}}\int_{\frac{\gamma-\mu}{\sigma}}^\infty e^{-(1/2)y^2}\, dy$$

$$= 1 - Pr\left(X \le \frac{\gamma-\mu}{\sigma}\right)$$

$$= 1 - F\left(\frac{\gamma-\mu}{\sigma}\right).$$

Although the parameters μ and σ are studied later, we can define them as follows: μ is the **mean** or **average** of the information, and σ^2 the **variance** of the random variable X, which is normally distributed. The σ or **standard deviation** indicates the amount of dispersion from μ of the values that X can assume. In actual applications of the normal distribution, the domain of X is finite and μ and σ^2 are estimated by statistical methods.

We illustrate some applications of the normal distribution with Examples 3.3.3 through 3.3.5.

Example 3.3.3

The scores, say, X, of an examination may be assumed to be a continuous random variable normally distributed with $\mu = 75$ and $\sigma^2 = 64$. Find the probability that:

a. A score chosen at random will be between 80% and 85%.

b. A score will be greater than 85%.

c. A score will be less than 90%.

Solution: The distribution of scores X is

$$f(x;\mu=75,\sigma^2=64)=\frac{1}{8\sqrt{2\pi}}\exp\left(-\frac{1}{2}\left(\frac{x-75}{8}\right)^2\right), -\infty<x<\infty.$$

Thus,

a. $Pr(80\leq X\leq 85)=\frac{1}{\sqrt{2\pi}8}\int_{80}^{85}\exp\left(-\frac{1}{2}\left(\frac{x-75}{8}\right)^2\right)dx.$

Let $y=(x-75)/8$, implying $8dy=dx$. Changing the limits accordingly, we have

$$Pr(80\leq X\leq 85)=\frac{1}{\sqrt{2\pi}}\int_{\frac{80-75}{8}}^{\frac{85-75}{8}}e^{-(1/2)y^2}dy=\frac{1}{\sqrt{2\pi}}\int_{0.625}^{1.25}e^{-(1/2)y^2}dy$$

$$=F(1.25)-F(0.625).$$

From Table 5, we obtain

$$Pr(80\leq X\leq 85)=0.8944-0.7340=0.1604.$$

Therefore, there is about 16% chance that the score will be a grade between 80 and 85.

b. $Pr(X>85)=1-Pr(X\leq 85)$

$$=1-\frac{1}{8\sqrt{2\pi}}\int_{-\infty}^{85}\exp\left(-\frac{1}{2}\left(\frac{x-75}{8}\right)^2\right)dx$$

$$=\frac{1}{\sqrt{2\pi}}\int_{-\infty}^{\frac{85-75}{8}}e^{-(1/2)y^2}dy$$

$$=1-F(1.25)$$

$$=1-0.8944$$

$$=0.1056$$

c. Similarly, we can calculate

$$Pr(X\leq 90)=1-\frac{1}{8\sqrt{2\pi}}\int_{-\infty}^{90}\exp\left(-\frac{1}{2}\left(\frac{x-75}{80}\right)^2\right)dx$$

$$=\frac{1}{\sqrt{2\pi}}\int_{-\infty}^{\frac{90-75}{8}}e^{-(1/2)y^2}dy$$

$$=F(1.875)$$

$$=0.9696.$$

Example 3.3.4

A noise voltage X, which assumes values of x volts, is normally distributed with $\mu = 1v$ and $\sigma = 0.80v$. Find the probability that:

a. $|X|$ will not exceed $2.0v$.

b. The noise voltage level is exceeded 95% of the time.

Solution:

a. $Pr(|X| \leq 2v) = Pr(-2v \leq X \leq 2v)$

$$= \frac{1}{\sqrt{2\pi}\,0.8v} \int_{-2v}^{2v} \exp\left(-\frac{1}{2}\left(\frac{xv-1v}{0.8v}\right)^2\right) dx$$

$$= \frac{1}{\sqrt{2\pi}} \int_{\frac{-2v-1v}{0.8v}}^{\frac{2v-1v}{0.8v}} e^{-(1/2)y^2} dy$$

$$= F(1.25) - [1 - F(3.75)] = 0.8943$$

b. $Pr(X > xv) = 1 - Pr(X \leq xv)$

$$= 1 - \frac{1}{\sqrt{2\pi}\,0.8v} \int_{-\infty}^{xv} \exp\left(-\frac{1}{2}\left(\frac{xv-1v}{0.8v}\right)^2\right) dx$$

$$= 1 - \frac{1}{\sqrt{2\pi}} \int_{-\infty}^{\frac{xv-1v}{0.8v}} e^{-(1/2)y^2} dy$$

$$= 1 - F\left(\frac{xv-1v}{0.8v}\right) = 0.05.$$

Thus,

$$F\left(\frac{xv-1v}{0.8v}\right) = 0.05,$$

and, from Table 5, we have

$$\frac{xv-1v}{0.8v} = -1.64 \text{ or } x = -0.312.$$

Hence,

$$Pr(X > -0.132v) = 0.95.$$

Example 3.3.5

Suppose that the diameters of golf balls manufactured by a certain company are normally distributed with $\mu = 1.96$ in. and $\sigma = 0.4$ in. A ball is considered defective if its diameter is less than 1.90 in. or greater than 2.02 in. What is the percentage of defective balls manufactured by the company?

Solution: Here, we want

$$Pr\left[(X < 90) \cup (X > 2.02)\right] = 1 - Pr(1.90 \leq X \leq 2.02)$$

$$= 1 - \frac{1}{0.04\sqrt{2\pi}} \int_{1.90}^{2.02} \exp\left(-\frac{1}{2}\left(\frac{x - 1.96}{0.04}\right)^2\right) dx$$

$$= 1 - \frac{1}{\sqrt{2\pi}} \int_{\frac{1.90 - 1.96}{0.04}}^{\frac{2.02 - 1.96}{0.04}} e^{-(1/2)y^2} dy$$

$$= 1 - \left[F\left(\frac{2.02 - 1.96}{0.04}\right) - F\left(\frac{1.90 - 1.96}{0.04}\right)\right]$$

$$= 1 - \left[F(1.5) - F(-1.5)\right]$$

$$= 1 - \left\{F(1.5) - \left[1 - F(1.5)\right]\right\}$$

$$= 1 - \left[2F(1.5) - 1\right]$$

$$= 2 \cdot \left[1 - F(1.5)\right]$$

$$= 2 \cdot (1 - 0.9332)$$

$$= 2(0.0668)$$

$$= 0.1336.$$

Therefore, the company manufactures about 13.4% defective golf balls.

Although few random phenomena precisely obey a normal probability law, the laws that they do obey can, under certain conditions, be closely approximated by the normal probability law. In this sense, the normal distribution is of paramount importance. Some densities that may be approximated by the normal distribution under certain conditions are the *binomial*, the *hypergeometric*, the *Poisson*, and the *gamma*. But it is of particular importance that, under certain conditions, as the size of a sample becomes large, the distribution of the sample mean,

$$\bar{x} = \frac{1}{n}\sum x_i,$$

3.3 Continuous Probability Distributions

may be justifiably approximated by a normal distribution. The last fact, called the **central limit theorem**, is discussed in greater detail later.

Gamma Probability Distribution We now define another density function that plays an important role in various physical phenomena.

Definition 3.3.3

A random variable X is said to be distributed as the **gamma probability distribution** if it has the following function for its probability density:

$$f(x;\alpha,\beta) = \begin{cases} \dfrac{1}{\Gamma(\alpha)\beta^\alpha} x^{\alpha-1} e^{-x/\beta}, & x > 0, \alpha, \beta > 0, \\ 0, & \text{elsewhere.} \end{cases} \quad (3.3.10)$$

We proceed by showing that this two-parameter function satisfies the conditions of a probability density. It is clear that $f(x;\alpha,\beta) \geq 0$ because

$$\Gamma(\alpha) = \int_0^\infty y^{\alpha-1} e^{-y} dy, \text{ for } \alpha > 0,$$

is a positive function. To show that

$$\int_0^\infty f(x;\alpha,\beta)dx = \int_0^\infty \frac{1}{\Gamma(\alpha)\beta^\alpha} x^{\alpha-1} e^{-x/\beta} dx = 1, \quad (3.3.11)$$

let $y = x/\beta$ and $\beta dy = dx$ in Equation (3.3.11). We then have

$$\int_0^\infty f(x;\alpha,\beta)dx = \frac{1}{\Gamma(\alpha)} \int_0^\infty y^{\alpha-1} e^{-y} dy = \frac{\Gamma(\alpha)}{\Gamma(\alpha)} = 1,$$

and the conditions of a probability density have been verified.

The parameters α and β determine the shape of the density function, which is skewed to the right for all values of α and β; however, the skewness decreases as α increases. Figure 3.3.5 shows the graph of the gamma density for a fixed $\beta = 1$ and $\alpha = 1, 2, 3$.

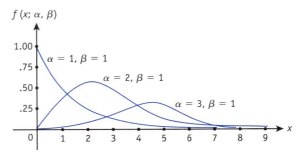

Figure 3.3.5 Gamma Probability Density Function

Figure 3.3.6 shows the graph of the gamma density for a fixed $\alpha = 2$ and $\beta = 1/2, 1, 2$.

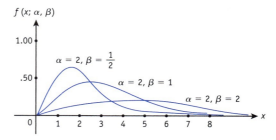

Figure 3.3.6 Gamma Probability Density Function

We can obtain the value of x at which the density attains its mode by differentiating $f(x;\alpha,\beta)$ and setting it equal to zero:

$$f'(x;\alpha,\beta) = \frac{1}{(\alpha-1)!\beta^\alpha}\left\{(\alpha-1)x^{\alpha-2}e^{-(x/\beta)} - \frac{1}{\beta}x^{\alpha-1}e^{-(x/\beta)}\right\} = 0.$$

This implies that $x = (\alpha-1)\beta$. Therefore, we can obtain a family of gamma densities by varying α and β, with each density attaining its maximum at the same value of x. Furthermore, as $x \to \infty$, $f(x;\alpha,\beta) \to 0$ for $\alpha, \beta > 0$. If $\alpha \geq 1$, then the gamma density curve starts at the origin. For $0 < \alpha < 1$, which implies that $\Gamma(\alpha) = (\alpha-1)!$ in Equation (3.3.10) is negative, $f(x;\alpha,\beta)$ becomes infinite as x approaches zero; that is, it has the $f(x;\alpha,\beta)$ axis as an asymptote.

When $\alpha = 1$ in the gamma probability density, we obtain an important density function, defined in Definition 3.3.4.

■ **Definition 3.3.4**

A random variable X is said to have an **exponential distribution** if it has the following function as its probability density:

$$f(x;\beta) = \begin{cases} \dfrac{1}{\beta}e^{-(x/\beta)}, & x \geq 0, \beta > 0, \\ 0, & \text{elsewhere.} \end{cases} \qquad (3.3.12)$$

It is not difficult to verify that the one-parameter function in Equation (3.3.12) satisfies the conditions of a probability density. The exponential density is a decaying type of function whose rate of decay depends on the parameter β. Figure 3.3.7 illustrates the exponential distribution for $\beta = 1/2, 1, 2$.

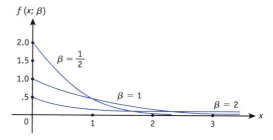

Figure 3.3.7 Exponential Probability Density Function

The cumulative distribution function of the gamma density is

$$F(x) = Pr(X \leq x) = \begin{cases} 0, & x \leq 0, \\ \dfrac{1}{\Gamma(\alpha)\beta^\alpha} \int_0^x t^{\alpha-1} e^{-(t/\beta)} dt, & x > 0. \end{cases} \quad (3.3.13)$$

The function $F(x)$ can be integrated exactly if $(\alpha-1)$ is a positive integer or zero; numerical integration is necessary if it is not an integer. Equation (3.3.13), which is known as the **incomplete gamma function**, has been tabulated for various values of α and β.[1]

Example 3.3.6

The daily consumption of fuel in millions of gallons at a certain airport can be treated as a random variable that has a gamma density with $\alpha=3$ and $\beta=1$. Suppose that the airport can store 2 million gallons of fuel; find the probability that this fuel supply will be inadequate on any given day.

Solution: The gamma distribution with $\alpha=3$ and $\beta=1$ is

$$f(x; \alpha=3, \beta=1) = \begin{cases} \dfrac{1}{2!} x^2 e^{-x}, & x \geq 0, \\ 0, & \text{elsewhere,} \end{cases}$$

and

$$Pr(X > 2) = \frac{1}{2!} \int_2^\infty x^2 e^{-x} dx = 5e^{-2}$$

$$= 0.6767.$$

Example 3.3.7

The time in hours during which an electrical generator is operational is a random variable that follows the exponential distribution with $\beta=160$. What is the probability that a generator of this type will be operational for (a) less than 40 hours, (b) between 60 and 160 hours, and (c) more than 200 hours?

Solution: The exponential density function is

$$f(x; 160) = \begin{cases} \dfrac{1}{160} e^{-(x/160)}, & x \geq 0, \\ 0, & \text{elsewhere.} \end{cases}$$

The cumulative distribution function of $f(x; 160)$ for $x \geq 0$ is given by

$$F(x) = Pr(X \leq x) = \int_0^x \frac{1}{160} e^{-(t/160)} dt$$

$$= 1 - e^{-(x/160)}.$$

Therefore,

$$F(x) = \begin{cases} 0, & x < 0, \\ 1 - e^{-(x/160)}, & x \geq 0. \end{cases}$$

Thus, the results are:

a. $Pr(X \leq 40) = 1 - e^{-0.25} = 0.22119$

b. $Pr(60 \leq X \leq 160) = F(160) - F(60) = e^{-0.375} - e^{-1} = 0.3194$

c. $Pr(X > 200) = 1 - F(200) = e^{-1.25} = 0.2865$

Students commonly ask, "When do we know that the gamma density is applicable in a given physical problem?" There is no direct answer to this question; however, you proceed to determine the answer by constructing a probability histogram from the information at hand. From the shape of this histogram, you decide whether the random variable follows the gamma density. Also, the parameters α and β must be statistically estimated. In Chapter 4, we discuss some additional properties of the gamma distribution.

Beta Probability Distribution A frequently employed distribution, for example, when the random variable assumes values that are percentages or when we are concerned with physical phenomena of the continuous type and have values lying between 0 and 1, is the beta probability distribution.

■ **Definition 3.3.5**

A random variable X is said to have a **beta distribution** if it has the following function for its probability density:

$$f(x; \alpha, \beta) = \begin{cases} \dfrac{\Gamma(\alpha+\beta)}{\Gamma(\alpha)\Gamma(\beta)} x^{\alpha-1}(1-x)^{\beta-1}, & 0 < x \leq 1,\, \alpha, \beta > 0, \\ 0, & \text{elsewhere.} \end{cases} \quad (3.3.14)$$

To verify that Equation (3.3.14) shows a density function, we must show that $f(x; \alpha, \beta) \geq 0$, for $0 \leq x \leq 1$ and $\alpha, \beta > 0$, and that the function integrates to one over the whole domain of X. The first condition can be easily shown, but for the second condition, we proceed as follows: We can write

$$(\alpha - 1)(\beta - 1)! = \Gamma(\alpha)\Gamma(\beta) \quad (3.3.15)$$

$$= \left(\int_0^\infty x^{\alpha-1} e^{-x} dx\right)\left(\int_0^\infty x^{\beta-1} e^{-y} dy\right)$$

$$= \int_0^\infty \int_0^\infty x^{\alpha-1} y^{\beta-1} e^{-(x+y)} dx dy.$$

Applying the transformation $z = x/(x+y)$ or $x = zy/(1-z)$ to Equation (3.3.15), with $dx = \{y/(1-z)^2\} dz$ and $0 \leq z \leq 1$, we have

$$\Gamma(\alpha)\Gamma(\beta) = \int_0^\infty \int_0^1 \left(\frac{zy}{1-z}\right)^{\alpha-1} y^{\beta-1} e^{-\{y/(1-z)\}} \frac{y}{(1-z)^2} dz dy. \quad (3.3.16)$$

Let $t = y/(1-z)$ and $(1-z)dt = dy$ in Equation (3.3.16). Thus,

$$\Gamma(\alpha)\Gamma(\beta) = \int_0^\infty \int_0^1 z^{\alpha-1}(1-z)^{\beta-1} t^{\alpha+\beta-1} e^{-t} dz dt$$

$$= \left(\int_0^\infty t^{\alpha+\beta-1} e^{-t} dt\right)\left(\int_0^1 z^{\alpha-1}(1-z)^{\beta-1} dz\right)$$

$$= \Gamma(\alpha+\beta) \int_0^1 z^{\alpha-1}(1-z)^{\beta-1} dz. \quad (3.3.17)$$

This implies that

$$\frac{\Gamma(\alpha+\beta)}{\Gamma(\alpha)\Gamma(\beta)} \int_0^1 z^{\alpha-1}(1-z)^{\beta-1} dz = 1,$$

which must be shown.

The two parameters α and β determine the shape of the beta distribution. Figure 3.3.8 shows the graph of the density for a fixed value of β while α varies.

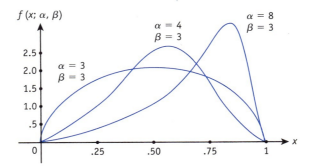

Figure 3.3.8 Beta Distribution

In Figure 3.3.9, we show the graph of the beta density for a fixed value of α while β varies.

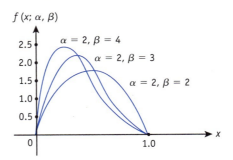

Figure 3.3.9 Beta Distribution

When $\alpha < \beta$ the density is skewed to the right, and when $\beta < \alpha$ it is skewed to the left. The beta distribution is symmetrical when $\alpha = \beta$. The graph of the distribution has a U shape when $(\alpha-1)$ and $(\beta-1)$ are negative and a J shape if only one of them is negative. The density function attains its mode at $x = (\alpha-1)/(\alpha+\beta-2)$ when $(\alpha-1)$ and $(\beta-1)$ are positive; the same point is its minimum when they are negative. It is possible to generate a family of beta density functions, all of which have the same mode, by selecting the values of the parameters α and β properly.

> Some special cases of the beta distribution are:
>
> 1. When $\alpha = \beta = 1$, we have the rectangular distribution,
>
> $$f(x) = \begin{cases} 1, & 0 \leq x \leq 1, \\ 0, & \text{elsewhere.} \end{cases}$$
>
> 2. When $\alpha = 1$, $\beta = 2$, and $\alpha = 2$, $\beta = 1$, we have the **triangular distributions**,
>
> $$f(x; \alpha = 1, \beta = 2) = \begin{cases} 2(1-x), & 0 \leq x \leq 1, \\ 0, & \text{elsewhere,} \end{cases}$$
>
> and
>
> $$f(x; \alpha = 2, \beta = 1) = \begin{cases} 2x, & 0 \leq x \leq 1, \\ 0, & \text{elsewhere.} \end{cases}$$

The cumulative distribution of the beta density is

$$F(x) = Pr(X \leq x) = \begin{cases} 0, & x \leq 0, \\ \dfrac{\Gamma(\alpha+\beta)}{\Gamma(\alpha)\Gamma(\beta)} \int_0^x t^{\alpha-1}(1-t)^{\beta-1}\,dt, & 0 < x < 1, \\ 1, & x \geq 1. \end{cases} \quad (3.3.18)$$

This $F(x)$ is known as the **incomplete beta function**, which cannot be directly computed if α and β are large. Tables have been compiled for various values of α and β.[2] The incomplete beta function and the cumulative binomial function are related as follows:

$$\sum_{x=\alpha}^{N} \binom{N}{x} p^x (1-p)^{N-x} = \frac{\Gamma(N+1)}{\Gamma(\alpha)\Gamma(N-\alpha+1)} \times \int_0^x x^{\alpha-1}(1-x)^{N-\alpha}\,dx,\; 0 \leq x < N.$$

◀

Example 3.3.8

The proportion of students who pass a standard examination may be treated as a continuous random variable, say, X, that has a beta distribution with $\alpha = 15$ and $\beta = 3$. Find the probability that, in a certain school, less than 80% of the students pass the test.

Solution:

$$f(x; \alpha = 15, \beta = 3) = \begin{cases} \dfrac{\Gamma(18)}{\Gamma(15)\Gamma(3)} x^{14}(1-x)^2, & x < x \leq 1, \\ 0, & \text{elsewhere,} \end{cases}$$

and

$$Pr(X < 0.8) = \frac{17 \cdot 16 \cdot 15}{2} \int_0^{0.8} x^{14}(1-x)^2 \, dx$$

$$= 17 \cdot 15 \cdot 8 \int_0^{0.8} x^{14}(1 - 2x + x^2) \, dx$$

$$= 0.30962.$$

Thus, we are 31% certain that less than 80% of the students will pass this examination. ◀

Example 3.3.9

The daily proportion of major automobile accidents across the United States can be treated as random variable that has a beta distribution with $\alpha = 6$ and $\beta = 4$. Find the probability that, on a certain day, the percentage of major accidents is less than 80% but greater than 60%.

Solution: Applying Equation (3.3.14), we have

$$Pr(0.60 \leq X \leq 0.80) = \frac{\Gamma(10)}{\Gamma(6)\Gamma(4)} \int_{0.6}^{0.8} x^5 (1-x)^3 \, dx$$

$$= 0.4317.$$ ◀

Cauchy Probability Distribution A classical distribution of minor importance is the Cauchy density, defined in Definition 3.3.6.

Definition 3.3.6

A random variable X is said to have a **Cauchy probability distribution** if it has the following function as its density:

$$f(x; \alpha, \mu) = \frac{1}{\pi} \frac{\alpha}{\alpha^2 + (x-\mu)^2}, \quad -\infty < x < \infty, \; -\infty < \mu < \infty, \; \alpha > 0. \qquad (3.3.19)$$

We can verify that function (3.3.19) is a probability density. By inspection, the first condition is satisfied; by letting $y = (x - \mu)/\alpha$ and $\alpha \, dy = dx$, $f(x; \alpha, \mu)$ becomes

$$\frac{1}{\pi} \int_{-\infty}^{\infty} \frac{dy}{1 + y^2} = \frac{1}{\pi} \left[\tan^{-1} y \right]_{-\infty}^{\infty} = \frac{1}{\pi} \left\{ \frac{\pi}{2} - \left(-\frac{\pi}{2} \right) \right\} = 1.$$

The parameter μ locates the center of symmetry of the Cauchy density, and α is a shape parameter. Under the linear transformation $z = x - \mu$, μ becomes zero and Equation (3.3.19) becomes

$$f(z; \alpha) = \frac{1}{\pi} \frac{\alpha}{(\alpha^2 + z^2)}. \qquad (3.3.20)$$

The cumulative distribution function of Equation (3.3.19) is

$$F(x) = Pr(X \leq x) = \frac{1}{\pi \alpha} \int_{-\infty}^{x} \frac{1}{1 + \left(\frac{t-\mu}{\alpha}\right)^2} dt = \frac{1}{\pi} \int_{-\infty}^{\frac{x-\mu}{\alpha}} \frac{1}{1+y^2} dy$$

$$= \frac{1}{\pi} \left[\tan^{-1} y\right]_{-\infty}^{(x-\mu)/\alpha} = \frac{1}{2} + \frac{1}{\pi} \tan^{-1}\left(\frac{x-\mu}{\alpha}\right), \quad \alpha > 0.$$

Figure 3.3.10 shows the graph of $f(x;\alpha,\mu)$ for $\mu = 0$ and $\alpha = 1, 2, 3$ and the corresponding graph of $F(x)$.

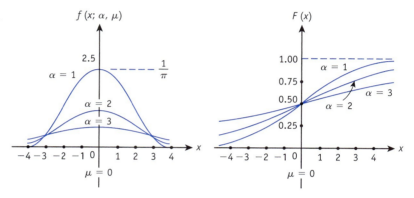

Figure 3.3.10 Cauchy Probability Density

The development of the Cauchy density, Equation (3.3.19), is as follows: In Figure 3.3.11, assume that a randomly chosen straight line is drawn through the point (μ, α). It hits the x-axis at the point $(x, 0)$, making an angle of θ with the α-axis. Assume that θ varies from $-(\pi/2)$ to $+(\pi/2)$. Then,

$$f(\theta) \begin{cases} \frac{1}{\pi}, & -\frac{\pi}{2} \leq \theta \leq \frac{\pi}{2}, \\ 0, & \text{elsewhere.} \end{cases}$$

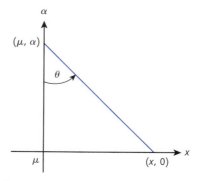

Figure 3.3.11 Developing the Cauchy Probability Distribution

If we wish to find the density of the random variable X, where $(x-\mu)=\alpha\tan\theta$, we proceed as follows:

$$h(x)=f(\theta)\left|\frac{d\theta}{dx}\right|,$$

where

$$\frac{d\theta}{dx}=\frac{1}{\frac{dx}{d\theta}}=\frac{1}{\alpha\sec^2\theta}=\frac{1}{\alpha\frac{\alpha^2+(x-\mu)^2}{\alpha^2}}=\frac{\alpha}{\alpha^2+(x-\mu)^2}.$$

When $\theta=-(\pi/2)$, $x=-\infty$; when $\theta=-\pi/2$, $x=+\infty$. Therefore,

$$h(x)=\frac{1}{\pi}\frac{\alpha}{\alpha^2+(x-\mu)^2},\quad-\infty<x<\infty.$$

◀

Example 3.3.10

A BB gun is mounted at the point $(0,\alpha)$ on a bench so that it can freely swing 180°. The gun fires BBs on a straight line marked on a board directly in front of the bench. If all angular positions of the gun are equally likely, what is the probability that a shot will hit the line between b_1 and b_2?

Solution: The physical description of this problem is similar to the development of the Cauchy density. If we assume that $\alpha=0.5$ and $\mu=0$, we can use Equation (3.3.19) to calculate the desired probability. Thus,

$$Pr(b_1\leq X\leq b_2)=\frac{1}{\pi}\int_{b_2}^{b_1}\frac{5}{(0.5)^2+(x)^2}dx$$

$$=\frac{1}{\pi}\left[\tan^{-1}\frac{b_2}{0.5}-\tan^{-1}\frac{b_1}{0.5}\right].$$

The probability that a shot is k ft or more to the right of vertical is

$$Pr(X>k)=1-Pr(X\leq k)=1-F(k)$$

$$=\frac{1}{2}-\frac{1}{\pi}\tan^{-1}\frac{k}{0.5}.$$

We discuss this distribution again in a later chapter.   ◀

Laplace Distribution A distribution of some importance to engineering problems is defined in Definition 3.3.7.

Definition 3.3.7

A random variable X is said to have a **Laplace distribution** if it has the following function for its density:

$$f(x;\mu,\sigma)=\frac{1}{2\sigma}e^{-|x-\mu|/\sigma},\quad-\infty<x<\infty,-\infty<\mu<\infty,\sigma>0.\quad(3.3.21)$$

The parameter μ located the mode of $f(x;\mu,\sigma)$, and σ serves as the shape parameter. The density is symmetrical with respect to μ. It is clear that $f(x;\mu,\sigma)$ is greater than zero for all μ, all x, and σ greater than zero. Next, let $y=(x-\mu)/\sigma$, $dx=\sigma dy$, and

$$\frac{1}{2\sigma}\int_{-\infty}^{\infty}e^{-(|x-\mu|)/\sigma}dx = \frac{1}{2\sigma}\int_{-\infty}^{\mu}e^{(x-\mu)/\sigma}dx + \frac{1}{2\sigma}\int_{\mu}^{\infty}e^{-(x-\mu)/\sigma}dx$$

$$= \frac{1}{2}\int_{-\infty}^{0}e^{y}dy + \frac{1}{2}\int_{0}^{\infty}e^{-y}dy$$

$$= \frac{1}{2} + \frac{1}{2} = 1.$$

Thus, Equation (3.3.21) is a probability density function. The cumulative distribution function is

$$F(x) = Pr(X \le x) = \frac{1}{2\sigma}\int_{-\infty}^{x}e^{-(|t-\mu|/\sigma)}dt. \tag{3.3.22}$$

For $x \le \mu$, Equation (3.3.22) becomes

$$F(x) = \frac{1}{2}e^{-(x-\mu)/\sigma};$$

for $x > \mu$, it becomes

$$F(x) = 1 - \frac{1}{2}e^{-(x-\mu)/\sigma}.$$

Therefore,

$$F(x)\begin{cases} \dfrac{1}{2}e^{-(x-\mu)/\sigma}, & x \le \mu, \\ 1 - \dfrac{1}{2}e^{-(x-\mu)/\sigma}, & x > \mu. \end{cases}$$

Figure 3.3.12 shows the graph of the Laplace density for $\mu=1$ and $\sigma=\dfrac{1}{2}, 1, 2$; it also shows the Laplace cumulative distribution function.

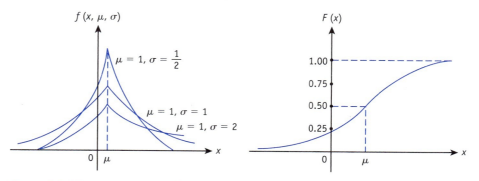

Figure 3.3.12 Laplacian f(x) and F(x)

Example 3.3.11

A company produces ball bearings, the diameters of which may vary. Let X be a variate that assumes for values the diameters of the bearings, and let it be distributed according to the Laplace density with $\mu=4$ units and $\sigma=0.008$. Customer specification requires the bearing diameter to lie in the interval 4.00 ± 0.016. Bearings outside this interval are considered defective. What fraction of the total production of the ball bearings will be acceptable to the customer?

Solution: The Laplace distribution for $\mu=4$ and $\sigma=0.008$ is

$$f(x,4,0.008)=\frac{1}{(0.008)^2}e^{-(|x-\mu|/0.008)}, \quad -\infty<x<\infty,$$

$$Pr(4.000-0.016 \le X \le 4.000+0.016)=\frac{1}{0.016}\int_{3.984}^{4.016} e^{(|x-4|/0.008)}dx$$

$$=F(4.016)-F(3.984)$$

$$=1-\frac{1}{2}e^{(4.016-4.000/0.008)}-\frac{1}{2}e^{(3.984-4.000/0.008)}$$

$$=1-\frac{1}{2}\left(e^{-2}+e^{-2}\right)$$

$$=0.864665.$$

Thus, 86.47% of the produced bearings will be acceptable to the customer.

Example 3.3.12

What should the limits be on the diameter of the ball bearing in Example 3.3.11 so that the proportion of acceptable bearings will be 90%?

Solution: We want to find x so that

$$Pr(4-x \le X \le 4+x)=\frac{1}{(0.008)^2}\int_{4-x}^{4+x} e^{(|x-4|/0.008)}dx=0.90.$$

Thus,

$$F(4+x)-F(4-x)=0.9,$$

which implies

$$1-\frac{1}{2}e^{(x/0.008)}-\frac{1}{2}e^{(x/0.008)}=0.9,$$

or

$$e^{-(x/0.008)}=0.1,$$

which implies

$$\frac{-x}{0.008} = \ln 0.1.$$

As a result,

$$x = 0.0184.$$

Therefore, the limits on the diameter of the ball bearings should be ± 0.0184 to make the unacceptable number of bearings equal to 10%.

Lognormal Probability Distribution The lognormal distribution in its simplest form may be defined as the distribution of a random variable whose logarithm obeys the normal probability density. Let Y be a random variable that is normally distributed, with parameters μ_y and σ_y. If $Y = \ln X$ or $X = e^Y$, then X is said to have lognormal probability distribution. This distribution arises in physical problems when the domain of the variate X is greater than zero and its histogram is markedly skewed. This skewing occurs when X is affected by random causes that produce small effects proportional to the variate X. The outcome of these random causes, each producing a small constant effect, is normally distributed. The lognormal probability distribution has been used in economics, sociology, biology, and anthropometry and in various physical and industrial processes.

The distribution of the variate X is defined in Definition 3.3.8.

■ **Definition 3.3.8**

A random variable X is distributed as **lognormal** if it has the following function for its probability density:

$$f(x) = \begin{cases} \dfrac{1}{x\sigma_y\sqrt{2\pi}} \exp\left\{-\dfrac{1}{2\sigma_y^2}(\ln x - \mu_y)^2\right\}, & x > 0, \sigma_y > 0, -\infty < \mu_y < \infty, \\ 0, & \text{elsewhere.} \end{cases} \qquad (3.3.23)$$

The two parameters that define the density completely, μ_y and σ_y, locate the relative position of the mean and the amount of dispersion of the information with respect to μ_y, respectively. These parameters are related to the parameters of the random variable X as follows:

$$\mu_y = \ln\left(\sqrt{\frac{\mu_x^4}{\mu_x^2 + \sigma_x^2}}\right), \quad \sigma_y = \ln\left(\sqrt{\frac{\mu_x^2 + \sigma_x^2}{\mu_x^2}}\right). \qquad (3.3.24)$$

(See Problem 3.29.)

To verify that the function in Equation (3.3.23) is a probability density, it is sufficient to show that $f(x; \mu_x, \sigma_x)$ integrates into one. Let

$$z = \frac{1}{\sigma_y}(\ln x - \mu_y), \quad \sigma_y x \, dz = dx.$$

Then,

$$\frac{1}{\sigma_y\sqrt{2\pi}} \int_0^\infty \frac{1}{x} \exp\left\{-\frac{1}{2\sigma_y^2}(\ln x - \mu_y)^2\right\} dx = \frac{1}{\sqrt{2\pi}} \int_{-\infty}^\infty e^{-(1/2)z^2} dz = 1.$$

The cumulative distribution of the lognormal is

$$F(x) = Pr(X \le x) = \frac{1}{\sigma_y \sqrt{2\pi}} \int_0^x \frac{1}{t} \exp\left\{-\frac{1}{2\sigma_y^2}(\ln t - \mu_y)^2\right\} dt. \qquad (3.3.25)$$

Let

$$z = \frac{\ln t - \mu_y}{\sigma_y}, \qquad \sigma_y t\, dz = dt,$$

and Equation (3.3.25) becomes

$$F(x) = \frac{1}{\sqrt{2\pi}} \int_{-\infty}^{\frac{\ln x - \mu_y}{\sigma_y}} e^{-(1/2)z^2}\, dz$$

$$= F\left(\frac{\ln x - \mu_y}{\sigma_y}\right).$$

Thus, the probability that the random variable X assumes its value from an interval greater than zero and less than x can be obtained from the normal tables. Similarly, if X is lognormally distributed with parameters, as given in Equation (3.3.24), and $0 < \alpha < \beta$, then

$$Pr(\alpha \le X \le \beta) = Pr(\alpha \le e^Y \le \beta)$$

$$= Pr(\ln \alpha \le Y \le \ln \beta)$$

$$= Pr\left(\frac{\ln \alpha - \mu_y}{\sigma_y} \le \frac{Y - \mu_y}{\sigma_y} \le \frac{\ln \beta - \mu_y}{\sigma_y}\right)$$

$$= F\left(\frac{\ln \beta - \mu_y}{\sigma_y}\right) - F\left(\frac{\ln \alpha - \mu_y}{\sigma_y}\right).$$

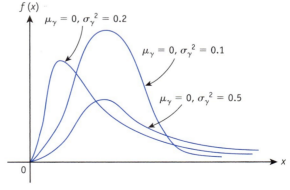

Figure 3.3.13 Lognormal Probability Density

Figures 3.3.13 and 3.3.14 show graphs of the lognormal density for fixed μ_y and varying σ_y^2 and for fixed σ_y^2 and varying μ_y, respectively. These figures illustrate that the distribution is positively skewed and that the greater the value of the parameter σ_y, the greater the skewness.

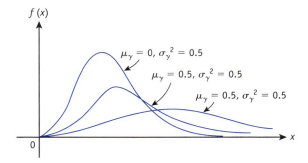

Figure 3.3.14 Lognormal Probability Density

The median of the lognormal is at $x = e^{\mu_y}$. The relative positions of the mean, meridian, and mode are at $e^{\mu_y + (1/2)\sigma_y^2}$, e^{μ_y}, and $e^{\mu_y - \sigma_y^2}$, respectively, where the mean is the average value of X.

$$\int_{-\infty}^{\infty} x f(x) dx,$$

the **median** is that value of m that satisfies the equation

$$\int_{-\infty}^{m} f(x) dx = \frac{1}{2} = \int_{m}^{\infty} f(x) dx,$$

and the mode is the value of x at which the probability density function attains its maximum.

The lognormal distribution possesses a number of interesting properties, most of which are immediate consequences of the normal distribution and are discussed in a later chapter.

When is the lognormal distribution applicable in a given physical problem if a certain amount of data has been obtained? This question can be answered by plotting the cumulative distribution of $\ln X$ on **normal probability paper**; if the resulting curve is nearly a straight line, then X has a lognormal distribution. In such a problem, the parameters μ_y and σ_y are estimated from the given information. ◀

Example 3.3.13

In an effort to establish a suitable height for the controls of a moving vehicle, information was gathered about X, the amount that the heights of the operators vary from 60 in., which is the minimum height of the operators. It was verified that the collected data followed the lognormal distribution by plotting the distribution of $\ln X$ on normal probability paper.

a. If we assume that $\mu_x = 6$ in. and $\sigma_x = 2$ in., what percentage of operators would have a height less than 65.5 in.?

b. If an operator is chosen at random, what is the probability that his height will be between 64 and 66 in.?

Solution:

a. $X = 65.5 - 60 = 5.5$

$$Pr(X \leq 5.5) = Pr(e^Y \leq 5.5) = Pr(Y \leq \ln 5.5)$$

$$= Pr\left(Z \leq \frac{\ln 5.5 - \mu_y}{\sigma_y}\right) = F\left(\frac{\ln 5.5 - \mu_y}{\sigma_y}\right).$$

But

$$\mu_y = \ln\sqrt{\frac{6^4}{6^2 + 2^2}} = \ln\frac{36}{\sqrt{40}}$$

$$= \ln\left(\frac{36}{6.32}\right) = \ln 5.69 = 1.74,$$

$$\sigma_y = \ln\sqrt{\frac{6^2 + 2^2}{6^2}} = \ln\sqrt{\frac{40}{36}}$$

$$= \ln\sqrt{1.11} = \ln 1.054 = 0.053.$$

Thus,

$$F\left(\frac{\ln 5.5 - \mu_y}{\sigma_y}\right) = F\left(\frac{1.70 - 1.74}{0.053}\right)$$

$$= F\left(\frac{-0.04}{0.053}\right) = F(-0.75)$$

$$= 1 - F(0.75) = 1 - 0.7734$$

$$= 0.2266.$$

b. $X_1 = 64 - 60 = 4.$

$$X_2 = 66 - 60 = 6.$$

Here, we want

$$Pr(4 \leq X \leq 6) = Pr(4 \leq e^Y \leq 6) = Pr(\ln 4 \leq Y \leq \ln 6)$$

$$= Pr\left[\left(\frac{\ln 6 - \mu_y}{\sigma_y}\right) \leq Z \leq \left(\frac{\ln 4 - \mu_y}{\sigma_y}\right)\right]$$

$$= F\left(\frac{1.79 - 1.74}{0.053}\right) - F\left(\frac{1.39 - 1.74}{0.053}\right)$$

$$= F\left(\frac{0.05}{0.053}\right) - F\left(\frac{-0.33}{0.053}\right)$$

$$= F(0.94) - F(-6.60)$$

$$= F(0.94) - [1 - F(6.60)]$$

$$= F(0.94) - 1 + F(6.60)$$

$$= 0.8264 - 1 + 1 = 0.8264.$$

For additional information, the book by J. Aitchison and J. A. C. Brown is devoted to the importance and usefulness of the lognormal distribution.[3]

Weibull Probability Distribution In 1951, Waloddi Weibull introduced a probability density function that is applicable to many physical phenomena.[1] It is extremely useful in studies of failure models.

Definition 3.3.9

A random variable X is said to be distributed as the **Weibull probability distribution** if it has the following function for its probability density:

$$f(x; \alpha, \beta, \gamma) = \begin{cases} \frac{\beta}{\alpha}(x-\gamma)^{\beta-1} e^{\{(x-\gamma)^\beta/\alpha\}}, & x > \gamma, \alpha, \beta, \gamma > 0, \\ 0, & \text{elsewhere.} \end{cases} \quad (3.3.26)$$

The three parameters $\alpha, \beta,$ and γ, which completely describe the Weibull density, are of significant importance. Here, α is a scale parameter, β is a shape parameter, and γ is a location or threshold parameter. The density function attains its maximum point when

$$x = \gamma + \frac{\gamma(\beta-1)^{1/\beta}}{\beta}.$$

Figure 3.3.15 shows a graphical representation of the Weibull density for $\alpha = 1$, $\gamma = 0$, and varying β, with $\beta = 1/2, 1, 2, 5$.

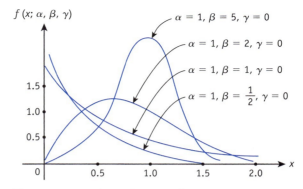

Figure 3.3.15 Weibull Probability Density

The density is skewed to the right for all values of the parameters. However, the skewness decreases as the shape parameter β increases.

It is clear that $f(x;\alpha,\beta,\gamma) \geq 0$ for $x > \gamma$, and $\alpha, \beta, \gamma > 0$. If we let

$$y = \frac{(x-\gamma)^\beta}{\alpha}, \quad \frac{\alpha}{\beta} \frac{1}{\beta(x-\gamma)^{\beta-1}} dy = dx,$$

we have

$$\frac{\beta}{\alpha} \int_\gamma^\infty (x-\gamma)^{\beta-1} e^{-\{(x-\gamma)/\alpha\}} dx = \int_0^\infty e^{-y} dy = 1.$$

Thus, Equation (3.3.26) satisfies the conditions of a probability density function.

The cumulative distribution of the Weibull distribution for $x > \gamma$ is given by

$$F(x) = Pr(X \leq x) = \frac{\beta}{\alpha} \int_\gamma^x (t-\gamma)^{\beta-1} e^{-\{(t-\gamma)^\beta/\alpha\}} dt,$$

which reduces to

$$1 - e^{-\{(x-\gamma)^\beta/\alpha\}}.$$

Thus,

$$F(x) = \begin{cases} 0, & x \leq 0, \\ 1, -e^{-\{(x-\gamma)^\beta/\alpha\}}, & x > \gamma, \alpha, \beta > 0, \gamma \geq 0. \end{cases}$$

The Weibull density function is flexible enough to be applicable to a number of problems. This flexibility can be displayed by deriving the following distributions as special cases of the Weibull:

1. When $\beta = 1$ and $\gamma = 0$ in the Weibull density, we obtain the exponential distribution,

$$f(x;\alpha) = \begin{cases} \dfrac{1}{\alpha} e^{-(x/\alpha)}, & x \geq 0, \alpha > 0, \\ 0, & \text{elsewhere,} \end{cases}$$

which we discussed earlier.

2. When we let $\beta = 2$ and $\gamma = 0$ in Equation (3.3.26), we obtain an important probability distribution, which is defined in Definition 3.3.10.

Definition 3.3.10

A random variable X is said to have a **Rayleigh distribution** if it has the following function for its probability density:

$$f(x) = \begin{cases} \dfrac{2}{\alpha} x e^{-(x^2/\alpha)}, & x \geq 0,\ \alpha > 0, \\ 0, & \text{elsewhere.} \end{cases} \qquad (3.3.27)$$

This distribution is of significant importance in the theory of sound. It meaning and derivation can be obtained by considering the following problem: In trying to locate an object on the x-y plane, we determine its distance from the origin by measuring the distance along the x- and y-axes and applying the Pythagorean formula: $r^2 = x^2 + y^2$. If the measurements are subject to random error, with X and Y representing errors in measurement that are assumed to be independent and normally distributed with $\mu = 0$ and $\sigma^2 = \alpha/2$, then the distribution of $R = \sqrt{X^2 + Y^2}$ is the Rayleigh distribution.

Figure 3.3.16 shows a graphical representation of the behavior of the shape parameter α in Equation (3.3.27).

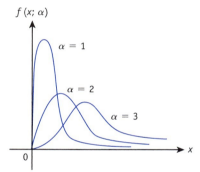

Figure 3.3.16 Rayleigh Probability Density

3. The Weibull distribution offers a close approximation to the normal distribution for certain values of parameters. It was found in a particular problem that, for $\alpha = 1$, $\beta = 3.2589$, and $\gamma = 0$ of the Weibull density and for $\mu = 0.8964$ and $\sigma^2 = 0.0924$ of the Gaussian density, the two are practically identical. This comparison is shown by Figure 3.3.17.

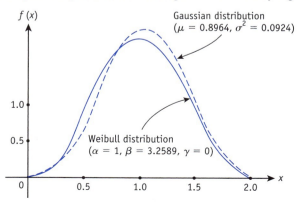

Figure 3.3.17 Gaussian vs. Weibull Probability Density

In Chapter 4, we discuss some additional properties of the Weibull and Rayleigh distributions.

Example 3.3.14

The amplitude of a signal generated by a certain source is usually assumed to be a random variable of the continuous type that is Rayleigh distributed. Given that the receiver can detect a signal of specified minimum amplitude x_0, find the probability that the receiver will be processing a signal.

Solution: Applying Equation (3.3.27), we have

$$Pr(X \geq x_0) = \frac{2}{\alpha} \int_{x_0}^{\infty} x e^{-(x^2/\alpha)} dx = e^{-(x_0^2/\alpha)}.$$

The shaded area of Figure 3.3.18 represents this probability.

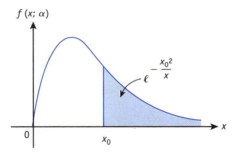

Figure 3.3.18 Rayleigh Probability Density

Other Continuous Probability Distributions The following probability density functions are important in various physical phenomena. We define the functions here and study their properties in the forthcoming chapters.

Definition 3.3.11

A random variable X is said to be distributed as the **Maxwell distribution** if it has the following function for its density:

$$f(x; \sigma) = \begin{cases} \frac{2\sigma^3}{\sqrt{2\pi}} x^2 e^{-(x^2\sigma^2/2)}, & x \geq 0, \sigma > 0, \\ 0, & \text{elsewhere.} \end{cases} \quad (3.3.28)$$

Definition 3.3.12

A random variable X is said to be distributed as the **extreme-value distribution** if it has the following function for its probability density:

$$f(x) = e^{-e^{-(x-\theta)}}, \quad -\infty < \theta < \infty. \quad (3.3.29)$$

Definition 3.3.13

A random variable X is said to be distributed as the **arcsine distribution** if it has the following function for its probability density:

$$f(x) = \begin{cases} \dfrac{1}{\pi}\{x(1-x)\}^{-(1/2)}, & 0 < x < 1, \\ 0, & \text{elsewhere.} \end{cases} \qquad (3.3.30)$$

Definition 3.3.14

A random variable X is said to be distributed as the **Pareto distribution** if it has the following function for its probability density:

$$f(x) = \begin{cases} r A^r (x^{r+1})^{-1}, & 0 \leq A \leq x, r \geq 0, \\ 0, & \text{elsewhere.} \end{cases} \qquad (3.3.31)$$

The Maxwell distribution, sometimes referred to as Maxwell's velocity distribution, is applicable to problems in physics. The extreme-value density has been used in the areas of hydrology, failure models, and meteorology. The Pareto and arcsine distributions have been applied, respectively, to various problems in econometrics and electrical engineering.

Summary: Important Concepts

Continuous random variables
Continuous probability density function
Continuous cumulative probability distribution
Uniform distribution
Gaussian or normal distribution
Gamma probability density function
Exponential probability density function
Beta probability density function
Cauchy probability density function
Laplace probability density function
Lognormal probability density function
Weibull probability density function
Rayleigh probability density function
Maxwell probability density function
Extreme-value probability density function
Arcsine probability density function

Theoretical Exercises

3.1 Suppose that $f(x = k \sin x), 0 < x < \pi$.
  a. Evaluate k so that $f(x)$ will be the probability density function of the random variable X.
  b. Find $F(x)$.
  c. Graph $f(x)$ and $F(x)$.

3.2 A continuous random variable X assumes its values from the set $S_x = \{x : 1 \leq x \leq 4\}$ and has on this set a probability density function $f(x)$, which is proportional to x.
  a. Determine the probability density function $f(x)$.
  b. Determine $Pr[X < x]$.
  c. Sketch $f(x)$ and $F(x)$.

3.3 The probability density function of the random variable X is given by

$$f(x) = \begin{cases} \frac{3}{2}x^2, & -1 \leq x \leq 1, \\ 0, & \text{elsewhere.} \end{cases}$$

  a. Find the cumulative distribution function of the random variable X.
  b. Sketch $f(x)$ and $F(x)$ on the same graph.

3.4 The cumulative distribution function of the random variable Y is given by

$$F(y) = \begin{cases} 0, & y < 0, \\ \frac{1}{2}, & y = 0, \\ \frac{1}{2} + \frac{y}{4}, & 0 < y < 2, \\ 1, & y \geq 2. \end{cases}$$

Find the probability density function of the random variable Y.

3.5 The cumulative distribution function of the random variable X is given by

$$F(x) = \begin{cases} 0, & x \leq 0, \\ x, & 0 < x \leq \frac{1}{2}, \\ \frac{1}{2}, & \frac{1}{2} < x \leq \frac{3}{4}, \\ 2x - 1, & \frac{3}{4} < x \leq 1, \\ 1, & x > 1. \end{cases}$$

  a. Find $f(x)$, being sure to include any points at which $f(x)$ is undefined.
  b. Show that the properties of a cumulative distribution are satisfied by $F(x)$.

3.6 Let the cumulative distribution function of the random variable Y be given by

$$F(y) = \begin{cases} 0, & y \leq 0, \\ y(3-2\sqrt{y}), & 0 < y < 1, \\ 1, & y \geq 1. \end{cases}$$

a. Find the probability density function of the random variable Y.
b. Sketch $f(y)$.

3.7 The cumulative distribution function of the random variable Z is given by

$$F(z) = \begin{cases} 1-(1+z)e^{-z}, & z \geq 0, \\ 0, & \text{elsewhere.} \end{cases}$$

Show that the properties of a cumulative distribution are satisfied.

3.8 The probability density function of the random variable X is given by

$$f(x) = \begin{cases} x^{11}, & 0 < x < 1, \\ \dfrac{1}{12}, & x = 1, 4, 5, 9.5, 10 < x < 11, \\ \dfrac{1}{4}(3-x)(x-1), & 2 < x < 3, \\ \dfrac{5}{3x^2}, & 4 < x < 5, \\ \dfrac{1}{12}\sin x, & 2\pi < x < 3\pi, \\ \dfrac{\sqrt{2}}{6}\cos 2x, & 4\pi < x < \dfrac{33\pi}{8}, \\ 0, & \text{elsewhere.} \end{cases}$$

a. Graph $f(x)$.
b. Derive $F(x)$.
c. Graph $F(x)$.

3.9 In the following functions, evaluate c so that $f(x)$ will be a probability density function:

a. $f(x) = \dfrac{c}{x^4}, \ x \geq 1$

b. $f(x;\theta) = c(\ln\theta)\theta^{-x}, \ x \geq 0, \ \theta > 1$

c. $f(x;\theta) = \dfrac{c}{3}(\theta+1)x^\theta, \ 0 \leq x \leq 1, \ \theta \geq 0$

d. $f(x;\theta) = 2c(1+2\theta x - 3\theta x^2), \ 0 \leq x \leq 1$

e. $f(x) = ce^{-(x/\theta)}, \ \theta \geq 0$

f. $f(x) = \dfrac{c}{1+x^2}, \ -\infty < x < \infty$

Theoretical Exercises

g. $f(x) = \begin{cases} \dfrac{c}{2}, & 0 \le x < \dfrac{1}{3}, \\ \dfrac{3c}{2}, & \dfrac{1}{3} \le x < \dfrac{2}{3}, \\ \dfrac{c}{2}, & \dfrac{2}{3} \le x < 1, \\ 0, & \text{elsewhere} \end{cases}$

h. $f(x) = c(1 - x^2), \; -1 < x \le 1$

i. $f(z) = \begin{cases} \dfrac{c}{2} z^2, & 0 \le z < 1, \\ c\left(-z^2 + 3z - \dfrac{3}{2}\right), & 1 \le z < 2, \\ \dfrac{c}{2}(z^2 - 6z + 9), & 2 \le z \le 3, \\ 0, & \text{elsewhere} \end{cases}$

j. $f(r) = c(1 - e^r)^{n-2} e^{-r}, \; -\infty < r < \infty, \; n \ge 0$

k. $f(x) = \dfrac{c}{\theta^2} x e^{-(2x/\theta)}, \; x > 0, \; \theta > 0$

l. $f(x) = c(9 - 2x), \; 2 < x \le 4$

m. $f(x) = \begin{cases} \dfrac{1}{20 x^2}, & -\infty < x \le -1, \\ \dfrac{3}{25}(2x^2 + x + 1), & -1 < x \le 1, \\ c, & 1 < x \le 3, \\ \dfrac{1}{20}(6 - x), & 3 < x \le 5, \\ c(10x - x^2 - 21), & 5 < x \le 7, \\ 0, & \text{elsewhere} \end{cases}$

n. $f(x) = \dfrac{c e^{-x}}{1 - e^{-1}}, \; 0 < x \le 1$

o. $f(x) = c \theta^{-(3/2)} x^2 e^{x^2/2\theta}, \; x > 0, \; \theta > 0$

p. $f(x) = \dfrac{c}{\theta^2}(\theta - x), \; 0 < x < \theta$

q. $f(x) = \dfrac{c(1 + \theta)}{(x + \theta)^2}, \; x \ge 1, \; \theta \ge -1$

r. $f(x) = \dfrac{c \theta}{\alpha} \left(\dfrac{\alpha}{x}\right)^{1 + \theta}, \; x > \alpha, \; \alpha > 0, \; \theta > 0$

s. $f(x) = \dfrac{c x^3}{\theta^4} e^{-(x/\theta)}, \; x > 0, \; \theta > 0$

3.10 **a.** Find the cumulative distribution functions of the variables whose probability density functions are given in Exercise 3.9.
b. Sketch $f(x)$ and $F(x)$.

3.11 Show that the following distributions satisfy the conditions of a probability density function, and determine the cumulative distribution for each density:
a. Maxwell distribution, given by Equation (3.3.28)
b. Extreme-value distribution, given by Equation (3.3.29)
c. Arcsine distribution, given by Equation (3.3.30)
d. Pareto distribution, given by Equation (3.3.31)

3.12 The probability that the random variable X will assume a value between α and β, with $1 \leq \alpha \leq \beta$ is given by

$$Pr(\alpha \leq X \leq \beta) = \frac{\beta - \alpha}{\alpha \beta}.$$

a. What is the probability density function of the random variable X?
b. Determine and sketch the cumulative distribution function $F(x)$.

3.13 Suppose that the probability density function of the random variable X is the gamma distribution with the parameters α and β. Show that, for $0 < a < b < \infty$,

$$Pr(a \leq X \leq b) = \frac{1}{\Gamma(n)} \left[\sum_{i=0}^{n-1} \frac{d^i}{dy^i}(y^{n-1}) \right]_{\frac{a}{\alpha}}^{\frac{b}{\beta}},$$

where

$$\frac{d^0}{dy^0}(y^{n-1}) = y^{n-1},$$

and if $a = 0$, then, for $c > 0$,

$$Pr(X \leq c) = 1 - \frac{1}{\Gamma(n)} \left[\sum_{i=0}^{n-1} \frac{d^i}{dy^i}(y^{n-1}) \right]^{\frac{c}{\beta}}.$$

3.14 A certain vacuum tube has been observed to fail uniformly over the time interval $[t_1, t_2]$.
a. Determine the reliability function for such a tube at time t for $t_1 \leq t \leq t_2$.
b. Calculate the failure rate of the tube.
c. If $180 \leq t \leq 220$, what is the reliability of such a tube at 200 hours?

3.15 An electrical component was studied in the laboratory, and it was determined that the time to failure of the component was characterized by the following failure density:

$$f(t;n,p) = \binom{n}{t} p^t (1-p)^{n-t}, \quad 0 \leq p \leq 1,$$

with

$$t = 0, 1, \ldots, n.$$

a. Determine the reliability function of the component at t.
b. Calculate the failure rate of the electrical component.

A needle of length a is tossed onto a flat surface with equidistant parallel lines, the distance between each of which is d, where $d > a$. We want to find the probability that the needle intersects one of the ruled lines. (This problem is known as Buffon's needle problem.)

Assume the needle to be positioned as in Figure 3.E.1. Let X be the perpendicular distance from the midpoint of the needle to the nearest parallel line, and let θ be the angle that the needle makes with this perpendicular line, Thus, X takes on some value x from 0 to $d/2$, and θ takes on some value θ from $-(\pi/2)$ to $\pi/2$. From Figure 3.E.1, we can see that if $x < a/2 \cos\theta$, then the needle intersects one of the lines; otherwise, there is no intersection. Figure 3.E.2 depicts the situation: The rectangle with base π and height $d/2$ represents the total probability involved in this situation, and the shaded area represents that portion of the total probability where intersection of a parallel line with the needle occurs. Thus, our desired probability equals the shaded area divided by the total area.

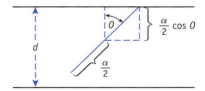

Figure 3.E.1

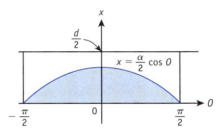

Figure 3.E.2

Use this information to solve Exercises 3.16 and 3.17.

3.16 Show that the result is $2a/\pi d$.

3.17 Show that, if $a > d$, the probability that the needle intersects a parallel line in

$$\frac{2}{\pi}\left[\frac{(a-\sin\emptyset)}{d+\emptyset}\right],$$

where

$$\emptyset = \cos^{-1}\left(\frac{d}{a}\right).$$

3.18 The base X and the height Y of a triangle are obtained by picking points x and y at random on two line segments of length a and b, respectively, where x is the distance from the left-hand end of the segment a to x; y is similar with respect to b. Find the probability that the area of the triangle is less than $ab/4$.

3.19 The numbers a and b are chosen at random between 1 and 2 and -1 and 1, respectively, all values being equally likely. Find the probability that the solution to $ax + b = 0$ is less than $1/4$.

3.20 The base of a triangle is 2 units long. The length of one of its sides X, where X was chosen at random from 0 to 4, all numbers being equally likely. The angle between the base and x is θ, where θ is chosen at random from 0 to π, all values being equally likely. Find the probability that the area of the triangle is less than 2.

Applied Problems

3.1 Let X be a random variable, the probability density function of which is given by

$$f(x) = \begin{cases} x, & 0 \le x \le 1, \\ 2-x, & 1 < x \le 2, \\ 0, & \text{elsewhere.} \end{cases}$$

We define the following events:

$$S_1 = \left\{ x : -4 < x \le \frac{1}{2} \right\}$$

$$S_2 = \left\{ x : 0 < x \le 1 \right\}$$

$$S_3 = \left\{ x : \frac{1}{2} \le x \le \frac{3}{2} \right\}$$

$$S_4 = \left\{ x : \frac{1}{2} \le x < \infty \right\}$$

Evaluate the probability of the following:
a. $S_1 \cup S_2$
b. $S_1 \cap S_3$
c. $\overline{S_1} \cup S_4$
d. $\overline{S_4 \cap S_3}$
e. $(S_1 \cap S_2) \cup S_3$
f. $(\overline{S_1} \cup S_2) \cap \overline{S_4}$
g. $(S_1 \cap S_3) \cup (S_2 \cap S_4)$
h. $(S_1 \cup S_2) \cap (S_3 \cup S_4)$
i. $(\overline{S_1} \cup \overline{S_3}) \cup S_2$
j. $\overline{(S_1 \cup S_3)} \cap S_4$

3.2 Determine the cumulative distribution function of the random variable X, whose density is given in Problem 3.1. Sketch $F(x)$.

3.3 The probability density function of the random variable X is given by

$$f(x) = \begin{cases} 3x^2, & 0 \le x \le 1, \\ 0, & \text{elsewhere.} \end{cases}$$

What is the probability that neither of two independent observations that the random variable X may assume falls in the interval $[1/3, 2/3]$?

3.4 In Exercise 3.1, what is the probability that the random variable X will assume a value from $\pi/6$ to $\pi/3$, exclusively?

3.5 The probability density function of the random variable Y is given by

$$F(y) = \begin{cases} y, & 0 \leq y \leq 1, \\ 2-y, & 1 < y < 2, \\ 0, & \text{elsewhere.} \end{cases}$$

What is the probability that the random variable Y will assume a value that fits the following description?
a. It is greater than $1/2$.
b. It is less than $3/2$?.
c. It is between $1/2$ and $3/2$.

3.6 In Exercise 3.2, what is the probability that X assumes a value from the set $S_X^* \cap S_X, S_X = \{x : 3/2 \leq x \leq 3\}$?

3.7 The amount of lamb meat in hundreds of pounds that a certain slaughterhouse is able to sell in a day is found to be a numerically valued random phenomenon described by the function

$$f(x) = \begin{cases} cx, & 0 \leq x < 2, \\ c(10-x), & 2 \leq x \leq 7. \end{cases}$$

a. Find c so that $f(x)$ will be a probability density function.
b. What is the probability that the amount of lamb sold the day before Easter will be between 150 and 550 pounds?

3.8 In Exercise 3.4, evaluate the following probabilities:
a. $Pr\left(Y > \dfrac{1}{2}\right)$ b. $Pr\left(\dfrac{3}{4} \leq Y \leq 1\right)$

3.9 Let the probability density function of the random variable X be given by

$$f(x) = \begin{cases} \dfrac{1}{\pi - 2} x^2 \sin x, & 0 < x < \dfrac{\pi}{2}, \\ 0, & \text{elsewhere.} \end{cases}$$

a. Find the cumulative distribution function of the random variable X.
b. Evaluate $Pr(X \geq \pi/3)$ and $Pr(|X| \leq \pi/6)$.

3.10 The graph of the cumulative distribution function of the random variable X is shown in the accompanying diagram.

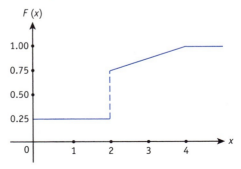

a. Find the probability density function and cumulative distribution function of the random variable X. Be sure to include all points at which $f(x)$ is undefined.
b. Evaluate $Pr(X \geq 5/2)$ and $Pr(1/2 \leq X \leq 5/2)$.

3.11 In Exercise 3.7, find the probability that the random variable Z will assume a value from $[1/2, 10]$ and $[16, \infty]$.

3.12 In Exercise 3.8, evaluate the following probabilities:

a. $Pr\left(\dfrac{3\pi}{2} \leq X \leq 10.5\right)$ b. $Pr(1 \leq X \leq 4.5)$

c. $Pr(X \geq 11.5)$ d. $Pr(|X| \leq 3)$

3.13 The hardness of a piece of pottery is proportional to the firing time. Assume that a rating system has been devised to rate the hardness of a ceramic dish and that this measure of hardness is a random variable that is uniformly distributed between 0 and 10. If a hardness in the interval $[5, 9]$ is desirable for kitchenware, what is the probability that a piece chosen at random will be suitable for kitchen use?

3.14 A geologist defines granite as a rock containing quartz, feldspar, and small amounts of other materials provided that it contains no more than 75% quartz. If all percentages are equally likely, what proportion of granite samples that the geologist collects during his lifetime will contain from 50% to 65% quartz?

3.15 A life insurance company has insured a group of 1000 people who are 28 years of age. Their insurance went into effect August 31; the fiscal year ended December 31. From the mortality table, it was found that the probability of dying at the age of 28 is approximately 0.15. If the company pays $5000 in benefits per claim on this group, assuming that the deaths are uniformly distributed throughout the year, how much reserve should the company hold for claims for this group for the rest of the fiscal year?

3.16 The thickness of the nickels minted in the Denver mint varies. Assume that the random variable that describes such a phenomenon is normally distributed with a mean of 1/16 in. and a variance of 1/40 in. A nickel activates a candy machine if its thickness is between 1/32 and 3/32 in. What is the probability that if a person randomly selects a nickel from her pocket and puts it into a machine, she will be able to get candy?

3.17 A company annually uses thousands of electrical lamps, which burn continuously day and night. Assume that, under such conditions, the life of a lamp may be regarded as a variable normally distributed about a mean of 60 days with a standard deviation of 20 days. On January 1, the company put 10,000 new lamps into service. Approximately how many lamps would be expected to need replacement by February 1 of the same year?

3.18 In the "time-term" method of refraction seismology, the depth from the surface to a refracting layer is estimated using the time of arrival of vibrations from a distant explosion. Suppose that, for a particular survey, the estimate of depth at each point is subject to a random error, which is assumed to be normally distributed with a mean of 0 ft and a standard deviation of 3 ft. If, at a certain point, the true depth is 2 ft, what is the probability that the estimated "depth" will be negative?

3.19 Assume the grades received by a class of students on an examination are normally distributed with a mean of 75 and a standard deviation of 10 and that the highest 10% of these grades are As. Find the lowest A.

3.20 A certain industrial process yields a large number of steel cylinders whose lengths are approximately normally distributed with a mean of 3.25 in. and a variance of 0.008 in^2. If two cylinders are chosen at random and placed end to end, what is the probability that their combined length is less than 6.55 in.?

3.21 To etch an aluminum tray successfully, the pH of the acid solution used must be between 1 and 4. This acid solution is made by mixing a fixed quantity of etching compound in powder form with a given volume of water. The actual pH of the solution obtained by this method is affected by the potency of the etching compound, by slight variations in the volume of water used, and perhaps by the pH of the water. Thus, the pH of the solution varies. Assume that the random variable that describes the random phenomenon is gamma distributed with $\alpha=2$ and $\beta=1$.
  a. What is the probability that an acid solution made by the preceding procedure will satisfactorily etch a tray?
  b. What would the answer to Problem 3.21(a) be if $\alpha=1$ and $\beta=2$?

3.22 In a certain country, the distribution of incomes in thousands of dollars is described by a gamma distribution with $\alpha=2$ and $\beta=8$. What is the probability that a man chosen at random will have the following incomes?
  a. More than $14,000
  b. At least $12,000

3.23 A cement mixing company has facilities for only a limited amount of storage space for the sand it uses in making ready-mix cement. The demand for sand varies from day to day. Suppose that the random variable that characterizes the problem follows the gamma distribution with $\alpha=2$ and $\beta=3$. If the storage space holds only 6000 tons of sand, what is the probability that on any given day the company will not run out of sand?

3.24 A climber is stranded on the side of a mountain and is signaling steadily with his flashlight, hoping to attract a rescue team. The time in hours that his flashlight is capable of sending out a beam that can be observed from below is a random variable following the exponential distribution with $\beta=6$.
  a. What is the probability (assuming other conditions are favorable) that his signal could be seen for at least 3 hours?
  b. Construct a table showing the answer to Problem 3.24(a) as β varies: $\beta=1, 2, 3, 4, 5, 6$.

3.25 Telephone conversation time follows an exponential distribution with $\beta=5$. If you are in a hurry to make a phone call and a woman reaches the phone booth and starts to dial before you arrive, what is the probability that you will have to wait less than 3 minutes before she completes her call? Show how your answer will vary as you vary β.

3.26 The life of lightbulbs used in traffic lights in a certain town is exponentially distributed with a mean of 500 hours. The lights are in operation 24 hours per day, and it is assumed for practical purposes that the red and green signals are on for equal intervals of time. On July 25, the maintenance crew installed new bulbs in a traffic light at a major intersection. If Labor Day is September 5 and the bulbs are all burning properly on September 1, what is the probability one of the bulbs will burn out during the holiday weekend of September 1 to 5?

3.27 Assume that the daily proportion of trailers and "wheeled campers" seeking accommodations in Great Smoky Mountains National Park can be treated as a random variable X that has a beta distribution with $\alpha=5$ and $\beta=7$. Find the probability that on a given day at least half of the campers seeking accommodations will have this type of equipment.

3.28 The percentage of marketable oysters that are collected from a large oyster bed varies from day to day. Assume that the random variable that describes this phenomenon follows the beta distribution with $\alpha=3$ and $\beta=2$. What is the probability that on any given day the percentage of marketable oysters is between 50% and 70%?

3.29 A spotlight is mounted 100 ft from the ground on a wall, and it is placed in such a way that it can swing freely in an 180° arc. Its purpose is to illuminate another wall that is directly opposite and 173 ft from the one on which the light is mounted. The light also is capable of illuminating a portion of the wall on which the light is mounted 10 ft to either side of the center spot. Assume that all angular positions of the light are equally likely. If a man is standing 100 ft to the right of center along the wall opposite the light, what is the probability that, when the light is turned on, he will be within the lighted portion of the wall?

3.30 In nineteenth-century thermodynamics, a "Maxwell's demon" was a hypothetical creature that could observe individual molecules in a gas and could operate a shutter over a pinhole in a partition to permit or prevent the molecules from entering an evacuated chamber. In the chamber was a screen that was parallel to the partition, with a center 5 in. from the pinhole and directly opposite it. The molecules behind the shutter had random direction, but the demon was instructed only to allow molecules with no vertical component of velocity through the hole. What is the probability that the next molecule to hit the screen would be between 5 and 6 in. from the center?

3.31 The telescopes on the west side of the observation deck of the Empire State Building are mounted so that they swing freely in 180° arcs, aimed along the New Jersey bank of the Hudson River, which lies due west. For a quarter, an observer receives 10 seconds of viewing time, not enough time to see from the south to the north end of the telescope's arc. Assume that Palisades Amusement Park lies 2 miles north of the point, directly opposite the Empire State Building, which is approximately 1 mile from New Jersey. Also assume that the observer can see Palisades Park if the telescope is aimed within 1/4 mile on either side of it. What is the probability that an observer will see the park in any given observation, assuming that each portion of the shore is equally likely to be observed?

3.32 A physical fitness test was given to a large number of college freshmen. In part of the test, each student was asked to run as far as he or she could in 10 minutes. The distance each student ran in miles was recorded and can be considered to be the random variable X. The data showed that X followed the lognormal distribution with $\mu_Y = 0.25$ and $\sigma_Y = 0.5$. A student is considered physically fit if he or she is able to run at least 1.5 miles in the time allowed. What percentage of college freshmen would be considered physically fit if you consider only this part of the test?

3.33 An experimenter is designing an experiment to test tetanus toxoid in guinea pigs. The survival of the animal following the dose of the toxoid is a random phenomenon. Past experience has shown that the random variable that describes such a situation follows the lognormal distribution with $\mu_Y = 0$ and $\sigma_Y = 0.70$. As a requirement of good design, the experimenter must choose doses at which the probability of surviving is 20%, 50%, and 80%. What three doses should she choose?

3.34 A potter produces plates using a potter's wheel. Because this is a hand process, considerable variation is possible in the diameters of the plates produced. A plate is considered to be satisfactory for sale if its diameter lies between 10 and 10.25 in. Assume that the random variable X represents the diameters of the plates and that X follows a Laplace distribution with $\mu = 10.125$ and $\sigma = 0.1$. What percentage of the potter's output will be salable? If the potter cannot make a profit unless at least 85% of his articles can be sold, what value must σ assume to allow a profit?

3.35 A company manufacturing seat belts for automobiles must keep the breaking strength above 2000 psi for the belts to be acceptable by automobile manufacturers. The breaking strength of the belts varies enough to be considered a random phenomenon characterized by X. If the probability distribution of the random variable X is *Laplacian* with $\mu = 2004$ psi and $\sigma^2 = 4$, what percentage of the belts will be rejected by the automobile manufacturers because their breaking strength is too low?

3.36 A geologist has collected a sample of a certain kind of soil and, by sifting it through a screen, has determined that the diameters of soil particles are distributed *lognormally* with a mean of 0.060 and a variance of 0.0016. If the geologist's sample consists of 5 ft³, from which he extracts 1 ft³ at random to conduct a test, what is the probability that the diameter of the soil particles will be between 0.0050 and 0.070 in.?

3.37 The amount of salt in kilograms that can be extracted from a given volume of water drawn from Great Salt Lake, Utah, can be considered to be the random variable X. Assume that X follows the Weibull distribution with $\alpha = 1$, $\beta = 2$, and $\gamma = 0$. To make it profitable to obtain salt by this particular extraction process, the yield per sample must be at least 0.5 kg. What is the probability that a given water sample will produce a satisfactory yield?

3.38 The time in seconds that the wheels in a slot machine are allowed to spin after a coin is dropped in is randomly distributed with a Weibull distribution that has a threshold of $\gamma = 5$ seconds, a scale parameter of 100, and a shape parameter of $\beta = 3$. What is the probability that, on a given trial, the wheels will spin for more than 10 seconds?

References

[1] Weibull, W. "A Statistical Distribution Function of Wide Applicability." *J. Appl. Mech.*, 18, pp. 293–297, 1957.

[2] Pearson, K. *Tables of the Incomplete Gamma Function.* London: Biometrika, 1922.

[3] Aitchison, J. and J. R. C. Brown. *The Lognormal Distribution.* London: Cambridge University Press, 1951.

"Probable impossibilities are to preferred to improbable possibilities."
—*Aristotle*

CHAPTER 4

Functions of a Random Variable

OBJECTIVES:

4.1 Distribution of a Continuous Function of a Discrete Random Variable
4.2 Distribution of a Continuous Function of a Continuous Random Variable
4.3 Other Types of Derived Distribution

Pierre Simon Laplace (1749–1827)

"It is remarkable that a science which begun with the consideration of games of chance should have become the most important object of human knowledge." "The most important questions of life are indeed, for the most part, really only problems of probability."
—*P. S. Laplace*

Pierre Simon Laplace was born in Beaumont-en-Auge, Normandy, on March 23, 1749, and died in Paris on March 5, 1827. He was the son of a farm laborer, but he gained the attention and assistance of some wealth neighbors through his abilities and engaging presence. Little is known of Laplace's early life except that he procured a letter of introduction to Jean Le Rond d'Alembert and went to Paris to pursue his fortune. On d'Alembert's recommendation, he obtained a professorship in mathematics at the Éncole Militaire.

Secure in his academic position, Laplace applied himself to original research. In 1773, he showed that the planetary motions were stable, and he carried the proof as far as the cubes of the eccentricities and inclinations. He followed this with several papers on points in integral calculus, finite differences, differential equations, and astronomy.

Laplace's career was one of almost uninterrupted prosperity. In 1784 he succeeded Étienne Bézout as examiner to the royal artillery, and the following year he became a member of the Academy of Sciences. He was made president of the Bureau of Longitude, assisted in

continued

the introduction of the decimal system, and taught mathematics with Joseph-Louis Lagrange in the Éncole Normale. After the 18 Brumaire, when Napoleon Bonaparte came to power, Laplace's ardor for republican principles suddenly gave way to a great devotion to the emperor. Napoleon rewarded this devotion by appointing him to the post of minister of the interior but dismissed him after 6 months for incapacity. Desiring to retain Laplace's allegiance, Napoleon elevated him to the senate and bestowed various other honors upon him. Nevertheless, in 1814 Laplace cheerfully supported the dethronement of his patron and hastened to tender his services to the Bourbons, thereby earning the title of marquis.

Though politically supple and servile, Laplace never misrepresented or concealed his own convictions in religion and science. In mathematics and astronomy, his genius shines with a luster exceeded by few. He gave three great works to the scientific world: *Mécanique Céleste, Exposition du Systéme du Monde,* and *Théorie Analytique des Probabilitiés.*[1] Besides these, he contributed important memoirs to the French Academy.

Laplace did more toward advancing the subject of probability than did any other investigator. The introduction to *Théorie Analytique des Probabilitiés* was published separately under the title *Essai Philosophique sur les Probabilitiés*[2] and is a masterly exposition, without the aid of analytical formulas, of the theory of generating functions. In his work on probability, Laplace reduced the solution of linear differential equations to definite integrals. His introduction of partial differential equations was concurrent with that of Lagrange. One of the most important parts of the work is the application of probability to the method of least squares, which is shown to give the most probable, as well as the most convenient, results.

[1] *Celestial Mechanics, Exposition of the System of the World,* and *Analytical Theory of Probabilities.*
[2] *Philosophical Essay on Probability.*

■ INTRODUCTION

We have seen that, basically, a *random variable* is a function (rule) that assigns a number to every outcome of an experiment; once we know the probability density function of the variable, we know the laws governing it. An important deductive problem in the study of random variables is the derivation of the probability density function of a functional form of the initial variate. That is, we are given the probability density function $f(x)$ of a random variable X, and we wish to find the probability distribution of a functional form of this variate:

$$Y = g(x).$$

In many physical problems, the derivation of the probability density of the functional form of the given variate is extremely important. For example, the velocity V of a gas molecule (Maxwell-Boltzmann law) behaves as a random variable that is gamma distributed. We want to derive the distribution of $E = mV^2$, the kinetic energy of the gas molecule. Because the value of V is the outcome of a random experiment, so is the value of E. Thus, the functional form of a random variable is a random variable.

In this chapter, we study the problem, beginning with a function of a single variate; in Chapter 6, we extend our investigation to multivariate functional forms of the random variable.

4.1 Distribution of a Continuous Function of a Discrete Random Variable

Let X be a discrete random variable that can assume the values $x_1, x_2, \ldots, x_n$, with the probabilities $f(x_1), f(x_2), \ldots, f(x_n)$ such that

$$\sum_{i=1}^{n} f(x_i) = 1.$$

Suppose that the random variable Y is a function of $X: Y = g(x)$ and we wish to determine the probability density of Y. To do this, we proceed as follows:

1. Determine the possible values that the variate Y can assume by substituting all values the random variable X can take on into $g(x)$, thus obtaining $y_1, y_2, \ldots, y_n$.

2. Obtain the probability that the variate Y assumes these values; that is, $h(y_1), h(y_2), \ldots, h(y_n)$ such that $\sum_{i=1}^{n} h(y_i) = 1$.

The probability $h(y_i)$ is obtained by summing $f(x)$ over all values of x for which $g(x) = y_i$. Several values of x may give rise to the same value of y. This situation occurs when the function g is not a one-to-one function; that is, when there exist points x_1 and x_2 that belong to the domain of g such that $x_1 \neq x_2$ and $g(x_1) = g(x_2)$.

We illustrate the procedure by considering the following examples.

Example 4.1.1

Suppose that the random variable X can assume the values 0, 1, 2, 3, and 4 with the following probabilities: $Pr(X=0) = p_0, Pr(X=1) = p_1, Pr(X=2) = p_2, Pr(X=3) = p_3,$ and $Pr(X=4) = p_4$. Find the probability density of the variate $Y = g(x) = x^2 - 2x + 1$. Here, Y can assume the values 0, 1, 4, and 9. Its density is given by Table 4.1.1.

Table 4.1.1

y	0	1	4	9
$h(y)$	p_1	$p_0 + p_2$	p_3	p_4

In this example, the values 0 and 2 of x give rise to the value 1 of y; thus,

$$h(1) = Pr(X=0) + Pr(X=2) = p_0 + p_2.$$

Now, assume that the function $Y = g(x)$ is a strictly monotone function in the interval over which X can assume its values; that is, $g(x)$ is a strictly monotone increasing or decreasing function of X, as shown in Figure 4.1.1. Recall that such a function guarantees that the mapping of $g^{-1}(y)$ on the set of ranges (or images) of $g(x)$ to the domain of $g(x)$ is a one-to-one function.

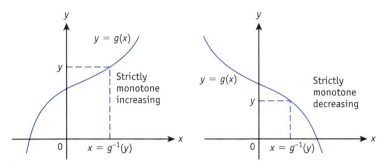

Figure 4.1.1 One-to-One Function

Thus, when the random variable X assumes the value x, Y assumes the value $y = g(x)$ and the inverse function $x = g^{-1}(y)$ is also strictly monotone. In the discrete case, let $f(x)$ be the probability density function of the variate X. Whenever X assumes a particular value x with the probability $f(x)$, then the variate Y assumes the corresponding value of $y = g(x)$, also with the same $f(x)$ probability. Therefore,

$$h(y) = f(g^{-1}(y)).$$

More precisely, the inverse of a function is derived by expressing the independent variable as a function of the dependent variable; that is, the inverse of $y = g(x)$ is $x = g^{-1}(y)$. For example; suppose that $y = g(x) = (x+3)/(2-x)$; then, $x = g^{-1}(y) = (2y-3)/(y+1)$.

Example 4.1.2

Suppose that the probability density function of the random variable X is given by

$$f(x) = \begin{cases} \dfrac{x}{8}|x-3|, & x = 0, 1, 2, 3, 4, \\ 0, & \text{elsewhere.} \end{cases}$$

We want to find the probability density of the random variable $Y = g(X) = (1/2)X + 1$.

Solution: The function $y = g(x) = (1/2)x + 1$ is an increasing function, $g^{-1}(y) = 2(y-1)$, and

$$h(y) = f(g^{-1}(y)) = \begin{cases} \dfrac{y-1}{4}|2y-5|, & y = 1, \dfrac{3}{2}, 2, \dfrac{5}{2}, 3, \\ 0, & \text{elsewhere.} \end{cases}$$

It is easy to verify that $h(y)$ is a probability density by showing that $\sum h(y) = 1$ and $h(y) \geq 0$ for all y.

To illustrate the preceding remarks, the probability that the random variable X assumes the value 4 is

$$Pr[X = 4] = f(4) = \frac{1}{2},$$

and the corresponding value of $y=(1/2)(4)+1=3$, which gives the same probability, is

$$Pr[Y=3]=h(3)=\frac{1}{2}.$$

Figure 4.1.2 shows the probability density function of the variate X and that of the derived density of the variate Y.

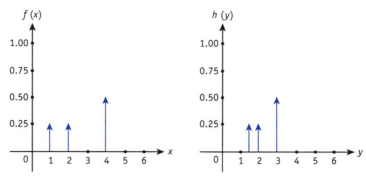

Figure 4.1.2   f(x) and h(y)

Example 4.1.3

Suppose that the random variable X has a Poisson distribution:

$$f(x;\lambda)=\begin{cases} \dfrac{e^{-\lambda}\lambda^x}{x!}, & x=0,1,2,\ldots, \\ 0, & \text{elsewhere.} \end{cases}$$

Find the probability density function of the random variable Y, where the following is true:

a. $Y=aX+b$

b. $Y=aX^2$

c. $Y=\sqrt{X}$

Solution:

a. Because $y=g(x)=ax+b$ is a strictly monotone function over the positive integers and $g^{-1}(y)=(y-b)/a$, we have

$$h(y;\lambda)=f(g^{-1}(y);\lambda)=\begin{cases} \dfrac{e^{-\lambda}\lambda^{\{(y-b)/a\}}}{\left(\dfrac{y-b}{a}\right)!}, & \begin{array}{l}\dfrac{y-b}{a}=0,1,2,\cdots \\ \text{or} \\ y=b, a+b, 2a+b,\ldots,\end{array} \\ 0, & \text{elsewhere.} \end{cases}$$

b. Here, $g^{-1}(y) = \sqrt{y^*/a}$, and the probability density of the variate Y is

$$h(y;\lambda) = f(g^{-1}(y);\lambda) = \begin{cases} \dfrac{e^{-\lambda}\lambda^{\sqrt{y/a}}}{\left(\sqrt{\dfrac{y}{a}}\right)!}, & \sqrt{\dfrac{y}{a}} = 0, 1, 2, \cdots \\ & \text{or} \\ & y = 0, a, 4a, 9a, \ldots, \\ 0, & \text{elsewhere.} \end{cases}$$

c. Similarly, $g^{-1}(y) = y^2$ and

$$h(y;\lambda) = f(g^{-1}(y);\lambda) = \begin{cases} \dfrac{e^{-\lambda}\lambda^{y^2}}{(y^2)!}, & y^2 = 0, 1, \cdots, n \\ & \text{or} \\ & y = 0, \sqrt{1}, \ldots, \sqrt{n}, \\ 0, & \text{elsewhere.} \end{cases}$$

◀

Example 4.1.4

Let the random variable X have the binomial distribution

$$f(x;n,p) = \begin{cases} \binom{n}{x} p^x (1-p)^{n-x}, & x = 0, 1, \ldots, n, \\ & 0 \leq p \leq 1, \\ 0, & \text{elsewhere.} \end{cases}$$

If $Y = g(X) = (X - np)^2$, find the probability density function of the random variable Y.

Solution: Here,

$$g^{-1}(y) = np \pm \sqrt{y}$$

and

$$h(y) = f(g^{-1}(y)) = \begin{cases} \binom{n}{np - \sqrt{y}} p^{np-\sqrt{y}} (1-p)^{n(1-p)+\sqrt{y}}, \\ \qquad y = g(x), \\ \qquad \text{where } x = 0, 1, \ldots, [np]; \\ \binom{n}{np + \sqrt{y}} p^{np+\sqrt{y}} (1-p)^{n(1-p)-\sqrt{y}}, \\ \qquad y = g(x), \\ \qquad \text{where } x = [np]+1, [np]+2, \ldots, n; \\ 0, \qquad \text{elsewhere,} \end{cases}$$

where $[np]$ denotes the largest integer not exceeding np.

---

* We choose the plus sign in all cases because $x = g^{-1}(y)$ can never be negative.

4.1 Distribution of a Continuous Function of a Discrete Random Variable

The cumulative distribution function of a random variable $Y = g(x)$ of the discrete type can be derived from the discrete probability density or by directly applying the definition of $F(y)$. For example, if we want to find the cumulative distribution function of $Y = aX + b$, where the random variable X has a Poisson distribution, we proceed as follows: For

$$x \leq \left[\frac{y-b}{a}\right],$$

$$F(y) = Pr(Y \leq y) = Pr(aX + b \leq y) = Pr\left(X \leq \frac{y-b}{a}\right) = \sum_{x=0}^{\left[\frac{y-b}{a}\right]} \frac{e^{-\lambda}\lambda^x}{x!}.$$

Therefore,

$$F(y) = \begin{cases} 0, & y < b, \\ \sum_{x=0}^{\left[\frac{y-b}{a}\right]} \frac{e^{-\lambda}\lambda^x}{x!}, & y \geq b. \end{cases}$$

The cumulative distribution function of the variate X is

$$F(x) = Pr(X \leq x) = \begin{cases} 0, & x < 0, \\ \sum_{0 \leq x_i \leq [x]} \frac{e^{-\lambda}\lambda^{x_i}}{x_i!}, & x \geq 0. \end{cases}$$

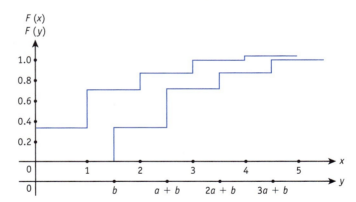

Figure 4.1.3 Cumulative Distribution Function of x

We conclude the discrete cases by summarizing the preceding developments.
Let X be a discrete random variable that has the probability density function

$$Pr(X = x) = \begin{cases} f(x), & x = x_1, x_2, x_3, \ldots, \\ 0, & \text{elsewhere}, \end{cases}$$

and let the random variable Y be a function of X: $Y = g(X)$. Then, the probability density of the variate Y is

$$h(y) = \begin{cases} f(g_{x_i}^{-1}(y)), & y = g(x_i), i = 1, 2, 3, \ldots, \\ 0, & \text{elsewhere}, \end{cases}$$

where $g_{x_i}^{-1}(y)$ is the inverse functional form of $g(x)$ that maps y back to x_i. Note: One $g(x_i)$ may be equal to another. If so, add the density parts corresponding to those functions that are equal. For example, if $g(x_{j1}) = g(x_{j2}) = \ldots = g(x_{jn}) = g(x_i)$, then the density part of $y = g(x_i)$ is

$$\sum_{i=1}^{n} f\left(g_{x_{j_i}}^{-1}(y)\right).$$

4.2 Distribution of a Continuous Function of a Continuous Random Variable

Let X be a continuous random variable with probability density function $f(x)$. We are concerned in this section with deriving the density of the continuous variate $Y = g(x)$. In practice, this situation arises quite frequently in engineering problems. The general procedure for deriving such distributions is given by Theorems 4.2.1 and 4.2.2.

Theorem 4.2.1

Let X be a random variable of the continuous type with probability density $f(x)$. If $Y = g(x)$ is a **strictly monotone increasing function** (its inverse exists for all values within the range of X) and if it is differentiable for all x, then the probability density function of the random variable $Y = g(x)$ is given by

$$h(y) = f\left[g^{-1}(y)\right] \frac{dg^{-1}(y)}{dy}. \quad (4.2.1)$$

Proof From the definition of the cumulative distribution function, we have

$$H(y) = Pr(Y \le y) = Pr(g(X) \le y)$$
$$= Pr(X \le g^{-1}(y)) = F(g^{-1}(y)). \quad (4.2.2)$$

Differentiating Equation (4.2.2) with respect to y, we obtain by, using the chain rule,

$$\frac{dH(y)}{dy} = \frac{dF(g^{-1}(y))}{dg^{-1}(y)} \frac{dg^{-1}(y)}{dy}$$

or

$$\frac{dH(y)}{dy} = f(g^{-1}(y)) \frac{dg^{-1}(y)}{dy}.$$

Thus, the density of the variate Y is given by Equation (4.2.1).

Example 4.2.1

Let X be a continuous random variable whose probability density is given by

$$f(x) = \begin{cases} \frac{1}{12}(x^2 + 1), & 0 \le x \le 3, \\ 0, & \text{elsewhere.} \end{cases}$$

If $Y = g(x) = 2x^2 - 1$, find the probability density of the variate Y.

4.2 Distribution of a Continuous Function of a Continuous Random Variable

Solution: We see that $g(X)$ is a strictly monotone increasing and differentiable function for $0 \le x \le 3$. Thus,

$$g^{-1}(y) = \sqrt{\frac{y+1}{2}}, \quad \frac{dg^{-1}(y)}{dy} = \frac{\sqrt{2}}{4\sqrt{y+1}}$$

and, applying Equation (4.2.1), we have

$$h(y) = \begin{cases} \dfrac{\sqrt{2}}{96} \dfrac{y+3}{\sqrt{y+1}}, & -1 \le y \le 17, \\ 0, & \text{elsewhere.} \end{cases}$$

It is easy to show that the conditions for a probability function have been met—that is, $h(y) \ge 0$ for all y in the interval $[-1, 17]$ and

$$\frac{\sqrt{2}}{96} \int_{-1}^{17} \frac{y+3}{\sqrt{y+1}} dy = 1.$$

Figure 4.2.1 shows a graphical comparison of $f(x)$, $g(x)$, and $h(y)$.

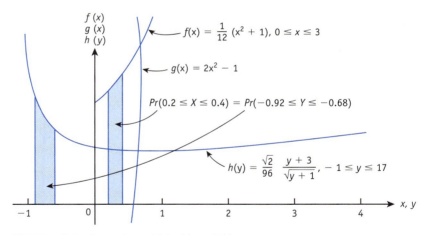

Figure 4.2.1 Comparison of f(x), g(x), and h(y)

Example 4.2.2

Let X be a continuous random variable whose probability density function is given by

$$f(x) = \begin{cases} e^{-x}, & x \ge 0, \\ 0, & \text{elsewhere.} \end{cases}$$

If $Y = g(X) = e^x$, find the probability density of the variate Y.

Solution: It can be easily shown that the conditions of Theorem 4.2.1 are satisfied. Hence,

$$g^{-1}(y) = \ln y, \quad \frac{dg^{-1}(y)}{dy} = \frac{1}{y}.$$

Applying Equation (4.2.1), we obtain

$$h(y) = \begin{cases} \dfrac{1}{y} e^{-\ln y}, & y \geq 1, \\ 0, & \text{elsewhere.} \end{cases}$$

To verify that $h(y)$ is a density function, we show that substituting $u = \ln y$ and $y\,du = dy$ into

$$\int_1^\infty \frac{1}{y} e^{-\ln y} dy$$

gives

$$\int_0^\infty e^{-u} du,$$

which equals 1, and $h(y) \geq 0$ for $y \geq 1$.

The cumulative distribution of $Y = e^X$ for $y \geq 1$ is given by

$$H(y) = Pr(Y \leq y) = \int_1^y \frac{1}{t} e^{-\ln t} dt = 1 - e^{-\ln y}$$

so that

$$H(y) = \begin{cases} 0, & y < 1, \\ 1 - e^{-\ln y}, & y \geq 1. \end{cases}$$

Figures 4.2.2 and 4.2.3 show the graphs of $g(x)$, $f(x)$, and $h(y)$ and of $F(x)$ and $H(y)$, respectively. ◀

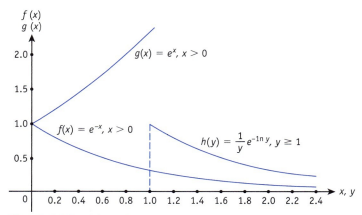

Figure 4.2.2 g(x), f(x), and h(y)

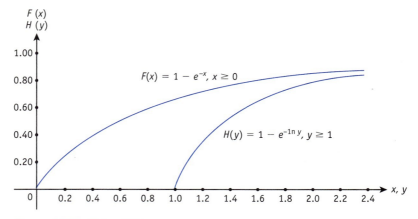

Figure 4.2.3 F(x) and H(y)

Example 4.2.3

The probability density of the velocity V of a gas molecule, according to the Maxwell-Boltzmann law, is given by

$$f(v;\beta) = \begin{cases} cv^2 e^{-\beta v^2}, & v > 0, \\ 0, & \text{elsewhere,} \end{cases}$$

where c is an appropriate constant and β depends on the mass of the molecule and the absolute temperature. We can find the density function of the kinetic energy E, which is related to V by $E = g(v) = (1/2)mv^2$.

Solution: For $v > 0$, the function $g(v)$ is a strictly monotone increasing and differentiable function. Thus,

$$g^{-1}(\varepsilon) = \sqrt{\frac{2\varepsilon}{m}}, \quad \frac{dg^{-1}(\varepsilon)}{d\varepsilon} = \frac{1}{m\sqrt{\frac{2\varepsilon}{m}}} = \frac{1}{\sqrt{2m\varepsilon}}$$

and

$$h(\varepsilon; m, \beta) = \begin{cases} \dfrac{c\sqrt{2\varepsilon}}{(m)^{3/2}} e^{-\frac{2\beta\varepsilon}{m}}, & \varepsilon > 0, \\ 0, & \text{elsewhere.} \end{cases}$$

The probability density of the kinetic energy $h(\varepsilon; m, \beta)$ is a gamma probability density function for a properly chosen c.

Example 4.2.4

Let the random variable X be normally distributed with the parameters μ_x and σ_x. If $Y = g(x) = e^x$, find the probability density function of the variate Y.

Solution: Here, $g(x)$ is a strictly monotone increasing and differentiable function. Thus, replacing x with $g^{-1}(y) = \ln y$ in

$$f(x;\mu_x,\sigma_x) = \frac{1}{\sqrt{2\pi}\,\sigma_x} \exp\left\{-\frac{1}{2}\left(\frac{x-\mu_x}{\sigma_x}\right)^2\right\}, \qquad -\infty < x < \infty,\; -\infty < \mu_x < \infty,\; \sigma_x > 0$$

and multiplying it by

$$\frac{dg^{-1}(y)}{dy} = \frac{1}{y},$$

we obtain

$$h(y;\mu_x,\sigma_x) = \begin{cases} \dfrac{1}{\sqrt{2\pi}\,\sigma_x y} \exp\left\{-\dfrac{1}{2}\left(\dfrac{\ln y - \mu_x}{\sigma_x}\right)^2\right\}, & y > 0, \\ 0, & \text{elsewhere.} \end{cases}$$

This is the lognormal probability density function, which was developed in Section 3.3. ◀

Example 4.2.5

The temperature recording equipment used to determine the melting point T of a given substance in degrees Celsius has an error that can be treated as a random variable that has a normal distribution with the parameters μ and σ. The temperature F, measured in degrees Fahrenheit, is related to T by $F = g(T) = (9/5)T + 32$. Find the probability density function that describes the behavior of F.

Solution: The conditions of Theorem 4.2.1 hold. Thus,

$$g^{-1}(f) = \frac{5}{9}(F - 32),\; -\infty < f < \infty,$$

$$\frac{dg^{-1}(f)}{df} = \frac{5}{9},$$

and

$$h(f) = \frac{5}{\sqrt{2\pi}\,9\sigma} \exp\left\{-\frac{1}{2}\left[\frac{f - (32 + (9/5)\mu)}{9\sigma/5}\right]^2\right\}, \qquad -\infty < f < \infty,\; -\infty < \mu < \infty,\; \sigma > 0.$$

Therefore, the temperature F is also normally distributed with a mean of $(32 + 9/5\mu)$ and a standard deviation of $(9/5)\sigma$. ◀

4.2 Distribution of a Continuous Function of a Continuous Random Variable

■ Theorem 4.2.2

Let X be a random variable of the continuous type with the probability density $f(x)$. If $Y = g(X)$ is a **strictly monotone decreasing function** (its inverse exists for all values within the range of X), and if it is differentiable for all x, then the probability density function of the variate $Y = g(x)$ is given by

$$h(y) = f(g^{-1}(y)) \left| \frac{dg^{-1}(y)}{dy} \right|. \quad (4.2.3)$$

Proof Because $g(x)$ is a decreasing function, we have

$$H(y) = Pr(Y \leq y) = Pr(g(X) \leq y)$$

$$= Pr(X > g^{-1}(y))$$

$$= 1 - Pr(X \leq g^{-1}(y))$$

$$= 1 - F(g^{-1}(y)). \quad (4.2.4)$$

Differentiating Equation (4.2.4) with respect to y, we obtain

$$\frac{dH(y)}{dy} = \frac{dH(y)}{dg^{-1}(y)} \cdot \frac{dg^{-1}(y)}{dy} = \frac{d}{dg^{-1}(y)} \left[1 - F(g^{-1}(y)) \right] \frac{dg^{-1}(y)}{dy}$$

or

$$h(y) = f(g^{-1}(y)) \frac{dg^{-1}(y)}{dy}.$$

Because $y = g(x)$ is a strictly monotone decreasing function,

$$\frac{dg^{-1}(y)}{dy} < 0.$$

Therefore,

$$-f(g^{-1}(y)) \frac{dg^{-1}(y)}{dy} = f(g^{-1}(y)) \left| \frac{dg^{-1}(y)}{dy} \right|,$$

giving Equation (4.2.3).

Example 4.2.6

Let X be a continuous random variable whose probability density function is given by

$$f(x) = \begin{cases} \dfrac{1}{4} x e^{-(x/2)}, & x \geq 0, \\ 0, & \text{elsewhere.} \end{cases} \quad (4.2.5)$$

If $Y = g(X) = -(1/2)X + 2$, find the probability density function of the variate Y.

Solution: Here, $g(X)$ is a decreasing and differentiable function, and Theorem 4.2.2 is applicable. Thus,

$$g^{-1}(y) = 4 - 2y, \quad \left|\frac{dg^{-1}(y)}{dy}\right| = |-2| = 2.$$

Applying Equation (4.2.3), we have

$$h(y) = \begin{cases} (2-y)e^{y-2}, & y \leq 2, \\ 0, & \text{elsewhere}, \end{cases} \quad (4.2.6)$$

and it can be shown that the conditions of a probability density function are satisfied. In Figure 4.2.4, we show a graphical comparison of $g(x)$, $f(x)$, and $h(y)$.

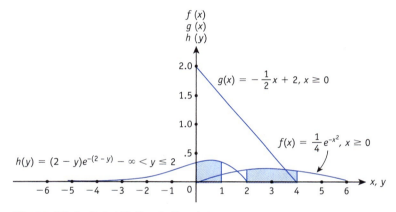

Figure 4.2.4 Probability Density Functions

We can use Equation (4.2.5) to calculate $Pr(2 \leq X \leq 4)$ or, correspondingly, $Pr(0 \leq Y \leq 1)$, which can be obtained from Equation (4.2.6). These two probabilities are shown as areas under their respective probability density functions in Figure 4.2.4.

Example 4.2.7

Let $Y = g(\theta) = A\cos\theta$, where the amplitude A is a known positive value and the phase θ is a random variable whose probability density is given by

$$f(\theta) = \begin{cases} \dfrac{2}{\pi}, & 0 \leq \theta \leq \dfrac{\pi}{2}, \\ 0, & \text{elsewhere}. \end{cases}$$

Find the distribution function of the random variable Y.

Solution: The functional form of θ is a decreasing and differentiable function for all θ such that $0 \leq \theta \leq \pi/2$. Hence,

$$g^{-1}(y) = \cos^{-1}\left(\frac{y}{A}\right)$$

and

$$\left|\frac{dg^{-1}(y)}{dy}\right| = \left|\frac{d}{dy}\cos^{-1}\left(\frac{y}{A}\right)\right| = \left|-\frac{1}{A\sqrt{1-\frac{y^2}{A^2}}}\right| = \left(A^2 - y^2\right)^{-1/2}.$$

Applying Theorem 4.2.2, we have

$$h(y) = \begin{cases} \frac{2}{\pi}\left(A^2 - y^2\right)^{-1/2}, & 0 \le y < A, \\ 0, & \text{elsewhere.} \end{cases}$$

To verify that $h(y)$ is a probability density, we must show that

$$\frac{2}{\pi}\int_0^A \frac{1}{\sqrt{A^2 - y^2}}\,dy = \frac{2}{\pi A}\int_0^A \frac{1}{\sqrt{1-\left(\frac{y}{A}\right)^2}}\,dy = 1. \tag{4.2.7}$$

Let $u = y/A$ and $A\,du = dy$ in Equation (4.2.7). Thus,

$$\frac{2}{\pi A}\int_0^A \frac{1}{\sqrt{1-\left(\frac{y}{A}\right)^2}}\,dy = \frac{2}{\pi}\int_0^1 \frac{1}{\sqrt{1-u^2}}\,du$$

$$= \frac{2}{\pi}\left[\sin^{-1} u\right]_0^1$$

$$= 1.$$

Because $h(y) \ge 0$ for all y such that $0 \le y < A$, the conditions of a probability density have been verified.

The cumulative distribution function of the variate Y is

$$H(y) = \Pr(Y \le y) = \begin{cases} 0, & y < 0, \\ \frac{2}{\pi}\sin^{-1}\left(\frac{y}{A}\right), & 0 \le y < A, \\ 1, & y \ge A. \end{cases}$$

◀

Example 4.2.8

Let the random variable X have the following function for its probability density:

$$f(x) = \begin{cases} \dfrac{1}{\beta - \alpha}, & \alpha \le x \le \beta, \\ 0, & \text{elsewhere.} \end{cases}$$

Find:

a. $Pr(x_1 \le -aX + b \le x_2)$, $a > 0$, $b - a\beta \le x_1 \le x_2 \le b - a\alpha$

b. $Pr(-x_3 \le -X^2 \le -x_4)$, $\alpha^2 < x_4 < x_3 < \beta^2$

Solution: To obtain the preceding probabilities, we must first determine the probability densities of the functions $Y = g(X) = -aX + b$ and $Y = g(X) = -X^2$. Both functions are differentiable and monotonically decreasing in the range of x.

a. Thus,

$$g^{-1}(y) = \frac{b-y}{a}, \quad \left|\frac{dg^{-1}(y)}{dy}\right| = \left|-\frac{1}{a}\right| = \frac{1}{a}.$$

Applying Equation (4.2.3), we have

$$h(y) = \begin{cases} \dfrac{1}{a} \dfrac{1}{\beta - \alpha}, & b - a\beta \le y \le b - a\alpha, \\ 0, & \text{elsewhere.} \end{cases}$$

Therefore,

$$Pr(x_1 \le -aX + b \le x_2) = Pr(x_1 \le Y \le x_2)$$

$$= \int_{x_1}^{x_2} \frac{1}{a} \frac{1}{\beta - \alpha} dy$$

$$= \frac{1}{a} \frac{1}{\beta - \alpha}(x_2 - x_1).$$

The same result can be obtained by writing

$$Pr\left(\frac{b - x_2}{a} \le X \le \frac{b - x_1}{a}\right) = \int_{\frac{b-x_2}{a}}^{\frac{b-x_1}{a}} \frac{1}{\beta - \alpha} dy.$$

b. Here, $g^{-1}(y) = \sqrt{-y}$ and

$$\left|\frac{d}{dy} g^{-1}(y)\right| = \left|-\frac{1}{2\sqrt{-y}}\right| = \frac{1}{2\sqrt{-y}}.$$

Substituting these results into Equation (4.2.1) gives

$$h(y) = \begin{cases} \dfrac{2}{2(\beta - \alpha)\sqrt{-y}}, & -\beta^2 < y < -\alpha^2, \\ 0, & \text{elsewhere.} \end{cases}$$

Here,

$$Pr\left(-x_3 \leq -X^2 \leq -x_4\right) = Pr\left(-x_3 \leq Y \leq -x_4\right)$$

$$= \int_{-x_3}^{-x_4} \frac{1}{2(\beta-\alpha)\sqrt{-y}} dy$$

$$= \frac{1}{\beta-\alpha}\left(\sqrt{x_3} - \sqrt{x_4}\right).$$

We can obtain the same results by writing

$$Pr\left(-x_3 \leq -X^2 \leq -x_4\right) = Pr\left(x_4 \leq X^2 \leq x_3\right)$$

$$= Pr\left(\sqrt{x_4} \leq X \leq \sqrt{x_3}\right)$$

$$= \int_{\sqrt{x_4}}^{\sqrt{x_3}} \frac{1}{\beta-\alpha} dy$$

$$= \frac{1}{\beta-\alpha}\left(\sqrt{x_3} - \sqrt{x_4}\right).$$

In a number of cases, the function of a random variable, the probability density function of which we know, is not a *one-to-one function;* that is, it is not a *strictly monotone function.*

When we encounter such a problem, we partition it into subintervals so that each resulting interval is strictly monotone and differentiable and so that no point over the interval equals a constant. We illustrate this approach with the following examples. ◀

Example 4.2.9

Let the continuous random variable X have the following function for its probability density:

$$f(x) = \begin{cases} \frac{1}{18}(x+2), & -2 \leq x \leq 4, \\ 0, & \text{elsewhere.} \end{cases}$$

Find the density function of the random variable $Y = g(X) = X^2$.

Solution: Here, we partition the interval $[-2, 4]$ into two subintervals, $[-2, 0]$ and $[0, 4]$, giving us

$$y = g(x) = \begin{cases} g_1(x) = x^2, & -2 \leq x \leq 0, \\ g_2(x) = x^2, & 0 < x \leq 4. \end{cases}$$

(See Figure 4.2.5.)

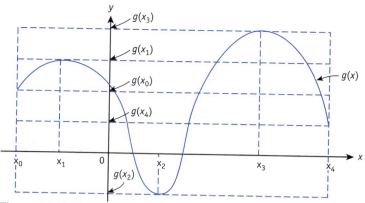

Figure 4.2.5 Probability Density Function

Also,

$$x = g^{-1}(y) = \begin{cases} g_1^{-1}(y) = -\sqrt{y}, & 0 \le y \le 4, \\ g_2^{-1}(y) = \sqrt{y}, & 4 < y \le 16, \end{cases}$$

and

$$\left|\frac{dg_2^{-1}(y)}{dy}\right| = \left|\frac{1}{2\sqrt{y}}\right| = \frac{1}{2\sqrt{y}} = \frac{dg_2^{-1}(y)}{dy}.$$

Therefore,

$$h(y) = \begin{cases} f(g_1^{-1}(y))\left|\dfrac{dg_1^{-1}(y)}{dy}\right| + f(g_2^{-1}(y))\left|\dfrac{dg_2^{-1}(y)}{dy}\right|, & g(0) \le y \le g(-2), \\ f(g_2^{-1}(y))\left|\dfrac{dg_2^{-1}(y)}{dy}\right|, & g(-2) < y \le g(4), \end{cases}$$

and

$$h(y) = \begin{cases} \dfrac{1}{9\sqrt{y}}, & 0 \le y \le 4, \\ \dfrac{1}{36\sqrt{y}}(2+\sqrt{y}), & 4 < y \le 16, \\ 0, & \text{elsewhere.} \end{cases}$$

The function $h(y)$ is nonnegative, and

$$\int_4^{16} \frac{1}{36\sqrt{y}}(2+\sqrt{y})\,dy + \int_0^4 \frac{1}{9\sqrt{y}}\,dy = 1.$$

Example 4.2.10

Let the random variable θ be distributed uniformly:

$$f(\theta)=\begin{cases}\dfrac{1}{\pi}, & -\dfrac{\pi}{2}\le\theta\le\dfrac{\pi}{2},\\ 0, & \text{elsewhere.}\end{cases}$$

Find the probability density function of the variate $\phi = g(\theta) = A\cos\theta$.

Solution: The function $g(\theta)$ is not a single-valued function, but by partitioning the interval $[-(\pi/2),\pi/2]$ into two intervals, $[-(\pi/2),0]$ and $[0,\pi/2]$, we get

$$\phi = g(\theta) = \begin{cases} g_1(\theta) = A\cos\theta, & -\dfrac{\pi}{2}\le\theta<0,\\ g_2(\theta) = A\cos\theta, & 0\le\theta\le\dfrac{\pi}{2},\end{cases}$$

and

$$h(\phi)f\left(g_1^{-1}(\phi)\right)\left|\dfrac{dg_1^{-1}(\phi)}{d\phi}\right| + f\left(g_2^{-1}(\phi)\right)\left|\dfrac{dg_2^{-1}(\phi)}{d\phi}\right|, \quad 0\le\phi\le A.$$

Also,

$$\left|\dfrac{dg_1^{-1}(\phi)}{d\phi}\right| = \left|\dfrac{dg_2^{-1}(\phi)}{d\phi}\right| = \dfrac{1}{A}\dfrac{1}{\sqrt{1-\left(\dfrac{\phi}{A}\right)^2}} = \left(A^2-\phi^2\right)^{-(1/2)}.$$

Hence,

$$h(\phi) = \begin{cases}\dfrac{2}{\pi}\left(A^2-\phi^2\right)^{-(1/2)}, & 0\le\phi\le A,\\ 0, & \text{elsewhere.}\end{cases}$$

The function $h(\phi)$ is nonnegative, and

$$\dfrac{2}{A\pi}\int_0^A \dfrac{1}{\sqrt{1-\left(\dfrac{\phi}{A}\right)^2}}\,d\phi = \dfrac{2}{\pi}\int_0^1 \dfrac{1}{\sqrt{1-u^2}}\,du = 1.$$

4.3 Other Types of Derived Distribution

Another case that arises is one in which the random variable X is continuous and its functional form $y = g(x)$ is piecewise uniformly continuous. For example, let X be a continuous variate with probability density $f(x)$. Suppose that

$$y = g(x) = \begin{cases} \alpha, & x > k, \\ -\alpha, & x \leq k. \end{cases}$$

We want to find the distribution of Y. Here, Y can assume only two values, $+\alpha$ and $-\alpha$, and

$$Pr(Y = -\alpha) = Pr(X \leq k) = \int_{-\infty}^{k} f(x) dx = F(k)$$

$$Pr(Y = \alpha) = Pr(X > k) = 1 - Pr(X \leq k)$$

$$= 1 - \int_{-\infty}^{k} f(x) dx = 1 - F(k).$$

Thus,

$$F(y) = \begin{cases} 0, & y < -\alpha, \\ F_x(k), & -\alpha \leq y < \alpha, \\ 1, & y \geq \alpha, \end{cases}$$

and

$$f(y) = \begin{cases} F_x(\mu), & y = -\alpha, \\ 1 - F_x(\mu), & y = \alpha, \\ 0, & \text{elsewhere}. \end{cases}$$

Example 4.3.1

Let the random variable X have the following function for its density:

$$f(x) = \begin{cases} e^{-(x-1)}, & x \geq 1, \\ 0, & \text{elsewhere}. \end{cases}$$

Find the distribution of the variate $y = g(x)$, which is the step function,

$$y = g(x) = \begin{cases} 1, & X > 2, \\ -1, & X \leq 2. \end{cases}$$

Solution: Thus,

$$Pr(Y = -1) = Pr(X \leq 2) = \int_{1}^{2} e^{-(x-1)} dx$$

$$= \int_{0}^{1} e^{-u} du = 0.632$$

and

$$Pr(Y=1) = Pr(X>2) = 1 - Pr(X \leq 2)$$

$$= 1 - \int_1^2 e^{-(x-1)} dx = 0.368.$$

Figure 4.3.1 illustrates $g(x)$, $F(x)$, and $F(y)$.

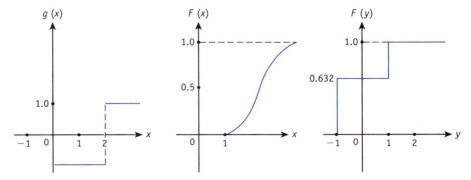

Figure 4.3.1 Graphical Representation of Functions

Thus,

$$h(y) = \begin{cases} 0.632, & y = -1, \\ 0.368, & y = 1, \\ 0, & \text{elsewhere,} \end{cases}$$

and

$$H(y) = \begin{cases} 0, & y < -1, \\ 0.632, & -1 \leq y < 1, \\ 1, & y \geq 1. \end{cases}$$

Consider also the case in which the random variable X is of the continuous type and $Y = g(X)$ is discrete. For example, let X be exponentially distributed with parameter β,

$$f(x;\beta) = \begin{cases} \beta e^{-\beta x}, & x \geq 0, \beta > 0, \\ 0, & \text{elsewhere,} \end{cases}$$

and $Y = g(X) = [X]$, where $[X]$ is the greatest integer function. $Y = 0$, $0 \leq x < 1$; $Y = 1$, $1 \leq x < 2$; ...; $Y = n$, $n \leq x < n+1$; and so on. The probability density of Y is given by

$$Pr(Y = n) = Pr(n \leq X < n+1) = \int_n^{n+1} \beta e^{-\beta x} dx$$

$$= e^{-\beta n} - e^{-\beta(n+1)}.$$

Thus,

$$h(y;\beta) = \begin{cases} e^{-y\beta}(1 - e^{-\beta}), & y = 0, 1, 2, \ldots, \beta > 0, \\ 0, & \text{elsewhere.} \end{cases}$$

It is clear that

$$\sum_{y=0}^{\infty} h(y) = \sum_{y=0}^{\infty} \left(1-e^{-\beta}\right) e^{-y\beta}$$

$$= \left(1-e^{-\beta}\right) \sum_{y=0}^{\infty} \left(e^{-\beta}\right)^y$$

$$= \left(1-e^{-\beta}\right) \frac{1}{\left(1-e^{-\beta}\right)} = 1,$$

◀

because $e^{-\beta} < 1 \ (\beta > 0)$ is a geometric series.

In this chapter, we have studied functional forms of a single random variable. More precisely, we have discussed the manner in which we can derive the probability distribution of a function of a given variate. In Chapter 6, we extend our investigation to multivariate functional forms of a random variable. In the study of probability theory and its applications, the *expected value* or *mean* of a random variable is perhaps the most important parameter. It is the aim of the next chapter to analyze the expected value and higher moments of a variate, along with the *moment-generating function* of a probability density function of a random variable.

Summary

The following cases were studied for deriving the probability density of a functional form of a single random variable:

1. Distribution of a continuous function of a discrete random variable
2. Distribution of a continuous function of a continuous random variable
3. Distribution of a piecewise uniformly continuous function of a continuous variate
4. Distribution of a discrete function of a continuous random variable

Theoretical Exercises

4.1 Let X be a discrete random variable that can assume the values $-1, 0, 1, 2, 3$, and 4 with the probabilities $1/6, 1/12, 1/6, 1/4, 1/12$, and $1/4$. Find the probability densities of the following random variables:

    a. $Y = \frac{1}{2}X - 1$ **b.** $Z = X^2 + 1$

4.2 The random variable X can assume the values $1, 2, 3, 4, 5$, and 6 with the probabilities P_1, P_2, P_3, P_4, P_5, and P_6. Find the probability density function of the following random variables:

    a. $Y = (X-1)^2$ **b.** $Z = \log X$

    c. $W = e^{-X}$ **d.** $Y = \frac{1}{X}$

4.3 The random variable X is Poisson distributed. Find the density functions of the following variates:
   a. $Y = -X$
   b. $Z = \sqrt{X}$
   c. $W = 2X - 1$
   d. $Y = X^\alpha$ (with α as an integer)

4.4 The random variable X assumes the values 1, 2, 3, and 4, with equal probability. Find the density functions of the following variates:
   a. $Y = 1 - 2X$
   b. $Z = \dfrac{X}{X+1}$
   c. $W = \cos 2\pi X$
   d. $Y = X^2 - X - 2$

4.5 Let the random variable X have the binomial distribution. Find the probability density functions of the following variates:
   a. $Y = aX^2$
   b. $Y = e^X$

4.6 Suppose that the random variable X is uniformly distributed over the interval (α, β). Find and sketch the probability density function of the following variates:
   a. $Z = X^2 + 2,\ 0 \leq \alpha \leq \beta < \infty$
   b. $W = \dfrac{1}{X-1},\ 2 < \alpha \leq \beta < \infty$
   c. $Y = aX + b,\ \alpha \neq 0$
   d. $T = \cos X,\ 0 \leq X \leq \pi$
   e. Graphically compare the densities $f(x)$, $h(z)$, and $h(w)$.

4.7 The probability density of the variate X is given by
$$f(x) = \begin{cases} e^{-(x-6)}, & x \geq 6, \\ 0, & \text{elsewhere.} \end{cases}$$
   a. Find the cumulative distribution function of the random variable $Y = X^2$.
   b. Compare $F(x)$ and $F(y)$ graphically.

4.8 The cumulative distribution function of the random variable X is given by
$$F(x) = \begin{cases} 0, & x < 0, \\ 1 - e^{-\frac{(x-\alpha)^\beta}{\alpha}}, & x \geq 0,\ \beta > 0,\ \alpha \geq 0. \end{cases}$$
Find the probability density function of the following variates:
   a. $Y = X^2$
   b. $Z = \ln X$

4.9 The density of the variate X is given by
$$f(x) = \begin{cases} \dfrac{1}{4} x e^{-(x/2)}, & x > 0, \\ 0, & \text{elsewhere.} \end{cases}$$
Find the probability distribution function of the following random variables:
   a. $Y = \alpha X + \beta,\ \alpha > 0$
   b. $Z = e^X$
   c. $Y = \dfrac{1}{X}$
   d. $Z = \sqrt{X}$

4.10 Suppose that the density of the random variable X is given by

$$f(x) = \begin{cases} \frac{2}{15}(x+2), & -1 < x < 2, \\ 0, & \text{elsewhere.} \end{cases}$$

Find the probability density function of the variate $Y = X^2$. Compare $f(x)$ and $h(y)$ graphically.

4.11 Let the random variable θ be uniformly distributed over the interval $(-\pi/2, \pi/2)$. Show that the variate $X = \mu + \alpha \tan \theta$ has a Cauchy distribution whose density function is given by Equation (3.3.19), where $\alpha = 1$.

4.12 Let the random variable X have the normal distribution with the parameters μ and σ^2. Find the probability density of the variate $Y = aX^2 + bX + c$.

4.13 Let the variate X be normally distributed with the parameters $\mu = 0$ and $\sigma = 1$. Find and sketch the probability density functions of the following random variables:

a. $Y = 2X + 1$ b. $Z = 2X^2 - 1$ c. $Y = \sqrt{|X|}$

4.14 Let the probability density function of the random variable X be given by

$$f(x) = \begin{cases} \dfrac{x^{(n-2)/2}}{2^{n/2} \Gamma\left(\frac{n}{2}\right)} e^{-(x/2)}, & x > 0, n = 1, 2, 3, \ldots, \\ 0, & \text{elsewhere,} \end{cases}$$

which is known as the *chi-square distribution*, with n degrees of freedom. Suppose that $S^2 = (\sigma^2/n)X$, $\sigma > 0$. Find the distribution of S^2. For what value of α and of β does the new density $h(s^2)$ become a gamma probability density function?

4.15 Let the variate X be gamma distributed with the parameters $\alpha = 1$ and $\beta = 2$. If

$$Y = \begin{cases} 2, & X \geq 3, \\ -2, & X < 3, \end{cases}$$

what is the cumulative distribution of the random variable Y?

4.16 In a sequence of Bernoulli trials, let X be a random variable representing the number of trials required to achieve k successes. Then, the probability density function of X is

$$f(x) = \begin{cases} \binom{x-1}{k-1} p^k (1-p)^{x-k}, & x = k, k+1, k+2, \ldots, \\ 0, & \text{elsewhere,} \end{cases}$$

where p is the probability of a success on any given trial. Show that the probability density function of $Y = X - k$ is

$$f(y) = \begin{cases} (-1)^y \binom{-k}{y} (1-p)^y p^k, & y = 0, 1, 2, \ldots, \\ 0, & \text{elsewhere.} \end{cases}$$

4.17 Let the random variable X have the continuous probability density function $f(x)$. Find the probability density function of $Y = F(x)$.

4.18 The probability density function of the random variable X is given by

$$f(x) = \begin{cases} \dfrac{3}{32}(2+x)(2-x), & -2 \leq x \leq 2, \\ 0, & \text{elsewhere.} \end{cases}$$

Find the probability density function of the variate $Y = X(X+1)(X-1)$.

4.19 If the probability density function of the variate T is given by

$$f(x) = \dfrac{\Gamma\left(\dfrac{n+1}{2}\right)}{\sqrt{n}\,\Gamma\left(\dfrac{n}{2}\right)\Gamma\left(\dfrac{1}{2}\right)\left(1+\dfrac{t^2}{n}\right)^{(n+1)/2}}, \quad -\infty < t < \infty,$$

what is the probability density function of the random variable $Y = T^2$?

4.20 If the probability density function of the variate X is given by

$$f(x) = \begin{cases} \dfrac{1}{2\pi}, & -\pi \leq x \leq \pi, \\ 0, & \text{elsewhere,} \end{cases}$$

what is the probability density function of the random variable $Y = 2\sin X$?

Applied Problems

4.1 In Exercise 4.6, find the probability that:
  a. The random variable Z assumes a value from $1/3$ to $2/3$.
  b. The random variable W assumes a value greater than 3.
  c. The variate Y assumes a value less than 4.
  d. The variate T assumes a value from 0 to $\pi/2$.

4.2 In Exercise 4.8, evaluate the following probabilities:
  a. $\Pr(1 \leq Y \leq 3.4)$ b. $\Pr(|X| \leq 4)$ c. $\Pr(1 < Z \leq 2)$

4.3 In Exercise 4.14, for $n = 4$ and $\sigma^2 = 1$, find the probability that the random variable S^2 assumes:
  a. A value less than or equal to 1
  b. A value between 3 and 4
  c. A value greater than 6

4.4 In Example 4.2.3, find c so that the probability density function of the kinetic energy of a gas molecule is actually a gamma probability density.

4.5 The temperature recording equipment used to determine the melting point T of a given substance in degrees Fahrenheit has an error that can be treated as a random variable that has a normal distribution with the parameters μ and σ. The temperature C, measured in degrees Celsius, is related to T by $C = g(T) = (5/9)T - 32(5/9)$.
  a. Find the probability density function that describes the behavior of C.
  b. Graphically compare this distribution with that of Example 4.2.5.

4.6 Suppose that X is normally distributed with $\mu=0$ and $\sigma=1$. Find the following probabilities:
  a. $Pr(0 \leq \sin \pi X \leq 0.2)$ b. $Pr(-0.25 \leq \sin \pi X \leq 0.25)$

4.7 In an electrical circuit, the voltage source V is a random variable uniformly distributed between 10 and 20 V. Find the probability density function of the power that is related to V and resistance R by $P = V^2/R$.

4.8 If a random sine wave is given by $X = A \sin \Phi$, where the amplitude A is a known positive constant and the argument Φ behaves as a random variable that is uniformly distributed on the interval $-(\pi/2)$ to $\pi/2$, what are the probability density and cumulative distribution functions of the variate X?

4.9 In Exercise 4.18, evaluate the following probabilities:
  a. $Pr(-2 \leq Y \leq 0.2)$ b. $Pr(0.2 \leq Y \leq 2)$

4.10 In Exercise 4.20, find the probability that the random variable Y assumes one of the following values:
  a. From $1/4$ to $3/4$
  b. From $-3/2$ to $1/4$

"The theory of probability as a mathematical discipline can and should be developed from geometry and algebra."
—A. N. Kolmogorov

CHAPTER 5

Expected Values, Moments, and the Moment-Generating Function

OBJECTIVES:
5.1 Mathematical Expectation
5.2 Properties of Expectation
5.3 Moments
5.4 Moment-Generating Function

Andrey Nikolayevich Kolmogorov (1903–1987)

"Have you ever noticed that those who have a natural capacity for calculation are, generally speaking, naturally quick at all kinds of study; while men of slow intellect, if they are trained and exercised in arithmetic, if they get nothing else from it, at least all improve and become sharper than they were before?"
—Plato, "The Republic," 360 BC

Andrey Nikolayevich Kolmogorov laid the mathematical foundations of probability theory and the theory of randomness. His monograph *Grundbegriffe der Wahrscheinlichketsrechnung,* published in 1933, introduced probability theory in a rigorous way from fundamental axioms. He later used probability theory to study the motion of the planets and the turbulent flow of air from a jet engine. He also made important contributions to stochastic processes, information theory, statistical mechanics, and nonlinear dynamics. Kolmogorov had numerous interests outside mathematics. In particular, he was interested in the form and structure of the poetry of the Russian author Aleksandr Sergeyevich Pushkin.

Chapter 5 Expected Values, Moments, and the Moment-Generating Function

INTRODUCTION

In the study of the theory and application of probability, the *expected value*, or *mean*, of a random variable is the most important parameter. The phrase "expected value" dates back to the development of probability theory in relation to games of chance. Consider a specific example:

A gambler tosses a fair die. If the numbers one, three, or five appear, he will be paid $2; if a two or a four turns up, he will lose $2; and if a six occurs, he will receive $4. Naturally, the gambler wants to know his *average* or *expected* winnings. The payoffs are $-2, 2$, and 4, and their respective probabilities are $1/3, 1/2$, and $1/6$. The expected value is obtained by multiplying each payoff by its chance of occurrence and adding the results; thus,

$$-2\left(\frac{1}{3}\right)+2\left(\frac{1}{2}\right)+4\left(\frac{1}{6}\right)=1.$$

This is the expected value of the game. In other words, if the player plays this game a great number of times, his average winnings will be $1 per toss.

When we speak of the expected value of a random variable X, we refer to an ideal or a theoretical average. However, if a given experiment is repeated many times, we would expect the average value of the random variable involved to be near its expected value.

In this chapter, in addition to studying the expected value of a random variable, we study moments and the moment-generating function of a probability density function of a random variable.

5.1 Mathematical Expectation

In this section, we give a precise definition of the mathematical expectation of a distribution function and illustrate the calculations using various examples.

Definition 5.1.1

Let X be a random variable of the discrete type, with the probability density function $p(x)$. The series

$$E(X)=\sum_i x_i p(x_i) \qquad (5.1.1)$$

is called the **expected value** of the random variable X if the following inequality is satisfied:

$$\sum_i p(x_i)|x_i|<\infty. \qquad (5.1.2)$$

Example 5.1.1

Let X be a discrete random variable whose probability density function is given by Table 5.1.1.

Table 5.1.1

x	−1	0	1	2	3	4	5
$P(x)$	$\frac{1}{7}$	$\frac{1}{7}$	$\frac{1}{14}$	$\frac{2}{7}$	$\frac{1}{14}$	$\frac{1}{7}$	$\frac{1}{7}$

Solution: The expected value of the random variable X is given by

$$E(X) = \sum_{x=-1}^{5} xp(x) = -1\left(\frac{1}{7}\right) + 0\left(\frac{1}{7}\right) + \ldots + 5\left(\frac{1}{7}\right) = 2.$$

It is clear that, in this example, the condition in Equation (5.1.2) is satisfied.

Example 5.1.2

Let the random variable X be binomially distributed with the parameters n and p. The expected value of the variate X is computed as follows:

$$E(X) = \sum_{x=0}^{n} x \binom{n}{x} p^x (1-p)^{n-x}$$

$$= \sum_{x=0}^{n} \frac{xn(n-1)!}{(n-x)!x(x-1)!} p^x (1-p)^{n-x}$$

$$= n \sum_{x=1}^{n} \frac{(n-1)!}{(n-x)!x(x-1)!} pp^{x-1} (1-p)^{n-x}$$

$$= np \sum_{x=1}^{n} \binom{n-1}{x-1} p^{x-1} (1-p)^{n-x}$$

$$= np,$$

where the sum is over all possible values that the random variable can assume. Arriving at the last step, we write

$$\sum_{x=1}^{n} \binom{n-1}{x-1} p^{x-1} (1-p)^{n-x} = \sum_{y=0}^{m} \binom{m}{y} p^y (1-p)^{m-y} = 1,$$

which follows by replacing $x-1$ with y and $n-1$ with m. One interpretation of the foregoing could be that, if we inspect n components of a system in which each has the probability p of being operable, we could *expect* that the total number of operable components would be close to np.

Example 5.1.3

Let the variate X be Poisson distributed with parameter λ. Find the expected value of the random variable X.

Solution: Here,

$$p(x) = \begin{cases} \dfrac{e^{-\lambda}\lambda^x}{x!}, & x = 0, 1, 2, \ldots, \\ 0, & \text{elsewhere}, \end{cases}$$

and

$$E(X) = \sum_{x=0}^{\infty} \frac{x e^{-\lambda} \lambda^x}{x!} = e^{-\lambda} \sum_{x=0}^{\infty} \frac{x \lambda^x}{x(x-1)!}$$

$$= \lambda e^{-\lambda} \sum_{x=1}^{\infty} \frac{\lambda^{x-1}}{(x-1)!} = \lambda e^{-\lambda} \sum_{k=0}^{\infty} \frac{\lambda^k}{k!}$$

$$= \lambda e^{-\lambda} e^{\lambda} = \lambda.$$

$E(X) = np$ for the binomial probability density function, so we expect $E(X)$ will equal λ for the Poisson probability density function because, in the derivation of the Poisson distribution, we set λ equal to np.

◀

Example 5.1.4

If the random variable X is uniformly distributed over the interval (α, β), what is its expected value?

Solution:

$$E(X) = \int_{-\infty}^{\infty} x f(x) dx = \frac{1}{\beta - \alpha} \int_{\alpha}^{\beta} x \, dx$$

$$= \frac{\beta^2 - \alpha^2}{2} \cdot \frac{1}{\beta - \alpha} = \frac{\beta + \alpha}{2}$$

◀

Definition 5.1.2

Let X be a random variable of the continuous type with the probability density function $f(x)$. The integral

$$E(X) = \int_{-\infty}^{\infty} x f(x) dx \qquad (5.1.3)$$

is the expected value of the random variable X if the following inequality is satisfied:

$$\int_{-\infty}^{\infty} f(x) |x| dx < \infty. \qquad (5.1.4)$$

Example 5.1.5

Let X be a continuous random variable whose probability density function is given by

$$f(x) = \begin{cases} e^{-(x-4)}, & x > 4, \\ 0, & \text{elsewhere.} \end{cases}$$

Find the expected value of X.

Solution: Applying Equation (5.1.3), we obtain the expected value of the random variable X:

$$E(X) = \int_4^\infty x e^{-e(x-4)} dx$$

$$= \int_4^\infty (u+4) e^{-u} du = 5.$$

◄

Example 5.1.6

Let X be normally distributed with the parameters μ and σ^2. Show that the expected value of the random variable X is μ.

Solution:

$$E(X) = \int_{-\infty}^\infty x f(x) dx = \frac{1}{\sqrt{2\pi}\sigma} \int_{-\infty}^\infty x \exp\left(-\frac{1}{2}\left(\frac{x-\mu}{\sigma}\right)^2\right) dx$$

Let $t = \frac{x-\mu}{\sigma}$, $\sigma t + \mu = x$, $\sigma dt = dx$, and

$$E(X) = \frac{1}{\sqrt{2\pi}} \int_{-\infty}^\infty (\sigma t + \mu) e^{-(1/2)t^2} dt$$

$$= \frac{1}{\sqrt{2\pi}} \int_{-\infty}^\infty \sigma t e^{-(1/2)t^2} dt + \frac{\mu}{\sqrt{2\pi}} \int_{-\infty}^\infty e^{-(1/2)t^2} dt$$

$$= 0 + \mu = \mu.$$

◄

The expected value of a random variable $E(X)$ is sometimes called the *mathematical expectation*, the *expectation*, or the *mean value* of the variate. The expected value of a random variable does not always exist. If the right side of Equation (5.1.1) or (5.1.3) exists but the inequality in Equation (5.1.2) or (5.1.4) is not satisfied, then the expected value of the variate does not exist. Thus, if the series in Equation (5.1.2) or the integral in Equation (5.1.4) is absolutely convergent, then the mathematical expectation of the random variable exists.

We demonstrate the preceding remarks with Examples 5.1.7 and 5.1.8.

Example 5.1.7

Let X be a random variable of the discrete type whose probability density function is given by

$$f(x) = \begin{cases} \dfrac{6}{\pi^2} \dfrac{1}{x^2}, & x = 1, -2, 3, -4, \ldots, \\ 0, & \text{elsewhere.} \end{cases}$$

If the expected value of X exists, then we would have

$$E[X] = \sum_i x_i f(x_i) = \frac{6}{\pi^2}\left[\frac{1}{1} - \frac{1}{2} + \frac{3}{9} - \frac{4}{16} + \frac{5}{25} - \cdots\right]$$

$$= \frac{6}{\pi^2} \sum_{x=1}^{\infty} \frac{(-1)^{x-1} x}{x^2} = \frac{6}{\pi^2} \sum_{x=1}^{\infty} \frac{(-1)^{x-1}}{x}.$$

It can be shown that $\sum 1/x^2 = \pi^2/6$, $x = 1, -2, 3, -4, \ldots$.
Recall that

$$\log_e(1+x) = x - \frac{x^2}{2} + \frac{x^3}{3} - \frac{x^4}{4} + \cdots + (-1)^{n-1} \frac{x^n}{n} + \cdots$$

for $-1 < x < 1$, which implies that

$$1 - \frac{1}{2} + \frac{1}{3} - \frac{1}{4} + \cdots + (-1)^{x-1}\frac{1}{x} + \cdots = \log_e(2).$$

Hence,

$$E[X] = \frac{6}{\pi^2} \log_e 2 < \infty.$$

However,

$$E[|X|] = \frac{6}{\pi^2} \sum_{x=1}^{\infty} \frac{\left|(-1)^{x-1} x\right|}{x^2} = \frac{6}{\pi^2} \sum_{x=1}^{\infty} \frac{x}{x^2}$$

$$= \frac{6}{\pi^2} \sum_{x=1}^{\infty} \frac{1}{x} = \infty.$$

Thus, even though $E[X]$ is calculated to be finite, we must conclude that the expectation of X does not exist because $E[|X|]$ is not finite. ◀

Example 5.1.8

Let X be a continuous random variable whose probability density function is given by

$$f(x) = \frac{1}{\pi}\frac{1}{1+x^2}, \quad -\infty < x < \infty.$$

Find the expected value of X.

Solution: Applying Equation (5.1.3), we have

$$E[X] = \int_{-\infty}^{\infty} f(x)\,dx = \frac{1}{\pi}\int_{-\infty}^{\infty}\frac{x}{1+x^2}\,dx$$

$$= \lim_{h\to\infty}\left[\frac{1}{2\pi}\int_{-h}^{h}\frac{2x}{1+x^2}\,dx\right]$$

$$= \lim_{h\to\infty}\left[\frac{1}{2\pi}\log(1+x^2)\Big|_{-h}^{h}\right]$$

$$= \lim_{h\to\infty}\left\{\frac{1}{2\pi}\left[\log(1+h^2) - \log(1+(-h)^2)\right]\right\}$$

$$= \lim_{h\to\infty}\left\{\frac{1}{2\pi}\cdot 0\right\} = 0.$$

However,

$$E[|X|] = \int_{-\infty}^{\infty} x|f(x)|\,dx = \frac{1}{\pi}\int_{-\infty}^{\infty}\frac{|x|}{(1+x^2)}\,dx$$

$$= \frac{1}{\pi}\left[\int_{-\infty}^{0}\frac{-x}{1+x^2}\,dx + \int_{0}^{\infty}\frac{x\,dx}{1+x^2}\right]$$

$$= \frac{2}{\pi}\int_{0}^{\infty}\frac{x}{1+x^2}\,dx = \frac{1}{\pi}\int_{0}^{\infty}\frac{2x\,dx}{1+x^2}$$

$$= \lim_{h\to\infty}\frac{1}{\pi}\int_{0}^{h}\frac{2x}{1+x^2}\,dx = \lim_{h\to\infty}\left[\log(1+x^2)\Big|_{0}^{h}\right]$$

$$= \lim_{h\to\infty}\frac{1}{\pi}\left[\log(1+h^2) - \log 1\right]$$

$$= \frac{1}{\pi}\lim_{n\to\infty}\left[\log(1+h^2)\right] = \infty.$$

Therefore, although the value of the integral in Equation (5.1.3) is finite, the condition in Equation (5.1.4) of Definition 5.1.2 is not satisfied. Thus, the expected value of the preceding distribution does not exist. This example is actually the Cauchy distribution, with $a=1$ and $\mu=0$, as defined in Definition 3.3.6. ◀

5.2 Properties of Expectation

We now state some useful theorems regarding expectation.

Theorem 5.2.1

Let X be a random variable with the probability density function $f(x)$. If $X = c$, where c is a constant, then $E(X) = c$.

Proof If the variate X is of the continuous type, then

$$E(c) = \int_{-\infty}^{\infty} cf(x)dx = c\int_{-\infty}^{\infty} f(x)dx = c.$$

The discrete case is found similarly. Because $X = c$,

$$F(x) = Pr(X \leq x) = \begin{cases} 0, & x < c, \\ 1, & x \geq c. \end{cases}$$

Theorem 5.2.2

Let X be a random variable with the probability density function $f(x)$. If $g(X)$ is a one-to-one functional form of the variate X, then the expected value of $g(X)$ is

$$E[g(X)] = \sum_{i=1}^{\infty} g(x_i) f(x_i) \qquad (5.2.1)$$

if X is discrete and

$$E[g(X)] = \int_{-\infty}^{\infty} g(x) f(x) dx \qquad (5.2.2)$$

if X is continuous. For the expected value of $g(X)$ to exist, Equations (5.2.1) and (5.2.2) must be absolutely convergent:

$$\sum_{i=1}^{\infty} |g(x_i)| f(x_i) < \infty$$

and

$$\int_{-\infty}^{\infty} |g(x)| f(x) dx < \infty.$$

Proof We prove the theorem for the discrete case. From the functional form, we have $y_i = g(x_i)$ or $x_i = g^{-1}(y_i)$. Applying the method of deriving distributions (Section 4.1), we can write the probability density of the random variable Y as $h(y_i) = f(g^{-1}(y_i))$. Thus, by definition,

$$E(Y) = \sum_{y_i} y_i h(y_i)$$

$$= \sum_{y_i} y_i f\left(g^{-1}(y_i)\right).$$

But $y_i = g(x_i)$ and $E(Y) = E[g(X)] = \sum_{x_i} g(x_i) f(x_i)$.

For example, the expected value of the discrete functional form given in Example 4.1.1 is

$$E(Y) = \sum_{i=1}^{4} y_i h(y_i)$$

$$= y_1 h(y_1) + y_2 h(y_2) + y_3 h(y_3) + y_4 h(y_4)$$

$$= y_1 f(x_1) + y_2 \{f(x_0) + f(x_2)\} + y_3 f(x_3) + y_4 f(x_4)$$

$$= g(x_1) f(x_1) + g(x_0) f(x_0) + g(x_2) f(x_2) + g(x_3) f(x_3) + g(x_4) f(x_4)$$

$$= \sum_{i=0}^{5} g(x_i) f(x_i) = E[g(X)].$$

Example 5.2.1

Let X be a discrete random variable that can assume the values, 0, 1, 2, and 3 with probabilities 1/6, 1/3, 1/6, and 1/3, respectively. Find the expected value of $g(X) = (X-2)^2$.

Solution: Applying Equation (5.2.1), we have

$$E[g(X)] = \sum_{x=0}^{3} p(x)(x-2)^2$$

$$= \frac{1}{6}(-2)^2 + \frac{1}{3}(-1)^2 + \frac{1}{6}(0)^2 + \frac{1}{3}(1)^2$$

$$= \frac{4}{3}.$$

Example 5.2.2

Let the random variable X be uniformly distributed over the interval (α, β). Find the expected value of $g(X) = e^{-(x/2)} + 1$.

Solution:

$$E[g(X)] = \int_{-\infty}^{\infty} g(x)f(x)\,dx$$

$$= \frac{1}{\beta-\alpha}\int_{\alpha}^{\beta}\left(e^{-(x/2)} + 1\right)dx$$

$$= \frac{1}{\beta-\alpha}\left[x - 2e^{-(x/2)}\right]_{\alpha}^{\beta}$$

$$= 1 - \frac{1}{\beta-\alpha}\left(e^{-\beta/2} - e^{-\alpha/2}\right)$$

◀

Example 5.2.3

The random variable X has for its probability density the function:

$$f(x) = \begin{cases} \dfrac{1}{2}e^{-(x-2)}, & x > 2, \\ \dfrac{1}{2}e^{x-2}, & x \leq 2. \end{cases}$$

Find $E[g(X)]$, where $g(X) = |x-2|$.

Solution: By definition,

$$E[g(X)] = \int_{-\infty}^{\infty} |x-2|f(x)\,dx$$

$$= \frac{1}{2}\int_{-\infty}^{2}-(x-2)e^{x-2}\,dx + \frac{1}{2}\int_{2}^{\infty}(x-2)e^{-(x-2)}\,dx$$

$$= 1.$$

◀

Examples 5.2.4 and 5.2.5 illustrate that the expected value of a functional form of a random variable exists if and only if $E[|g(X)|]$ converges.

Example 5.2.4

Let the discrete random variable have for its probability density:

$$f(x) = \begin{cases} \dfrac{1}{2}2^{-|x|}, & x = \pm 1, \pm 2, \pm 3, \ldots, \\ 0, & \text{elsewhere.} \end{cases}$$

Show that the expected value of the functional form,

$$g(X) = (-1)^{|X|-1} \frac{2^{|X|}}{2|X|-1},$$

does not exist, although its $E[g(X)]$ is finite.

Solution: Here,

$$E[g(X)] = \frac{1}{2}\sum_{x=-1}^{-\infty}(-1)^{|x|-1}\frac{2^{|x|}}{2|x|-1}2^{-|x|} + \frac{1}{2}\sum_{x=1}^{\infty}(-1)^{x-1}\frac{2^x}{2x-1}2^{-x}$$

$$= \frac{1}{2}\sum_{x=1}^{\infty}(-1)^{x-1}\frac{1}{2x-1} + \frac{1}{2}\sum_{x=1}^{\infty}(-1)^{x-1}\frac{1}{2x-1}$$

$$= \sum_{x=1}^{\infty}(-1)^{x-1}\frac{1}{2x-1}$$

$$= 1 - \frac{1}{3} + \frac{1}{5} - \frac{1}{7} + \dots$$

$$= \frac{\pi}{4} < \infty.$$

However,

$$E[|g(X)|] = \frac{1}{2}\sum_{x=-1}^{-\infty}\left|(-1)^{|x|-1}\frac{2^{|x|}}{2|x|-1}\right|2^{-|x|} + \frac{1}{2}\sum_{x=1}^{\infty}\left|(-1)^{|x|-1}\frac{2^{|x|}}{2^{|x|}-1}\right|2^{-|x|}$$

$$= \frac{1}{2}\sum_{x=1}^{\infty}\left|(-1)^{x-1}\frac{2^x}{2x-1}\right|2^{-x} + \frac{1}{2}\sum_{x=1}^{\infty}\left|(-1)^{x-1}\frac{2^x}{2x-1}\right|2^{-x}$$

$$= \frac{1}{2}\sum_{x=1}^{\infty}\frac{1}{2x-1} + \frac{1}{2}\sum_{x=1}^{\infty}\frac{1}{2x-1}$$

$$= \sum_{x=1}^{\infty}\frac{1}{2x-1} = 1 + \frac{1}{3} + \frac{1}{5} + \frac{1}{7} + \dots .$$

Note that

$$1 + \frac{1}{3} + \frac{1}{5} + \frac{1}{7} + \dots > \frac{1}{3} + \frac{1}{6} + \frac{1}{9} + \frac{1}{12} + \dots$$

$$= \sum_{x=1}^{\infty}\frac{1}{3x}$$

$$= \infty.$$

Therefore,

$$E[|g(X)|] = \sum_{x=1}^{\infty}\frac{1}{2x-1} = \infty,$$

and the expected value of $g(X) = (-1)^{|X|-1}2^{|X|}/(2|X|-1)$ does not exist.

Example 5.2.5

The probability density function of the continuous random variable is given by

$$f(x) = \frac{1}{2}e^{-|x|}, \quad -\infty < x < \infty.$$

As in the previous example, show that the expectation of the functional form of the continuous variate $Y = g(X) = Xe^{|X|}$ does not exist.

Solution: Here,

$$E(Y) = E[g(X)] = \int_{-\infty}^{\infty} g(x)f(x)\,dx$$

$$= \frac{1}{2}\int_{-\infty}^{\infty} xe^{|x|}e^{-|x|}\,dx$$

$$= \frac{1}{2}\int_{-\infty}^{\infty} x\,dx = 0 \quad \text{(odd function)}.$$

But

$$E[|Y|] = E[|g(X)|]\int_{-\infty}^{\infty}|g(x)|f(x)\,dx$$

$$= -\frac{1}{2}\int_{-\infty}^{0} x\,dx + \frac{1}{2}\int_{0}^{\infty} x\,dx$$

$$= \int_{0}^{\infty} x\,dx = \infty \quad \text{(even function)}.$$

Therefore, the expected value of $g(X) = Xe^{|X|}$ does not exist.

Theorem 5.2.3

Let X be a random variable with the probability density function $f(x)$. If $g(X) = \alpha X + \beta$, then

$$E[g(X)] = \alpha E(X) + \beta.$$

Proof For the continuous case, we have

$$E[g(X)] = \int_{-\infty}^{\infty} g(x)f(x)\,dx$$

$$= \int_{-\infty}^{\infty}(\alpha x + \beta)f(x)\,dx$$

$$= \alpha\int_{-\infty}^{\infty} xf(x)\,dx + \beta\int_{-\infty}^{\infty} f(x)\,dx$$

$$= \alpha E(X) + \beta.$$

Example 5.2.6

Let X be a continuous random variable whose probability density function is given by

$$f(x) = \begin{cases} e^{-\{x-(1/2)\}}, & x > \frac{1}{2}, \\ 0, & \text{elsewhere.} \end{cases}$$

Find the expected value of the variate $Y = (1/2)X - 2$.

Solution:

$$E(Y) = E\left(\frac{1}{2}X - 2\right)$$

$$= \int_{1/2}^{\infty} \left(\frac{1}{2}X - 2\right) e^{-\{x-(1/2)\}} dx$$

$$= \frac{1}{2} \int_{1/2}^{\infty} x e^{-\{x-(1/2)\}} dx - 2 \int_{1/2}^{\infty} e^{-\{x-(1/2)\}} dx$$

$$= \frac{3}{4} - 2 = -\frac{5}{4}.$$

Example 5.2.7

Let X be normally distributed with the parameters μ and σ. Find the expected value of the variate $g(X) = (1/2)X - 3$.

Solution: In Example 5.1.6, we showed $E(X) = \mu$. Thus, applying Theorem 5.2.3, we have

$$E[g(X)] = E\left(\frac{1}{2}X - 3\right) = \frac{1}{2}E(X) - 3 = \frac{1}{2}\mu - 3.$$

Theorem 5.2.4

Let X be a random variable with the probability density function $f(x)$. If $f(x)$ is symmetrical about the point $x = a$, $f(a-x) = f(a+x)$ for all x, then

$$E(X) = a,$$

provided $E(X)$ exists.

Proof If X is a continuous random variable, then

$$E(a - X) = \int_{-\infty}^{\infty} (a - t) f(t) dt. \tag{5.2.3}$$

In Equation (5.2.3), let $t = a - x$ and $dx = -dt$. Then,

$$\int_{-\infty}^{\infty} (a-t) f(t) dt = \int_{\infty}^{-\infty} x f(a-x)(-dx) = \int_{-\infty}^{\infty} x f(a-x) dx.$$

continued

Thus,
$$E(a-X) = \int_{-\infty}^{\infty} x f(a-x)\,dx. \tag{5.2.4}$$

But if we let $t = a+x$ and $dx = dt$ in Equation (5.2.3), Then
$$\int_{-\infty}^{\infty}(a-t)f(t)\,dt = \int_{-\infty}^{\infty} -x f(a+x)\,dx.$$

Thus,
$$E(a-X) = \int_{-\infty}^{\infty} -x f(a+x)\,dx. \tag{5.2.5}$$

Substituting $f(a+x) = f(a-x)$ into Equation (5.2.5), we have
$$E(a-X) = \int_{-\infty}^{\infty} -x f(a-x)\,dx. \tag{5.2.6}$$

Equating Equations (5.2.4) and (5.2.6) gives
$$E(a-X) = \int_{-\infty}^{\infty} x f(a-x)\,dx = -\int_{-\infty}^{\infty} x f(a-x)\,dx,$$

which implies $2\int_{-\infty}^{\infty} x f(a-x)\,dx = 0$ or $E(a-x) = 0$.
But,
$$E(a-X) = 0 = \int_{-\infty}^{\infty}(a-x)f(x)\,dx$$
$$= a\int_{-\infty}^{\infty} f(x)\,dx - \int_{-\infty}^{\infty} x f(x)\,dx,$$

which implies that $0 = a - E(X)$ or $E(X) = a$.

If we let $a = 0$ in the theorem, we get $E(X) = 0$. Also, $f(x) = f(-x)$. In this case, we say that $f(x)$ is *even*; if we wish, we may write
$$\int_{-\infty}^{\infty} f(x)\,dx = 2\int_{0}^{\infty} x f(x)\,dx.$$

Example 5.2.8

Show that, if the variate X is distributed according to the standard normal distribution, $E(X) = 0$.

Solution: Here, $a = 0$ and the density function is even, $f(x) = f(-x)$. Thus,
$$E(X) = \frac{1}{\sqrt{2\pi}} \int_{-\infty}^{\infty} x e^{-(1/2)x^2}\,dx$$
$$= -\frac{1}{\sqrt{2\pi}}\left[e^{-(x^2/2)}\right]_{-\infty}^{\infty} = 0.$$

◀

Theorem 5.2.5

Let X be a random variable with the probability density function $f(x)$. If $g(X) = g_1(X) + g_2(X)$, the sum of the two functional forms of the variate X, then

$$E[g(X)] = E[g_1(X)] + E[g_2(X)].$$

Proof If the variate X is continuous, then

$$E[g(X)] = E[g_1(X) + g_2(X)]$$

$$= \int_{-\infty}^{\infty} [g_1(X) + g_2(X)] f(x) dx$$

$$= \int_{-\infty}^{\infty} g_1(X) f(x) dx + \int_{-\infty}^{\infty} g_2(X) f(x) dx$$

$$= E[g_1(X)] + E[g_2(X)].$$

Example 5.2.9

Let X be Rayleigh distributed with $\alpha = 2$. Find the expected value of the random variable

$$g(X) = \frac{1}{X} + e^{-(1/3)X^2}.$$

Solution: Applying Theorem 5.2.5, we have

$$E[g(X)] = E(X^{-1}) + E\left(e^{-(1/3)X^2}\right),$$

$$= \int_{-\infty}^{\infty} \frac{1}{x} f(x) dx + \int_{-\infty}^{\infty} e^{(-1/3)x^2} f(x) dx$$

$$= \int_{0}^{\infty} \frac{1}{x} \cdot x e^{-x^2/2} dx + \int_{0}^{\infty} e^{-(1/3)x^2} x e^{-(1/2)x^2} dx$$

$$= \frac{\sqrt{2\pi}}{2} \cdot \frac{2}{\sqrt{2\pi}} \int_{0}^{\infty} e^{-(1/2)x^2} dx + \int_{0}^{\infty} x e^{-(5/6)x^2} dx$$

$$= \frac{\sqrt{2\pi}}{2} - \frac{3}{5} \int_{0}^{\infty} e^{-(5/6)x^2} \left(-\frac{5}{3} x\right) dx$$

$$= \frac{\sqrt{2\pi}}{2} - \frac{3}{5} \left[e^{-(5/6)x^2} \right]_{0}^{\infty}$$

$$= \sqrt{\frac{\pi}{2}} + \frac{3}{5}.$$

Theorem 5.2.6

Let X be a random variable with the probability density function $f(x)$. If $g_1(X) \leq g_2(X)$ for all possible values that the variate X can assume, then:

1. $E[g_1(X)] \leq E[g_2(X)]$
2. $|E[g_1(X)]| \leq E[|g_1(X)|]$

Proof

1. $E[g_1(X)] = \int_{-\infty}^{\infty} g_1(x) f(x) \leq \int_{-\infty}^{\infty} g_2(x) f(x) dx \leq E[g_2(X)]$

2. $|E[g_1(X)]| = \left| \int_{-\infty}^{\infty} g_1(x) f(x) dx \right| \leq \int_{-\infty}^{\infty} |g_1(x) f(x)| dx$

$$= \int_{-\infty}^{\infty} |g_1(x)| f(x) dx$$

$$= E[|g_1(x)|]$$

The discrete case can be handled similarly.

5.3 Moments

The moments of a distribution are a collection of descriptive constants that can be used for measuring its properties and, under some conditions, for specifying it.

The kth moment of the distribution of a random variable is the expected value of the variate involved when it is raised to the kth power.

Definition 5.3.1

Let X be a random variable with the density function $f(x)$. The **kth moment** or **kth ordinary moment** of the distribution of the variate X is given by

$$\mu_k = E(X^k) = \sum_{i=1}^{\infty} x_i^k f(x_i) \qquad (5.3.1)$$

if X is discrete and by

$$\mu_k = E(X^k) = \int_{-\infty}^{\infty} x^k f(x) dx \qquad (5.3.2)$$

if X is continuous.

The series in Equation (5.3.1) and the integral in Equation (5.3.2) must be absolutely convergent for μ_k to exist. In addition, if $k = 1$, we obtain the mean of the random variable, which we denote using μ.

Example 5.3.1

Let X be gamma distributed with the parameters α and β. Compute the rth moment of this gamma density.

Solution: Here,

$$\mu_r = E(X^r) = \int_{-\infty}^{\infty} x^r f(x) dx$$

$$= \frac{1}{\Gamma(\alpha)\beta^\alpha} \int_{-\infty}^{\infty} x^r x^{\alpha-1} e^{-x/\beta} dx$$

$$= \frac{1}{\Gamma(\alpha)\beta^\alpha} \int_{-\infty}^{\infty} x^{\alpha+r-1} e^{-x/\beta} dx$$

$$= \frac{\Gamma(\alpha+r)\beta^{\alpha+r}}{\Gamma(\alpha)\beta^\alpha} = \frac{\Gamma(\alpha+r)\beta^r}{\Gamma(\alpha)}.$$

When $r=1$, we obtain the mean of the gamma distribution $E(X) = \alpha\beta$.

Theorem 5.3.1

Let X have $f(x)$ for its probability density function. If a is some constant, then

$$E\left[(aX)^k\right] = a^k E(X^k).$$

Proof If X is continuous, then

$$E\left[(aX^k)\right] = \int_{-\infty}^{\infty} a^k x^k f(x) dx$$

$$= a^k \int_{-\infty}^{\infty} x^k f(x) dx$$

$$= a^k E(X^k).$$

Definition 5.3.2

Let X be a random variable with the probability density $f(x)$. The **kth moment of $f(x)$, with respect to any point b**, is given by

$$E\left[(X-b)^k\right] = \sum_{i=1}^{\infty} (x_i - b)^k f(x_i) \tag{5.3.3}$$

if X is discrete and by

$$E\left[(X-b)^k\right] = \int_{-\infty}^{\infty} (x-b)^k f(x) dx \tag{5.3.4}$$

if X is continuous.

Example 5.3.2

Let X be uniformly distributed over the interval (α, β). Compute the rth moment of this uniform distribution with respect to any point b.

Solution: Applying Equation (5.3.4), we have

$$E\left[(X-b)^r\right] = \int_{-\infty}^{\infty}(x-b)^r f(x)dx$$

$$= \frac{1}{\beta-\alpha}\int_{\alpha}^{\beta}(x-b)^r\, dx$$

$$= \frac{1}{\beta-\alpha}\left[\frac{(\beta-b)^{r+1} - (\alpha-b)^{r+1}}{r+1}\right]$$

$$= \frac{1}{r+1}\sum_{i=0}^{r}(\alpha-b)^i(\beta-b)^{r-i}.$$

If $b=0$ and $r=1$, we have

$$E(X) = \frac{\beta+\alpha}{2},$$

which is the mean of the uniform distribution over the interval (α, β), as was obtained in Example 5.1.4.

Definition 5.3.3

Let X be a random variable with the probability density function $f(x)$. The kth moment of $f(x)$, with respect to $E(X) = \mu$, n_k, is called the **kth central moment** of X. Thus, we have

$$\eta_k = E\left[X - E(X)\right]^k = \sum_{i=1}^{\infty}(x_i - \mu)^k f(x_i) \tag{5.3.5}$$

if X is discrete and

$$\eta_k = E\left[X - E(X)\right]^k = \int_{-\infty}^{\infty}(x-\mu)^k f(x)dx \tag{5.3.6}$$

if X is continuous.

If $k=1$, then $n_1 = E[X - \mu] = 0$; that is,

$$\eta_1 = E[X-\mu]^k = \int_{-\infty}^{\infty}(x-\mu)f(x)dx$$

$$= \int_{-\infty}^{\infty}x f(x)dx - \mu\int_{-\infty}^{\infty}f(x)dx$$

$$= \mu - \mu = 0.$$

◀

Example 5.3.3

Compute the second central moment of the Poisson distribution with parameter λ.

Solution: Because $E(X) = \lambda$,

$$E\left[X - E(X)\right]^2 = \sum_{x=0}^{\infty} (x-\lambda)^2 \frac{e^{-\lambda}\lambda^x}{x!}$$

$$= \sum_{x=0}^{\infty} \frac{x^2 e^{-\lambda}\lambda^x}{x!} - \sum_{x=0}^{\infty} \frac{2\lambda x e^{-\lambda}\lambda^x}{x!} + \sum_{x=0}^{\infty} \frac{\lambda^2 e^{-\lambda}\lambda^x}{x!}$$

$$= \sum_{x=0}^{\infty} \frac{[x(x-1)+x]}{x(x-1)!} e^{-\lambda}\lambda x - 2\lambda \sum_{x=0}^{\infty} \frac{x e^{-\lambda}\lambda^x}{x(x-1)!} + \lambda^2 \sum_{x=0}^{\infty} \frac{e^{-\lambda}\lambda^x}{x!}$$

$$= \lambda^2 \sum_{x=2}^{\infty} \frac{e^{-\lambda}\lambda^{x-2}}{(x-2)!} + \lambda \sum_{x=1}^{\infty} \frac{e^{-\lambda}\lambda^{x-1}}{(x-1)!} - 2\lambda^2 \sum_{x=1}^{\infty} \frac{e^{-\lambda}\lambda^{x-1}}{(x-1)!} + \lambda^2$$

$$= \lambda^2 + \lambda - 2\lambda^2 + \lambda^2 = \lambda .$$

Thus, the second central moment of the Poisson distribution is equal to its mean.

Example 5.3.4

Express the second and third central moments of the distribution of a random variable X in terms of ordinary moments.

Solution: For the second central moment, we have

$$\eta_2 = E\left[X - E(X)\right]^2 = E\left[X - \mu\right]^2$$

$$= E\left[X^2\right] - 2\mu E\left[X\right] + \mu^2$$

$$= \mu_2 - \mu^2 .$$

Also,

$$\eta_3 = E\left[X - E(X)\right]^3 + E\left[X - \mu\right]^3$$

$$= E\left[X^3\right] - 3\mu E\left[X^2\right] + 3\mu^2 E\left[X\right] - \mu^3$$

$$= \mu_3 - 3\mu\mu_2 + 2\mu^3 .$$

Thus, from the preceding example, we can conclude that central moments of a random variable can be expressed in terms of ordinary moments.

Definition 5.3.4

The second central moment η_2 of a random variable X is the variance of X and is denoted by σ^2.

Thus, if X has a discrete distribution,

$$\operatorname{Var}(X) = \sigma^2 = E\left[(X-\mu)^2\right] = \sum_{i=1}^{\infty}(x_i - \mu)^2 f(x_i). \tag{5.3.7}$$

If X is a continuous random variable,

$$\sigma^2 = E\left[(X-\mu)^2\right] = \int_{-\infty}^{\infty}(x_i - \mu)^2 f(x)dx. \tag{5.3.8}$$

The variance σ^2 is a measure of the dispersion of the random variable around its expected value. The smaller the value of σ^2, the more concentrated the distribution around the mean.

Definition 5.3.5

The square root of the variance of the distribution of a random variable is the standard deviation.

To compute the variance, Theorem 5.3.2 is useful.

Theorem 5.3.2

If X is a random variable with the probability density $f(X)$, then

$$\sigma^2 = E\left[X - E(X)\right]^2 = E\left[X^2\right] - \left[E(X)\right]^2. \tag{5.3.9}$$

Proof If X is continuous, then

$$\sigma^2 = \int_{-\infty}^{\infty}(x-\mu)^2 f(x)dx$$

$$= \int_{-\infty}^{\infty} x^2 f(x) - 2\mu \int_{-\infty}^{\infty} x f(x)dx + \mu^2$$

$$= E\left[X^2\right] - \mu^2.$$

Example 5.3.5

Compute the variance of the binomial distribution with the parameters n and p.
We saw in Example 5.1.2 that $E(X)=np$. Next, we compute

$$E(X^2) = \sum_{x=0}^{\infty} x^2 \binom{n}{x} p^x (1-p)^{n-x}$$

$$= \sum_{x=0}^{\infty} [x(x-1)+x] \frac{n(n-1)!}{(n-x)!x(x-1)!} p^x (1-p)^{n-x}$$

$$= n(n-1)p^2 \sum_{x=2}^{n} \binom{n-2}{x-2} p^{x-2} (1-p)^{n-x} + np \sum_{x=1}^{n} \binom{n-1}{x-1} p^{x-1} (1-p)^{n-x}$$

$$= n(n-1)p^2 + np.$$

Thus, applying Equation (5.3.9), we obtain

$$\sigma^2 = E[X^2] - [E(X)]^2$$

$$= n(n-1)p^2 + np - n^2 p^2$$

$$= np(1-p).$$

Theorem 5.3.3

If a functional form of a random variable X is given by $Y = \alpha X + \beta$, then
$$\text{Var}(Y) = \alpha^2 \text{Var}(X).$$

Proof Applying Definition 5.3.5, we have

$$\text{Var}(Y) = \text{Var}(\alpha X + \beta) = E\left[(\alpha X + \beta) - E(\alpha X + \beta)\right]^2$$

$$= E\left[\alpha X + \beta - \alpha E(X) - \beta\right]^2$$

$$= E\left[\alpha X - \alpha E(X)\right]^2$$

$$= E\left\{\alpha[X - E(X)]\right\}^2 = E\left\{\alpha^2 [X - E(X)]^2\right\}$$

$$= \alpha^2 E\left[X - E(X)\right]^2$$

$$= \alpha^2 \text{Var}(X).$$

Theorem 5.3.4

For every constant $c \neq \mu$, the variance of a random variable X satisfies the following inequality:

$$\mathrm{Var}(X) \leq E\left[(X-c)^2\right]. \tag{5.3.10}$$

Proof The quantity $E\left[(X-c)^2\right]$ can be written as follows:

$$E\left[(X-c)^2\right] = E\left\{\left[(X-\mu)+(\mu-c)\right]^2\right\}$$

$$= E\left[(X-\mu)^2\right] + 2(\mu-c)E(X-\mu) + (\mu-c)^2$$

$$= \mathrm{Var}(X) + (\mu-c)^2.$$

Because $(\mu-c)^2 > 0$, from the hypothesis of the theorem, we obtain the inequality in Equation (5.3.10).

Definition 5.3.6

A random variable X is called a **standardized random variable** if

$$E(X) = 0$$

and

$$\mathrm{Var}(X) = 1.$$

Thus, if the expected value of a random variable X is μ and its standard deviation is σ, then the random variable defined by

$$Z = \frac{X-\mu}{\sigma}$$

is a standardized random variable; that is,

$$E(Z) = E\left(\frac{X-\mu}{\sigma}\right) = \frac{1}{\sigma}\left[E(X) - \mu\right] = 0.$$

Applying Theorem 5.3.3 to the variate Z, we have

$$\mathrm{Var}(Z) \leq \frac{1}{\sigma^2} \mathrm{Var}(X) = 1.$$

Definition 5.3.7

The ratio of the standard deviation of a random variable X to its expected value,

$$V = \frac{\sqrt{E[X-\mu]^2}}{E[X]} = \frac{\sigma}{\mu},$$

is called the **coefficient of variation**.

The coefficient of variation, V, expresses the magnitude of variation of a random variable relative to its expected value. Thus, if the mean of the variate is one, then V equals the standard deviation, which is a measure of the dispersion of the variate around one.

Example 5.3.6

Compute the coefficient of variation of the random variable X, which is binomially distributed with the parameters n and p.

Solution: We saw in Example 5.1.2 that $E(X) = np$ and in Example 5.3.5 that $\text{Var}(X) = np(1-p)$. Therefore,

$$V = \frac{\sqrt{np(1-p)}}{np} = \sqrt{\frac{1-p}{np}}.$$

If the probability density function of a random variable is symmetrical, then its expected value, if it exists, equals the center of symmetry. Furthermore, the odd central moments of a symmetrical distribution, if they exist, are equal to zero. In certain cases, we are interested in establishing the extent to which such a distribution departs from symmetry. Such a measure of asymmetry is given by the following definition.

Definition 5.3.8

The ratio of the third central moment of the distribution of a random variable X to the cube of its standard deviation,

$$\gamma = \frac{E\left[(X - E(X))^3\right]}{\left(\sqrt{E\left[(X - E(X))^2\right]}\right)^3} = \frac{\eta_3}{\sigma^3},$$

is called the **coefficient of skewness**.

Thus, if $\gamma = 0$, the distribution is symmetrical; if $\gamma > 0$, we speak of positive (skewness) asymmetry; and if $\gamma < 0$, we refer to negative asymmetry.

Example 5.3.7

Compute the coefficient of skewness for the distribution of the random variable X if it is Poisson distributed.

Solution: We have seen that if X is Poisson distributed, then $E(X) = \text{Var}(X) = \lambda$. Thus,

$$\gamma = \frac{E\left[(X - E(X))^3\right]}{\left[\sqrt{E[X - E(X)]^2}\right]^3} = \frac{E\left[(X - \lambda)^3\right]}{\left(\sqrt{\lambda}\right)^3}$$

and
$$E\left[(X-\lambda)^3\right] = E\left[X^3 - 3X^2\lambda + 3\lambda^2 X - \lambda^3\right]$$
$$= E\left[X^3\right] - 3\lambda E\left[X^2\right] + 3\lambda^2 E\left[X\right] - \lambda^3.$$

Here,
$$E\left[X^3\right] = \lambda^3 + 3\lambda^2 + \lambda,$$
$$E\left[X^2\right] = \lambda(\lambda+1)$$

and
$$E\left[X\right] = \lambda.$$

Thus,
$$E\left[(X-\lambda)^3\right] = \lambda^3 + 3\lambda^2 + \lambda - 3\lambda^3 - 3\lambda^2 + 3\lambda^3 - \lambda^3$$
$$= \lambda.$$

Therefore,
$$\gamma = \frac{\lambda}{\lambda^{3/2}} = \frac{1}{\sqrt{\lambda}}.$$
◀

We conclude this section by briefly introducing several additional terms that describe the peakedness of a distribution. The degree of peakedness of a distribution is referred to as the *kurtosis* of the distribution. A distribution with a relatively high peak is called *leptokurtic,* a flat-topped distribution is called *platykurtic,* and a distribution that is not very peaked or very flat topped is called *mesokurtic.* Figure 5.3.1 illustrates these types of kurtosis.

A usual measure of kurtosis is expressed by a dimensionless quantity given by the ratio of the fourth central moment to the second central moment squared, minus three:
$$\xi = \frac{\eta_4}{\eta_2^2} - 3.$$

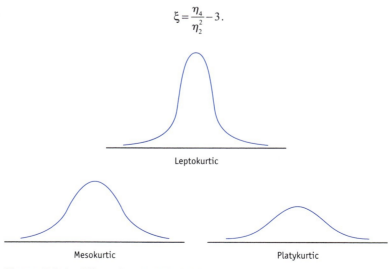

Figure 5.3.1 Different Distributions' Shapes

Distributions for which $\xi=0$ are mesokurtic. Those for which $\xi>0$ and $\xi<0$ are leptokurtic and platykurtic, respectively. For the normal distribution with μ and σ^2, ξ is zero.

Example 5.3.8

The kurtosis of the binomial distribution is

$$\xi = \frac{1-6pq}{npq}.$$

As $n \to \infty$, $\xi \to 0$, which is in agreement with the binomial distribution tending to the normal distribution as $n \to \infty$. This result is shown in a later chapter.

5.4 Moment-Generating Function

In the theory and application of probability, moments of probability density functions play an important role. In the preceding section, we saw by working various examples that the manner of computing the moments of various probability distributions can be extremely tedious; it would be useful to obtain an expression that would generate for us all the moments, if they exist, of a density function. Such an expression is called a moment-generating function, and we define it as follows.

Definition 5.4.1

Let X be a random variable with the probability density $f(x)$. The **moment-generating function (MGF)** of $f(x)$ is the expected value of e^{tX}, if it exists, for every value of t in some interval $|t| \leq T$, $T > 0$. We denote the MGF by

$$m_X(t) = E(e^{tX}) = \sum_{i=1}^{\infty} e^{tx_i} f(x_i) \qquad (5.4.1)$$

if the random variable is discrete and by

$$m_X(t) = E(e^{tX}) = \int_{-\infty}^{\infty} e^{tx} f(x) dx \qquad (5.4.2)$$

if the random variable is continuous. Thus, if the random variable X is discrete, its MGF is defined as the infinite series in Equation (5.4.1); if it is continuous, its MGF is defined as the improper integral in Equation (5.4.2). Such a series or integral may not always converge to a finite value for all values of t. Hence, the MGF may not be defined for all values of t. However, we say that a probability distribution possesses an MGF if a positive number T exists such that $m_X(t)$ is finite in the interval $-T \leq t \leq T$.

For illustrations of the manner in which the MGF is evaluated, see Examples 5.4.1 through 5.4.5.

Example 5.4.1

Let X be binomially distributed with parameters p and n. Applying Equation (5.4.1).

Solution

$$m_X(t) = E(e^{tX}) = \sum_{x=0}^{n} e^{tx} \binom{n}{x} p^x (1-p)^{n-x}$$

$$= \sum_{x=0}^{n} \binom{n}{x} (pe^t)^x (1-p)^{n-x}$$

$$= [pe^t + (1-p)]^n.$$

◀

Example 5.4.2

Let X be Poisson distributed with parameters λ. Find the MGF.

Solution: The MGF is

$$m_X(r) = E(e^{tX})$$

$$= \sum_{x=0}^{\infty} e^{tx} \frac{e^{-\lambda} \lambda^x}{x!}$$

$$= e^{-\lambda} \sum_{x=0}^{\infty} \frac{(\lambda e^t)^x}{x!}.$$

Recall from the Maclaurin series expansion of e^x that

$$e^{\lambda e^t} = 1 + \lambda e^t + \frac{\lambda^2 e^{2t}}{2!} + \ldots + \frac{\lambda^n e^{nt}}{n!} + \ldots$$

$$= \sum_{x=0}^{\infty} \frac{(\lambda e^t)^x}{x!}.$$

Thus,

$$m_X(t) = e^{-\lambda} e^{\lambda e^t} = e^{\lambda(e^t - 1)}.$$

◀

Example 5.4.3

Suppose that the random variable X is uniformly distributed over the interval (α, β). Find the MGF.

Solution: Applying Equation (5.4.2), we obtain the following MGF:

$$m_X(t) = E(e^{tX}) = \frac{1}{\beta - \alpha} \int_{\alpha}^{\beta} e^{tx} dx$$

$$= \frac{1}{t(\beta - \alpha)} (e^{\beta t} - e^{\alpha t}).$$

◀

Example 5.4.4

Find the MGF of the random variable X, which is gamma distributed with the parameters α and β.

Solution: Here,

$$m_X(t) = E(e^{tX}) = \int_{-\infty}^{\infty} e^{tx} f(x) dx$$

$$= \frac{1}{\Gamma(\alpha)\beta^\alpha} \int_0^\infty e^{tx} x^{\alpha-1} e^{-x/\beta} dx$$

$$= \frac{1}{\Gamma(\alpha)\beta^\alpha} \int_0^\infty x^{\alpha-1} e^{-x/\beta + tx} dx$$

$$= \frac{\Gamma(\alpha)\left(\frac{\beta}{1-t\beta}\right)^\alpha}{\Gamma(\alpha)\beta^\alpha}$$

$$= \left(\frac{1}{1-t\beta}\right)^\alpha = (1-t\beta)^{-\alpha},$$

provided that $t < 1/\beta$, because the expression in the brackets represents the area under a gamma density function, with the parameters α and $\beta/(1-t\beta)$, and is by definition equal to one if $t=0$.

Example 5.4.5

Let the random variable X be normally distributed with the parameters μ and σ. Find the MGF.

Solution: The MGF of $f(x)$ is obtained as follows:

$$m_X(t) = E(e^{tX}) = \frac{1}{\sqrt{2\pi}\sigma} \int_{-\infty}^{\infty} e^{tx} \exp\left\{-\frac{1}{2}\left(\frac{x-\mu}{\sigma}\right)^2\right\} dx$$

$$= \frac{e^{t\mu}}{\sqrt{2\pi}\sigma} \int_{-\infty}^{\infty} e^{t(x-\mu)} e^{-(1/2\sigma^2)(x-\mu)^2} dx$$

$$= \frac{e^{t\mu}}{\sqrt{2\pi}\sigma} \int_{-\infty}^{\infty} e^{-(1/2\sigma^2)\left[(x-\mu)^2 - 2\sigma^2 t(x-\mu)\right]} dx.$$

Completing the square inside the bracket of the exponent, we have

$$\left[(x-\mu)^2 - 2\sigma^2 t(x-\mu)\right] = (x-\mu)^2 - 2\sigma^2 t(x-\mu) + \sigma^4 t^2 - \sigma^4 t^2$$

$$= (x-\mu-\sigma^4 t)^2 - \sigma^4 t^2.$$

Thus,

$$m_X(t) = e^{t\mu+(\sigma^2 t^2/2)} \frac{1}{\sqrt{2\pi}\sigma} \int_{-\infty}^{\infty} \exp\left\{-\frac{1}{2}\left(\frac{x-\mu-\sigma^2 t}{\sigma}\right)^2\right\} dx.$$

Let

$$y = \frac{x-\mu-\sigma^2 t}{\sigma}, \quad \sigma dy = dx,$$

and

$$m_X(t) = e^{t\mu+(\sigma^2 t^2/2)} \left[\frac{1}{\sqrt{2\pi}} \int_{-\infty}^{\infty} e^{-(1/2)y^2} dy\right].$$

Therefore, because the integral in the bracket equals one, we have

$$m_X(t) = e^{t\mu+(\sigma^2 t^2/2)}.$$

If the variate X is a standard normal distribution—that is, $\mu=0$ and $\sigma^2=1$—then its MGF is simply

$$m_X(t) = e^{t^2/2}.$$

◀

5.4.1 Properties of the Moment-Generating Function
The Maclaurin series expansion of e^x is

$$e^x = 1 + x + \frac{x^2}{2!} + \ldots + \frac{x_n}{n!} + \ldots,$$

which converges for all values of x. If the MGF $m_X(t)$ is finite for $|t| \leq T, T > 0$, it possesses a power-series expansion, which is valid for $-T \leq t \leq T$. Thus,

$$m_X(t) = E(e^{tX})$$

$$= E\left(1 + tX + \frac{t^2 X^2}{2!} + \ldots + \frac{t^n X^n}{n!} + \ldots\right).$$

Taking the expected value of the series by treating t as a constant, we have

$$m_X(t) = 1 + tE(X) + \frac{t^2}{2!}E(X^2) + \ldots + \frac{t^n}{n!}E(X^n) + \ldots$$

$$= 1 + t\mu + \frac{t^2}{2!}\mu_2 + \ldots + \frac{t^n}{n!}\mu_n + \ldots.$$

Differentiating the MGF with respect to t and letting $t=0$, we obtain the first moment. That is,

$$\left.\frac{dm_X(t)}{dt}\right|_{t=0} = \mu + \frac{2t^2}{2!}\mu_2 + \ldots + \frac{nt^{n-1}}{n!}\mu_n + \ldots \bigg|_{t=0}$$

$$= \mu = E(X).$$

Similarly,

$$\left.\frac{d^2 m_X(t)}{dt^2}\right|_{t=0} = \mu_2 + t\mu_{3+} + \ldots \left.\frac{n(n-1)t^{n-2}}{n!}\mu_n + \ldots\right|_{t=0}$$

$$= \mu_2 = E(X^2).$$

$$\left.\frac{d^n m_X(t)}{dt^n} = \mu_n + t\mu_{n+1} + \ldots\right|_{t=0} = \mu_n = E(X^n).$$

Thus, we have obtained Theorem 5.4.1.

■ **Theorem 5.4.1**

If the MGF of a random variable exists for $|t| \leq T, T > 0$, then

$$\left.\frac{d^n m_X(t)}{dt^n}\right|_{t=0} = E(X^n) = \mu_n.$$

We can also develop the theorem as follows: Let X be a continuous variate with the probability density function $f(x)$. Then, differentiating its MGF with respect to t, we have

$$\frac{dm_X(t)}{dt} = \int_{-\infty}^{\infty} \frac{d}{dt}\{e^{tx}\} f(x) dx$$

$$= \int_{-\infty}^{\infty} x e^{tx} f(x) dx = E(X e^{tX})$$

and

$$\left.\frac{dm_X(t)}{dt}\right|_{t=0} = E(X) = \mu.$$

Similarly,

$$\frac{d^n m_X(t)}{dt^n} = \int_{-\infty}^{\infty} \frac{d^n}{dt^n}(e^{tx}) f(x) dx$$

$$= \int_{-\infty}^{\infty} x^n e^{tx} f(x) dx = E(X^n e^{tX})$$

and

$$\left.\frac{d^n m_X(t)}{dt^n}\right|_{t=0} = E(X^n) = \mu_n.$$

Thus, we see that the preceding function, $m_X(t)$, if it exists, generates all ordinary moments of a given probability distribution.

Example 5.4.6

We saw in Example 5.4.1 that the MGF of a binomial distribution, with the parameters n and p, is given by

$$m_X(t) = \left[pe^t + (1-p)\right]^n.$$

Find the mean and variance of X.

Solution: Applying Theorem 5.3.1, we obtain the following:

$$E(X) = \left.\frac{dm_X(t)}{dt}\right|_{t=0} = n\left[pe^t + (1-p)\right]^{n-1}(pe^t)\Big|_{t=0} = np$$

and

$$E(X^2) = \left.\frac{d^2 m_X(t)}{dt^2}\right|_{t=0} = n(n-1)\left[pe^t + (1-p)\right]^{n-2} p^2 e^{2t}$$

$$+ n\left[pe^t + (1-p)\right]^{n-1} pe^t \Big|_{t=0} = n(n-1)p^2 + np.$$

Thus,

$$\text{Var}(X) = E(X^2) - \left[E(X)\right]^2$$

$$= n(n-1)p^2 + np - n^2 p^2 = np(1-p).$$

◀

Example 5.4.7

Using Theorem 5.3.1, show that the mean and variance of the Poisson distribution, with parameter λ, is equal to λ.

Solution: In Example 5.4.2, we saw that

$$m_X(t) = e^{\lambda(e^t - 1)}.$$

The mean is obtained by

$$E(X) = \left.\frac{dm_X(t)}{dt}\right|_{t=0} = e^{\lambda(e^t-1)} \lambda e^t \Big|_{t=0} = \lambda.$$

Also,

$$E(X^2) = \left.\frac{d^2 m_X(t)}{dt^2}\right|_{t=0} = e^{\lambda(e^t-1)} \lambda^2 e^{2t} + e^{\lambda(e^t-1)} \lambda e^t \Big|_{t=0} = \lambda^2 + \lambda$$

and

$$\text{Var}(X) = E(X^2) - \left[E(X)\right]^2 = \lambda^2 + \lambda - \lambda^2 = \lambda.$$

◀

Example 5.4.8

Suppose that the variate X is normally distributed with the parameters μ and σ. Find the mean and variance of X.

Solution: In Example 5.4.5, the MGF of X is

$$m_X(t) = e^{t\mu + (\sigma^2 t^2/2)}.$$

Thus,

$$E(X) = \frac{dm_X(t)}{dt}\bigg|_{t=0} = e^{t\mu + (\sigma^2 t^2/2)}(\mu + \sigma^2 t)\bigg|_{t=0} = \mu,$$

$$E(X^2) = \frac{d^2 m_X(t)}{dt^2}\bigg|_{t=0} = e^{t\mu + (\sigma^2 t^2/2)}(\mu + \sigma^2 t)^2 + e^{t\mu + (\sigma^2 t^2/2)}\sigma^2\bigg|_{t=0}$$

$$= \mu^2 + \sigma^2,$$

and

$$\text{Var}(X) = E(X^2) - [E(X)]^2 = \mu^2 + \sigma^2 - \mu^2 = \sigma^2.$$

Theorem 5.4.2

Let the random variable X have the MGF $m_X(t)$. If $Y = \alpha X + \beta$, then the MGF of the variate Y is given by

$$m_Y(t) = e^{\beta t} m_X(\alpha t).$$

Proof Here,

$$m_Y(t) = E(e^{Yt}) = E\left(e^{t(\alpha X + \beta)}\right)$$

$$= E\left(e^{\alpha t X + \beta t}\right) = e^{\beta t} E\left(e^{\alpha t X}\right)$$

$$= e^{\beta t} m_X(\alpha t).$$

We have seen that $m(t)$, if it exists, generates the moments of a probability distribution. In addition, the MGF answers the important question: If we are given a set of moments, what is the probability density function from which these moments came? In relation to this question, we state Theorem 5.4.3.

Theorem 5.4.3

Let X and Y be two random variables with the probability densities $f(x)$ and $g(y)$, respectively. If $m_X(t)$ and $m_Y(t)$ exist and are equal for all t in the interval $|t| \leq T$, $T > 0$, then X and Y have the same probability distributions; that is, $f(x) = g(y)$.

The proof of this theorem is beyond the scope of this book, but we illustrate the meaning of the theorem with Example 5.4.9.

Example 5.4.9

Suppose that the random variable X has the MGF

$$m_X(t) = \frac{\alpha}{\alpha - t}, \quad t < \frac{1}{\alpha}.$$

Let the random variable Y have the following function for its probability density:

$$g(y) = \begin{cases} \alpha e^{-\alpha y}, & y > 0, \alpha > 0, \\ 0, & \text{elsewhere.} \end{cases}$$

Can we obtain the probability density of the variate X with the preceding information?

Solution: The MGF of Y is

$$m_Y(t) = E(e^{tY}) = \int_0^\infty \alpha e^{ty} e^{-\alpha y} dy$$

$$= \alpha \int_0^\infty e^{-y(\alpha - t)} dy$$

$$= \frac{-\alpha}{\alpha - t} e^{-y(\alpha - t)} \Big|_0^\infty$$

$$= \frac{\alpha}{\alpha - t}, \quad t < \frac{1}{\alpha}.$$

Thus, applying Theorem 5.3.3, we have

$$m_X(t) = \frac{\alpha}{\alpha - t} = m_Y(t), \quad t < \frac{1}{\alpha},$$

Therefore,

$$f(x) \equiv g(y) = \begin{cases} \alpha e^{-\alpha x}, & \alpha > 0, x > 0, \\ 0, & \text{elsewhere.} \end{cases}$$

The theorem says that if two variables have the same MGF, they have the same probability density function. The MGF uniquely determines the probability distribution of the random variable.

An MGF that sometimes simplifies the problem of find the moments of discrete probability distributions is the *factorial moment-generating function*, given by

$$g_x(t) = E\left[(1+t)^X\right] = E\left[e^{X \log(1+t)}\right], \quad |t| < 1.$$

The nth derivative, evaluated at $t=0$, is

$$\left.\frac{d^n g_X(t)}{dt^n}\right|_{t=0} = E\left[X(X-1)(X-2)\ldots(X-n+1)\right]$$

This is the nth factorial moment of the probability distribution of X. Knowing the first n factorial moments of a probability distribution, we can obtain the first n moments of the distribution.

For example,

$$\left.\frac{dg_X(t)}{dt}\right|_{t=0} = \left.E\left[X(1+t)^{X-1}\right]\right|_{t=0} = E[X],$$

$$\left.\frac{d^2 g_X(t)}{dt^2}\right|_{t=0} = \left.E\left[X(X-1)(1+t)^{X-1}\right]\right|_{t=0} = E[X(X-1)],$$

and

$$E[X(X-1)] = E(X^2) - E(X)$$

or

$$E[X^2] = E[X(X-1)] + E[X].$$

We can also obtain the factorial moments if we know the moments of a probability distribution.

Example 5.4.10

Find the first three moments of the Poisson distribution using factorial moments.

Solution:

$$g_X(t) = E\left[(1+t)^X\right] = \sum_{x=0}^{\infty} \frac{(1+t)^x e^{-\lambda} \lambda^x}{x!}$$

$$= e^{-\lambda} \sum_{x=0}^{\infty} \frac{[\lambda(1+t)]^x}{x!}$$

$$= e^{-\lambda} e^{\lambda(1+t)} = e^{\lambda t},$$

$$\left.\frac{dg_X(t)}{dt}\right|_{t=0} = E[X] = \left.\lambda e^{\lambda t}\right|_{t=0} = \lambda,$$

$$\left.\frac{d^2 g_X(t)}{dt^2}\right|_{t=0} = E[X(X-1)] = \left.\lambda^2 e^{\lambda t}\right|_{t=0} = \lambda^2,$$

and

$$\left.\frac{d^3 g_X(t)}{dt^3}\right|_{t=0} = E[X(X-1)(X-2)] = \lambda^3 e^{\lambda t}\bigg|_{t=0} = \lambda^3.$$

Therefore,

$$E[X] = \lambda,$$

$$E[X^2] = E[X] + E[X(X-1)] = \lambda + \lambda^2,$$

and

$$E[X^3] = \lambda^3 + 3E[X^2] - 2E[X] = \lambda^3 + 3\lambda^2 + \lambda.$$

Up to this point, we have been concerned with one-dimensional random variables and their probability distributions. However, in many physical situations, we have to deal with two or more random variables. The next chapter aims to extend the concepts of one random variable and its probability distribution function to the case of two random variables.

Summary

Let X be a random variable with the probability density function $f(x)$. The series

$$E(X) = \sum_i x_i f(x_i)$$

is the *expected value* of the discrete variate X if

$$\sum_i f(x_i)|x_i| < \infty$$

is satisfied. The integral

$$E(X) = \int_{-\infty}^{\infty} x f(x) dx$$

is the *expected value* of the continuous variate X if

$$\int_{-\infty}^{\infty} f|x| dx < \infty$$

is satisfied.

The following properties of expectation were discussed. Let X be a random variable with the probability density $f(x)$.

1. If c is a constant, then $E(c) = c$.
2. If $g(X)$ is a functional form of the variate X, then

$$E[g(X)] = \sum_{i=1}^{\infty} g(x_i) f(x_i)$$

if X is discrete and

$$E[g(X)] = \int_{-\infty}^{\infty} g(x) f(x) dx$$

if X is continuous, provided that

$$\sum_{i=1}^{\infty} |g(x_i)| f(x_i) < \infty$$

and

$$\int_{-\infty}^{\infty} |g(x)| f(x) dx < \infty.$$

3. If $g(X) = \alpha X + \beta$, then

$$E[g(X)] = \alpha E(X) + \beta.$$

4. If $f(x)$ is symmetrical about the point $x = a$ for all x, then $E(X) = a$.

5. If $g(X) = g_1(X) + g_2(X)$, then

$$E[g(X)] = E[g_1(X)] + E[g_2(X)].$$

6. If $g_1(X) \le g_2(X)$ for all possible values that the random variable X can assume, then

$$E[g_1(X)] \le E[g_2(X)]$$

and

$$|E[g_1(X)]| \le E[|g_1(X)|].$$

The k*th ordinary moment* of the distribution of the variate X is

$$\mu_k = E(X^k) = \sum_{i=1}^{\infty} x_i^k f(x_i)$$

if X is discrete and

$$\mu_k = E(X^k) = \int_{-\infty}^{\infty} x^k f(x) dx$$

if X is continuous, provided that the series and the integral are absolutely convergent.

The k*th moment of* $f(x)$, with respect to any point b, is given by

$$E\left[(X-b)^k\right] = \sum_{i=1}^{\infty} (x_i - b)^k f(x_i)$$

if X is discrete and

$$E\left[(X-b)^k\right] = \int_{-\infty}^{\infty} (x-b)^k f(x) dx$$

if X is continuous.

If b is replaced with $E(X) = \mu$, then $E\left[(X-\mu)^k\right]$ is called the *k*th *central moment* of X. The second central moment of X is the *variance* of the variate X, which is a measure of the dispersion of the variable around its expected value. The square root of the variance is called the *standard deviation*.

Let X be a random variable with the probability density $f(x)$. Then

1. $\text{Var}(X) = E(X^2) - [E(X)]^2$
2. $\text{Var}(Y) = \alpha^2 \text{Var}(X)$, where $Y = \alpha X + \beta$
3. $\text{Var}(X) \leq E\left[(X-c)^2\right]$, $c \neq \mu$

The ratio σ/μ, which expresses the magnitude of variation of a variate relative to its expectation, is called the *coefficient of variation*. The *coefficient of skewness* is the ratio of the third central moment of the distribution of a variate to the cube of its standard deviation.

The *moment-generating function* of a probability density $f(x)$, $m_X(t)$, is the expected value of e^{tX}, if it exists, for every value of t in some interval $|t| \leq T, T > 0$. It possesses the following properties:

1. $\left.\dfrac{d^n m_X(t)}{dt^n}\right|_{t=0} = E(X^n) = \mu_n$

2. $M_Y(t) = e^{\beta t} M_X(\alpha t)$, where $Y = \alpha X + \beta$

The *factorial moment-generating function* $g_X(t)$ is given by

$$g_X(t) = E\left[(1+t)^X\right] = E\left[e^{x\log(1+t)}\right], |t| < 1.$$

Theoretical Exercises

5.1 In an experiment, the probability of success is equal to p. Find the expected number of trials required for k successes.

5.2 Find the expected value and variance of the random variable X for the following two probability density functions:

a. $f(x) = \begin{cases} \dfrac{(x-3)^2}{5}, & x = 3, 4, 5, \\ 0, & \text{elsewhere} \end{cases}$

b. $f(x) = \begin{cases} \dfrac{14-3x}{15}, & x = 2, 3, 4, \\ 0, & \text{elsewhere} \end{cases}$

5.3 Let the probability density function of the random variable X be

$$f(x) = \begin{cases} \dfrac{3}{16}(x^2 + 5), & 0 \leq x \leq 1, \\ 0, & \text{elsewhere.} \end{cases}$$

Find
a. the expected value of X
b. the variance of X

5.4 The probability density function of the random variable X is given by

$$f(x) = \begin{cases} 0, & x < -\alpha, \\ \dfrac{\alpha + x}{\alpha^2}, & -\alpha \leq x \leq 0, \\ \dfrac{\alpha - x}{\alpha^2}, & 0 < x \leq \alpha, \\ 0, & \alpha > x. \end{cases}$$

Compute
a. $E(X)$
b. $\text{Var}(X)$

5.5 Suppose that

$$f(x) = \begin{cases} \dfrac{cx}{2}, & 0 \leq x \leq 1, \\ c - \dfrac{cx}{2}, & 1 < x \leq 2, \\ 0, & \text{elsewhere.} \end{cases}$$

a. Find c so that $f(x)$ is a probability density function.
b. Find $E(X)$.
c. Find $\text{Var}(X)$.

5.6 Suppose that

$$f(x) = \begin{cases} c \sin x e^{-x}, & 0 \leq x \leq \pi, \\ 0, & \text{elsewhere.} \end{cases}$$

a. Find c so that $f(x)$ is a probability density function.
b. Compute $E(X)$.
c. Compute $\text{Var}(X)$.

5.7 The probability of a success in an experiment is p. Show that the expected number of trials for two successes in a row is $(p+1)/p^2$.

5.8 The probability density of the random variable X is given by

$$f(x) = \begin{cases} \dfrac{x^2}{2}, & 0 < x \leq 1, \\ \dfrac{6x - 2x^2 - 3}{2}, & 1 < x \leq 2, \\ \dfrac{(x-3)^2}{2}, & 2 < x \leq 3, \\ 0, & \text{elsewhere.} \end{cases}$$

Find the expected value of the random variable X.

5.9 Let

$$f(x) = ce^{-|x|}, \quad -\infty < x < \infty.$$

a. Find c so that $f(x)$ is the probability density function of the random variable X.
b. Evaluate $E(X)$.
c. Evaluate $\text{Var}(X)$.

5.10 In Exercise 5.1, find the mean value of the square of the number of trials required for k successes and the variance of the number of trials required for k successes.

5.11 If

$$f(x) = \begin{cases} cx(2x - 1), & x = 1, 2, 3, 4, 5, \\ 0, & \text{elsewhere,} \end{cases}$$

Find
a. c so that $f(x)$ is the probability density function of the random variable X
b. $E(X)$
c. $E(X^2)$
d. $E(1/2 X^2 - 3)$
e. $E\{X - E(X)\}^2$
f. $E(X^2) - \mu^2$
g. Show that the result in Exercise 5.11(f) is the same as the result in Exercise 5.11(e)

5.12 Show that the expected value of the random variable X whose probability density is given by

$$f(x) = \frac{1}{\pi(x^2 + 1)}, \quad -\infty < x < \infty,$$

does not exist.

5.13 Let

$$f(x) = \frac{c}{x^2}, \quad x = \pm 1, \pm 2, \pm 3, \ldots.$$

a. Find c so that $f(x)$ is a probability density.
b. Show that the expected value of the random variable X does not exist, that is, that

$$E(X) = 0$$

but
$$E(|X|) = \infty.$$

5.14 Show that
$$\{E(X-c)\}^2 \leq E(X-c)^2,$$
where c is an arbitrary constant.

5.15 The probability density function of the discrete random variable X is given by
$$f(x) = \begin{cases} \frac{1}{6}(3x^2 - 5x + 2), & 0, 1, 2, \\ 0, & \text{elsewhere.} \end{cases}$$

Find
a. μ b. η_2 c. γ d. V e. ξ

5.16 The probability density function of the random variable X is given by
$$f(x) = \begin{cases} \frac{1}{2\sqrt{x}}, & 0 < x < 1, \\ 0, & \text{elsewhere.} \end{cases}$$

Find
a. μ b. σ^2 c. V d. γ

5.17 The probability density function of the random variable X is given by
$$f(x) = \begin{cases} \frac{1}{\beta - \alpha}, & \alpha \leq x \leq \beta, \\ 0, & \text{elsewhere.} \end{cases}$$

Find
a. μ b. η_2 c. σ^2 d. γ e. ξ

5.18 In Exercise 5.3, find
a. μ_2 b. σ^2 c. η d. γ e. ξ

5.19 Show that
a. $m_{X+a}(t) = e^{at} m_X(t)$
b. $m_{bX}(t) = e^{at} m_X(bt)$

5.20 The probability density function of the random variable X is given by
$$f(x) = \begin{cases} \dfrac{\Gamma[(n_1 + n_2)/2](n_1/n_2)^{n_1/2}}{\Gamma(n_1/2)\Gamma(n_2/2)} x^{n_1/2 - 1} (1 + (n_1/n_2)x)^{-(n_1+n_2)/2}, & x > 0, \\ 0, & \text{elsewhere.} \end{cases}$$

Show that

a. $E(X) = n_2/(n_2 - 2)$, for $n_2 > 2$

b. $\text{Var}(X) = \dfrac{2n_2^2(n_1 + n_2 - 2)}{n_1(n_2 - 2)^2(n_2 - 4)}$

Hint: Use the transformation $y = \dfrac{(n_1/n_2)x}{1 + (n_1/n_2)x}$.

5.21 The probability density function of the random variable X is given by

$$f(x) = \frac{\Gamma[(n+1)/2]}{\sqrt{n\pi}\,\Gamma(n/2)}\left(1 + \frac{x^2}{n}\right)^{-(n+1)/2}, \quad -\infty < x < \infty.$$

Show that

a. $E(X) = 0$
b. $\text{Var}(X) = n/(n-2)$, for $n > 2$

Hint: To integrate $K\int_{-\infty}^{\infty} x^2\left(1 + \dfrac{x^2}{n}\right)^{-(n+1)/2} dx$, use integration by parts, where $u = (n/2)Kx$

and $dv = \left(\dfrac{2x}{n}\right)\left(1 + \dfrac{x^2}{n}\right)^{-(n+1)/2}$.

5.22 Show that the kth central moment can be expressed as

$$\eta_k = \sum_{i=0}^{k}(-1)^i\binom{k}{i}\mu^i \mu_{k-i}.$$

5.23 Derive an expression similar to that in Exercise 5.22 that expresses ordinary moments in terms of moments about the mean.

Applied Problems

5.1 Find the average number of spots showing on a throw of k fair dice, where $k = 1, 2, 3$.

5.2 Suppose that five cards are dealt from a deck containing only the aces, kings, and queens. Find the expected number of aces dealt.

5.3 Let the random variable X assume the number of spots that come up on a single roll of a die. Find the expected value and variance of

a. X **b.** $Y = 2X - 6$ **c.** $Z = 6 - X$ **d.** $W = X + Z$

5.4 Let the random variable X be normally distributed with the parameters 0 and σ^2. Show that $E(X^{2k+1}) = 0$, where $k = 0, 1, 2, \ldots$.

5.5 Show that in the game of craps the expected number of throws required for the thrower to win is $13{,}144/9075$, which is approximately equal to 1.448 throws.

5.6 In Problem 5.1,

a. Find the average value of the square of the sum of the spots showing.
b. Compute the variance of the number of spots showing.

5.7 In Problem 5.2, find
  a. the expected value of the square of the number of aces dealt
  b. the variance of the number of aces dealt

5.8 Let X be normally distributed with the parameters 0 and σ^2. Find
  a. $E(X^2)$ **b.** $E(\alpha X^2 + \beta)$ **c.** $\text{Var}(\alpha X - \beta)$

5.9 Calculate the mean and variance of the
  a. geometric distribution
  b. negative binomial distribution

5.10 Find the average value of the difference of the number of spots showing on a pair of fair dice if they are tossed repeatedly.

5.11 In Exercises 5.2 and 5.6 and Problem 5.4, find
  a. $E(3X^2 + 2X + 1)$ **b.** $\text{Var}(X - 2)$

5.12 Find the expected value of $g(X) = 3X^2 + 2X + 1$ if X is characterized by the
  a. binomial distribution
  b. Poisson distribution
  c. negative binomial distribution
  d. geometric distribution
  e. hypergeometric distribution
  f. gamma distribution with the parameters α and β
  g. normal distribution with the parameters μ and σ^2

5.13 Find the expected value of $g(X) = X^k$ for the
  a. gamma distribution
  b. beta distribution
  c. lognormal distribution.

5.14 Use the results of Problem 5.13 to find the mean and variance of these probability density functions.

5.15 In Exercise 5.7, find the kth moment of the random variable X about
  a. 0
  b. the mean
  c. 3
  d. Use the results of Problem 5.15(a) to find the mean and variance of X.

5.16 In Exercise 5.1 and Problem 5.6, determine the mean and variance of the
  a. geometric distribution
  b. negative binomial distribution

5.17 Determine Problem 5.15(a) through (d) for the probability density functions given in Problem 5.4 and Exercises 5.2 and 5.6.

5.18 Find the moment-generating function of the
  a. geometric probability density function
  b. negative binomial probability density function

5.19 Using the result of Problem 5.18, find the mean and variance of the
  a. geometric distribution
  b. negative binomial distribution

5.20 Find the expected value of $g(X) = 3X^2 + 2X + 1$ for the probability density functions given in Exercise 3.9.

5.21 Find the factorial moment-generating function of the
  a. binomial distribution
  b. geometric distribution
  c. negative binomial distribution

5.22 Use the results of Problem 5.21 to find the mean and variance of the following distributions:
  a. binomial
  b. geometric
  c. negative binomial

5.23 Compute for the point binomial, binomial, Poisson, hypergeometric, geometric, negative binomial, gamma, and beta distributions for:
  a. γ
  b. V
  c. ξ

Check your answers with the accompanying tables.

5.24 Calculate the mean and variance of the lognormal probability distribution; see Equation (3.3.24).

Table 5.P.1 Discrete Distributions

Name	Probability Mass Function	Conditions on Parameters
	$\sum_{i=1}^{\infty} f(x_i) = 1, \quad f(x_i) \geq 0,$ for all x_i	
Point binomial	$f(x;p) = \begin{cases} p^x(1-p)^{1-x}, & x=0,1, \\ 0, & \text{elsewhere} \end{cases}$	$0 \leq p \leq 1$
Binomial	$f(x;n,p) = \begin{cases} \binom{n}{x} p^x q^{n-x}, & x=0,1,\ldots,n, \\ 0, & \text{elsewhere} \end{cases}$	$0 \leq p \leq 1$ $q = 1-p$
Poisson	$f(x;\lambda) = \begin{cases} \dfrac{e^{-\lambda}\lambda^x}{x!}, & x=0,1,2,\ldots, \\ 0, & \text{elsewhere} \end{cases}$	$\lambda = np > 0$
Geometric	$f(x;p) = \begin{cases} p(1-p)^{x-1}, & x=1,2,3,\ldots, \\ 0, & \text{elsewhere} \end{cases}$	$0 \leq p \leq 1$

Name	Probability Mass Function	Conditions on Parameters
Hypergeometric	$f(x;r,n_1,n_2) = \begin{cases} \dfrac{\binom{n_1}{x}\binom{n_2}{r-x}}{\binom{n_1+n_2}{r}}, & \begin{matrix} x=1,2,\ldots,r \leq n_1 \\ \text{or} \\ x=1,2,\ldots,n_1 \leq r, \end{matrix} \\ 0, & \text{elsewhere} \end{cases}$	$r \leq n_1 + n_2$
Negative binomial	$f(x;p,r) = \begin{cases} \binom{x-1}{r-1} p^r (1-p)^{x-r}, & x=r, r+1, r+2, \ldots, \\ 0, & \text{elsewhere} \end{cases}$	$0 \leq p \leq 1$ $r > 0$

Expected Value	Variance	Coefficient of Variation
$\mu = E(X)$	$\mathrm{Var}(X) = E(X-\mu)^2$	$V = \dfrac{\sigma}{\mu}$
$E(X) = p$	$\mathrm{Var}(X) = p(1-p)$	$V = \dfrac{\sqrt{p(1-p)}}{p}$
$E(X) = np$	$\mathrm{Var}(X) = npq$	$V = \dfrac{\sqrt{npq}}{np}$
$E(X) = \lambda$	$\mathrm{Var}(X) = \lambda$	$V = \dfrac{\sqrt{\lambda}}{\lambda}$
$E(X) = \dfrac{1}{p}$	$\mathrm{Var}(X) = \dfrac{q}{p^2}$	$V = \sqrt{1-p}$
$E(X) = \dfrac{m_1}{n_1 + n_2}$	$\mathrm{Var}(X) = \dfrac{n_1 n_2}{(n_1+n_2)^2}\left(\dfrac{n_1+n_2-r}{n_1+n_2-1}\right)$	$V = \dfrac{\sqrt{n_1 n_2}}{m_1}\sqrt{\dfrac{n_1+n_2-2}{n_1-n_2-1}}$
$E(X) = \dfrac{r}{p}$	$\mathrm{Var}(X) = \dfrac{r(1-p)}{p^2}$	$V = \dfrac{\sqrt{(1-p)}}{r}$

continued

Table 5.P.1 Discrete Distributions (continued)

Name	Probability Mass Function	Conditions on Parameters
	$\sum_{i=1}^{\infty} f(x_i) = 1, \quad f(x_i) \geq 0$, for all x_i	
Point binomial	$f(x;p) = \begin{cases} p^x(1-p)^{1-x}, & x=0,1, \\ 0, & \text{elsewhere} \end{cases}$	$0 \leq p \leq 1$
Binomial	$f(x;n,p) = \begin{cases} \binom{n}{x} p^x q^{n-x}, & x=1,2,\ldots,n, \\ 0, & \text{elsewhere} \end{cases}$	$0 \leq p \leq 1$
Poisson	$f(x;\lambda) = \begin{cases} \dfrac{e^{-\lambda}\lambda^x}{x!}, & x=0,1,2,\ldots, \\ 0, & \text{elsewhere} \end{cases}$	$\lambda = np > 0$
Geometric	$f(x;p) = \begin{cases} p(1-p)^{x-1}, & x=0,1,2,3,\ldots, \\ 0, & \text{elsewhere} \end{cases}$	$0 \leq p \leq 1$
Hypergeometric	$f(x;r,n_1,n_2) = \begin{cases} \dfrac{\binom{n_1}{x}\binom{n_2}{r-x}}{\binom{n_1+n_2}{r}}, & \begin{array}{l} x=0,1,\ldots,r \leq n_1 \\ \text{or} \\ x=0,1,\ldots,n_1 \leq r, \end{array} \\ 0, & \text{elsewhere} \end{cases}$	$r \leq n_1 + n_2$
Negative binomial	$f(x;p,r) = \begin{cases} \binom{x-1}{r-1} p^r (1-p)^{x-r}, & x=r,r+1,r+2,\ldots, \\ 0, & \text{elsewhere} \end{cases}$	$0 \leq p \leq 1$ $r > 0$

Moment-Generating Function	Coefficient of Skewness
$m(t) = E(e^{tX})$	$\gamma = \dfrac{E\left[(X-\mu)^3\right]}{\sigma^3}$
$m(t) = pe^t + q$ $q = 1-p$	$\gamma = \dfrac{q-p}{\sqrt{pq}}$

Moment-Generating Function	Coefficient of Skewness
$m(t) = (pe^t + q)^n$ $q = 1 - p$	$\gamma = \dfrac{q - p}{\sqrt{npq}}$
$m(t) = e^{\lambda(e^t - 1)}$	$\gamma = \dfrac{\lambda}{\lambda^{2/3}}$
$m(t) = \dfrac{pe^t}{1 - qe^t}$ $q = 1 - p$	$\gamma = \dfrac{2 - p}{(1 - p)^{1/2}}$
$m(t) = \dfrac{(n_2)^{[r]}}{(n_1 + n_2)^{[r]}} \sum_{j=0}^{r} \dfrac{(n_1)^{[j]} r^{[j]} e^t}{(n_2 - r + j)^{[j]} j!}$ *	Does not exist
$m(t) = \dfrac{(pe^t)r}{(1 - qe^t)r}$ $q = 1 - p$	$\gamma = 0$

* [r] denotes the largest integer not exceeding r.

Table 5.P.2 Continuous Distributions

Name	Probability Density Function	Conditions on Parameters
	$\int_{-\infty}^{\infty} f(x) dx = 1,$ $f(x) \geq 0$, for all x	
Uniform	$f(x) = \begin{cases} \dfrac{1}{\beta - \alpha}, & \alpha < x < \beta, \\ 0, & \text{elsewhere} \end{cases}$	$-\infty < \alpha < \beta < \infty$
Normal or Gauss	$f(x; \mu, \sigma) = \dfrac{1}{\sqrt{2\pi}\sigma} \exp\left\{-\dfrac{1}{2}\left(\dfrac{x - \mu}{\sigma}\right)^2\right\}, \; -\infty < x < \infty$	$-\infty < \mu < \infty$ $\sigma > 0$

continued

Table 5.P.2 Continuous Distributions (continued)

Name	Probability Density Function	Conditions on Parameters
Gamma	$f(x;\alpha,\beta) = \begin{cases} \dfrac{1}{\Gamma(\alpha)\beta^\alpha} x^{\alpha-1} e^{-x/\beta}, & x > 0, \\ 0, & \text{elsewhere} \end{cases}$	$\alpha > 0$ $\beta > 0$
Exponential	$f(x;\beta) = \begin{cases} \dfrac{1}{\beta} e^{-x/\beta}, & x > 0, \\ 0, & \text{elsewhere} \end{cases}$	$\beta > 0$
Beta	$f(x;\alpha,\beta) = \begin{cases} \dfrac{\Gamma(\alpha+\beta)}{\Gamma(\alpha)\Gamma(\beta)} x^{\alpha-1}(1-x)^{\beta-1}, & 0 < x \le 1, \\ 0, & \text{elsewhere} \end{cases}$	$\alpha > 0$ $\beta > 0$
Triangular	$f(x) = \begin{cases} 2(1-x), & 0 \le x \le 1, \\ 0, & \text{elsewhere} \end{cases}$ $f(x) = \begin{cases} 2x, & 0 \le x \le 1, \\ 0, & \text{elsewhere} \end{cases}$	
Cauchy	$f(x;\alpha,\mu) = \begin{cases} \dfrac{1}{\pi} \dfrac{\alpha}{\alpha^2 + (x-\mu)^2}, & -\infty < x < \infty, \\ 0, & \text{elsewhere} \end{cases}$	$\alpha > 0$ $-\infty < \mu < \infty$

Expected Value	Variance	Coefficient of Variation
$\mu = E(x)$	$\text{Var}(X) = E(X-\mu)^2$	$V = \dfrac{\sigma}{\mu}$
$E(X) = \dfrac{\alpha+\beta}{2}$	$\text{Var}(X) = \dfrac{(\beta-\alpha)^2}{12}$	$V = \dfrac{\beta-\alpha}{\sqrt{3}(\alpha+\beta)}$
$E(X) = \mu$	$\text{Var}(X) = \sigma^2$	$V = \dfrac{\sigma}{\mu}$

Expected Value	Variance	Coefficient of Variation
$E(X) = \alpha\beta$	$\text{Var}(X) = \alpha\beta^2$	$V = \dfrac{\sqrt{\alpha}}{\alpha}$
$E(X) = \beta$	$\text{Var}(X) = \beta^2$	$V = 1$
$E(X) = \dfrac{\alpha}{\alpha+\beta}$	$\text{Var}(X) = \dfrac{\alpha\beta}{(\alpha+\beta+1)(\alpha+\beta)^2}$	$V = \dfrac{\sqrt{\alpha\beta}}{\alpha\sqrt{\alpha+\beta+1}}$
$E(X) = \dfrac{1}{3}$	$\text{Var}(X) = \dfrac{1}{18}$	$V = \dfrac{1}{\sqrt{2}}$
$E(X) = \dfrac{2}{3}$	$\text{Var}(X) = \dfrac{1}{18}$	$V = \dfrac{1}{2\sqrt{2}}$
Does not exist	Does not exist	Does not exist

Name	Probability Density Function	Conditions on Parameters
	$\int_{-\infty}^{\infty} f(x)\,dx = 1,$ $f(x) \geq 0,$ for all x	
Laplace	$f(x;\mu,\sigma) = \begin{cases} \dfrac{1}{2\sigma} e^{-(|x-\mu|/\sigma)}, & -\infty < x < \infty, \\ 0, & \text{elsewhere} \end{cases}$	$\sigma > 0$ $-\infty < \mu < \infty$
Lognormal	$f(x) = \begin{cases} \dfrac{1}{x\sigma_y \sqrt{2\pi}} \exp\left\{-\dfrac{1}{2\sigma_y^2}(\ln x - \mu_y)^2\right\}, & x < 0, \\ 0, & \text{elsewhere} \end{cases}$	$\sigma_x > 0$ $-\infty < \mu_y < \infty$ $\sigma_y \ln\sqrt{\dfrac{\mu_x^2 + \sigma_x^2}{\mu_x^2}}$ $\mu_y \ln\sqrt{\dfrac{\mu_x^4}{\mu_x^2 + \sigma_x^2}}$

continued

Table 5.P.2 Continuous Distributions (continued)

Name	Probability Density Function	Conditions on Parameters
Weibull	$f(x;\alpha,\beta,\gamma) = \begin{cases} \dfrac{\beta}{\alpha}(x-\gamma)^{\beta-1} e^{\{(x-\gamma)\beta/\alpha\}}, & x \geq \gamma, \\ 0, & \text{elsewhere} \end{cases}$	$\alpha > 0$ $\beta > 0$ $\gamma \geq 0$
Maxwell	$f(x;\sigma) = \begin{cases} \dfrac{2\sigma^3}{\sqrt{2\pi}} x^2 e^{-(x^2\sigma^2/2)}, & x \geq 0, \\ 0, & \text{elsewhere} \end{cases}$	$\sigma > 0$
Rayleigh	$f(x;\alpha) = \begin{cases} \dfrac{2}{\alpha} x e^{-(x^2/\alpha)}, & x \geq 0, \\ 0, & \text{elsewhere} \end{cases}$	$\alpha > 0$
Pareto	$f(x;r) = \begin{cases} rA^r (x^{r+1})^{-1}, & 0 < A \leq x, \\ 0, & \text{elsewhere} \end{cases}$	$A > r$ $r \leq 0$
Extreme value	$f(x) = e^{-x-e^{-x}}, \quad -\infty < x < \infty$	

Expected Value	Variance	Coefficient of Variation
$\mu = E(x)$	$\text{Var}(X) = E(X-\mu)^2$	$V = \dfrac{\sigma}{\mu}$
$E(X) = \mu$	$\text{Var}(X) = 2\sigma^2$	$V = \dfrac{\sigma\sqrt{2}}{\mu}$
$E(X) = e^{\mu_y + \sigma_y}$	$\text{Var}(X) = e^{2(\mu_y + \sigma_y)}\left(e^{2\sigma_{y-1}}\right)$	$V = \sqrt{e^{2\sigma_y} - 1}$
$E(X) = \alpha^{1/\beta}$, $\Gamma\left(1+\dfrac{1}{\beta}\right) + \gamma$	$\text{Var}(X) = \alpha^{2/\beta}\left\{\Gamma\left(1+\dfrac{2}{\beta}\right) - \left[\Gamma\left(1+\dfrac{1}{\beta}\right)\right]^2\right\}$	$V = \dfrac{\sqrt{\Gamma\left(1+\dfrac{2}{\beta}\right) - \left[\Gamma\left(1+\dfrac{1}{\beta}\right)\right]}}{\Gamma\left(1+\dfrac{1}{\beta}\right) + \gamma^{\alpha^{-1/\beta}}}$

Expected Value	Variance	Coefficient of Variation
$E(X) = \dfrac{4}{\sigma\sqrt{2\pi}}$	$\text{Var}(X) = \dfrac{1}{\sigma^2}\left(3 - \dfrac{8}{\pi}\right)$	$V = \sqrt{\dfrac{3}{8}\pi - 1}$
$E(X) = \alpha\dfrac{\sqrt{\pi}}{2}$	$\text{Var}(X) = \alpha^2\left(1 - \dfrac{\pi}{2}\right)$	$V = \sqrt{\alpha}\left(\dfrac{1-\pi}{\pi}\right)^{1/2}$
$E(X) = A\left(\dfrac{r}{r-1}\right)$, $r > 1$	$\text{Var}(X) = \dfrac{r^3 - 3r^2 + r}{(r-2)(r-1)^2}A^2$, $r > 2$	$V = \dfrac{\sqrt{r^3 - 3r^2 + r}}{r\sqrt{r-1}}$
$E(X) = 0$	$\text{Var}(X) = 1$	Infinite

Name	Probability Density Function	Conditions on Parameters
	$\int_{-\infty}^{\infty} f(x)\,dx = 1$, $f(x) \geq 0$, for all x	
Uniform	$f(x) = \begin{cases} \dfrac{1}{\beta - \alpha}, & \alpha < x < \beta, \\ 0, & \text{elsewhere} \end{cases}$	$-\infty < \alpha < \beta < \infty$
Normal or Gauss	$f(x;\mu,\sigma) = \dfrac{1}{\sqrt{2\pi}\sigma}\exp\left\{-\dfrac{1}{2}\left(\dfrac{x-\mu}{\sigma}\right)^2\right\}$, $-\infty < x < \infty$	$-\infty < \mu < \infty$ $\sigma > 0$
Gamma	$f(x;\alpha,\beta) = \begin{cases} \dfrac{1}{\Gamma(\alpha)\beta^\alpha}x^{\alpha-1}e^{(-x/\beta)}, & x > 0, \\ 0, & \text{elsewhere} \end{cases}$	$\alpha > 0$ $\beta > 0$
Exponential	$f(x;\beta) = \begin{cases} \dfrac{1}{\beta}e^{(-x/\beta)}, & x \geq 0, \\ 0, & \text{elsewhere} \end{cases}$	$\beta > 0$

continued

Table 5.P.2 Continuous Distributions (continued)

Name	Probability Density Function	Conditions on Parameters
Beta	$f(x;\alpha,\beta) = \begin{cases} \dfrac{\Gamma(\alpha+\beta)}{\Gamma(\alpha)\Gamma(\beta)} x^{\alpha-1}(1-x)^{\beta-1}, & 0 < x \leq 1, \\ 0, & \text{elsewhere} \end{cases}$	$\alpha > 0$ $\beta > 0$
Triangular	$f(x) = \begin{cases} 2(1-x), & 0 \leq x \leq 1, \\ 0, & \text{elsewhere} \end{cases}$ $f(x) = \begin{cases} 2x, & 0 < x \leq 1, \\ 0, & \text{elsewhere} \end{cases}$	
Cauchy	$f(x;\alpha,\mu) = \begin{cases} \dfrac{1}{\pi} \dfrac{\alpha}{\alpha^2 + (x-\mu)^2}, & -\infty < x < \infty, \\ 0, & \text{elsewhere} \end{cases}$	$\alpha > 0$ $-\infty < \mu < \infty$

Moment-Generating Function	Coefficient of Skewness
$m(t) = E(e^{tX})$	$\gamma = \dfrac{E\left[(X-\mu)^3\right]}{\sigma^3}$
$m(t) = \dfrac{e^{t\beta} - e^{t\alpha}}{t(\beta-\alpha)}$	$\gamma = 0$
$m(t) = e^{t\mu + (1/2)t^2\sigma^2}$	$\gamma = 0$
$m(t) = (1-t\beta)^{-\alpha}, \; t < \dfrac{1}{\beta}$	$\gamma = \dfrac{2\sqrt{\alpha}}{\alpha}$
$m(t) = (1-t\beta)^{-1}, \; t < \dfrac{1}{\beta}$	$\gamma = 2$

Moment-Generating Function	Coefficient of Skewness
$m(t) = \dfrac{\Gamma(\alpha+\beta)}{\Gamma(\alpha)}, \quad \displaystyle\sum_{j=0}^{\infty} \dfrac{\Gamma(\alpha+j)(t)^j}{\Gamma(\alpha+\beta+j)\Gamma(j+1)}$	$\gamma = \dfrac{2(\beta-\alpha)\sqrt{\alpha+\beta+1}}{(\alpha+\beta+2)\sqrt{\alpha\beta}}$
$m(t) = \dfrac{2}{t^2}(e^t - t - 1)$	$\gamma = \dfrac{2\sqrt{2}}{5}$
$m(t) = \dfrac{2}{t^2}(te^t - e^t + 1)$	$\gamma = \dfrac{-2\sqrt{2}}{5}$
Exists only for $t = 0$	

Name	Probability Density Function	Conditions on Parameters
	$\displaystyle\int_{-\infty}^{\infty} f(x)\,dx = 1,$ $f(x) \geq 0,$ for all x	
Laplace	$f(x;\mu,\sigma) = \begin{cases} \dfrac{1}{2\sigma} e^{-(|x-\mu|/\sigma)}, & -\infty < x < \infty, \\ 0, & \text{elsewhere} \end{cases}$	$\sigma > 0$ $-\infty < \mu < \infty$
Lognormal	$f(x) = \begin{cases} \dfrac{1}{x\sigma_y\sqrt{2\pi}} \exp\left\{-\dfrac{1}{2\sigma_y^2}(\ln x - \mu)^2\right\}, & x < 0, \\ 0, & \text{elsewhere} \end{cases}$	$-\infty < \mu_y < \infty$ $\mu_y = \ln\sqrt{\dfrac{\mu_x^4}{\mu_x^2 + \sigma_x^2}}$ $\sigma_y = \mathrm{n}\sqrt{\dfrac{\mu_x^2 + \sigma_x^2}{\mu_x^2}}$
Weibull	$f(x;\alpha,\beta,\gamma) = \begin{cases} \dfrac{\beta}{\alpha}(x-\gamma)^{\beta-1} \cdot e^{\{(x-\gamma)\beta/\alpha\}}, & x \geq \gamma, \\ 0, & \text{elsewhere} \end{cases}$	$\alpha > 0$ $\beta > 0$ $\gamma \geq 0$

continued

Table 5.P.2   Continuous Distributions (continued)

Name	Probability Density Function	Conditions on Parameters
Maxwell	$f(x;\sigma) = \begin{cases} \dfrac{2\sigma^3}{\sqrt{2\pi}} x^2 e^{-(x^2\sigma^2/2)}, & x \geq 0, \\ 0, & \text{elsewhere} \end{cases}$	$\sigma > 0$
Rayleigh	$f(x;\alpha) = \begin{cases} \dfrac{2}{\alpha} x e^{-(x^2/\alpha)}, & x \geq 0, \\ 0, & \text{elsewhere} \end{cases}$	$\alpha > 0$
Pareto	$f(x;r) = \begin{cases} rA^r (x^{r-1})^{-1}, & 0 < A \leq x, \\ 0, & \text{elsewhere} \end{cases}$	$r \geq 0$ $A > r$
Extreme value	$f(x) = e^{-e^{-(x-\theta)}}, \quad -\infty < x < \infty$	$-\infty < \theta < \infty$

Moment-Generating Function	Coefficient of Skewness
$m(t) = E(e^{tX})$	$\gamma = \dfrac{E\left[(X-\mu)^3\right]}{\sigma^3}$
$m(t) = \dfrac{e^{t\mu}}{1-\sigma^2 t^2}$	
$m(t) = e^{t\gamma} \displaystyle\int_0^\infty e^{-y + t(\sigma y)^{1/\beta}} \, dy$	$\gamma = \dfrac{\left\{\Gamma\!\left(1+\dfrac{3}{\beta}\right) - 3\Gamma\!\left(1+\dfrac{1}{\beta}\right)\cdot\Gamma\!\left(1+\dfrac{2}{\beta}\right) + 2\left[\Gamma\!\left(1+\dfrac{1}{\beta}\right)\right]^3\right\}}{\left\{\Gamma\!\left(1+\dfrac{2}{\beta}\right) - \left[\Gamma\!\left(1+\dfrac{1}{\beta}\right)\right]^2\right\}^{3/2}}$

Moment-Generating Function	**Coefficient of Skewness**
$m(t) = \dfrac{e^{t^2/2\sigma^2}}{\sqrt{2\pi\sigma^2}} \cdot \left[\sqrt{2\pi}\left(\sigma^2 + t^2\right) + 4t\sigma \right]$	$\gamma = \dfrac{2\sqrt{2}\left(16 - 5\pi\right)}{\left(3\pi - 8\right)^{3/2}}$
$m(t) = e^{\alpha t^2/4}\left[1 + t\dfrac{\sqrt{\alpha\pi}}{2}\right]$	
$m(t) = e^{tA}\sum\limits_{j=1}^{r-1}\dfrac{(r-j-1)!\,A^{j}t^{j}}{(r-1)!}$	
$m(t) = \Gamma(1-t),\ t < 1$	

> "When he was asked, 'What is a friend?' he said, 'One soul inhabiting two bodies.'"
> —Aristotle

CHAPTER 6

Two Random Variables

OBJECTIVES:

6.1 Joint Probability Density Function
6.2 Bivariate Cumulative Distribution Function
6.3 Marginal Probability Distributions
6.4 Conditional Probability Density and Cumulative Distribution Functions
6.5 Independent Random Variables
6.6 Function of Two Random Variables
6.7 Expected Value and Moments
6.8 Conditional Expectations
6.9 Bivariate Normal Distribution

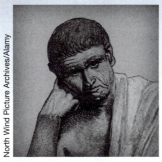

Aristotle (384–322 BC)

> "What you leave behind is not what is engraved in stone monuments but what is woven into the lives of others."
> —Pericles

> "Education is an ornament in prosperity and a refuge in adversity."
> —Aristotle

Chapter 6 Two Random Variables

INTRODUCTION

We have until now been concerned with studying the situation of one-dimensional random variables and their distributions. In many physical phenomena, however, we must deal with random variables of more than one dimension. For example, we might be interested in an experiment involving the weight W and height H of a certain group of people, in which case the sample space consists of 2-pules. In this situation, we need the two-dimensional random variable (W, H) to describe the numerical characteristics of the experiment. Many other situations require the use of the two-dimensional random variable. However, as in the univariate case, it is necessary to develop the joint probability density function of the two-dimensional variate; once we know the distribution of the variate, we know the laws that govern it.

In this chapter, we extend the concept of a random variable and its probability distribution function to the concept of two random variables. Once the conceptual extension for the univariate to the bivariate case has been accomplished, you will note that the further generalization to the n-dimensional case follows essentially the same idea. The n-dimensional random variable is the subject of Chapter 7.

6.1 Joint Probability Density Function

We define a two-dimensional discrete random variable in Definition 6.1.1.

Definition 6.1.1

Let S be a sample space and $X(s)$ and $Y(s)$ be two real valued functions (defined on S) that assign a real number to each element of the sample space. The pair $(X(s), Y(s))$ is then referred to as a **two-dimensional random variable**. For convenience, we denote $X(s) = X$ and $Y(s) = Y$. Hence, the range space $R_{X,Y}$ of the two-dimensional discrete variate (X, Y) is a subset of the Euclidean plane (x-y plane).

Definition 6.1.2

Let (X, Y) be a two-dimensional random variable. If the number of possible values of (X, Y) is finite or countably infinite, then (X, Y) is called a **two-dimensional discrete random variable**.

We denote the possible values that (X, Y) may assume by

$$(x_i, y_j), \quad i = 1, 2, \ldots, n, \ldots$$

and

$$j = 1, 2, \ldots, k, \ldots.$$

As it was in the one-dimensional case, our aim is to calculate the probability that the two-dimensional random variable will assume certain values. Some of these probabilities are $Pr(X = x_1, Y = y_1)$, the probability that the values that X and Y assume simultaneously are x_1 and y_1, respectively, and $Pr(x_1 < X < x_2, y_1 < Y < y_2)$, the probability that the values that X

and Y assume fall in the intervals (x_1, x_2) and (y_1, y_2), respectively. More precisely, if $E_1 = \{(x,y): x_1 < x < x_2, y_1 < y < y_2\}$, then

$$Pr(x_1 < X < x_2, y_1 < Y < y_2) = Pr[(X,Y) \text{ in } E_1],$$

the probability that the values that the two-dimensional variate assumes lie in a certain set $E_1 \subset R_{X,Y}$, where $R_{X,Y}$ is the range space.

A mathematical function used to calculate the preceding probabilities is the joint probability density function of the random variable (X,Y).

Definition 6.1.3

Let (X,Y) be a two-dimensional discrete random variable. A function $f(x_i, y_j)$ is called the **joint probability density function** of the random variable (X,Y) if the following conditions are satisfied:

1. $f(x_i, y_j) \geq 0$, for all $(x_i, y_j) \in R_{X,Y}$
2. $\sum\sum_{(x_i, y_j) \in R_{x,y}} f(x_i, y_j) = 1$

Here, the summation is over all possible values (x_i, y_j), $i = 1, 2, \ldots, n, \ldots$, $j = 1, 2, \ldots, k, \ldots$. Furthermore, for any real valued pair (x_i, y_j), we have $f(x_i, y_j) = Pr(X = x_i, Y = y_j)$.

Example 6.1.1

An experiment consists of drawing four objects from a container that holds eight operable, six defective, and 10 semioperable objects. Let X be the number of operable objects drawn and Y the number of defective objects drawn. Find the probability density function.

Solution: The probability density function of the two-dimensional random variable (X,Y) is

$$f(x,y) = \begin{cases} \dfrac{\binom{8}{x}\binom{6}{y}\binom{10}{4-(x+y)}}{\binom{24}{4}}, & \begin{array}{l} x, y = 0, 1, \ldots, 4 \\ \text{such that} \\ 0 \leq x + y \leq 4, \end{array} \\ 0, & \text{elsewhere.} \end{cases}$$

It can be shown that $f(x,y) \geq 0$ for $0 \leq x + y \leq 4$ and that

$$\sum_{x=0}^{4-y}\sum_{y=0}^{4} \dfrac{\binom{8}{x}\binom{6}{y}\binom{10}{4-x-y}}{\binom{24}{4}} = 1.$$

Now we can use $f(x,y)$ to calculate various probabilities. For example,

$$Pr(X=3, Y=0) = f(3,0) = \frac{\binom{8}{3}\binom{6}{0}\binom{10}{1}}{\binom{24}{4}} = 0.0527$$

is the probability that the sample will consist of three operable components and one semi-operable component, and

$$Pr(X<3, Y=1) = \sum_{x=0}^{2} f(x,1) = \sum_{x=0}^{2} \frac{\binom{8}{x}\binom{6}{1}\binom{10}{3-x}}{\binom{24}{4}} = 0.4291$$

is the probability that the random sample will consist of two or fewer operable components and one defective component.

Figure 6.1.1 shows the graph of $f(x,y)$; the length of the vertical lines indicates the probabilities for all possible values that the random variable (X,Y) can assume.

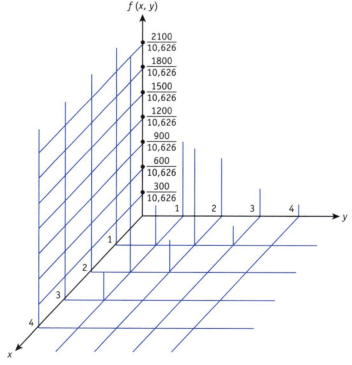

Figure 6.1.1

Example 6.1.2

Let X denote the number of aces and Y the number of kings occurring in a bridge hand of 13 cards. The values that the random variable (X,Y) can assume are $(0,0),(0,1),...,(4,4)$. Find the joint probability density function.

Solution: If we assume that the $\binom{52}{13}$ possible bridge hands are equally likely to occur, the joint probability density function is given by

$$f(x,y) = \begin{cases} \dfrac{\binom{4}{x}\binom{4}{y}\binom{44}{13-x-y}}{\binom{52}{13}}, & x,y = 0,1,\ldots,4, \\ 0, & \text{elsewhere.} \end{cases}$$

It can be shown that $f(x,y) \geq 0$ for all x and y and that

$$\sum_{y=0}^{4}\sum_{x=0}^{4} \dfrac{\binom{4}{x}\binom{4}{y}\binom{44}{13-x-y}}{\binom{52}{13}} = 1.$$

Now, we can use $f(x,y)$ to calculate various probabilities. For example, the probability of having neither an ace nor a king in a bridge hand is

$$Pr(X=0, Y=0) = f(0,0) = \dfrac{\binom{4}{0}\binom{4}{0}\binom{44}{13}}{\binom{52}{13}} = \dfrac{51{,}915{,}526{,}432}{635{,}013{,}559{,}600} = 0.0818.$$

Also, the probability of having no ace or one ace and exactly four kings in such a hand is

$$Pr(X<2, Y=4) = \sum_{x=0}^{1} f(x,4) = \sum_{x=0}^{1} \dfrac{\binom{4}{x}\binom{4}{4}\binom{44}{9-x}}{\binom{52}{13}} = 0.0022.$$

◀

■ **Definition 6.1.4**

Let (X,Y) be a two-dimensional random variable. If (X,Y) can assume all values in some uncountable (nondenumerable) subset of the Euclidean plane, then (X,Y) is called a **two-dimensional continuous random variable**.

The range space of the two-dimensional variate (X,Y) is the set of all possible values of (X,Y), which is an uncountable subset of the Euclidean plane. For example, if (X,Y) assumes its values from the set $S_1 = \{(x,y): x_1 \leq x \leq x_2, y_1 \leq y \leq y_2\}$, we would say that (X,Y) is a two-dimensional continuous random variable, the range space of which is a rectangular subset of the x-y plane. Furthermore, for any two real valued pairs (a,b) and (c,d) such that $a<b$ and $c<d$, respectively, we have

$$Pr(a \leq X \leq b, c \leq Y \leq d) = \int_a^b \int_c^d f(x,y)\,dy\,dx.$$

Definition 6.1.5

Let (X,Y) be a two-dimensional continuous random variable. A function $f(x,y)$ is the joint probability density function of the variate (X,Y) if the following conditions are satisfied:

1. $f(x,y) \geq 0$, for all $(x,y) \in R_{X,Y}$
2. $\iint_{R_{X,Y}} f(x,y)\,dxdy = 1$

Here, the integration is over $R_{X,Y}$, the range space in which (X,Y) assumes all values.

For the double integral to exist in a given plane, the joint function $f(x,y)$ must be continuous almost everywhere on the plane. (We assume that this condition exists throughout this book.) The second condition states that the total volume under the surface given by $f(x,y)$ is 1, which is analogous to the total area under $f(x)$.

Given the joint probability density $f(x,y)$ of the continuous variate (X,Y) and letting $E_1 = \{(x,y): a \leq x \leq b, c \leq y \leq d\}$, we can define

$$Pr(E_1) = Pr(a \leq X \leq b, c \leq Y \leq d) = \iint_{E_1} f(x,y)\,dxdy$$

$$= \int_c^d \int_a^b f(x,y)\,dxdy.$$

The probability is represented as the volume above the shaded rectangle, shown in Figure 6.1.2.

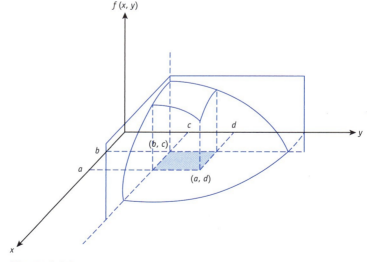

Figure 6.1.2

Example 6.1.3

Use the function $g(x,y) = 4x + 2y + 1$, defined over the rectangle $0 \leq x \leq 2, 0 \leq y \leq 2$, to define a joint probability density function over that region.

Solution: Here,

$$\int_0^2 \int_0^2 (4x+2y+1)\,dx\,dy = 28.$$

Thus,

$$f(x,y) = \begin{cases} \dfrac{1}{28}(4x+2y+1), & 0 \le x \le 2, 0 \le y \le 2, \\ 0, & \text{elsewhere,} \end{cases}$$

is the joint probability density function of the random variable (X,Y). The probability that the variate (X,Y) will assume a value in the region $x \le 1$, $y < 1/2$ is

$$Pr\left(X \le 1, Y < \frac{1}{2}\right) = \int_{-\infty}^{1/2} \int_{-\infty}^{1} f(x,y)\,dx\,dy$$

$$= \int_0^{1/2} \int_0^1 \frac{1}{28}(4x+2y+1)\,dy\,dx$$

$$= \frac{1}{16}.$$

Furthermore, the probability that $x+y$ will be less than 1 is

$$Pr(X+Y<1) = \int_0^1 \int_0^{1-x} \frac{1}{28}(4x+2y+1)\,dx\,dy$$

$$= \frac{3}{56}.$$

The graph of the joint probability density $f(x,y)$ is shown in Figure 6.1.3. In addition, $Pr(X \le 1, Y < 1/2)$ is shown as the volume above the area of the rectangle in the x-y plane, shaded ▭; the volume above the area of the triangle in the x-y plane, shaded ▭, represents $Pr(X+Y<1)$.

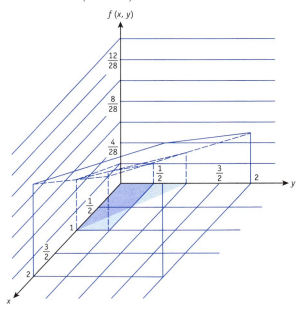

Figure 6.1.3

Example 6.1.4

Find k so that $f(x,y) = kxy$, $1 \leq x \leq y \leq 2$, will be a probability density function.

Solution: Here,

$$\frac{1}{k} = \int_1^2 \int_x^2 xy\,dy\,dx = \frac{9}{8}.$$

Thus, $k = 8/9$ and

$$f(x,y) = \begin{cases} \dfrac{8}{9}xy, & 1 \leq x \leq y \leq 2, \\ 0, & \text{elsewhere,} \end{cases}$$

is a probability density function. Figure 6.1.4 clearly shows the region in which $f(x,y)$ is defined.

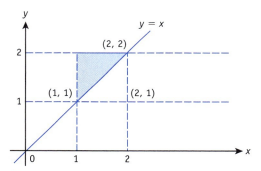

Figure 6.1.4

Thus,

$$Pr\left(X \leq \frac{3}{2}, Y \leq \frac{3}{2}\right) = \frac{8}{9}\int_1^{3/2}\int_x^{3/2} xy\,dy\,dx = \frac{25}{144}$$

and

$$Pr\left(X + Y \leq \frac{3}{2}\right) = 0,$$

because both X and Y are greater than or equal to 1.

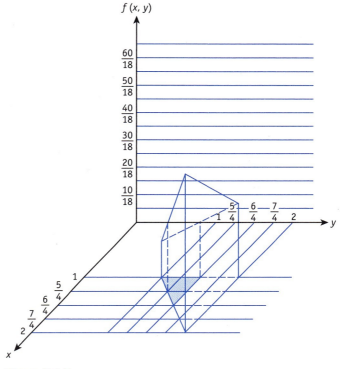

Figure 6.1.5

Figure 6.1.5 shows the graph of $f(x,y)$ and $Pr(X \leq 3/2, Y \leq 3/2)$, which is the volume above the shaded area in the x-y plane.

6.2 Bivariate Cumulative Distribution Function

The cumulative distribution function played an important role in our study of the one-dimensional random variable. In this section, we extend the discussion of the cumulative distribution function to the two-dimensional random variable.

Definition 6.2.1

Let (X, Y) be a two-dimensional random variable. The bivariate function, defined by

$$F(x,y) = Pr(X \leq x, Y \leq y),$$

is called the *cumulative distribution function* of the variate (X, Y).

continued

If (X, Y) is of the discrete type, then

$$F(x,y) = Pr(X \leq x, Y \leq y) = \sum_{y_j \leq y} \sum_{x_i \leq x} f(x_i, y_j);$$

if (X, Y) is of the continuous type, then

$$F(x,y) = Pr(X \leq x, Y \leq y) = \int_{-\infty}^{y} \left\{ \int_{-\infty}^{x} f(t,s) dt \right\} ds,$$

where $f(x,y)$ is the joint probability density function of the random variable (X, Y).

Note: The comma in the probability parentheses means "and also."

The two-dimensional cumulative distribution has a number of properties that resemble those of the one-dimensional case. We state these properties, omitting the proofs of the first four properties because these properties are analogous to those of the univariate cumulative distribution function.

Theorem 6.2.1

If $F(x,y)$ is the bivariate cumulative distribution function of the random variable (X, Y), then $F(x,y)$ has the following properties:

1. $F(\infty, \infty) = 1$.
2. $F(-\infty, y) = F(x, -\infty) = 0$.
3. $F(x, y)$ is a monotone nondecreasing function in each variable separately.
4. $F(x, y)$ is continuous, at least from the right, in each variable.
5. For all values x_1, x_2 $(x_2 > x_1)$ and y_1, y_2 $(y_2 > y_1)$, the relation

$$F(x_2, y_2) - F(x_1, y_2) - F(x_2, y_1) + F(x_1, y_1) \geq 0$$

is satisfied.

We verify the final condition by showing that it is equal to $Pr(x_1 < X \leq x_2, y_1 < Y \leq y_2)$, which is certainly nonnegative. As you can observe from Figure 6.2.1, it is possible to write

$$\{X \leq x_2, Y \leq y_2\} = \{X \leq x_1, Y \leq y_1\}$$
$$\cup \{X \leq x_1, y_1 < Y \leq y_2\}$$
$$\cup \{x_1 < X \leq x_2, Y \leq y_1\}$$
$$\cup \{x_1 < X \leq x_2, y_1 < Y \leq y_2\}.$$

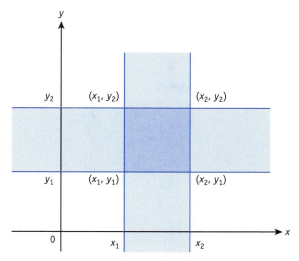

Figure 6.2.1

Thus, the event $\{X \le x_2, Y \le y_2\}$ is expressed in terms of mutually exclusive events. Upon taking the probability of both sides, we have

$$Pr(x_1 < X \le x_2, < Y \le y_2) = Pr(X \le x_2, Y \le y_2)$$
$$-Pr\{X \le x_1, Y \le y_1\}$$
$$-Pr(X \le x_1, y_1 < Y \le y_2)$$
$$-Pr(x_1 < X \le x_2, Y \le y_1). \tag{6.2.1}$$

But

$$\{X \le x_1, Y \le y_2\} = \{X \le x_1, Y \le y_1\} \cup \{X \le x_1, y_1 < Y \le y_2\} \tag{6.2.2}$$

and

$$\{X \le x_2, Y \le y_1\} = \{X \le x_1, Y \le y_1\} \cup \{x_1 < X \le x_2, Y \le y_1\}. \tag{6.2.3}$$

Taking the probabilities of both sides of Equations (6.2.2) and (6.2.3), we have

$$Pr(X \le x_1, y_1 < Y \le y_2) = Pr(X \le x_1, Y \le y_2) - Pr\{X \le x_1, Y \le y_1\} \tag{6.2.4}$$

and

$$Pr(x_1 < X \le x_2, Y \le y_1) = Pr(X \le x_2, Y \le y_2) - Pr\{X \le x_1, Y \le y_1\}. \tag{6.2.5}$$

Substituting Equations (6.2.4) and (6.2.5) into Equation (6.2.1) results in

$$Pr(x_1 < X \le x_2, y_1 < Y \le y_2) = Pr(X \le x_2, Y \le y_2) - Pr(X \le x_1, Y \le y_1)$$
$$-Pr(X \le x_1, Y \le y_2) + Pr(X \le x_1, Y \le y_1)$$
$$-Pr(X \le x_2, Y \le y_1) + Pr(X \le x_1, Y \le y_1). \tag{6.2.6}$$

Applying the definition of the function $F(x,y)$ to Equation (6.2.6), we obtain the desired result:
$$Pr(x_1 < X \leq x_2, y_1 < Y \leq y_2) = F(x_2, y_2) - F(x_1, y_2) - F(x_2, y_1) + F(x_1, y_1) \geq 0. \quad (6.2.7)$$

Furthermore, if $F(x,y)$ is the bivariate cumulative distribution function of the two-dimensional continuous random variable (X, Y), whose first and second partial derivatives exist, then
$$\frac{\partial^2 F(x,y)}{\partial x \partial y} = f(x,y),$$

where $f(x,y)$ is the joint probability density function of the variate (X, Y). This statement is true because

$$\frac{\partial^2 F(x,y)}{\partial x \partial y} = \frac{\partial}{\partial x}\left\{\lim_{\Delta y \to 0} \frac{F(x, y+\Delta y) - F(x,y)}{\Delta y}\right\} = \lim_{\Delta x \to 0} \lim_{\Delta y \to 0}$$

$$\left\{\frac{F(x+\Delta x, y+\Delta y) - F(x, y+\Delta y) - F(x+\Delta x, y) + F(x,y)}{\Delta x \Delta y}\right\},$$

which from the inequality in Equation (6.2.7) we can write as

$$\lim_{\Delta x \to 0} \lim_{\Delta y \to 0} \left\{\frac{Pr(x < X \leq x+\Delta x, y < Y \leq y+\Delta y)}{\Delta x \Delta y}\right\} = f(x,y).$$

Hence,
$$Pr(x < X \leq x+dx, y < Y \leq y+dy) = f(x,y) dx\, dy,$$

and because the left-hand side is nonnegative, we have
$$f(x,y) \geq 0.$$

Example 6.2.1

The joint probability mass of the discrete random variable (X, Y) is given in Table 6.2.1.

Table 6.2.1

x \ y	0	1	2	
0	0	$\frac{1}{4}$	0	$\frac{1}{4}$
1	$\frac{1}{4}$	$\frac{3}{8}$	0	$\frac{5}{8}$
2	0	0	$\frac{1}{8}$	$\frac{1}{8}$
	$\frac{1}{4}$	$\frac{5}{8}$	$\frac{1}{8}$	1.0

The cumulative distribution function of the discrete variate (X,Y) is

$$F(x,y) = \begin{cases} 0, & x<0 \text{ or } y<0, \\ 0, & x<1 \text{ or } y<1, \\ \dfrac{1}{4}, & x<1 \text{ or } y<2, \\ \dfrac{1}{4}, & x<1 \text{ or } 1 \le y < \infty, \\ \dfrac{1}{4}, & x<2 \text{ or } y<1, \\ \dfrac{1}{4}, & 1 \le x < \infty \text{ or } y<1, \\ \dfrac{7}{8}, & x<2 \text{ or } 1 \le y < \infty, \\ \dfrac{7}{8}, & 1 \le x < \infty \text{ or } y<2, \\ 1, & 2 \le x < \infty \text{ or } 2 \le y < \infty. \end{cases}$$

◀

Example 6.2.2

Determine the cumulative distribution of the probability density function given in Example 6.1.3.

Solution: For $0 \le x \le 2$ and $0 \le y \le 2$,

$$F(x,y) = \Pr(X \le x, Y < y) = \int_0^y \left\{ \int_0^x \frac{1}{28}(4t + 2s + 1)\,dt \right\} ds$$

$$= \int_0^y \frac{1}{28}(2x^2 + 2sx + x)\,ds$$

$$= \frac{1}{28}(2x^2 y + y^2 x + xy)$$

$$= \frac{xy}{28}(2x + y + 1).$$

Thus,

$$F(x,y) = \begin{cases} 0, & x \le 0 \text{ or } y \le 0, \\ \dfrac{y}{28}(2x + y + 1), & 0 < x \le 2, 0 < y \le 2, \\ \dfrac{x}{14}(2x + 3), & 0 < x \le 2, y > 2, \\ \dfrac{y}{14}(y + 5), & x > 2, 0 < y \le 2, \\ 1, & x, y > 2. \end{cases}$$

The graph of $F(x,y)$ is shown in Figure 6.2.2.

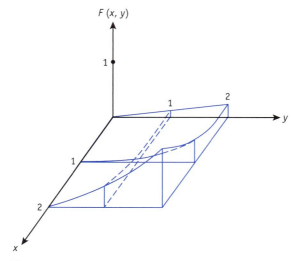

Figure 6.2.2

Example 6.2.3

Determine the cumulative distribution function of the bivariate density function

$$f(x,y) = \begin{cases} \dfrac{8}{9}xy, & 1 \leq x \leq y \leq 2, \\ 0, & \text{elsewhere}, \end{cases}$$

given in Example 6.1.4.

Solution: To find $F(x,y)$, we must integrate $f(x,y)$ first with respect to y and then with respect to x. We use this technique because we are trying to find various volumes under $f(x,y)$, the graph of which is shown in Figure 6.2.3. The density $f(x,y)$ is simply a surface in three-dimensional space and may be thought of as a family of hyperbolas. Satisfaction of the restriction $1 \leq x \leq y \leq 2$ means that we are interested only in the volume under $f(x,y)$ and over the shaded area between the lines $x = y$, $x = 1$, and $y = 2$. Thus, finding $F(x,y)$ may be thought of as finding various volumes under the surface $f(x,y)$, subject to the restriction $1 \leq x \leq y \leq 2$. To find any volume under a surface, we take the double integral of the equation of the surface between appropriate limits of integration for the variables involved. The shaded area of Figure 6.2.3(a), above which we wish to find various volumes, is shown more clearly in Figure 6.2.3(b) and (c).

6.2 Bivariate Cumulative Distribution Function

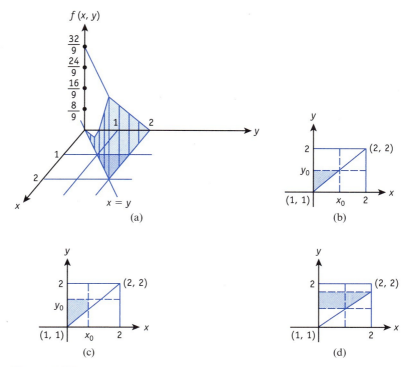

Figure 6.2.3

In calculating $F(x,y)$, we must find volumes over the shaded areas, as shown in Figure 6.2.3(b) and (c); however, we will never have a shaded area similar to that in Figure 6.2.3(d). Thus, to ensure that we involve the correct area, we must always integrate first with respect to y. For the shaded area in Figure 6.2.3(b), it would not matter with which variable we integrated first, but for the area shown in Figure 6.2.3(c), we must integrate first with respect to y. Proceeding in this manner, we derive $F(x,y)$ for the various intervals:

a. For $x<1$ or $y<1$: $F(x,y)=0$.

b. For $1 \leq x \leq y \leq 2$: $F(x,y) = \Pr(X \leq x, Y \leq y)$

$$= \int_1^x \left\{ \int_t^y \frac{8}{9} ts \, ds \right\} dt$$

$$= \frac{4}{9} \int_1^x t(y^2 - t^2) \, dt$$

$$= \frac{1}{9}(x^2 - 1)(2y^2 - x^2 - 1).$$

c. For $x > y$, $1 \leq y \leq 2$: $F(x,y) = Pr(X \leq x, Y \leq y)$

$$= \int_1^y \left\{ \int_t^y \frac{8}{9} ts\, ds \right\} dt$$

$$= \frac{4}{9} \int_1^y t\left(y^2 - t^2\right) dt$$

$$= \frac{1}{9}\left(y^2 - 1\right)^2.$$

Example 6.2.3(c) could have been obtained simply by replacing x with y (the upper limit of x) in Example 6.2.3(b).

d. For $1 \leq x \leq 2$, $y > 2$: $F(x,y) = Pr(X \leq x, Y \leq y)$

$$= \int_1^x \left\{ \int_t^2 \frac{8}{9} st\, ds \right\} dt$$

$$= \frac{4}{9} \int_1^x t\left(4 - t^2\right) dt$$

$$= \frac{1}{9}\left(x^2 - 1\right)\left(7 - x^2\right).$$

Example 6.2.3(d) could have been obtained by letting $y = 2$ (the upper limit of y) in Example 6.2.3(b).

e. For $x > 2$ and $y > 2$: $F(x,y) = Pr(X \leq x, Y \leq y)$

$$= \int_1^2 \left\{ \int_x^2 \frac{8}{9} ts\, ds \right\} dt$$

$$= \frac{4}{9} \int_1^2 t\left(4 - t^2\right) dt$$

$$= \frac{4}{9}\left[2x^2 - \frac{x^4}{4} \right]_1^2$$

$$= \frac{1}{9}\left[32 - 16 - 8 + 1 \right]$$

$$= \frac{1}{9}(9) = 1.$$

Example 6.2.3(e) could have been obtained in any of the following ways:

1. Let $x = y$ (x's upper limit) and $y = 2$ (y's upper limit) in Example 6.2.3(b), or let $y = 2$ (the upper limit of y) and $x = 2$ (x's upper limit), also in Example 6.2.3(b).

2. Let $y = 2$ (y's upper limit) in Example 6.2.3(c).

3. Let $x = 2$ (the upper limit of x) in Example 6.2.3(d).

Thus,

$$F(x,y) = \begin{cases} 0, & x<1 \text{ or } y<1, \\ \frac{1}{9}(x^2-1)(2y^2-x^2-1), & 1 \le x \le y \le 2, \\ \frac{1}{9}(y^2-1)^2, & x>y, 1 \le y \le 2, \\ \frac{1}{9}(x^2-1)(7-x^2), & 1 \le x \le 2, y>2, \\ 1, & x>2 \text{ and } y>2. \end{cases}$$

Example 6.2.4

The bivariate cumulative distribution function of the random variable (X,Y) is

$$F(x,y) = \begin{cases} 0, & x \le 0 \text{ or } y \le 0, \\ \dfrac{xy}{(1+x)(1+y)}, & x,y>0. \end{cases}$$

Obtain the joint probability density function.

Solution: The joint probability density function is obtained by

$$\frac{\partial F(x,y)}{\partial x} = \frac{y}{(1+x)^2(1+y)}, \quad x,y>0,$$

and

$$\frac{\partial F(x,y)}{\partial x \partial y} = \frac{1}{(1+x)^2(1+y)^2}, \quad x,y>0.$$

Thus,

$$f(x,y) = \begin{cases} \dfrac{1}{(1+x)^2(1+y)^2}, & x,y>0, \\ 0, & \text{elsewhere.} \end{cases}$$

6.3 Marginal Probability Distributions

Let $f(x,y)$ be the joint probability density function of the random variable (X,Y). In problems that involve more than one random variable, we are occasionally interested in obtaining the probability density of the variate X alone, say, $f_1(x)$, or that of Y, $f_2(y)$, the *marginal probability density functions* of the random variables X or Y, respectively.

First consider the discrete case. Suppose that the joint probability density function of the discrete random variable (X, Y) is given by Table 6.3.1.

Table 6.3.1

			y		
x	y_1	y_2	$\cdots$	y_m	Sum
x_1	$f(x_1, y_1)$	$f(x_1, y_2)$	$\cdots$	$f(x_1, y_m)$	$\sum_{j=1}^{m} f(x_1, y_j)$
x_2	$f(x_2, y_1)$	$f(x_2, y_2)$	$\cdots$	$f(x_2, y_m)$	$\sum_{j=1}^{m} f(x_2, y_j)$
$\vdots$	$\vdots$	$\vdots$	$\cdots$	$\vdots$	$\vdots$
x_n	$f(x_n, y_1)$	$f(x_n, y_2)$	$\cdots$	$f(x_n, y_m)$	$\sum_{j=1}^{m} f(x_n, y_j)$
Sum	$\sum_{i=1}^{n} f(x_i, y_1)$	$\sum_{i=1}^{n} f(x_i, y_2)$	$\cdots$	$\sum_{i=1}^{n} f(x_i, y_m)$	1.00

Thus, Table 6.3.1 represents

$$Pr(X = x_i, Y = y_i) = f(x_i, y_i)$$

such that

$$\sum_{i=1}^{n} \sum_{j=1}^{m} f(x_i, y_j) = 1.$$

Suppose that we are interested in finding the probability that $X = x_k$, $k = 1, 2, \ldots, n$. Because $X = x_k$ must occur with $Y = y_j$ for some $j = 0, 1, \ldots, m$ and can occur with $Y = y_j$ for only one j, we can write this probability as follows:

$$f_1(x_k) = Pr[X = x_k] = Pr[X = x_k, Y = y_1 \text{ or } X = x_k, Y = y_2 \text{ or } \ldots$$
$$\text{or } X = x_k, Y = y_m]$$

$$= \sum_{j=1}^{m} Pr(X = x_k, Y = y_j)$$

$$= \sum_{j=1}^{m} f(x_k, y_j).$$

6.3 Marginal Probability Distributions

The function $f_1(x_k)$, $k=1, 2,\ldots,n$ is called the *marginal probability density function* of the random variable X. Similarly,

$$f_2(y_j) = Pr(Y = y_j) = \sum_{i=1}^{n} Pr(X = x_i, Y = y_j)$$

$$= \sum_{i=1}^{n} f(x_i, y_j)$$

is the *marginal function* of the variate Y. Therefore, the last column and the last row of Table 6.3.1 give the marginal probability density functions of the variates X and Y, respectively. ◀

Example 6.3.1

Find the marginal probability density function of the random variables X and Y if their joint probability density function is given by Table 6.3.2.

Table 6.3.2

			y			
x	−2	0	1	4	Sum
−1	0.3	0.1	0	0.2	0.6
3	0	0.2	0.1	0	0.3
5	0.1	0	0	0	0.1
Sum	0.4	0.3	0.1	0.2	1

Solution: Table 6.3.3 is based on the information in Table 6.3.2 and gives the marginal densities of the variates X and Y.

Table 6.3.3

x_i	−1	3	5	Otherwise	y_j	−2	0	1	4	Otherwise
$f_1(x_i)$	0.6	0.3	0.1	0	$f_2(y_j)$	0.4	0.3	0.1	0.2	0

Note that $f_1(x_i)$ and $f_2(y_j)$ satisfy the conditions of a probability density function.

Definition 6.3.1

Let $f(x_i, y_j)$ be the discrete probability density function of the random variable (X,Y). The **marginal probability density functions** of the random variables X and Y are given by

$$f_1(x_i) = \sum_{j=1}^{\infty} f(x_i, y_j)$$

continued

and
$$f_2(y_j) = \sum_{i=1}^{\infty} f(x_i, y_j),$$

respectively. Furthermore, their **marginal cumulative distribution functions** are given by

$$F_1(x) = F(x, \infty) = Pr(X \leq x, Y \leq \infty) = \sum_{x_i \leq x} \sum_{j=1}^{\infty} f(x_i, y_j) = \sum_{x_i \leq x} f_1(x_i)$$

and

$$F_2(y) = F(\infty, y) = Pr(X \leq \infty, Y \leq y) = \sum_{y_j \leq y} \sum_{i=1}^{\infty} f(x_i, y_j) = \sum_{y_j \leq y} f_2(y_j).$$

Note that $f_1(x_i)$ and $f_2(y_j)$ are actually probability density functions. However, because they are derived from $f(x_i, y_j)$, they are, in this case, called *marginal* probability density functions.

Definition 6.3.2

Let $f(x, y)$ be the joint probability density function of the continuous random variable (X, Y). The marginal probability density functions of the continuous variates X and Y are given by

$$f_1(x) = \int_{-\infty}^{\infty} f(x, y) dy$$

and

$$f_2(y) = \int_{-\infty}^{\infty} f(x, y) dx,$$

respectively.

Example 6.3.2

The joint probability density function of the two-dimensional random variable (X, Y) is given by

$$f(x, y) = \begin{cases} \dfrac{x^3 y^3}{16}, & 0 \leq x \leq 2, 0 \leq y \leq 2, \\ 0, & \text{elsewhere.} \end{cases}$$

6.3 Marginal Probability Distributions

The marginal densities of X and Y are

$$f_1(x) = \int_{-\infty}^{\infty} f(x,y)\,dy = \int_0^2 \frac{x^3 y^3}{16}\,dy = \begin{cases} \dfrac{x^3}{4}, & 0 \le x \le 2, \\ 0, & \text{elsewhere,} \end{cases}$$

and

$$f_2(y) = \int_{-\infty}^{\infty} f(x,y)\,dx = \int_0^2 \frac{x^3 y^3}{16}\,dx = \begin{cases} \dfrac{y^3}{4}, & 0 \le y \le 2, \\ 0, & \text{elsewhere.} \end{cases}$$

Obtain the cumulative distribution of X.

Solution: For $0 \le x \le 2$,

$$F_1(x) = Pr(X \le x) = Pr(X \le x, Y \le \infty) = \int_{-\infty}^{x}\left\{\int_{-\infty}^{\infty} f(t,y)\,dy\right\}dt$$

$$= \int_{-\infty}^{x} f_1(t) = \int_0^x \frac{t^3}{4}\,dt = \frac{x^4}{16}.$$

Therefore,

$$F_1(x) = \begin{cases} 0, & x < 0 \\ \dfrac{x^4}{16}, & 0 \le x \le 2, \\ 1, & x > 2. \end{cases}$$

For $0 \le y \le 2$,

$$F_2(y) = Pr(Y \le y) = \int_{-\infty}^{y} f_2(s)\,ds = \int_0^y \frac{s^3}{4}\,ds = \frac{y^4}{16}.$$

Therefore,

$$F_2(y) = \begin{cases} 0, & y < 0 \\ \dfrac{y^4}{16}, & 0 \le y \le 2, \\ 1, & y > 2. \end{cases}$$

Figure 6.3.1 gives a graphical presentation of $f_1(x)$, $f_2(y)$, $F_1(x)$, and $F_2(y)$.

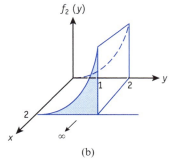
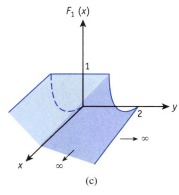
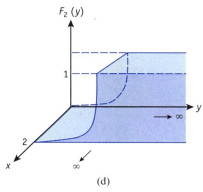

Figure 6.3.1

Example 6.3.3

Obtain the marginal densities of the variates X and Y from the bivariate density function of the two-dimensional random variable (X,Y), as given in Example 6.1.4.

Solution: Here,

$$f(x,y) = \begin{cases} \dfrac{8}{9}xy, & 1 \leq x \leq y \leq 2, \\ 0, & \text{elsewhere.} \end{cases}$$

For $1 \leq x \leq 2$,

$$f_1(x) = \int_{-\infty}^{\infty} f(x,y)\,dy = \int_x^2 \dfrac{8}{9}xy\,dy = \dfrac{4}{9}x(4-x^2);$$

for $1 \leq y \leq 2$,

$$f_2(y) = \int_{-\infty}^{\infty} f(x,y)\,dx = \int_1^y \dfrac{8}{9}xy\,dx = \dfrac{4}{9}y(y^2-1).$$

Thus,

$$f_1(x) = \begin{cases} \dfrac{4}{9}x(4-x^2), & 1 \leq x \leq 2, \\ 0, & \text{elsewhere,} \end{cases}$$

and

$$f_2(y) = \begin{cases} \frac{4}{9}y(y^2-1), & 1 \le y \le 2, \\ 0, & \text{elsewhere.} \end{cases}$$

Furthermore, for $1 \le x \le 2$,

$$F_1(x) = Pr(X \le x) = \int_1^x \left\{ \int_t^2 \frac{8}{9} ty \, dy \right\} dt \, ;$$

for $1 \le x \le 2$,

$$F_1(x) = \int_1^x f_1(t) \, dt = \int_1^x \frac{4}{9} t(4-t^2) \, dt$$

$$= \frac{4}{9} \left[2t^2 - \frac{t^4}{4} \right]_1^x = \frac{1}{9}(x^2-1)(7-x^2).$$

Thus,

$$F_1(x) = \begin{cases} 0, & x < 1, \\ \frac{1}{9}(x^2-1)(7-x^2), & 1 \le x \le 2, \\ 1, & x > 2. \end{cases}$$

Also, for $1 \le y \le 2$,

$$F_2(y) = Pr(Y \le y) = \int_1^y \left\{ \int_t^s \frac{8}{9} xs \, dx \right\} ds \, ;$$

for $1 \le y \le 2$,

$$F_2(y) = \int_1^y \left\{ \int_x^y \frac{8}{9} xs \, ds \right\} dx \, ;$$

and for $1 \le y \le 2$,

$$F_2(y) = \int_1^y f_2(s) \, ds = \int_1^y \frac{4}{9} s(s^2-1) \, ds$$

$$= \frac{4}{9} \left[\frac{s^4}{4} - \frac{s^2}{2} \right]_1^y = \frac{1}{9}(y^2-1)^2.$$

Hence,

$$F_2(y) = \begin{cases} 0, & y < 1, \\ \frac{1}{9}(y^2-1)^2, & 1 \le y \le 2, \\ 1, & y > 2. \end{cases}$$

6.4 Conditional Probability Density and Cumulative Distribution Functions

In Chapter 1, we discussed the conditional probability of events. That is, if S_1 and S_2 are events or subsets of the sample space S, the *conditional probability* of the event S_1, given that event S_2 has occurred, was defined as the probability $S_1 \cap S_2$ divided by the probability of S_2:

$$Pr(S_1|S_2) = \frac{Pr(S_1 \cap S_2)}{Pr(S_2)}, \quad Pr(S_2) > 0.$$

In this section, we investigate conditional distributions.

Let $f(x_i, y_j)$, $i = 1, 2, 3, \ldots$, $j = 1, 2, 3, \ldots$, be the bivariate probability density function of the discrete random variable (X, Y). Thus,

$$f(x_i, y_j) = Pr(X = x_i, Y = y_j).$$

We can obtain the marginal densities by

$$f_1(x_i) = Pr(X = x_i) = \sum_{j=1}^{\infty} f(x_i, y_j)$$

and

$$f_2(y_j) = Pr(Y = y_j) = \sum_{i=1}^{\infty} f(x_i, y_j).$$

Definition 6.4.1

For every i and j, we define the probabilities

$$h_1(x_i | y_j) = Pr(X = x_i | Y = y_j)$$

$$= \frac{Pr(X = x_i, Y = y_j)}{Pr(Y = y_j)} = \frac{f(x_i, y_j)}{f_2(y_j)} \quad (6.4.1)$$

and

$$h_2(y_j | x_i) = Pr(y = y_j | X = x_i)$$

$$= \frac{Pr(X = x_i, Y = y_j)}{Pr(X = x_i)} = \frac{f(x_i, y_j)}{f_1(x_i)}. \quad (6.4.2)$$

If y_j is fixed and x_i varies over all possible values, then the expression in Equation (6.4.1) is the **conditional probability density function** of the random variable X of the discrete type, under the condition $Y = y_j$.

Similarly, if x_i is fixed and y_j varies over all possible values, then the expression in Equation (6.4.2) is the conditional probability density function of the discrete variate Y, given that

$X = x_i$. Moreover, $f_1(x_i)$ and $f_2(y_j)$ must be greater than zero. The conditional densities in Equations (6.4.1) and (6.4.2) are nonnegative:

$$\sum_{i=1}^{\infty} h(x_i | y_j) = \frac{\sum_{i=1}^{\infty} f(x_i, y_j)}{f_2(y_j)} = \frac{f_2(y_j)}{f_2(y_j)} = 1$$

and

$$\sum_{j=1}^{\infty} h(y_j | x_i) = \frac{\sum_{j=1}^{\infty} f(x_i, y_j)}{f_1(x_i)} = \frac{f_1(x_i)}{f_1(x_i)} = 1.$$

Thus, the conditions of a probability density are satisfied.

Example 6.4.1

Consider Example 6.3.1; we want to determine the conditional probability $Pr(X = -1 | Y = 0)$.

Solution: Applying Equation (6.4.1), we obtain

$$h_1(-1 | 0) = Pr(X = -1 | Y = 0) = \frac{Pr(X = -1, Y = 0)}{Pr(Y = 0)}$$

$$= \frac{0.1}{0.1 + 0.2 + 0} = \frac{0.1}{0.3} = \frac{1}{3}.$$

Similarly,

$$h_2(4 | -1) = Pr(Y = 4 | X = -1) = \frac{Pr(X = -1, Y = 4)}{Pr(X = -1)}$$

$$= \frac{0.2}{0.3 + 0.1 + 0 + 0.2} = \frac{0.2}{0.6} = \frac{1}{3}.$$

◀

Definition 6.4.2

Let $f(x, y)$ be the joint probability density function of the continuous random variable (X, Y). The **conditional density function** of the random variable X for a given value $Y = y$ is defined by

$$h_1(x | y) = \frac{f(x, y)}{f_2(y)}, \quad f_2(y) > 0, \tag{6.4.3}$$

and the conditional probability density function of the variate Y for a given value $X = x$ is defined by

$$h_2(y | x) = \frac{f(x, y)}{f_1(x)}, \quad f_1(x) > 0. \tag{6.4.4}$$

Note that $f_1(x)$ and $f_2(y)$ are the marginal probability density functions of the variates X and Y:

$$f_1(x) = \int_{-\infty}^{\infty} f(x,y)\,dy$$

and

$$f_2(y) = \int_{-\infty}^{\infty} f(x,y)\,dx.$$

As in the discrete case, the expressions in Equations (6.4.3) and (6.4.4) satisfy all requirements of a one-dimensional probability density function. For a fixed y, $h_1(x|y) \geq 0$ and

$$\int_{-\infty}^{\infty} h_1(x|y)\,dx = \int_{-\infty}^{\infty} \frac{f(x,y)}{f_2(y)}\,dx = \frac{1}{f_2(y)} \int_{-\infty}^{\infty} f(x,y)\,dx = \frac{f_2(y)}{f_2(y)} = 1.$$

The situation is similar for the conditional density $h_2(y|x)$. Thus, if we want to obtain the probability that the value of X is in $S_1 = \{x : a \leq x \leq b\}$, given that Y is known, we have

$$Pr(a \leq X \leq b | y) = \int_a^b h_1(x|y)\,dx.$$

The conditional cumulative distribution functions of the continuous random variables X given Y and Y given X are

$$F_X(x|y) = Pr(X \leq x | Y \leq y) = \frac{Pr(X \leq x, Y \leq y)}{Pr(Y \leq y)}$$

$$= \frac{\int_{-\infty}^{x}\left\{\int_{-\infty}^{y} f(t,s)\,ds\right\}dt}{\int_{-\infty}^{y}\left\{\int_{-\infty}^{\infty} f(x,s)\,dx\right\}ds}$$

$$= \frac{\int_{-\infty}^{x}\left\{\int_{-\infty}^{y} f(t,s)\,ds\right\}dt}{\int_{-\infty}^{y} f_2(s)\,ds} \qquad (6.4.5)$$

and

$$F_Y(y|x) = Pr(Y \leq y | X \leq x) = \frac{Pr(X \leq x, Y \leq y)}{Pr(X \leq x)}$$

$$= \frac{\int_{-\infty}^{x}\left\{\int_{-\infty}^{y} f(t,s)\,ds\right\}dt}{\int_{-\infty}^{y} f_1(t)\,dt}, \qquad (6.4.6)$$

respectively. However, a more common way of introducing the concept of cumulative distribution function is

$$F_X(x|y) = \int_{-\infty}^{x} h(t|y)\,dt = \int_{-\infty}^{x} \frac{f(t,y)}{f(y)}\,dt. \qquad (6.4.7)$$

6.4 Conditional Probability Density and Cumulative Distribution Functions

We now discuss the development of Equation (6.4.7) in view of what has come before it. From Equation (6.4.5), we have

$$F_X(x|y) = \frac{F(x,y)}{F_2(y)}. \tag{6.4.8}$$

Suppose that $f_2(y) \neq 0$. We can then define $F_X(x|Y=y)$ as the limit

$$F_X(x|Y=y) = \lim_{\Delta y \to 0} F_X(x|y < Y \leq y + \Delta y).$$

Equation (6.4.8) can also be written as

$$F_X(x|Y=y) = \frac{\lim_{\Delta y \to 0} \dfrac{F(x, y+\Delta y) - F(x,y)}{\Delta y}}{\lim_{\Delta y \to 0} \dfrac{F(y+\Delta y) - F(y)}{\Delta y}}$$

$$= \frac{\dfrac{\partial}{\partial y} F(x,y)}{\dfrac{d}{dy} F_2(y)}.$$

Assuming that the partial derivative of $F(x,y)$ with respect to y and the derivative of $F_2(y)$ exist, then

$$F_X(x|Y=y) = \frac{\int_{-\infty}^{x} f(t,y)\,dt}{f_2(y)}, \quad f_2(y) > 0. \tag{6.4.9}$$

Similarly,

$$F_Y(y|X=x) = \frac{\int_{-\infty}^{y} f(x,s)\,ds}{f_1(x)}, \quad f_1(x) > 0.$$

We now illustrate that we can express marginal cumulative distributions as functions of conditional cumulative distributions. In view of Equation (6.4.9), we have

$$\int_{-\infty}^{x} f(t,y)\,dt = f_2(y) F_X(x|y).$$

It follows that

$$\int_{-\infty}^{x} \left\{ \int_{-\infty}^{\infty} f(t,y)\,dy \right\} dt = \int_{-\infty}^{\infty} f_2(y) F_X(x|y)\,dy,$$

$$\int_{-\infty}^{x} f_1(t)\,dt = \int_{-\infty}^{\infty} f_2(y) F_X(x|y)\,dy,$$

or

$$F_1(x) = \int_{-\infty}^{\infty} f_2(y) F_X(x|y)\,dy.$$

Similarly,

$$F_2(y) = \int_{-\infty}^{\infty} f_1(x) F_Y(y|x) dx.$$

The conditional distribution function satisfies the properties of ordinary distributions:

1. $F_X(\infty|y) = \dfrac{\int_{-\infty}^{\infty} f(x,y) dx}{f_2(y)} = \dfrac{f_2(y)}{f_2(y)} = 1.$

2. $F_X(-\infty|y) = \int_{-\infty}^{\infty} \dfrac{f(x,y) dx}{f_2(y)} = 0.$

3. $F_X(x_2|y) - F_X(x_1|y) = Pr(x_1 < X \leq x_2 |y)$

   $= \dfrac{Pr(x_1 < X \leq x_2, Y = y)}{f_2(y)} \geq 0,$ for $f_2(y) > 0.$

4. If the random variable X is of the continuous type, then

$$h(x|y) = \dfrac{d}{dx} F_X(x|y) = \lim_{h \to 0} \dfrac{Pr\{x \leq X \leq x+h | Y = y\}}{h}.$$

Example 6.4.2

The joint probability density function of the continuous random variable (X,Y) is

$$f(x,y) = \begin{cases} \dfrac{1}{28}(4x+2y+1), & 0 \leq x \leq 2, 0 \leq y \leq 2, \\ 0, & \text{elsewhere.} \end{cases}$$

We want to find the marginal densities of X and Y.

Solution: For $0 \leq x \leq 2$,

$$f_1(x) = \dfrac{1}{28} \int_0^2 (4x+2y+1) dy = \dfrac{1}{14}(4x+3);$$

therefore,

$$f_1(x) = \begin{cases} \dfrac{1}{14}(4x+3), & 0 \leq x \leq 2, \\ 0, & \text{elsewhere.} \end{cases}$$

For $0 \leq y \leq 2$,

$$f_2(y) = \dfrac{1}{28} \int_0^2 (4x+2y+1) dx = \dfrac{1}{14}(2y+5);$$

6.4 Conditional Probability Density and Cumulative Distribution Functions

therefore,

$$f_2(y) = \begin{cases} \dfrac{1}{14}(2y+5), & 0 \le y \le 2, \\ 0, & \text{elsewhere.} \end{cases}$$

From Equation (6.4.3), the conditional density of X for a given value of Y is as follows: For $0 \le x \le 2$ and $0 \le y \le 2$,

$$h_1(x|y) = \dfrac{\dfrac{1}{28}(4x+2y+1)}{\dfrac{1}{14}(2y+5)} = \dfrac{1}{2}\dfrac{(4x+2y+1)}{(2y+5)};$$

therefore,

$$h_1(x|y) = \begin{cases} \dfrac{1}{2}\dfrac{(4x+2y+1)}{(2y+5)}, & 0 \le x \le 2 \text{ and } 0 \le y \le 2, \\ 0, & \text{elsewhere.} \end{cases}$$

Also, for $0 \le y \le 2$ and $0 \le x \le 2$,

$$h_2(y|x) = \dfrac{\dfrac{1}{28}(4x+2y+1)}{\dfrac{1}{14}(4x+3)} = \dfrac{1}{2}\dfrac{(4x+2y+1)}{(4x+3)};$$

therefore,

$$h_2(y|x) = \begin{cases} \dfrac{1}{2}\dfrac{(4x+2y+1)}{(4x+3)}, & 0 \le y \le 2 \text{ and } 0 \le x \le 2, \\ 0, & \text{elsewhere.} \end{cases}$$

◀

Example 6.4.3

We have seen that the marginal densities of the bivariate density function given in Example 6.3.3 are

$$f_1(x) = \begin{cases} \dfrac{4}{9}x(4-x^2), & 1 \le x \le 2, \\ 0, & \text{elsewhere,} \end{cases}$$

and

$$f_2(y) = \begin{cases} \dfrac{4}{9}y(y^2-1), & 1 \le y \le 2, \\ 0, & \text{elsewhere.} \end{cases}$$

We want to find the conditional probability density function of X for a fixed $Y = y_0$.

Solution: For $1 \leq x \leq y_0$ and $1 \leq y_0 \leq 2$,

$$h_1(x|y_0) = \frac{f(x,y_0)}{f_2(y_0)} = \frac{\frac{8}{9}xy_0}{\frac{4}{9}y_0(y_0^2-1)} = \frac{2x}{y_0^2-1};$$

therefore,

$$h_1(x|y) = \begin{cases} \frac{2x}{y_0^2-1}, & 1 \leq x \leq y_0 \text{ and } 1 \leq y_0 \leq 2, \\ 0, & \text{elsewhere.} \end{cases}$$

Similarly, for $x_0 \leq y \leq 2$ and $1 \leq x_0 \leq 2$,

$$h_2(y|x_0) = \frac{f(x_0,y)}{f_1(x_0)} = \frac{\frac{8}{9}x_0 y}{\frac{4}{9}x_0(4-x_0^2)} = \frac{2y}{4-x_1^2};$$

therefore,

$$h_2(y|x_0) = \begin{cases} \frac{2y}{4-x_0^2}, & x_0 \leq y \leq 2 \text{ and } 1 \leq x_0 \leq 2, \\ 0, & \text{elsewhere.} \end{cases}$$

Furthermore, for $1 \leq x \leq y_0$ and $1 \leq y_0 \leq 2$,

$$F_X(x|y_0) = Pr(X \leq x|Y=y_0) = \frac{PrX \leq x, Y=y_0}{Pr(Y=y_0)}$$

$$= \frac{\int_1^x \frac{8}{9}ty_0 \, dt}{\int_1^{y_0} \frac{8}{9}xy_0 \, dx} = \frac{x^2-1}{y_0^2-1}.$$

Thus,

$$F_X(x|y_0) = \begin{cases} 0, & x < 1, \\ \frac{x^2-1}{y_0^2-1}, & 1 \leq x \leq y_0 \text{ and } 1 \leq y_0 \leq 2, \\ 1, & x > y_0. \end{cases}$$

The same result can be obtained for $1 \leq x \leq y_0$ and $1 \leq y_0 \leq 2$ by using

$$F_X(x|y_0) = \int_1^x h_1(t|y_0) \, dt = \int_1^x \frac{2t}{y_0^2-1} \, dt = \frac{x^2-1}{y_0^2-1}.$$

Also, for $x_0 \leq y \leq 2$ and $1 \leq x_0 \leq 2$,

$$F_Y(y|x_0) = \frac{\int_{x_0}^{y} \frac{8}{9} x_0 s \, ds}{\int_{x_0}^{2} \frac{8}{9} x_0 y \, dy},$$

or for $x_0 \leq y \leq 2$ and $1 \leq x_0 \leq 2$,

$$F_Y(y|x_0) = \int_{x_0}^{y} h_2(s|x_0) \, ds = \int_{x_0}^{y} \frac{2s}{4 - x_0^2} \, ds = \frac{y^2 - x_0^2}{4 - x_0^2}.$$

Thus,

$$F_Y(y|x_0) = \begin{cases} 0, & y < x_0, \\ \dfrac{y^2 - x_0^2}{4 - x_0^2}, & x_0 \leq y \leq 2 \text{ and } 1 \leq x_0 \leq 2, \\ 1, & y > 2. \end{cases}$$

6.5 Independent Random Variables

In Section 1.6, we studied the notion of independent events. That is, two random events S_1 and S_2 are *independent* of each other if

$$Pr(S_1 \cap S_2) = Pr(S_1) Pr(S_2).$$

In this section, we introduce the notion of independent random variables. Throughout the section, we let $F(x, y)$, $F_1(x)$, and $F_2(y)$ represent the cumulative distribution function of the two-dimensional random variable (X, Y) and the marginal cumulative distribution functions of the variates X and Y, respectively.

Definition 6.5.1

Two random variables X and Y are said to be *independent (mutually independent)* if

$$F(x, y) = F_1(x) F_2(y) \tag{6.5.1}$$

is satisfied for every real pair of numbers (x, y).

Thus, the random variables X and Y are independent if the events $\{X \leq x\}$ and $\{Y \leq y\}$ are independent for every real x and y. That is,

$$Pr(X \leq x, Y \leq y) = Pr(X \leq x) Pr(Y \leq y).$$

Furthermore, for all values x_1, x_2 $(x_2 > x_1)$ and y_1, y_2 $(y_2 > y_1)$, we show that, if X and Y are independent,

$$Pr(x_1 < X \le x_2, y_1 < Y \le y_2) = Pr(x_1 < X \le x_2)Pr(y_1 < Y \le y_2):$$

$$Pr(x_1 < X \le x_2, y_1 < Y \le y_2) = F(x_1, y_1) - F(x_1, y_2)$$
$$-F(x_2, y_1) + F(x_2, y_2)$$
$$= F_1(x_1)F_2(y_1) - F_1(x_1)F_2(y_2)$$
$$-F_1(x_2)F_2(y_1) + F_1(x_2)F_2(y_2)$$
$$= \{F_1(x_2) - F_1(x_1)\}\{F_2(y_2) - F_2(y_1)\}$$
$$= Pr(x_1 < X \le x_2)Pr(y_1 < Y \le y_2).$$

Thus, if X and Y are independent, the probability that the random variable (X,Y) assumes values in the rectangle shown in Figure 6.2.1 is equal to the product of the probabilities that X assumes its values in the vertical strip and that Y assumes its values in the horizontal strip, both shown in the figure.

Theorem 6.5.1

Let $f(x, y)$ be the joint probability density function of the random variable (X,Y).

1. If (X,Y) is of the discrete type, then X and Y are independent random variables if and only if

$$f(x_i, y_j) = f_1(x_i)f_2(y_j) \tag{6.5.2}$$

for all i and j.

2. If (X,Y) is of the continuous type, then X and Y are independent random variables if and only if

$$f(x, y) = f_1(x)f_2(y) \tag{6.5.3}$$

for all real (x, y).

Proof We now prove the second part of the theorem. Applying the relation between probability density and cumulative distribution functions, we have

$$f(x, y) = \frac{\partial^2 F(x, y)}{\partial x \partial y} = \frac{dF_1(x)}{dx} \frac{dF_2(y)}{dy} = f_1(x)f_2(y).$$

Conversely,

$$F(x,y) = \int_{-\infty}^{x}\left\{\int_{-\infty}^{y} f(t,s)\,ds\right\}dt$$

$$= \int_{-\infty}^{x}\left\{\int_{-\infty}^{y} f_1(t)f_2(s)\,ds\right\}dt$$

$$= \int_{-\infty}^{x} f_1(t)\,dt \int_{-\infty}^{y} f_2(s)\,ds$$

$$= Pr(X \leq x) Pr(Y \leq y)$$

$$= F_1(x) F_2(y).$$

Independence of random variables can be equivalently defined by using the notion of conditional probability density functions, as shown by Theorem 6.5.2.

Theorem 6.5.2

Let $f(x,y)$, $f_1(x)$, and $f_2(y)$ be the joint probability density function of the random variable (X,Y) and the marginal densities X and Y, respectively.

1. If (X,Y) is of the discrete type, then X and Y are independent if and only if

$$h(x_i | y_j) = f_1(x_i) \text{ or } h_2(y_j | x_i) = f_2(y_j) \tag{6.5.4}$$

for all i and j.

2. If (X,Y) is of the continuous type, then X and Y are independent if and only if

$$h_1(x|y) = f_1(x) \text{ or } h_2(y|x) = f_2(y) \tag{6.5.5}$$

for all real (x,y), such that $f_1(x)$ and $f_2(x)$ are positive.

Proof The proof is straightforward, applying the definition of conditional probability density function and Theorem 6.5.1.

Example 6.5.1

Suppose that a fair coin is tossed twice. Let the random variable X assume the value 0 or 1, depending on whether a head or a tail appears on the first toss. The random variable Y assumes the value 0 or 1, depending on whether a head or a tail appears on the second toss. Table 6.5.1 gives the joint probability density (X,Y).

Table 6.5.1

		y		
x		0	1	$f_1(x_i)$
0		$\frac{1}{4}$	$\frac{1}{4}$	$\frac{1}{2}$
1		$\frac{1}{4}$	$\frac{1}{4}$	$\frac{1}{2}$
$f_2(y_j)$		$\frac{1}{2}$	$\frac{1}{2}$	1

Thus, because

$$f(x_i, y_j) = f_1(x_i) f_2(y_j),$$

X and Y are independent random variables.

Example 6.5.2

The cumulative distribution function of the continuous random variable (X,Y) is given by

$$F(x,y) = \begin{cases} 1 - e^{-x} - e^{-y} + e^{-(x+y)}, & x, y \geq 0, \\ 0, & \text{elsewhere.} \end{cases}$$

The joint probability density function is

$$\frac{\partial^2 F(x,y)}{\partial x \partial y} = f(x,y) = \begin{cases} e^{-(x+y)} & x, y \geq 0, \\ 0, & \text{elsewhere.} \end{cases}$$

The marginal probability density and marginal cumulative functions of X and Y, are

$$f_1(x) = \begin{cases} e^{-x}, & x \geq 0, \\ 0, & \text{elsewhere,} \end{cases}$$

$$f_2(y) = \begin{cases} e^{-y}, & y \geq 0, \\ 0, & \text{elsewhere,} \end{cases}$$

$$F_1(x) = \begin{cases} 0, & y < 0, \\ 1 - e^{-x}, & x \geq 0, \end{cases}$$

and

$$F_2(y) = \begin{cases} 0, & x < 0, \\ 1 - e^{-y}, & y \geq 0. \end{cases}$$

In view of Equation (6.5.1), we have

$$F_1(x) F_2(y) = (1 - e^{-x})(1 - e^{-y}) = 1 - e^{-x} - e^{-y} + e^{-(x+y)} = F(x,y).$$

Hence, X and Y are independent. Their independence is also clear from Equations (6.5.3) and (6.5.5).

Example 6.5.3

Referring to Example 6.1.4,

$$f(x,y) = \begin{cases} \dfrac{8}{9}xy, & 1 \le x \le y \le 2, \\ 0, & \text{elsewhere,} \end{cases}$$

$$f_1(x) = \begin{cases} \dfrac{4}{9}x(4-x^2), & 1 \le x \le 2, \\ 0, & \text{elsewhere,} \end{cases}$$

and

$$f_2(y) = \begin{cases} \dfrac{4}{9}y(y^2-1), & 1 \le y \le 2, \\ 0, & \text{elsewhere.} \end{cases}$$

Thus,

$$f_1(x)f_2(y) = \frac{16}{81}xy(4-x^2)(y^2-1) \ne f(x,y),$$

and the random variables X and Y are not independent.

Theorem 6.5.3

If X and Y are independent random variables, then the one-to-one functional forms of X and Y, $W_1 = g_1(X)$ and $W_2 = g_2(Y)$, are also independent.

Proof Because W_1 and W_2 are single-valued functions, their inverses exist; thus,

$$F(w_1, w_2) = Pr(W_1 \le w_1, W_2 \le w_2)$$

$$= Pr(g_1(X) \le w_1, g_2(Y) \le w_2)$$

$$= Pr(X \le g_1^{-1}(w_1), Y \le g_2^{-1}(w_2))$$

$$= Pr(X \le g_1^{-1}(w_1)) Pr(Y \le g_2^{-1}(w_2))$$

$$= Pr(g_1(X) \le w_1) Pr(g_2(X) \le w_2)$$

$$= Pr(W_1 \le w_1) Pr(W_2 \le w_2)$$

$$= F_1(w_1) F_2(w_2).$$

The usefulness of this theorem is shown in later sections.

6.6 Function of Two Random Variables

We saw in Chapter 4 that a functional form of a random variable is also a random variable, and we discussed the manner in which we can derive such distribution functions. In this section, we study some functional forms of two random variables.

> **6.6.1 One Function of Two Discrete Random Variables** We begin by investigating one function of two discrete random variables. Let (X,Y) be a discrete two-dimensional random variable whose distribution is known; we wish to obtain the distribution of $H = g(X,Y)$. To obtain the probability mass function of H is quite easy, and we illustrate the function by considering Example 6.6.1.

Example 6.6.1

Consider an experiment in which there are two consecutive throws of a die. Let X and Y denote the results of the first and second throws, respectively. Thus, both X and Y take on the values 1, 2,…,6, each with the probability 1/6, because they are independent, as shown in Table 6.6.1.

Table 6.6.1

				y			
x	1	2	3	4	5	6	$f_1(x_i)$
1	$\frac{1}{36}$	$\frac{1}{36}$	$\frac{1}{36}$	$\frac{1}{36}$	$\frac{1}{36}$	$\frac{1}{36}$	$\frac{1}{6}$
2	$\frac{1}{36}$	$\frac{1}{36}$	$\frac{1}{36}$	$\frac{1}{36}$	$\frac{1}{36}$	$\frac{1}{36}$	$\frac{1}{6}$
3	$\frac{1}{36}$	$\frac{1}{36}$	$\frac{1}{36}$	$\frac{1}{36}$	$\frac{1}{36}$	$\frac{1}{36}$	$\frac{1}{6}$
4	$\frac{1}{36}$	$\frac{1}{36}$	$\frac{1}{36}$	$\frac{1}{36}$	$\frac{1}{36}$	$\frac{1}{36}$	$\frac{1}{6}$
5	$\frac{1}{36}$	$\frac{1}{36}$	$\frac{1}{36}$	$\frac{1}{36}$	$\frac{1}{36}$	$\frac{1}{36}$	$\frac{1}{6}$
6	$\frac{1}{36}$	$\frac{1}{36}$	$\frac{1}{36}$	$\frac{1}{36}$	$\frac{1}{36}$	$\frac{1}{36}$	$\frac{1}{6}$
$f_2(y_j)$	$\frac{1}{6}$	$\frac{1}{6}$	$\frac{1}{6}$	$\frac{1}{6}$	$\frac{1}{6}$	$\frac{1}{6}$	1

Solution:

a. We want to obtain the probability density of the one-dimensional random variable H, where $H = g(X,Y) = X+Y$. The variate H can assume the values $2, 3, \ldots, 12$, and because X and Y are independent, we can compute the probability function of H. That is,

$$Pr(H=2) = Pr(X=1)Pr(Y=1) = \frac{1}{36},$$

$$Pr(H=3) = Pr(X=1)Pr(Y=2)$$

$$+Pr(X=2)Pr(Y=1) = \frac{1}{36} + \frac{1}{36} = \frac{1}{18},$$

$$Pr(H=4) = Pr(X=1)Pr(Y=3) + Pr(X=3)Pr(Y=1)$$

$$+Pr(X=2)Pr(Y=2) = \frac{1}{36} + \frac{1}{36} + \frac{1}{36} = \frac{1}{12},$$

and so on. Table 6.6.2 shows the probability mass function of the variate H.

Table 6.6.2

h_i	2	3	4	5	6	7	8	9	10	11	12
$f(h_i)$	$\frac{1}{36}$	$\frac{2}{36}$	$\frac{3}{36}$	$\frac{4}{36}$	$\frac{5}{36}$	$\frac{6}{36}$	$\frac{5}{36}$	$\frac{4}{36}$	$\frac{3}{36}$	$\frac{2}{36}$	$\frac{1}{36}$

b. Similarly, we can obtain the probability density function of W, where $W = g(X,Y) = X - Y$. The variate W can assume the values $-5, -4, \ldots, 0, \ldots, 4, 5$. Its distribution is shown in Table 6.6.3.

Table 6.6.3

w_i	-5	-4	-3	-2	-1	0	1	2	3	4	5
$f(w_i)$	$\frac{1}{36}$	$\frac{2}{36}$	$\frac{3}{36}$	$\frac{4}{36}$	$\frac{5}{36}$	$\frac{6}{36}$	$\frac{5}{36}$	$\frac{4}{36}$	$\frac{3}{36}$	$\frac{2}{36}$	$\frac{1}{36}$

Therefore, to obtain the probability mass function of H, we have

$$f(h_k) = Pr(H = h_k) = \sum_{x_i + y_j = h_k} Pr(X = x_i, Y = y_j)$$

$$= \sum_{x_i + y_j = h_k} f(x_i, y_j),$$

and for the variate W, we have

$$f(w_r) = Pr(W = w_r) = \sum_{x_i - y_j = w_r} Pr(X = x_i, Y = y_j)$$

$$= \sum_{x_i - y_j = w_r} f(x_i, y_j).$$

The cumulative distribution function of H is simply

$$F(h) = Pr(H \leq h) = \sum_{x_i + y_j \leq h} f(x_i, y_j).$$

6.6.2 Two Functions of Two Continuous Random Variables Before we study the manner in which we can obtain the distribution of a function of two continuous random variables, we consider the joint distribution of two functional forms of such a variate. Suppose that we are given $f(x, y)$, the joint probability density function of the continuous random variable (X, Y), and we are interested in obtaining a new probability density function $z(u, v)$ of the random variable (U, V), where the relation between the new and the old variates is given by

$$U = g_1(X, Y) \text{ and } V = g_2(X, Y). \tag{6.6.1}$$

To obtain this new density function, we must assume that the functions g_1 and g_2 define a one-to-one transformation that maps a two-dimensional set S of the x-y plane onto a two-dimensional set E^* in the u-v plane and has continuous first partial derivatives with respect to x and y (as shown later in Figure 6.6.1). The conditions imply that we can uniquely solve for the inverse of the transformation:

$$x = z_1^{-1}(u, v) \text{ and } y = z_2^{-1}(u, v). \tag{6.6.2}$$

Furthermore, we assume that the inverse of the transformation has continuous first partial derivatives with respect to u and v.

Under the preceding assumptions, the joint probability density function of the new random variable (U, V) is given by

$$z(u, v) = f\left[z_1^{-1}(u, v), z_2^{-1}(u, v)\right]|J|, \tag{6.6.3}$$

where $|J|$ is the absolute value of the Jacobian of the transformation, defined by the determinant

$$J = \frac{\partial(x, y)}{\partial(u, v)} = \begin{vmatrix} \frac{\partial x}{\partial u} & \frac{\partial x}{\partial v} \\ \frac{\partial y}{\partial u} & \frac{\partial y}{\partial v} \end{vmatrix},$$

which cannot be equal to zero. For complete justification of the development of $z(u, v)$, we need certain concepts of advanced calculus, which are beyond the scope of this book; however, you can see how this function is obtained by considering the following formulation: Let $E_1 \subset E$ be a two-dimensional subset (an event of E) in the x-y plane, and let $E_2 \subset E^*$ be a two-dimensional subset (an event E^*) in the u-v plane. Denote the range, or image, of E_1 under a one-to-one transformation, as shown in Figure 6.6.1. Thus, the events E_1 and E_2, where $(X,Y) \in E_1$ and $(U,V) \in E_2$, are equivalent, and

$$Pr\left[(X,Y) \in E_1\right] = \iint_{E_1} f(x,y)\,dx\,dy = Pr\left[(U,V) \in S_2\right]$$

$$= \iint_{E_2} f\left[z_1^{-1}(u,v), z_2^{-1}(u,v)\right]|J|\,du\,dv,$$

where

$$x = z_1^{-1}(u,v) \text{ and } y = z_2^{-1}(u,v).$$

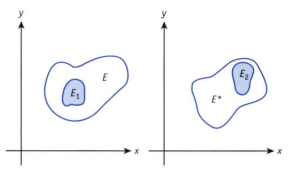

Figure 6.6.1

This situation implies that the joint probability density function of the random variable (U,V) is

$$z(u,v) = f\left[z_1^{-1}(u,v),\, z_2^{-1}(u,v)\right]|J|.$$

Hence, from the knowledge of $z(u,v)$, we can obtain the marginal distributions of U and V, say, $f_1(u)$ and $f_2(v)$, by integrating $z(u,v)$ with respect to v and u, respectively:

$$f_1(u) = \int_{-\infty}^{\infty} z(u,v)\,dv \text{ and } f_2(v) = \int_{-\infty}^{\infty} z(u,v)\,du.$$

We summarize the preceding discussion in Theorem 6.6.1.

■ Theorem 6.6.1

Let $f(x,y)$ be the joint probability density function of the continuous random variable (X,Y). Suppose that

$$U = g_1(X,Y)$$

and

$$V = g_2(X,Y),$$

continued

Chapter 6 Two Random Variables

such that g_1 and g_2 define a one-to-one transformation of the random variable (X,Y) and that this random variable has continuous first partial derivatives, with respect to x and y. The joint probability density function of the random variable (U,V) is given by

$$z(u,v) = f\left[z_1^{-1}(u,v), z_2^{-1}(u,v)\right]|J|,$$

where

$$x = z_1^{-1}(u,v),$$

$$y = z_2^{-1}(u,v),$$

and the Jacobian $|J| \neq 0$.

Example 6.6.2

Let

$$f(x,y;\sigma) = \frac{1}{2\pi\sigma^2} e^{-(1/2\sigma^2)(x^2+y^2)}, \quad -\infty < x, y < \infty,$$

be the joint probability density function of the variate (X,Y). Let

$$U = \sqrt{X^2 + Y^2}$$

and

$$V = \tan^{-1}\left(\frac{Y}{X}\right), \quad 0 \leq V \leq 2\pi.$$

We wish to determine the probability density function of the random variables U and V.

Solution: For the transformation be one to one, we write

$$f(x,y;\sigma) = \begin{cases} \dfrac{1}{2\pi\sigma^2} e^{-(1/2\sigma^2)(x^2+y^2)}, & x < 0, -\infty < y < \infty, \\ \dfrac{1}{2\pi\sigma^2} e^{-(1/2\sigma^2)(x^2+y^2)}, & x \geq 0, -\infty < y < \infty; \end{cases}$$

then

$$\begin{aligned} X &= U\cos V \\ Y &= U\sin V \end{aligned} \bigg| \begin{aligned} &U > 0 \\ &\frac{\pi}{2} < V < \frac{3\pi}{2} \end{aligned}$$

for the first part of $f(x,y;\sigma)$ and

$$\begin{aligned} X &= U\cos V \\ Y &= U\sin V \end{aligned} \bigg| \begin{aligned} &U > 0 \\ &-\frac{\pi}{2} < V \leq \frac{\pi}{2} \end{aligned}$$

for the second part of the joint probability density function $f(x, y; \sigma)$, as shown in Figure 6.6.2.

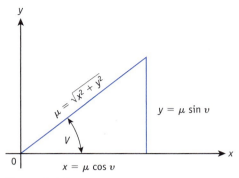

Figure 6.6.2

The Jacobian of the transformation is

$$J = \begin{vmatrix} \dfrac{\partial x}{\partial u} & \dfrac{\partial x}{\partial v} \\ \dfrac{\partial y}{\partial u} & \dfrac{\partial y}{\partial v} \end{vmatrix} = \begin{vmatrix} \cos v & -u \sin v \\ \sin v & u \cos v \end{vmatrix} = u \cos^2 v + u \sin^2 v = u.$$

Thus, applying Equation (6.6.3), we have

$$z(u, v) = \begin{cases} \dfrac{u}{2\pi\sigma^2} e^{-(u^2/2\sigma^2)}, & u > 0, \dfrac{\pi}{2} < v < \dfrac{3\pi}{2}, \\ \dfrac{u}{2\pi\sigma^2} e^{-(u^2/2\sigma^2)}, & u > 0, -\dfrac{\pi}{2} < v < \dfrac{\pi}{2}, \\ 0, & \text{elsewhere.} \end{cases}$$

The marginal probability density of the random variable U is obtained as follows:

$$z_1(u) = \dfrac{1}{2\pi\sigma^2} \int_{\frac{\pi}{2}}^{\frac{3\pi}{2}} u e^{-(u^2/2\sigma^2)} dv + \dfrac{1}{2\pi\sigma^2} \int_{-\frac{\pi}{2}}^{\frac{\pi}{2}} u e^{-(u^2/2\sigma^2)} dv$$

$$= \dfrac{1}{2\pi\sigma^2} u e^{-(u^2/2\sigma^2)} \int_0^{2\pi} dv = \dfrac{1}{\sigma^2} u e^{-(u^2/2\sigma^2)}.$$

Therefore,

$$z_1(u) = \begin{cases} \dfrac{1}{\sigma^2} u e^{-(u^2/2\sigma^2)}, & u > 0, \\ 0, & \text{elsewhere.} \end{cases}$$

The marginal probability density of the variate V is derived as follows: For $\pi/2 < v < 3\pi/2$, we obtain $\int_0^\infty z(u,v)\,du = 1/2\pi$. For $-(\pi/2) < v < \pi/2$, we also obtain the value $1/2\pi$. Combining these two results gives

$$z_2(v) = \begin{cases} \dfrac{1}{2\pi}, & 0 < v < 2\pi, \\ 0, & \text{elsewhere.} \end{cases}$$

Hence, the variate V is uniformly distributed over the interval from 0 to 2π. When we replace σ^2 with $\alpha/2$ in $z_1(u)$, we obtain the Rayleigh distribution, as defined in Section 3.3. ◂

Example 6.6.3

Let the probability density function of the random variable (X,Y) be

$$f(x,y) = \begin{cases} \beta^{-(1/2)} e^{-\{(x+y)/\beta\}}, & x, y > 0, \beta > 0, \\ 0, & \text{elsewhere.} \end{cases}$$

In addition, let

$$U = \frac{X - Y}{2}$$

and

$$V = Y.$$

We wish to determine the probability density of (U,V).

Solution: Here, $x = 2u + v$ and $y = v$ define a one-to-one transformation from event $E_1 = \{(x,y): x, y > 0\}$ onto event $E_2 = \{(u,v): -2u < v \text{ and } 0 < v, -\infty < u < \infty\}$. The Jacobian of the transformation is

$$J = \begin{vmatrix} \dfrac{\partial x}{\partial u} & \dfrac{\partial x}{\partial v} \\ \dfrac{\partial y}{\partial u} & \dfrac{\partial y}{\partial v} \end{vmatrix} = \begin{vmatrix} 2 & 1 \\ 0 & 1 \end{vmatrix} = 2.$$

Thus, in view of Equation (6.6.3), the joint probability density function of the random variable (U,V) is

$$z(u,v) = \begin{cases} \dfrac{|2|}{\beta^2} e^{-(2/\beta)(u+v)}, & -\infty < u < \infty, \quad v > -2u, \text{ if } u < 0, \\ & \qquad\qquad\qquad\quad\; v > 0, \quad\; \text{if } u \geq 0, \\ 0, & \text{elsewhere,} \qquad\qquad\qquad \beta > 0. \end{cases}$$

The marginal probability density function of U is

$$z_1 u = \begin{cases} \int_{-2u}^{\infty} \dfrac{2}{\beta^2} e^{-(2/\beta)(u+v)} dv, & u < 0, \\ \int_{0}^{\infty} \dfrac{2}{\beta^2} e^{-(2/\beta)(u+v)} dv, & u \geq 0, \end{cases}$$

or

$$z_1(u) = \begin{cases} \dfrac{1}{\beta} e^{2u/\beta}, & u < 0, \\ \dfrac{1}{\beta} e^{-(2u/\beta)}, & u \geq 0. \end{cases}$$

Thus,

$$z_1(u) = \dfrac{1}{\beta} e^{-(2/\beta)|u|}, \quad -\infty < u < \infty.$$

It can easily be shown that

$$\int_{-\infty}^{0} \dfrac{1}{\beta} e^{2u/\beta} du + \int_{-\infty}^{0} \dfrac{1}{\beta} e^{-(2u/\beta)} du = 1.$$

◀

Example 6.6.4

Let X and Y be two independent chi-square–distributed variates with n_1 and n_2 degrees of freedom, respectively. We wish to obtain the distribution of (U,V), where

$$U = X + Y$$

and

$$V = \dfrac{X}{Y}.$$

Solution: The joint probability density function of X and Y can be written as

$$f(x,y) = f_1(x) f_2(y) = \dfrac{1}{\Gamma(n_1/2)} \dfrac{1}{2^{n_1/2}} x^{(n_1/2)-1} e^{-(1/2)x}$$

$$\cdot \dfrac{1}{\Gamma(n_2/2)} \dfrac{1}{2^{n_2/2}} y^{(n_2/2)-1} e^{-(1/2)y} \quad x, y > 0$$

$$= \begin{cases} \dfrac{1}{\Gamma(n_1/2)\Gamma(n_2/2) 2^{(n_1+n_2)/2}} x^{(n_1/2)-1} y^{(n_2/2)-1} e^{-(1/2)(x+y)}, & x, y > 0, \\ 0, & \text{elsewhere.} \end{cases}$$

The functions z_1^{-1} and z_2^{-1} are given by $x = uv/(v+1)$ and $y = u/(v+1)$, respectively. The Jacobian is

$$J = \begin{vmatrix} \dfrac{\partial x}{\partial u} & \dfrac{\partial x}{\partial v} \\ \dfrac{\partial y}{\partial u} & \dfrac{\partial y}{\partial v} \end{vmatrix} = \begin{vmatrix} \dfrac{v}{v+1} & \dfrac{u}{(v+1)^2} \\ \dfrac{1}{v+1} & \dfrac{-u}{(v+1)^2} \end{vmatrix} = \dfrac{-u}{(v+1)^2}.$$

Thus,

$$z(u,v) = f\left(\dfrac{uv}{v+1}, \dfrac{u}{v+1}\right) \left| \dfrac{-u}{(v+1)^2} \right|, \quad u,v > 0,$$

and zero elsewhere or

$$z(u,v) = \begin{cases} \dfrac{u^{(n_1+n_2-2)/2} v^{(n_1-2)/2} e^{-u/2}}{\Gamma(n_1/2)\Gamma(n_2/2)\left[2(v+1)\right]^{(n_1+n_2)/2}}, & u,v \geq 0, \\ 0, & \text{elsewhere.} \end{cases}$$

◂

> **6.6.3 Transformations of Variables of the Discrete Type** Essentially, no difficulties are involved in transformations of variables of the discrete type. Let $f(x,y)$ be the joint probability density function of the discrete random variable (X,Y) defined on a two-dimensional set of points S. If the new random variable (U,V), given by
>
> $$U = g_1(X,Y)$$
>
> and
>
> $$V = g_2(X,Y),$$
>
> defines a one-to-one transformation that maps the set E_1 of the x-y plane onto E^*, the two-dimensional set of the u-v plane, then the joint probability density function of the new random variable is given by
>
> $$z(u,v) = \begin{cases} f\left(z_1^{-1}(u,v), z_2^{-1}(u,v)\right), & (u,v) \in E^*, \\ 0, & \text{elsewhere,} \end{cases}$$
>
> where $x = z_1^{-1}(u,v)$ and $y = z_2^{-1}(u,v)$ are the single-valued inverses of g_1 and g_2, respectively. By knowing $z(u,v)$, we can obtain the marginal probability mass functions of the discrete random variables U and V by summing on v and u, respectively.

Example 6.6.5

If X and Y are independent and Poisson distributed, with parameters λ_1 and λ_2, respectively, their joint probability density function is given by

$$f(x,y) = \begin{cases} \dfrac{e^{-(\lambda_1+\lambda_2)} \lambda_1^x \lambda_2^y}{x! y!}, & x = 0, 1, 2, \ldots, y = 0, 1, 2, \ldots, \\ 0, & \text{elsewhere.} \end{cases}$$

We wish to find the probability mass function of a new random variable (U,V), defined by

$$U = X + Y$$

and

$$V = Y.$$

Solution: Here, $f(x,y)$ is defined on the two-dimensional set $E = \{(x,y) : x = 0, 1, 2, \ldots$ and $y = 0, 1, 2, \ldots\}$, and the new random variable (U,V) represents a one-to-one transformation that maps the set S onto the set $E^* = \{(u,v) : u = 0, 1, 2, \ldots$ and $v = 0, 1, 2, \ldots\}$. The inverse functions are

$$x = u - v \text{ and } y = v.$$

Thus, the probability mass function of the variate (U,V) is

$$z(u,v) = \begin{cases} \dfrac{e^{-(\lambda_1 + \lambda_2)} \lambda_1^{u-v} \lambda_2^{v}}{(u-v)! v!}, & u, v = 0, 1, 2, \ldots, \text{ such that } 0 \leq v \leq u, \\ 0, & \text{elsewhere.} \end{cases}$$

The marginal density of U is obtained by

$$z_1(u) = \sum_{v=0}^{u} \frac{e^{-(\lambda_1 + \lambda_2)} \lambda_1^{u-v} \lambda_2^{v}}{(u-v)! v!}$$

$$= \frac{e^{-(\lambda_1 + \lambda_2)}}{u!} \sum_{v=0}^{u} \frac{u!}{(u-v)! v!} \lambda_1^{u-v} \lambda_2^{v}$$

$$= \frac{e^{-(\lambda_1 + \lambda_2)}}{u!} \sum_{v=0}^{u} \binom{u}{v} \lambda_2^{v} \lambda_1^{u-v}$$

or

$$z_1(u) = \begin{cases} \dfrac{e^{-(\lambda_1 + \lambda_2)} (\lambda_1 + \lambda_2)^{u}}{u!}, & u = 0, 1, 2, \ldots, \\ 0, & \text{elsewhere.} \end{cases}$$

Similarly,

$$z_2(v) = \begin{cases} \dfrac{e^{-\lambda_2} \lambda_2^{v}}{v!}, & v = 0, 1, 2, \ldots, \\ 0, & \text{elsewhere.} \end{cases}$$

Thus, we conclude that the new variable $U = X + Y$ is Poisson distributed, with the parameter $\lambda_1 + \lambda_2$, and $V = Y$ is Poisson distributed, with the parameter λ_2.

6.6.4 One Function of Two Continuous Random Variables: Auxiliary Variable

In many problems, we are interested in determining the probability density function of a single functional form of the two-dimensional random variable (X,Y). That is, if the probability distribution of the variate (X,Y) is known, we wish to find the distribution of a new random variable U, given by

$$U = g_1(X,Y).$$

To obtain the probability density of U, we introduce an *auxiliary variable*, either

$$V = g_2(X,Y) = X$$

or

$$V = g_2(X,Y) = Y,$$

and under the assumptions given in Theorem 6.6.1, we can obtain the joint density of the random variable (U,V). Thus, the probability density function of U is found by integrating out all V. We use the preceding technique to obtain the distribution of some of the most important functions of the random variable (X,Y)—namely, the sum, product, and quotient of the variates X and Y.

■ Theorem 6.6.2

Let $f(x,y)$ be the probability density function of the continuous random variable (X,Y). If

$$U = X + Y,$$

then the probability density function of the variate U is given by

$$h(u) = \int_{-\infty}^{\infty} f(u-v,v)\,dv.$$

Proof We introduce the auxiliary variable $V = Y$ so that the new random variable (U,V), defined by

$$U = X + Y$$

and

$$V = Y,$$

is a one-to-one mapping of the random variable (X,Y). Thus, $x = u - v$ and $y = v$. The Jacobian of the transformation is

$$J = \begin{vmatrix} 1 & -1 \\ 0 & 1 \end{vmatrix} = 1.$$

Hence, the joint probability density function of (U,V) is

$$z(u,v) = f(u-v,v)|J|,$$

and the marginal density of U is

$$h(u) = \int_{-\infty}^{\infty} f(u-v,v)\,dv.$$

Example 6.6.6

Let the probability density function of the random variable (X, θ) be

$$f(x,\theta) = \begin{cases} \dfrac{1}{\sigma\pi\sqrt{2\pi}} e^{-(1/2\sigma^2)x^2}, & -\infty < x < \infty, 0 < \theta < \pi, \sigma < 0, \\ 0, & \text{elsewhere.} \end{cases}$$

Suppose that

$$U = X + a\cos\theta,$$

where a is an arbitrary constant. Find the probability distribution of the variate U.

Solution: In view of Theorem 6.6.2, we have

$$h(u) = \dfrac{1}{\sigma\pi\sqrt{2\pi}} \int_0^\pi e^{-(1/2\sigma^2)(u-a\cos\phi)^2} \, d\phi, \quad -\infty < u < \infty.$$

Note that $h(u)$ can be obtained by introducing the auxiliary variable $\Phi = \theta$; together with $U = X + a\cos\theta$, we have the single-valued inverses

$$x = u - a\cos\theta$$

and

$$\theta = \Phi.$$

The Jacobian of the transformation is

$$J = \begin{vmatrix} 1 & -a\sin\phi \\ 0 & 1 \end{vmatrix} = 1.$$

Thus,

$$z(u,\phi) = \begin{cases} \dfrac{1}{\sigma\pi\sqrt{2\pi}} e^{-(1/2\sigma^2)(u-a\cos\phi)^2}, & -\infty < u < \infty, 0 < \phi < \pi, \sigma > 0, \\ 0, & \text{elsewhere.} \end{cases}$$

Hence, integrating out ϕ, we obtain $h(u)$. ◀

Theorem 6.6.3

Let $f(x, y)$ be the probability density function of the continuous random variable (X, Y). If

$$U = XY,$$

then the probability density function of the variate U is given by

$$h(u) = \int_{-\infty}^{\infty} f\left(\dfrac{u}{v}, v\right) \left|\dfrac{1}{v}\right| dv.$$

continued

> **Proof** Let
> $$U = XY$$
> and
> $$V = Y.$$
> Thus, the single-valued inverses are $x = u/v$ and $y = v$. The Jacobian is
> $$J = \begin{vmatrix} \frac{1}{v} & -\frac{u}{v^2} \\ 0 & 1 \end{vmatrix} = \frac{1}{v}.$$
> Therefore,
> $$z(u,v) = f\left(\frac{u}{v}, v\right)\left|\frac{1}{v}\right|,$$
> and the marginal density of U is
> $$h(u) = \int_{-\infty}^{\infty} f\left(\frac{u}{v}, v\right)\left|\frac{1}{v}\right| dv.$$

Example 6.6.7

The joint density of the random variable (X, Y) is
$$f(x, y) = \begin{cases} \theta e^{-(x + \theta y)}, & x, y > 0, \theta > 0, \\ 0, & \text{elsewhere}. \end{cases}$$

We wish to obtain the probability density function of the variate U, given by $U = XY$.

Solution: Applying Theorem 6.6.3, we have
$$h(u) = \int_{-\infty}^{\infty} f\left(\frac{u}{v}, v\right)\left|\frac{1}{v}\right| dv$$
$$= \int_{-\infty}^{\infty} \theta e^{-(u/v + \theta v)} \left|\frac{1}{v}\right| dv.$$

Because $v = y > 0$, we have $1/v > 0$; thus,
$$h(u) = \int_0^{\infty} \frac{1}{v} \theta e^{-(u/v + \theta v)} dv.$$

Theorem 6.6.4

Let $f(x,y)$ be the probability density function of the continuous random variable (X,Y). If

$$U = \frac{X}{Y},$$

then the probability density function of the variate U is given by

$$h(u) = \int_{-\infty}^{\infty} f(uv,v)|v|dv.$$

Proof Let

$$U = \frac{X}{Y}$$

and

$$V = Y.$$

Thus, the single-valued inverses are $x = uv$ and $y = v$. The Jacobian of the transformation is

$$J = \begin{vmatrix} v & u \\ 0 & 1 \end{vmatrix} = v$$

Therefore,

$$z(u,v) = f(uv,v)|v|$$

and

$$h(u) = \int_{-\infty}^{\infty} f(uv,v)|v|dv.$$

Example 6.6.8

The joint probability density of the random variable (X,Y) is given by

$$f(x,y) = \frac{1}{2\pi\sigma_x\sigma_y} \exp\left\{-\frac{1}{2}\left(\frac{x^2}{\sigma_x^2} + \frac{y^2}{\sigma_y^2}\right)\right\}, \quad -\infty < x, y < \infty, \sigma_x, \sigma_y > 0.$$

Determine the probability density function of the variate U, defined by

$$U = \frac{X}{Y}.$$

Solution: In view of Theorem 6.6.4, the probability density of U is given by

$$h(u) = \frac{1}{2\pi\sigma_x\sigma_y} \int_{-\infty}^{\infty} |v| \exp\left\{-\frac{1}{2}\left(\frac{u^2v^2}{\sigma_x^2} + \frac{v^2}{\sigma_y^2}\right)\right\} dv$$

$$= \frac{1}{2\pi\sigma_x\sigma_y} \int_{0}^{\infty} v \exp\left\{-\frac{1}{2}\left(\frac{u^2v^2}{\sigma_x^2} + \frac{v^2}{\sigma_y^2}\right)\right\} dv$$

$$- \int_{-\infty}^{0} v \exp\left\{-\frac{1}{2}\left(\frac{u^2+v^2}{\sigma_x^2} + \frac{v^2}{\sigma_y^2}\right)\right\} dv$$

$$= \frac{1}{2\pi\sigma_x\sigma_y} \int_{0}^{\infty} v \exp\left\{\frac{-\frac{1}{2}v^2}{\frac{\sigma_x^2\sigma_y^2}{\sigma_x^2 + u^2\sigma_y^2}}\right\} dv$$

$$- \int_{0}^{\infty} v \exp\left\{\frac{-\frac{1}{2}v^2}{\frac{\sigma_x^2\sigma_y^2}{\sigma_x^2 + u^2\sigma_y^2}}\right\} dv.$$

Let

$$t = \frac{v}{\frac{\sigma_x\sigma_y}{\left(\sigma_x^2 + u^2\sigma_y^2\right)^{1/2}}}, \quad \frac{\sigma_x\sigma_y}{\left(\sigma_x^2 + u^2\sigma_y^2\right)^{1/2}} dt = dv$$

and

$$v = \sigma_x\sigma_y \Big/ \left(\sigma_x^2 + u^2\sigma_y^2\right)^{1/2} t.$$

Thus,

$$h(u) = \frac{\sigma_x\sigma_y}{2\pi\left(\sigma_x^2 + u^2\sigma_y^2\right)} \int_{0}^{\infty} t e^{-(1/2)t^2} dt$$

$$- \frac{\sigma_x\sigma_y}{2\pi\left(\sigma_x^2 + u^2\sigma_y^2\right)} \int_{-\infty}^{0} t e^{-(t^2/2)} dt$$

$$= \frac{\sigma_x\sigma_y}{2\pi\left(\sigma_x^2 + u^2\sigma_y^2\right)} + \frac{\sigma_x\sigma_y}{2\pi\left(\sigma_x^2 + u^2\sigma_y^2\right)}$$

$$= \frac{\sigma_x\sigma_y}{\pi\left(\sigma_x^2 + u^2\sigma_y^2\right)}, \quad -\infty < u < \infty,$$

or

$$h(u) = \frac{1}{\pi\{\sigma_x/\sigma_y + [(\sigma_y/\sigma_x)u^2]\}}, \quad -\infty < u < \infty.$$

Hence, $h(u)$ has the form of the *Cauchy distribution*.

6.7 Expected Value and Moments

In this section, we extend the basic concepts of expectation and moments, both of which were introduced in Chapter 5, to two random variables and functions of those random variables.

Definition 6.7.1

Let (X, Y) be a two-dimensional random variable of the discrete type, with the probability mass function $p(x,y)$. The series

$$E(XY) = \sum_i \sum_j x_i y_j p(x_i, y_j) \tag{6.7.1}$$

is called the **expected value of the product of X and Y** if the inequality

$$\sum_i \sum_j |x_i y_j| p(x_i, y_j) < \infty \tag{6.7.2}$$

is satisfied.

Example 6.7.1

The joint probability density function of the discrete random variable (X, Y) is given in Table 6.7.1.

Table 6.7.1

		y		
x	1	2	3	$p_1(x)$
1	$\frac{1}{6}$	$\frac{1}{6}$	$\frac{1}{6}$	$\frac{1}{2}$
2	$\frac{1}{6}$	$\frac{1}{12}$	$\frac{1}{12}$	$\frac{1}{3}$
3	$\frac{1}{12}$	$\frac{1}{12}$	0	$\frac{1}{6}$
$p_2(y)$	$\frac{5}{12}$	$\frac{1}{3}$	$\frac{1}{4}$	1

The expected value of the product of the random variables X and Y is

$$E(XY) = \sum_{x=1}^{3}\sum_{y=1}^{3}(xy)p(x,y)$$

$$= \sum_{x=1}^{3}\{(x)(1)p(x,1)+(x)(2)p(x,2)+(x)(3)p(x,3)\}$$

$$= (1)(1)p(1,1)+(1)(2)p(1,2)+(1)(3)p(1,3)$$

$$+(2)(1)p(2,1)+(2)(2)p(2,2)+(2)(3)p(2,3)$$

$$+(3)(1)p(3,1)+(3)(2)p(3,2)+(3)(3)p(3,3)$$

$$= \frac{1}{6}+(2)\frac{1}{6}+(3)\frac{1}{6}+(2)\frac{1}{6}+(4)\frac{1}{12}$$

$$+(6)\frac{1}{12}+(3)\frac{1}{12}+(6)\frac{1}{12}+(9)0$$

$$= \frac{35}{12}.$$

◀

Example 6.7.2

The expected value of the product of the two discrete random variables X and Y, whose joint probability mass function is given in Example 6.1.1, is

$$E(X,Y) = \sum_{y=0}^{4-x}\sum_{x=0}^{4} xy\, f(x,y)$$

$$= \sum_{y=1}^{4-x}\sum_{x=1}^{4} \frac{xy\binom{8}{x}\binom{6}{y}\binom{10}{4-x-y}}{\binom{24}{4}}$$

$$= (1)f(1,1)+(2)f(1,2)+(3)f(1,3)+(2)f(2,1)$$

$$+(4)f(2,2)+(3)f(3,1)$$

$$= \frac{2160}{10,626}+\frac{2400}{10,626}+\frac{480}{10,626}+\frac{3360}{10,626}+\frac{1680}{10,626}+\frac{1008}{10,626}$$

$$= \frac{24}{23}.$$

◀

Definition 6.7.2

Let (X,Y) be a two-dimensional continuous random variable with the joint probability density function $f(x,y)$. The double integral

$$E(XY) = \int_{-\infty}^{\infty}\int_{-\infty}^{\infty} xy\, f(x,y)\, dx\, dy \qquad (6.7.3)$$

is called the **expected value of the product of the variates X and Y** if the inequality

$$\int_{-\infty}^{\infty}\int_{-\infty}^{\infty} |xy|\, f(x,y)\, dx\, dy < \infty \qquad (6.7.4)$$

is satisfied.

Example 6.7.3

The joint probability density function of the continuous random variable (X,Y) given in Example 6.1.3 is

$$f(x,y) = \begin{cases} \dfrac{1}{28}(4x+2y+1), & 0 \le x \le 2,\, 0 \le y \le 2, \\ 0, & \text{elsewhere.} \end{cases}$$

The expected value of the product of the variates X and Y is

$$E(XY) = \int_0^2 \left\{ \int_0^2 xy \frac{1}{28}(4x+2y+1)\, dx \right\} dy$$

$$= \int_0^2 \left(\frac{8}{21} y + \frac{y^2}{7} + \frac{y}{14} \right) dy$$

$$= \left. \frac{8y^2}{42} + \frac{y^3}{21} + \frac{y^2}{28} \right|_0^2$$

$$= \frac{9}{7}.$$

Example 6.7.4

The expected value of the product of the random variables X and Y, the joint density of which is given in Example 6.1.4, is

$$E(XY) = \int_{-\infty}^{\infty}\int_{-\infty}^{\infty} xy\, f(x,y)\, dx\, dy$$

$$= \int_1^2 \left\{ \int_1^y \frac{8}{9} x^2 y^2\, dx \right\} dy$$

$$= \int_1^2 \frac{8}{27}(y^5 - y^2)\, dy$$

$$= \frac{196}{81}.$$

The expected value of a random variable, as we mentioned in Chapter 5, does not always exist. That is, if the right side of Equation (6.7.1) or (6.7.3) exists but the inequality in Equation (6.7.2) or (6.7.4) is not satisfied, then the expected value of the product of two random variables does not exist.

> **6.7.1 Properties of Expectation** The properties of expectation that we discussed in Section 5.2 for the one-dimensional random variable are also present in a two-dimensional random variable. We present here some additional properties of the two-dimensional variate.

Theorem 6.7.1

Let (X,Y) be a two-dimensional random variable with the joint probability density function $f(x,y)$. If $U = g(X,Y)$ is a functional form of the variate (X,Y), then the expected value of U is

$$E(U) = E[g(X,Y)] = \sum_i \sum_j g(x_i, y_j) f(x_i, y_j) \qquad (6.7.5)$$

if (X,Y) is discrete and

$$E(U) = E[g(X,Y)] = \int_{-\infty}^{\infty} \int_{-\infty}^{\infty} g(x,y) f(x,y) \, dx \, dy \qquad (6.7.6)$$

if (X,Y) is continuous.

The proof of this theorem is similar to that of Theorem 5.2.2; however, the expected value of $g(X,Y)$ exists only if the right-hand side of Equations (6.7.5) and (6.7.6) are *absolutely convergent*. ◀

Example 6.7.5

If $U = g(X,Y) = X+Y$, then $E(U) = E(X) + E(Y)$. Let (X,Y) be a continuous two-dimensional random variable with the joint probability density function $f(x,y)$. Then, the expected value of U is

$$E(U) = E(X+Y) = \int_{-\infty}^{\infty} \int_{-\infty}^{\infty} (x+y) f(x,y) \, dx \, dy$$

$$= \int_{-\infty}^{\infty} x \int_{-\infty}^{\infty} f(x,y) \, dy \, dx$$

$$+ \int_{-\infty}^{\infty} y \int_{-\infty}^{\infty} f(x,y) \, dx \, dy$$

$$= \int_{-\infty}^{\infty} x f_1(x) \, dx + \int_{-\infty}^{\infty} y f_2(y) \, dy$$

$$= E(X) + E(Y).$$

The situation is similar in the discrete case. Thus, the expected value of the sum of any two random variables is equal to the sum of their expected values. The variance of U is

$$\operatorname{Var}(U) = \operatorname{Var}(X+Y) = E\left[(X+Y)^2\right] - \left[E(X+Y)\right]^2$$

$$= E(X^2) - \left[E(X)\right]^2 + E(Y^2) - \left[E(Y)\right]^2 + 2E(XY) - 2E(X)E(Y).$$

Then,

$$\operatorname{Var}(X+Y) = \operatorname{Var}(X) + \operatorname{Var}(Y) + 2E(XY) - 2E(X)E(Y).$$

Similarly,

$$\operatorname{Var}(X-Y) = \operatorname{Var}(X) + \operatorname{Var}(Y) - 2E(XY) + 2E(X)E(Y).$$

Example 6.7.6

The expected value of $U = X^2 + Y^2$, where the joint density of the two-dimensional random variable (X,Y) is given in Example 6.7.4 is

$$E(U) = E(X^2 + Y^2) = \int_{-\infty}^{\infty}\int_{-\infty}^{\infty}(x^2+y^2)f(x,y)\,dx\,dy$$

$$= \int_1^2\left\{\int_1^y \frac{8}{9}(x^2+y^2)xy\,dx\right\}dy$$

$$= \int_1^2 \frac{8}{9}\left(\frac{3y^5}{4} - \frac{y^3}{2} - \frac{y}{4}\right)dy$$

$$= \frac{45}{9}$$

$$= 5.$$

It can be shown that

$$= \int_1^2\left\{\int_1^y |x^2+y^2|\frac{8}{9}xy\,dx\right\}dy < \infty.$$

Thus, we are assured of the existence of $E(U)$.

Theorem 6.7.2

Let (X,Y) be a two-dimensional random variable with the joint probability density function $f(x,y)$. If

$$U = g_1(X,Y)$$

and

$$V = g_2(X,Y),$$

then

$$E(UV) = E\left[g_1(X,Y)g_2(X,Y)\right].$$

Thus, if (X,Y) is of the discrete type, we have

$$E(UV) = \sum_i \sum_j g_1(x_i, y_j) g_2(x_i, y_j) f(x_i, y_j),$$

and we have

$$E(UV) = \int_{-\infty}^{\infty} \int_{-\infty}^{\infty} g_1(x,y) g_2(x,y) f(x,y) \, dx \, dy$$

if the variate is of the continuous type. For $E(UV)$ to exist, the preceding expressions must be absolutely convergent.

We can also obtain $E(UV)$ by first finding the joint probability density function of (U,V), $z(u,v)$ (as illustrated in Section 6.6), if possible, and then applying either Equation (6.7.1) or (6.7.3). That is,

$$E(UV) = \int_{-\infty}^{\infty} \int_{-\infty}^{\infty} u v \, z(u,v) \, du \, dv$$

or

$$E(UV) = \sum_i \sum_j u_i v_j z(u_i, v_j).$$

We illustrate the preceding remarks with Example 6.7.7.

Example 6.7.7

In Example 6.6.3, we saw that if the joint probability density function of the variate (X,Y) is given by

$$f(x,y) = \begin{cases} \dfrac{1}{\beta^2} e^{-\{(x+y)/\beta\}}, & x, y > 0, \beta > 0, \\ 0, & \text{elsewhere,} \end{cases}$$

then the joint density of the random variable (U,V), where

$$U = \frac{X-Y}{2}$$

and

$$V = Y,$$

is

$$z(u,v) = \begin{cases} \dfrac{2}{\beta^2} e^{-(2/\beta)(u+v)}, & -\infty < u < \infty,\ v > -2u,\ \text{if}\ u < 0, \\ & v > 0,\ \text{if}\ u \geq 0, \\ 0, & \text{elsewhere},\ \beta > 0. \end{cases}$$

We now show that $E(UV) = E[g_1(X,Y)g_2(X,Y)]$.

Solution: Here,

$$E(UV) = \int_{-\infty}^{\infty} \int_{-\infty}^{\infty} uv\, z(u,v)\, du\, dv$$

$$= \int_{-\infty}^{0} \left\{ \int_{-2u}^{\infty} uv \left(\frac{2}{\beta^2}\right) e^{-(2/\beta)(u+v)} dv \right\} du$$

$$+ \int_{0}^{\infty} \left\{ \int_{2}^{\infty} uv \left(\frac{2}{\beta^2}\right) e^{-(2/\beta)(u+v)} dv \right\} du$$

$$= \frac{2}{\beta^2} \int_{-\infty}^{0} u e^{-(2u/\beta)} \left\{ -\frac{\beta^2}{4} e^{(4u/\beta)} \left(\frac{4u}{\beta} - 1\right) \right\} du$$

$$+ \frac{2}{\beta^2} \int_{0}^{\infty} u e^{-(2u/\beta)} \left\{ -\frac{\beta^2}{4}(-1) \right\} du$$

$$= -\frac{1}{\beta^2} \int_{-\infty}^{0} u^2 e^{-(2u/\beta)} du + \frac{1}{2} \int_{-\infty}^{0} u e^{-(2u/\beta)} du$$

$$+ \frac{1}{2} \int_{0}^{\infty} u e^{-(2u/\beta)} du$$

$$= 2\left[\frac{\beta}{4}(-1)\right] + \frac{1}{2}\left(\frac{\beta^2}{4}\right) + \frac{1}{2}\frac{\beta^2}{4}$$

$$= -\frac{\beta^2}{2}$$

and

$$E[g_1(X,Y)g_2(X,Y)] = E\left[\left(\frac{X-Y}{2}\right)(Y)\right]$$

$$= \int_0^\infty \int_0^\infty \left(\frac{x-y}{2}\right) y \frac{1}{\beta^2} e^{-[(x+y)/\beta]} dx\, dy$$

$$= \frac{1}{2}\left\{\int_0^\infty x\frac{1}{\beta}e^{-(x/\beta)}\left[\int_0^\infty y\frac{1}{\beta}e^{-(y/\beta)}dy\right]dx\right.$$

$$\left. -\int_0^\infty y^2 \frac{1}{\beta}e^{-(y/\beta)}\left[\int_0^\infty \frac{1}{\beta}e^{-(x/\beta)}dx\right]dy\right\}$$

$$= \frac{1}{2}\left\{\beta^2 - 2\beta\int_0^\infty \frac{1}{\beta}ye^{-(y/\beta)}dy\right\}$$

$$= -\frac{\beta^2}{2}.$$

It is also clear that the double integral

$$\int_0^\infty \int_0^\infty \left|\frac{x-y}{2}\right| |y| \frac{1}{\beta^2} e^{-[(x+y)/\beta]} dx\, xy$$

converges to $(5/8)\beta^2$, which ensures the existence of the expected value of the product of the two random variables U and V. ◀

Theorem 6.7.3

Let (X,Y) be a two-dimensional random variable with the joint probability density function $f(x,y)$. If X and Y are independent variates, then

$$E(XY) = E(X)E(Y).$$

Proof If (X,Y) is discrete, then

$$E(XY) = \sum_i \sum_j x_i y_j f(x_i, y_j)$$

$$= \sum_i \sum_j x_i y_j f_1(x_i) f_2(y_j)$$

$$= \sum_i x_i f_1(x_i) \sum_j y_j f_2(y_j)$$

$$= E(X)E(Y),$$

where

$$f_1(x_i) = \sum_j f(x_i, y_j) \text{ and } f_2(y_j) = \sum_i (x_i, y_j).$$

The situation is similar in regard to the continuous case.

Example 6.7.8

If, in Example 6.7.5, the random variables X and Y are independent, then

$$\text{Var}(X+Y) = \text{Var}(X)\text{Var}(Y) + 2E(XY) - 2E(X)E(Y)$$
$$= \text{Var}(X) + \text{Var}(Y).$$

Similarly,

$$\text{Var}(X-Y) = \text{Var}(X) + \text{Var}(Y).$$

◀

Example 6.7.9

Let (X,Y) be a discrete random variable whose joint probability mass function is given by

$$f(x,y;\lambda_1,\lambda_2) = \begin{cases} \dfrac{e^{-(\lambda_1+\lambda_2)}\lambda_1^x \lambda_2^y}{x!\,y!}, & \begin{array}{l} x=0,1,2,\ldots, \\ y=0,1,2,\ldots, \end{array} \\ 0, & \text{elsewhere.} \end{cases}$$

Because

$$f_1(x;\lambda_1) = \begin{cases} \dfrac{e^{-(\lambda_1)}\lambda_1^x}{x!}, & x=0,1,2,\ldots, \\ 0, & \text{elsewhere,} \end{cases}$$

and

$$f_2(y;\lambda_2) = \begin{cases} \dfrac{e^{-(\lambda_2)}\lambda_2^y}{y!}, & y=0,1,2,\ldots, \\ 0, & \text{elsewhere,} \end{cases}$$

X and Y are independently distributed. In view of Theorem 6.7.3, we have

$$E(X,Y) = \sum_{x=0}^{\infty}\sum_{y=0}^{\infty} xy \frac{e^{-(\lambda_1+\lambda_2)}\lambda_1^x \lambda_2^y}{x!\,y!}$$

$$= \sum_{x=0}^{\infty} \frac{xe^{-\lambda_1}\lambda_1^x}{x!} \sum_{y=0}^{\infty} \frac{ye^{-\lambda_2}\lambda_2^y}{y!}$$

$$= \lambda_1\lambda_2.$$

◀

Example 6.7.10

In Example 6.7.3, the marginal densities of the random variables X and Y are

$$f_1(x) = \begin{cases} \frac{2}{7}x + \frac{3}{14}, & 0 \leq x < 2, \\ 0, & \text{elsewhere,} \end{cases}$$

and

$$f_2(y) = \begin{cases} \frac{1}{7}y + \frac{5}{14}, & 0 \leq y < 2, \\ 0, & \text{elsewhere,} \end{cases}$$

respectively. Thus,

$$E(X) = \int_0^2 \left(\frac{2}{7}x^2 + \frac{3}{14}x \right) dx = \frac{25}{21}$$

and

$$E(Y) = \int_0^2 \left(\frac{1}{7}y^2 + \frac{5}{14}y \right) dy = \frac{23}{21}.$$

As expected, $E(XY) \neq E(X)E(Y)$ because X and Y are dependent variables. ◀

6.7.2 Moments
We proceed by defining various moments of bivariate probability distributions.

Definition 6.7.3

Let (X, Y) be a two-dimensional random variable with the joint density function $f(x, y)$. The **moment of order** $k+m$ of the distribution of the variate (X, Y) is defined by

$$E(X^k Y^m) = \sum_i \sum_j x_i^k y_j^m f(x_i, y_j) \tag{6.7.7}$$

if (X, Y) is discrete and by

$$E(X^k Y^m) = \int_{-\infty}^{\infty} \int_{-\infty}^{\infty} x^k y^m f(x, y) \, dx \, dy \tag{6.7.8}$$

if (X, Y) is continuous.

Again, the *moment of order* $k+m$ of the distribution of (X, Y) exists only if Equations (6.7.7) and (6.7.8) are absolutely convergent.

We denote the moment of order $k+m$ of a bivariate distribution by

$$\mu_{km} = E(X^k Y^m).$$

Thus, if $k=1$, $m=0$, and $k=0$, $m=1$, we have

$$\mu_{10} = E(X^1 Y^0) = E(X)$$

and

$$\mu_{01} = E(X^0 Y^1) = E(Y),$$

the expected value of the marginal distributions of X and Y, respectively.

Example 6.7.11

With reference to Example 6.7.1, we can calculate the moments for (a) μ_{10}, (b) μ_{01}, (c) μ_{20}, and (d) μ_{02}.

Solution:

a. $\mu_{10} = \sum_{x=1}^{3} x p_1(x) = (1)\frac{1}{2} + (2)\frac{1}{3} + (3)\frac{1}{6} = \frac{5}{3}$

b. $\mu_{01} = \sum_{y=1}^{3} y p_2(y) = (1)\frac{5}{12} + (2)\frac{1}{3} + (3)\frac{1}{4} = \frac{11}{6}$

c. $\mu_{20} = \sum_{x=1}^{3} x^2 p_1(x) = (1)^2 \frac{1}{2} + (2)^2 \frac{1}{3} + (3)^2 \frac{1}{6} = \frac{10}{3}$

d. $\mu_{02} = \sum_{y=1}^{3} y^2 p_2(y) = (1)^2 \frac{5}{12} + (2)^2 \frac{1}{3} + (3)^2 \frac{1}{4} = 4$

Example 6.7.12

In Example 6.7.4, we can calculate the moment of order $k+m$ as follows:

$$\mu_{km} = E(X^k Y^m) = \int_{-\infty}^{\infty} \int_{-\infty}^{\infty} x^k y^m f(x,y) \, dx \, dy$$

$$= \int_{1}^{2} \left\{ \int_{1}^{y} \frac{8}{9} x^{k+1} y^{m+1} \, dx \right\} dy$$

$$= \int_{1}^{2} \frac{8}{9} \left(\frac{y^{k+m+3} - y^{m+1}}{k+2} \right) dy$$

$$= \frac{8}{9(k+2)} \left[\frac{y^{k+m+4}}{k+m+4} - \frac{y^{m+2}}{m+2} \right]_{1}^{2}$$

$$= \frac{8}{9(k+2)} \left[\frac{2^{k+m+4} - 1}{k+m+4} - \frac{2^{m+2} - 1}{m+2} \right].$$

Thus, if we let $k=m=1$, we obtain $E(XY) = 196/81$, as expected.

> **Definition 6.7.4**
>
> Let (X,Y) be a two-dimensional random variable with the joint probability density function $f(x,y)$. The **central moment of order** $k+m$ of the distribution of the variate (X,Y) is given by
>
> $$\eta_{km} = E\left[(X-\mu_{10})^k (Y-\mu_{01})^m\right]$$
>
> $$= \sum_i \sum_j (x_i - \mu_{10})^k (y_j - \mu_{01})^m f(x_i, y_j)$$
>
> if (X,Y) is discrete and by
>
> $$\eta_{km} = E\left[(X-\mu_{10})^k (Y-\mu_{01})^m\right]$$
>
> $$= \int_{-\infty}^{\infty} \int_{-\infty}^{\infty} (x_i - \mu_{10})^k (y_j - \mu_{01})^m f(x,y) \, dx \, dy$$
>
> if (X,Y) is continuous.

Thus, if $k=1$, $m=0$, and $k=0$, $m=1$, we have

$$\eta_{10} = E\left[(X-\mu_{10})\right] = E(X) - \mu_{10} = 0$$

and

$$\eta_{01} = E\left[(Y-\mu_{10})\right] = E(Y) - \mu_{10} = 0.$$

Also, if $k=2$, $m=0$, and $k=0$, $m=2$, we have the variance of the marginal distributions of the random variables X and Y;

$$\eta_{20} = E\left[(X-\mu_{10})^2\right] = \sigma_x^2$$

and

$$\eta_{02} = E\left[(Y-\mu_{01})^2\right] = \sigma_y^2.$$

◀

Example 6.7.13

We can express central moments of a bivariate distribution in terms of ordinary moments:

$$\eta_{20} = E\left[(X-\mu_{10})^2\right] = E(X^2) - \mu_{10}^2$$

or

$$\eta_{20} = \mu_{20} - \mu_{10}^2$$

and

$$\eta_{02} = E\left[(Y-\mu_{10})^2\right] = E(Y^2) - \mu_{01}^2$$

or

$$\eta_{02} = \mu_{02} - \mu_{01}^2.$$

Thus, in reference to Example 6.7.11, we have

$$\eta_{20} = \mu_{20} - \mu_{10}^2 = \frac{10}{3} - \left(\frac{5}{3}\right)^2 = \frac{5}{9}$$

and

$$\eta_{02} = \mu_{02} - \mu_{01}^2 = 4 - \left(\frac{11}{6}\right)^2 = \frac{23}{36}.$$

◀

Definition 6.7.5

The second $(k=1, m=1)$ central moment of the distribution of the two-dimensional random variable (X,Y),

$$\eta_{11} = E\left[(X-\mu_{10})(Y-\mu_{01})\right],$$

is called the **covariance of X and Y**.

Hence, η_{11} can be expressed as follows:

$$\eta_{11} = E\left[(X-\mu_{10})(Y-\mu_{01})\right]$$
$$= E(XY) - \mu_{10}E(Y) - \mu_{01}E(X) + \mu_{10}\mu_{01}$$
$$= \mu_{11} - \mu_{10}\mu_{01} - \mu_{01}\mu_{10} + \mu_{10}\mu_{01}$$
$$= \mu_{11} - \mu_{10}\mu_{01}.$$

The covariance of the random variables X and Y is usually denoted by $\mathrm{Cov}(X,Y)$.

Example 6.7.14

The covariance of the variates X and Y in Example 6.7.1 is

$$\eta_{11} = \mathrm{Cov}(X,Y) = \mu_{11} - \mu_{10}\mu_{01}$$

$$= \frac{35}{12} - \left(\frac{5}{3}\right)\left(\frac{11}{6}\right)$$

$$= -\frac{5}{36}.$$

◀

Example 6.7.15

In reference to Example 6.7.10, we have

$$\mu_{10} = E(X) = \frac{25}{31}, \mu_{01} E(Y) = \frac{23}{21} \text{ and } \mu_{11} = E(XY) = \frac{9}{7}.$$

Thus, the covariance of the continuous random variables X and Y is

$$\eta_{11} = \text{Cov}(X,Y) = \eta_{11} - \mu_{10}\mu_{01}$$

$$= \frac{9}{7} - \left(\frac{25}{21}\right)\left(\frac{23}{21}\right)$$

$$= -\frac{8}{441}.$$

◀

Theorem 6.7.4

If X and Y are independent random variables, then

$$\text{Cov}(X,Y) = 0.$$

Proof

$$\text{Cov}(X,Y) = E\big[(X-\mu_{10})(Y-\mu_{01})\big]$$

$$= E\big[XY - \mu_{10}Y - \mu_{01}X + \mu_{10}\mu_{01}\big]$$

$$= E(X)E(Y) - \mu_{10}E(Y) - \mu_{01}E(X) + \mu_{10}\mu_{01}$$

$$= \mu_{10}\mu_{01} - \mu_{10}\mu_{01} - \mu_{01}\mu_{10} + \mu_{10}\mu_{01}$$

$$= 0.$$

The converse of this theorem is not true; that is, if $\text{Cov}(X,Y) = 0$, it does *not* follow that X and Y are independent.

Example 6.7.16

The joint mass function of the discrete random variable (X,Y) is given by Table 6.7.2.

Table 6.7.2

		y			
x	-1	0	1	2	$p_1(x)$
-2	0	$\frac{1}{6}$	0	0	$\frac{1}{6}$
-1	$\frac{1}{3}$	0	0	$\frac{1}{6}$	$\frac{1}{2}$
0	$\frac{1}{6}$	0	$\frac{1}{6}$	0	$\frac{1}{3}$
$p_2(y)$	$\frac{1}{2}$	$\frac{1}{6}$	$\frac{1}{6}$	$\frac{1}{6}$	1

Solution: Here,

$$\mu_{10} = \sum_{x=-2}^{0} x p_1(x) = (-2)\frac{1}{6} + (-1)\frac{1}{2} = -\frac{5}{6}.$$

$$\mu_{01} = \sum_{y=-1}^{2} y p_2(y) = (-1)\frac{1}{2} + (1)\frac{1}{6} + (2)\frac{1}{6} = 0$$

$$\mu_{11} = \sum_{x=-1}^{0} x \sum_{y=-1}^{2} y p(x,p)$$

$$= (-2)\left[(-1)0 + 0 + (1)0 + (2)0\right]$$

$$+ (-1)\left[(-1)\frac{1}{3} + 0 + (1)0 + (2)\frac{1}{6}\right]$$

$$= 0.$$

Thus,

$$\text{Cov}(X,Y) = \mu_{11} - \mu_{10}\mu_{01} = 0,$$

but

$$p_1(-2) \cdot p_2(-1) = \left(\frac{1}{6}\right)\left(\frac{1}{2}\right) = \frac{1}{12} \neq p(-2,-1) = 0,$$

$$p_1(-1) \cdot p_2(-1) = \left(\frac{1}{2}\right)\left(\frac{1}{2}\right) = \frac{1}{4} \neq p(-1,-1) = \frac{1}{3},$$

and so on. Hence, $\text{Cov}(X,Y) = 0$ does not imply independence.

Theorem 6.7.5

If X and Y are any random variables and

$$U = \alpha_1 X + \beta_1, \quad V = \alpha_2 Y + \beta_2,$$

then

$$\text{Cov}(U,V) = \alpha_1 \alpha_2 \text{Cov}(X,Y).$$

Proof

$$\text{Cov}(U,V) = E(UV) - E(U)E(V)$$
$$= E[(\alpha_1 X + \beta_1)(\alpha_2 Y + \beta_2)]$$
$$- E(\alpha_1 X + \beta_1) E(\alpha_2 Y + \beta_2)$$
$$= \alpha_1 \alpha_2 E(XY) + \alpha_2 \beta_1 \mu_{01} + \alpha_1 \beta_2 \mu_{10} + \beta_1 \beta_2$$
$$- \alpha_1 \alpha_2 \mu_{10} \mu_{01} - \alpha_2 \beta_1 \mu_{01} - \alpha_1 \beta_2 \mu_{10} - \beta_1 \beta_2$$
$$= \alpha_1 \alpha_2 E(XY) - \alpha_1 \alpha_2 \mu_{10} \mu_{01}$$
$$= \alpha_1 \alpha_2 \{ E(XY) - \mu_{10} \mu_{01} \}$$
$$= a_1 a_2 \text{Cov}(X,Y).$$

Thus, if the variates are equal, that is, $X = Y$, then,

$$\text{Cov}(U,V) = \alpha_1 \alpha_2 \text{Var}(X).$$

The *covariance* of two random variables is the expected value of the product of the deviations of the two random variables from their respective means. It is also referred to as the *first product-moment*.

One important parameter that characterizes the distribution of a two-dimensional random variable (X,Y) is the coefficient of correlation.

Definition 6.7.6

The **coefficient of correlation** ρxy between X and Y is defined by

$$\rho xy = \frac{E[(X - \mu_{10})(Y - \mu_{01})]}{\sqrt{\text{Var}(X)} \sqrt{\text{Var}(Y)}} = \frac{\eta_{11}}{\sigma_X \sigma_Y}.$$

Thus, when we define the coefficient of correlation between two random variables, we assume that their expected values and variances exist and that their standard deviations are different from zero. The correlation coefficient is a dimensionless quantity that measures the linear relationship between two (quantitative) variables. It is also known as the *Pearson product-moment coefficient of correlation*. Its value ranges from −1 to +1, where 0 indicates the absence of any linear relationship and where −1 and +1 indicate a perfect *negative (inverse)* and a perfect *positive (direct)* relationship, respectively. To better understand the meaning and interpretation of the correlation coefficient, we exhibit its properties in Theorems 6.7.6 through 6.7.10.

■ Theorem 6.7.6

If the random variables X and Y are independent, then $\rho_{xy} = 0$.

Proof The proof follows immediately from Theorem 6.7.4. The converse of this theorem is, in general, not true. That is, if $\rho_{xy} = 0$, the random variables X and Y need not be independent.

■ Theorem 6.7.7

For every pair of random variables, (X, Y), the coefficient of correlation, if it exists, assumes a value between −1 and +1:

$$-1 \leq \rho_{xy} \leq +1.$$

Proof Let

$$U = X - \mu_{10} \text{ and } V = Y - \mu_{01} \tag{6.7.9}$$

Consider the following quadratic function of the arbitrary real number c:

$$h(c) = E(U + cV)^2 \geq 0. \tag{6.7.10}$$

Expanding Equation (6.7.10), we have

$$h(c) = E(c^2V^2 + 2cVU + U^2)$$

$$= E(V^2)c^2 + 2E(UV)c + E(U^2). \tag{6.7.11}$$

Equation (6.7.11) is a quadratic function in c and, according to Equation (6.7.10), is greater than or equal to zero for all c. If the discriminant of Equation (6.7.11) were positive, there would exist some values that could be selected for c so that this equation would be negative. Therefore, the discriminant must be less than or equal to zero, ensuring that Equation (6.7.11) always remains nonnegative:

$$4[E(UV)]^2 - 4E(U^2)E(V^2) \leq 0$$

or

$$[E(UV)]^2 \leq E(U^2)E(V^2). \tag{6.7.12}$$

Dividing both sides of the inequality in Equation (6.7.12) by $E(U^2)E(V^2)$, we obtain

$$\frac{[E(UV)]^2}{E(U^2)E(V^2)} \leq 1. \qquad (6.7.13)$$

Substituting Equation (6.7.9) into the inequality in Equation (6.7.13), we have

$$\frac{\{E[(X-\mu_{10})(Y-\mu_{01})]\}^2}{E[(X-\mu_{10})^2]E[(Y-\mu_{01})^2]} \leq 1$$

or

$$\rho_{XY}^2 \leq 1.$$

Hence,

$$-1 \leq \rho XY \leq 1.$$

Theorem 6.7.8 shows that the correlation coefficient does not change with respect to a location change and a scale change.

Theorem 6.7.8

If X and Y are random variables and

$$U = \alpha_1 X + \beta_1, \quad V = \alpha_2 Y + \beta_2,$$

where $\alpha_i, \beta_i, i = 1, 2$, are arbitrary real numbers, then

$$\rho_{UV} = \rho_{XY}.$$

Proof

$$\rho_{UV} = \frac{\text{Cov}(U, V)}{\sqrt{\text{Var}(U)}\sqrt{\text{Var}(V)}} \qquad (6.7.14)$$

In view of Theorem 6.7.5 and because

$$\text{Var}(U) = \text{Var}(\alpha_1 X + \beta_1) = \alpha_1^2 \text{Var}(X)$$

and

$$\text{Var}(V) = \text{Var}(\alpha_2 Y + \beta_2) = \alpha_2^2 \text{Var}(Y),$$

Equation (6.7.14) becomes

$$\rho_{UV} = \frac{\alpha_1 \alpha_2 \text{Cov}(X, Y)}{\alpha_1 \sqrt{\text{Var}(X)} \alpha_2 \sqrt{\text{Var}(Y)}} = \frac{\text{Cov}(X, Y)}{\sqrt{\text{Var}(X)}\sqrt{\text{Var}(Y)}}.$$

Thus,

$$\rho_{UV} = \rho_{XY}.$$

Theorem 6.7.9

If X and Y are two random variables for which $Y = \alpha X + \beta$, where α and β are constants and $\alpha \neq 0$, then:

1. $\rho_{XY} = 1$, if $\alpha > 0$
2. $\rho_{XY} = -1$, if $\alpha < 0$

Proof From Theorem 6.7.5, we have

$$\text{Cov}(X, Y) = \text{Cov}(X, \alpha X + \beta) = \alpha \text{Var}(X).$$

Also,

$$\text{Var}(Y) = \text{Var}(\alpha X + \beta) = \alpha^2 \text{Var}(X).$$

Thus,

$$\rho_{XY} = \frac{\text{Cov}(X, Y)}{\sqrt{\text{Var}(X)}\sqrt{\text{Var}(Y)}} = \frac{\alpha \text{Var}(X)}{\sqrt{\sigma_x^2}\sqrt{\alpha^2 \sigma_x^2}}$$

or

$$\rho_{XY} = \frac{\alpha}{|\alpha|}.$$

Therefore,

$$\rho_{XY} = +1, \text{ if } \alpha > 0,$$

and

$$\rho_{XY} = -1, \text{ if } \alpha < 0.$$

Theorem 6.7.10 shows that if $|\rho XY| = 1$, then, with a probability of 1, Y is a linear function of X.

Theorem 6.7.10

If $|\rho XY| = 1$, then $Pr(Y = \alpha X + \beta) = 1$.

Proof Let

$$Z = \frac{X - \mu_{10}}{\sigma_X}$$

and

$$W = \frac{Y - \mu_{01}}{\sigma_Y}.$$

continued

Then
$$E(Z)=\frac{1}{\sigma_X}\left[E(X)-\mu_{10}\right]=0, \quad E(W)=0,$$

and
$$\operatorname{Var}(Z)=\frac{1}{\sigma_X^2}\operatorname{Var}(X)=1, \quad \operatorname{Var}(W)=1.$$

Thus,
$$\rho_{ZW}=\frac{\operatorname{Cov}(Z,W)}{\sigma_Z\sigma_W}=\frac{E(ZW)-E(Z)E(W)}{\sigma_Z\sigma_W}$$

or
$$\rho_{ZW}=E(ZW).$$

Let
$$G_1=Z+W$$

and
$$G_2=Z-W.$$

From Example 6.7.5, we have
$$\operatorname{Var}(G_1)=\sigma_{Z+W}^2=\sigma_Z^2+\sigma_W^2+2E(ZW)$$

and
$$\operatorname{Var}(G_2)=\sigma_{Z-W}^2=\sigma_Z^2+\sigma_W^2-2E(ZW).$$

Hence,
$$0\leq \sigma_Z^2+\sigma_W^2+2\rho ZW$$

and
$$0\leq \sigma_Z^2+\sigma_W^2-2\rho ZW.$$

Because $\sigma_Z^2=\sigma_W^2=1$, we have
$$0\leq 2+2\rho ZW$$

and
$$0\leq 2-2\rho ZW.$$

Assume that $|\rho ZW|=1$. Then from Theorem 6.7.8, we have

$$\rho XY = \rho ZW;$$

thus, $|\rho ZW|=1$. If $\rho_{ZW}=-1$, then $\sigma^2_{G_1}=\sigma^2_{Z+W}=2+2\rho_{ZW}=0$; thus $Z+W=0$, with a probability of 1. We conclude that, with a probability of 1, $Z=-W$. This condition exists because the distribution of G_1 has a variance of 0, which implies that its distribution is discrete; the total probability of the distribution is massed at its mean, which in this case, is 0. Such a distribution is called a *degenerate distribution*. We say that $G_1=0$, with a probability of 1. Therefore, we can write

$$\frac{X-\mu_{10}}{\sigma_X} = -\frac{Y-\mu_{01}}{\sigma_Y}$$

with a probability of 1 or

$$Y = \alpha X + \beta,$$

with a probability of 1, where

$$\alpha = \frac{\sigma_Y}{\sigma_X}$$

and

$$\beta = \mu_{01} + \frac{\sigma_Y}{\sigma_X}\mu_{10}.$$

Note that α is simply the slope of the line and β is its y-intercept. If $\rho_{ZW}=1$, then $\sigma^2_{G_2}=\sigma^2_{Z-W}=2-2\rho_{ZW}=0$. Similarly, we conclude that

$$Y = \alpha X + \beta$$

with a probability of 1, where

$$\alpha = \frac{\sigma_Y}{\sigma_X}$$

and

$$\beta = \mu_{01} - \frac{\sigma_Y}{\sigma_X}\mu_{10}.$$

See Figure 6.7.1.

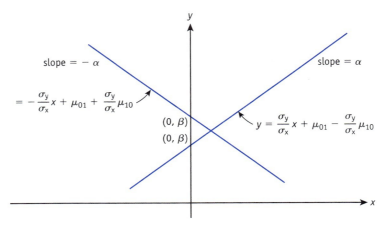

Figure 6.7.1

We conclude by giving an analogous interpretation of the correlation coefficient. If $\rho_{XY} = 0$, we say that the variates X and Y are *uncorrelated*; if $\rho_{XY} = \pm 1$, then X and Y are *perfectly correlated*. Thus, in a linear relation, we can predict Y if X has been observed, with ρ_{XY} indicating the accuracy of the predicted value.

Example 6.7.17

For the discrete bivariate density given in Example 6.7.1, we saw that

$$\mu_{10} = \frac{5}{3},$$

$$\mu_{01} = \frac{11}{6},$$

$$\mu_{11} = \frac{35}{12},$$

$$\mu_{20} = \frac{10}{3},$$

and

$$\mu_{02} = 4.$$

Thus, the correlation coefficient of X and Y is

$$P_{XY} = \frac{\text{Cov}(X,Y)}{\sigma_X \sigma_Y} = \frac{\eta_{11}}{\sqrt{\mu_{20} - \mu_{10}^2}\sqrt{\mu_{02} - \mu_{01}^2}}$$

$$= \frac{\mu_{11} - \mu_{10}\mu_{01}}{\sqrt{\eta_{20}}\sqrt{\eta_{02}}} = \frac{\left(\frac{35}{12}\right) - \left(\frac{5}{3}\right)\left(\frac{11}{6}\right)}{\sqrt{\frac{5}{9}}\sqrt{\frac{23}{36}}}$$

$$= \frac{-\frac{5}{36}}{\frac{1}{3}\sqrt{5}\cdot\frac{1}{6}\sqrt{23}} = \frac{-5}{2\sqrt{115}}.$$

6.8 | Conditional Expectations

In this section, we discuss conditional expectations, which are expected values computed with respect to conditional distributions.

Definition 6.8.1

Let (X, Y) be a two-dimensional random variable. The **conditional expected value** of the variate X for a given $Y = y$ is

$$E(X \mid Y = y_j) = \sum_i x_i h_1(x_i \mid y_j) \tag{6.8.1}$$

if (X, Y) is discrete and

$$E(X \mid Y = y) = \int_{-\infty}^{\infty} x h_1(x \mid y) dx \tag{6.8.2}$$

if (X, Y) is of the continuous type.

Similarly, we can define the conditional expected value of Y for a given $X = x$. We may write Equations (6.8.1) and (6.8.2) as

$$E(X \mid Y = y_j) = \sum_i x_i \frac{f(x_i, y_j)}{f_2(y_j)}, \quad f_2(y_j) > 0,$$

and

$$E(X \mid Y = y) = \int_{-\infty}^{\infty} x \frac{f(x, y)}{f_2(y)} dx, \quad f_2(y) > 0,$$

where

$$p_2(y_j) = \sum_i f(x_i, y_j)$$

and

$$f_2(y) = \int_{-\infty}^{\infty} f(x, y) dx.$$

For the conditional expected value to exist, the right-hand side of Equation (6.8.1) or (6.8.2) must be absolutely convergent. Furthermore, if X and Y are independent random variables, then

$$E(X|Y=y) = E(X)$$

and

$$E(Y|X=x) = E(Y).$$

Example 6.8.1

The conditional expectation of X given $Y = -2$ in Example 6.3.1, is

$$E(X|Y=-2) = \sum_{x=-1}^{5} x \frac{f(x,-2)}{f_2(-2)}$$

$$= \frac{1}{p_2(-2)}\{(-1)p(-1,-2) + (3)p(3,-2) + (5)p(5,-2)\}$$

$$= \frac{5}{2}\{(-1)(0.3) + (3)(0) + (5)(0.1)\}$$

$$= \frac{1}{2}.$$

Also,

$$E(Y|X=-1) = \sum_{x=-2}^{4} y \frac{p(-1,y)}{p_1(-1)}$$

$$= \frac{1}{p_1(-1)}\{(-2)p(-1,-2) + (1)p(-1,1) + (4)p(-1,4)\}$$

$$= \frac{5}{2}\{(-2)(0.3) + (1)(0) + (4)(0.2)\}$$

$$= \frac{1}{3}.$$

◀

Example 6.8.2

The conditional expected value of the random variable X given $Y = y_0$, whose conditional distribution is given in Example 6.4.3, is

$$E(X|Y=y_0) = \int_{-\infty}^{\infty} x h_1(x|y_0) dx$$

$$= \int_1^{y_0} x \frac{2x}{y_0^2 - 1} dx$$

$$= \frac{2}{3}\left(\frac{y_0^2 + y_0 + 1}{y_0 + 1}\right).$$

Also,

$$E(Y|X=x_0) = \int_{-\infty}^{\infty} y h_2(y|x_0) dy$$

$$= \int_{x_0}^{2} y \frac{2y}{4 - x_0^2} dy$$

$$= \frac{2}{3}\left(\frac{x_0^2 + 2x_0 + 4}{x_0 + 2}\right).$$ ◀

Again, for the conditional expectation of a random variable to exist, Equations (6.8.1) and (6.8.2) must be absolutely convergent. Also, the conditional expected value of one random variable is a constant for a fixed value of the second variate.

But $E(X|Y=y)$ is a function of y, where y represents a value of the random variable Y; thus, it is not a random variable. Similarly, $E(Y|X=x)$ is not a random variable because the conditional expectation is a function of x, where x represents a value that the variate X may assume. However, in general, $E[X|Y]$ is a function of the random variable Y and $E[Y|X]$ is a function of the random variable X.

Theorem 6.8.1

Let (X,Y) be a two-dimensional random variable. If the conditional expected values $E(X|Y=y)$ and $E(Y|X=x)$ exist, then:

1. $E[E(X|Y=y)] = E(X)$
2. $E[E(Y|X=x)] = E(Y)$

continue

Proof

1. For all discrete cases, we have

$$E\left[E(X|Y=y_j)\right] = \sum_{j=1}^{\infty} E(X|Y=y_j) f_2(y_j)$$

$$= \sum_{j=1}^{\infty} \sum_{i=1}^{\infty} x_i h(x_i|y_j) f_2(y_j)$$

$$= \sum_{j=1}^{\infty} \sum_{i=1}^{\infty} x_i \frac{f(x_i, y_j)}{f_2(y_j)} f_2(y_j)$$

$$= \sum_{i=1}^{\infty} x_i \sum_{j=1}^{\infty} f(x_i, y_j)$$

$$= \sum_{i=1}^{\infty} x_i f_1(x_i)$$

$$= E(X).$$

2. For the continuous case, we have

$$E\left[E(Y|X=x)\right] = \int_{-\infty}^{\infty} E(Y|X=x) f_1(x) dx$$

$$= \int_{-\infty}^{\infty} \left\{\int_{-\infty}^{\infty} y h_2(y|x) dy\right\} f_1(x) dx$$

$$= \int_{-\infty}^{\infty} \left\{\int_{-\infty}^{\infty} y \frac{f(x,y)}{f_1(x)} dy\right\} f_1(x) dx.$$

Because the expected values exist, we may change the order of integration in the preceding integral. Hence,

$$E\left[E(Y|X=x)\right] = \int_{-\infty}^{\infty} y \left\{\int_{-\infty}^{\infty} \frac{f(x,y)}{f_1(x)} f_1(x) dx\right\} dy$$

$$= \int_{-\infty}^{\infty} y \left\{\int_{-\infty}^{\infty} f(x,y) dx\right\} dy$$

$$= E(Y).$$

If $g(X,Y)$ is a functional form of the random variable (X,Y), which has a probability density of $f(x,y)$, then

$$E\left[g(X,Y)|x_1 \leq X \leq x_2\right] = \int_{-\infty}^{\infty} \int_{x_1}^{x_2} \frac{g(x,y) f(x,y)}{Pr(x_1 \leq X \leq x_2)} dx\, dy$$

$$= \int_{-\infty}^{\infty} \int_{x_1}^{x_2} \frac{g(x,y) f(x,y)}{F(x_2) - F(x_1)} dx\, dy.$$

This conditional expected value exists only if the preceding integral is absolutely convergent and $Pr(x_1 \leq X \leq x_2) \neq 0$:

$$E[g(X,Y)|Y=y] = \lim_{\Delta y \to 0} \{E[g(X,Y)|y \leq Y \leq y+\Delta y]\}$$

$$= \int_{-\infty}^{\infty} \left\{ \lim_{\Delta y \to 0} \int_{y}^{y+\Delta y} \frac{g(x,s)f(x,s)}{F(y+\Delta y)-F(y)} ds \right\} dx$$

$$= \int_{-\infty}^{\infty} \frac{g(x,y)f(x,y)}{f_2(y)} dx.$$

Applying an argument similar to that used in Theorem 6.8.1, we can show that

$$E\{E[g(X,Y)|X=x]\} = E\{E[g(X,Y)|Y=y]\} = E[g(X,Y)].$$

Example 6.8.3

If the probability density function of the random variable (X,Y) is given by

$$f(x,y) = \begin{cases} \dfrac{8}{9}xy, & 1 \leq x \leq y \leq 2, \\ 0, & \text{elsewhere,} \end{cases}$$

and

$$f_1(x) = \begin{cases} \dfrac{4}{9}x(4-x^2), & 1 \leq x \leq 2, \\ 0, & \text{elsewhere,} \end{cases}$$

compute $E[g(X,Y)|X=x]$, where

$$g(X,Y) = X^2 + Y^2.$$

Solution: We write

$$E[X^2+Y^2|X=x] = \int_{x}^{2}(x^2+y^2)\frac{f(x,y)}{f_1(x)}dy$$

$$= \int_{x}^{2} 2(x^2+y^2)\frac{y}{(4-x^2)}dy$$

$$= \frac{3x^2+4}{2}, \quad 1 \leq x \leq 2.$$

Thus, $E\left[g(X,Y)|X=x\right]$ is a random variable. Furthermore,

$$E\left\{E\left[X^2+Y^2|X=x\right]\right\}=E\left[X^2+Y^2\right]$$

$$=\int_1^2\left\{\int_1^y\frac{8}{9}(x^2+y^2)xy\,dx\right\}dy$$

$$=\frac{8}{9}\int_1^2\left(\frac{3}{4}y^5-\frac{1}{2}y^3-\frac{1}{4}y\right)dy$$

$$=5.$$

6.8.1 Regression Curve

An important geometric interpretation of conditional expectation is that of a regression curve. Let (X,Y) be a two-dimensional random variable with the joint probability density function $f(x,y)$. If (X,Y) is of the discrete type, we denote the conditional expected values of the variates X given $Y=y_j$ and Y given $X=x_i$ using $\xi_1(y_j)$ and $\xi_2(x_i)$, respectively. Thus,

$$\xi_1(y_j)=E(X|Y=y_j)=\sum_i x_i\frac{f(x_i,y_j)}{f_2(y_j)} \qquad (6.8.3)$$

and

$$\xi_2(x_i)=E(Y|X=x_i)=\sum_j y_j\frac{f(x_i,y_j)}{f_1(x_i)}. \qquad (6.8.4)$$

Similarly, for the continuous case,

$$\xi_1(y)=E(X|Y=y)=\int_{-\infty}^{\infty}x\frac{f(x,y)}{f_2(y)} \qquad (6.8.5)$$

and

$$\xi_2(x)=E(Y|X=x)=\int_{-\infty}^{\infty}y\frac{f(x,y)}{f_1(x)}dy. \qquad (6.8.6)$$

The graph of the collection of points $\left(x=\xi_1(y_j),\,y=y_j\right)$ from Equation (6.8.3) is known as the *regression curve* of X on Y. Analogously, the graph of the set of points $\left(x=x_i,y=\xi_2(x_i)\right)$ is called the *regression curve* of the random variable Y on X. Similar interpretations may be made for Equations (6.8.5) and (6.8.6). See Figure 6.8.1.

6.8 Conditional Expectations

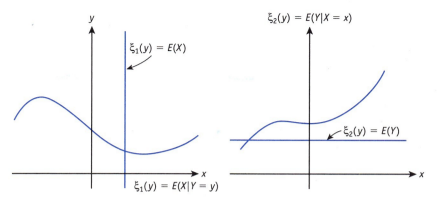

Figure 6.8.1

We formulate the preceding discussion in Definition 6.8.2

Definition 6.8.2

The graph of the collection of points $\left(\xi_1(y_j), y_j\right)$ or $\left(\xi_1(y), y\right)$ from Equations (6.8.3) and (6.8.5) is called the **regression curve of the random variable X on the random variable Y**. Similarly, the graph of the set of points $\left(x_i, \xi_2(x_i)\right)$ or $\left(x, \xi_2(x)\right)$ from Equation (6.8.4) or (6.8.6) is called the **regression curve of the random variable Y on the random variable X**.

Thus, as we stated previously, the conditional expectation of X, given Y, is a function of y; that is, it runs over all possible values that the random variable Y can assume, and the values of $E(X|Y=y)$ are located on the regression curve of X on Y. The situation is similar for the regression curve of Y on X.

If the random variables X and Y are independent, then

$$\xi_1(y) = E(X|Y=y) = E(X)$$

and

$$\xi_2(x) = E(Y|X=x) = E(Y).$$

Thus, the conditional expectation $\xi_1(y)$ does not depend on Y, and the regression curve is a line with infinite slope. Similarly, the conditional expectation $\xi_2(x)$ does not depend on X, and the regression curve is a line parallel to the x-axis, as shown in Figure 6.8.1.

Example 6.8.4

Compute the regression curves for the discrete bivariate distribution, given in Example 6.3.1.

Solution: Here,

$$\xi_1(y_j) = E(X|Y=y_j) = \sum_i x_i \frac{f(x_i, y_j)}{f_2(y_j)}$$

$$\xi_1(-2) = \frac{1}{f_2(-2)}\left[(-1)f(-1,-2) + (3)f(3,-2) + (5)f(5,-2)\right]$$

$$= \frac{1}{0.4}\left[(-1)(0.3) + (3)(0) + (5)(0.1)\right]$$

$$= \frac{1}{2};$$

$$\xi_1(0) = \frac{1}{f_2(0)}\left[(-1)f(-1,0) + (3)f(3,0) + (5)f(5,0)\right]$$

$$= \frac{1}{0.3}\left[(-1)(0.1) + (3)(0.2) + (5)(0)\right]$$

$$= \frac{5}{3};$$

$$\xi_1(1) = \frac{1}{f_2(1)}\left[(-1)f(-1,1) + (3)f(3,1) + (5)f(5,1)\right]$$

$$= \frac{1}{0.1}\left[(-1)(0) + (3)(0.1) + (5)(0)\right]$$

$$= 3;$$

$$\xi_1(4) = \frac{1}{f_2(4)}\left[(-1)f(-1,4) + (3)f(3,4) + (5)f(5,4)\right]$$

$$= \frac{1}{0.2}\left[(-1)(0.2) + (3)(0) + (5)(0)\right]$$

$$= -1;$$

$$\xi_2(x_i) = E[Y|X=x_i] = \sum_j y_j \frac{f(x_i, y_j)}{f_1(x_i)}$$

$$\xi_2(-1) = \frac{1}{f_1(-1)}\left[(-2)f(-1,-2)+(0)f(-1,0)+(1)f(-1,1)+(4)f(-1,4)\right]$$

$$= \frac{1}{0.6}\left[(-2)(0.3)+(4)(0.2)\right]$$

$$= \frac{1}{3};$$

$$\xi_2(3) = \frac{1}{f_1(3)}\left[(-2)f(3,-2)+(0)f(3,0)+(1)f(3,1)+(4)f(3,4)\right]$$

$$= \frac{1}{0.3}\left[(1)(0.1)\right]$$

$$= \frac{1}{3};$$

and

$$\xi_2(5) = \frac{1}{f_1(5)}\left[(-2)f(5,-2)+(0)f(5,0)+(1)f(5,1)+(4)f(5,4)\right]$$

$$= \frac{1}{0.1}\left[(-2)(0.1)\right]$$

$$= -2.$$

Thus, the regression curves of X on Y and Y on X consist of the following points:

$$X \text{ on } Y: \left(\frac{1}{2},-2\right), \left(\frac{5}{3},0\right), (3,1), \text{ and } (-1,4).$$

$$Y \text{ on } X: \left(-1,\frac{1}{3}\right), \left(3,\frac{1}{3}\right), \text{ and } (5,-2).$$

Example 6.8.5

Compute the regression curves for the continuous bivariate distribution, given in Example 6.4.2.

Solution: Here,

$$X \text{ on } Y: \xi_1(y) = E(X|Y=y) = \int_{-\infty}^{\infty} x\, h_1(x|y)\, dx$$

$$= \frac{1}{2(2y+5)} \int_0^2 x(4x+2y+1)\, dx$$

$$= \frac{1}{(2y+5)}\left(2y+\frac{19}{3}\right) = \frac{1}{3}\frac{6y+19}{2y+5}.$$

Y on X : $\xi_2(x) = E(Y|X=x) = \int_{-\infty}^{\infty} x\, h_2(y|x)\, dy$

$$= \frac{1}{2(4x+3)} \int_0^2 y(4x+2y+1)\, dy$$

$$= \frac{1}{(4x+3)}\left(4x + \frac{11}{3}\right) = \frac{1}{3}\cdot\frac{12x+11}{4x+3}.$$

Thus, the regression curve of X on Y is simply the location of the mean of X for various values of Y in the conditional of X given Y; the situation is similar for the regression curve Y on X. See Figure 6.8.2.

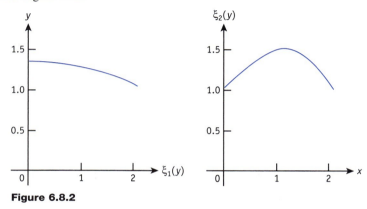

Figure 6.8.2

6.9 Bivariate Normal Distribution

We conclude our discussion of two random variables by illustrating with an important bivariate distribution most of the important probabilistic concepts we have studied.

Definition 6.9.1

A two-dimensional random variable (X, Y) is said to be a **bivariate normal distribution** if its joint probability density function is

$$f(x, y; \mu_X, \mu_Y, \sigma_X, \sigma_Y)$$

$$= K\exp\left\{-\frac{1}{2(1-\rho^2)}\left[\left(\frac{x-\mu_X}{\sigma_X}\right)^2 - 2\rho\left(\frac{x-\mu_X}{\sigma_X}\right)\right.\right.$$

$$\left.\left.\times\left(\frac{y-\mu_Y}{\sigma_Y}\right) + \left(\frac{y-\mu_Y}{\sigma_Y}\right)^2\right]\right\}, \quad -\infty < x, y < \infty,$$

where

$$K = \frac{1}{2\pi\sigma_X\sigma_Y\sqrt{1-\rho^2}}, \quad -\infty < \mu_X < \infty, -\infty < \mu_Y < \infty,$$

$$\sigma_X > 0, \sigma_Y > 0, -1 < \rho < 1.$$

6.9 Bivariate Normal Distribution

1. We can show that the conditions of a joint probability density function are satisfied. It is clear that $f(x, y; \mu_X, \mu_Y, \sigma_X, \sigma_Y) > 0$ for all (x, y). Thus, we must show that

$$\int_{-\infty}^{\infty}\int_{-\infty}^{\infty} f(x, y; \mu_X, \mu_Y, \sigma_X, \sigma_Y)\, dx\, dy = 1.$$

Consider

$$A = \int_{-\infty}^{\infty}\int_{-\infty}^{\infty} \frac{1}{2\pi\sigma_X\sigma_Y\sqrt{1-\rho^2}} \cdot \exp\left\{-\frac{1}{2(1-\rho^2)}\left[\left(\frac{x-\mu_X}{\sigma_X}\right)^2 - 2\rho\left(\frac{x-\mu_X}{\sigma_X}\right)\left(\frac{y-\mu_Y}{\sigma_Y}\right) + \left(\frac{y-\mu_Y}{\sigma_Y}\right)^2\right]\right\} dx\, dy.$$

Let

$$u = \frac{x-\mu_X}{\sigma_X},\ v = \frac{y-\mu_Y}{\sigma_Y},$$

or

$$x = \sigma_X u + \mu_X,\ y = \sigma_Y v + \mu_Y.$$

The Jacobian of the transformation is

$$J = \begin{vmatrix} \frac{\partial x}{\partial u} & \frac{\partial x}{\partial v} \\ \frac{\partial y}{\partial u} & \frac{\partial y}{\partial v} \end{vmatrix} = \begin{vmatrix} \sigma_X & 0 \\ 0 & \sigma_Y \end{vmatrix} = \sigma_X \sigma_Y.$$

We then have

$$A = \int_{-\infty}^{\infty}\int_{-\infty}^{\infty} \frac{1}{2\pi\sqrt{1-\rho^2}} \exp\left\{-\frac{1}{2(1-\rho^2)}(u^2 - 2\rho uv + v^2)\right\} du\, dv;$$

however, we can write

$$(u^2 - 2\rho uv + v^2)$$

as

$$u^2 - 2\rho uv + \rho^2 v^2 - \rho^2 v^2 + v^2,$$

which equals

$$(u - \rho v)^2 - \rho^2 v^2 + v^2$$

or

$$(u - \rho v)^2 + v^2(1 - \rho^2).$$

Thus,

$$A = \int_{-\infty}^{\infty}\int_{-\infty}^{\infty} \frac{1}{2\pi\sqrt{1-\rho^2}} \exp\left\{-\frac{1}{2(1-\rho^2)}(u-\rho v)^2 - \frac{v^2}{2}\right\} du\, dv$$

$$= \int_{-\infty}^{\infty}\int_{-\infty}^{\infty} \frac{1}{2\pi\sqrt{1-\rho^2}} \exp\left\{-\frac{1}{2}\left(\frac{u-\rho v}{\sqrt{1-\rho^2}}\right)^2\right\} e^{-(1/2)v^2} du\, dv.$$

Now, let

$$t_1 = \frac{u-\rho v}{\sqrt{1-\rho^2}}, \quad t_2 = v,$$

or

$$u = \sqrt{1-\rho^2}\, t_1 + \rho t_2, \quad v = t_2,$$

and

$$J = \begin{vmatrix} \sqrt{1-\rho^2} & \rho \\ 0 & 1 \end{vmatrix} = \sqrt{1-\rho^2}.$$

Hence,

$$A = \int_{-\infty}^{\infty}\left\{\int_{-\infty}^{\infty} \frac{1}{\sqrt{2\pi}} e^{-(1/2)t_1^2} dt_1\right\} \frac{1}{\sqrt{2\pi}} e^{-(1/2)t_2^2} dt_2 = 1.$$

2. We now compute the marginal probability densities of the random variables X and Y.

$$f_1(x;\mu_X,\sigma_X) = \int_{-\infty}^{\infty} f(x,y,\mu_X,\mu_y,\sigma_X,\sigma_Y) dy, \quad -\infty < x < \infty,$$

$$= \int_{-\infty}^{\infty} \frac{1}{2\pi\sigma_X\sigma_Y\sqrt{1-\rho^2}} \cdot \exp\left\{-\frac{1}{2(1-\rho^2)}\right.$$

$$\left[\left(\frac{x-\mu_X}{\sigma_X}\right)^2 - 2\rho\left(\frac{x-\mu_X}{\sigma_X}\right)\left(\frac{y-\mu_Y}{\sigma_Y}\right)\right.$$

$$\left.\left.+\frac{y-\mu_Y}{\sigma_Y}\right]^2\right\} dy, \quad -\infty < x < \infty.$$

Let

$$v = \frac{y-\mu_Y}{\sigma_Y}$$

and

$$dy = \sigma_Y\, dv.$$

6.9 Bivariate Normal Distribution

Then,

$$f_1(x;\mu_X,\sigma_Y) = \int_{-\infty}^{\infty} \frac{1}{2\pi\sigma_X\sqrt{1-\rho^2}} \cdot \exp\left\{-\frac{1}{2(1-\rho^2)}\right.$$

$$\left.\left[\left(\frac{x-\mu_X}{\sigma_X}\right)^2 - 2\rho\left(\frac{x-\mu_X}{\sigma_X}\right)v + v^2\right]\right\} dv, \quad -\infty < x < \infty.$$

We may write

$$\left(\frac{x-\mu_X}{\sigma_X}\right)^2 - 2\rho\left(\frac{x-\mu_X}{\sigma_X}\right)v + v^2$$

as

$$\left(\frac{x-\mu_X}{\sigma_X}\right)^2 - 2\rho\left(\frac{x-\mu_X}{\sigma_X}\right)$$

or

$$v + \rho^2\left(\frac{x-\mu_X}{\sigma_X}\right)^2 - \rho^2\left(\frac{x-\mu_X}{\sigma_X}\right)^2 + v^2,$$

which equals

$$\left[v - \rho\left(\frac{x-\mu_X}{\sigma_X}\right)\right]^2 + (1-\rho^2)\left(\frac{x-\mu_X}{\sigma_X}\right)^2.$$

Thus,

$$f_1(x;\mu_X,\sigma_X) = \int_{-\infty}^{\infty} \frac{1}{2\pi\sigma_X\sqrt{1-\rho^2}}$$

$$\cdot \exp\left\{-\frac{1}{2(1-\rho^2)}\left[v - \rho\left(\frac{x-\mu_X}{\sigma_X}\right)\right]^2\right\}$$

$$\cdot \exp\left\{-\frac{1}{2}\left(\frac{x-\mu_X}{\sigma_X}\right)^2\right\} dv, \quad -\infty < x < \infty,$$

$$= \frac{1}{\sqrt{2\pi}\sigma_X} \exp\left\{-\frac{1}{2}\left[\frac{x-\mu_X}{\sigma_X}\right]^2\right\} \int_{-\infty}^{\infty} \frac{1}{\sqrt{2\pi}\sqrt{1-\rho^2}}$$

$$\cdot \exp\left\{-\frac{1}{2}\left[\frac{v - \rho\left(\frac{x-\mu_X}{\sigma_X}\right)}{\sqrt{1-\rho^2}}\right]^2\right\} dv, \quad -\infty < x < \infty.$$

Now, let

$$t = \frac{v - \rho[(x-\mu_X)/\sigma_X]}{\sqrt{1-\rho^2}};$$

then,

$$\sqrt{1-\rho^2}\, dt = dv.$$

Hence,

$$f_1(x;\mu_X,\sigma_X) = \frac{1}{\sqrt{2\pi}\sigma_X} \exp\left\{-\frac{1}{2}\left(\frac{x-\mu_X}{\sigma_X}\right)^2\right\}$$

$$\int_{-\infty}^{\infty} \frac{1}{\sqrt{2\pi}} e^{-(1/2)t^2}\, dt, \quad -\infty < x < \infty.$$

Therefore, the marginal probability density of the random variable X is

$$f_1(x;\mu_X,\sigma_X) = \frac{1}{\sqrt{2\pi}\sigma_X} \exp\left\{-\frac{1}{2}\left(\frac{x-\mu_X}{\sigma_X}\right)^2\right\}, \quad -\infty < x < \infty,\ \sigma_X > 0.$$

Similarly, the marginal probability density function of the variate Y is

$$f_2(y;\mu_Y,\sigma_Y) = \frac{1}{\sqrt{2\pi}\sigma_Y} \exp\left\{-\frac{1}{2}\left(\frac{y-\mu_Y}{\sigma_Y}\right)^2\right\}, \quad -\infty < y < \infty,\ \sigma_Y > 0.$$

3. The conditional probability density function of the random variable X, given that $Y = y$, is

$$h_1(x|y) = h_1(x;\mu_X,\sigma_X|y;\mu_Y,\sigma_Y), \quad -\infty < x < \infty,$$

$$= \frac{f(x,y;\mu_X,\mu_Y,\sigma_X,\sigma_Y)}{f_2(y,\mu_Y,\sigma_Y)}$$

$$= \frac{\dfrac{1}{2\pi\sigma_X\sigma_Y\sqrt{1-\rho^2}} \exp\left\{-\dfrac{1}{2(1-\rho^2)}\left[\left(\dfrac{x-\mu_X}{\sigma_X}\right)^2 - 2\rho\left(\dfrac{x-\mu_X}{\sigma_X}\right)\left(\dfrac{y-\mu_Y}{\sigma_Y}\right) + \left(\dfrac{y-\mu_Y}{\sigma_Y}\right)^2\right]\right\}}{\dfrac{1}{\sqrt{2\pi}\sigma_Y}\exp\left\{-\dfrac{1}{2}\left[\dfrac{y-\mu_Y}{\sigma_Y}\right]^2\right\}}, \quad -\infty < x < \infty,\ -\infty < y < \infty,$$

$$= \frac{1}{2\pi\sigma_X\sqrt{1-\rho^2}} \exp\left\{-\frac{1}{2(1-\rho^2)}\left[\left(\frac{x-\mu_X}{\sigma_X}\right)^2 - 2\rho\left(\frac{x-\mu_X}{\sigma_X}\right)\right.\right.$$

$$\left.\left.\cdot\left(\frac{y-\mu_Y}{\sigma_Y}\right)+\rho^2\left(\frac{y-\mu_Y}{\sigma_Y}\right)^2\right]\right\}, \quad \begin{array}{l}-\infty<x<\infty,\\-\infty<y<\infty,\end{array}$$

$$=\frac{1}{2\pi\sigma_X\sqrt{1-\rho^2}}$$

$$\cdot\exp\left\{-\frac{1}{2\sigma_X^2(1-\rho^2)}\left[(x-\mu_X)^2-2\rho\sigma_X(x-\mu_X)\right.\right.$$

$$\left.\left.\cdot\left(\frac{y-\mu_Y}{\sigma_Y}\right)+\rho^2\sigma_X^2\left(\frac{y-\mu_Y}{\sigma_Y}\right)^2\right]\right\}, \quad -\infty<x<\infty, -\infty<y<\infty$$

Therefore,

$$h_1(x|y)=\frac{1}{\sqrt{2\pi}\sigma_X\sqrt{1-\rho^2}}$$

$$\cdot\exp\left\{-\frac{1}{2\sigma_X^2(1-\rho^2)}\left[(x-\mu_X)-\rho\sigma_X\left(\frac{y-\mu_Y}{\sigma_Y}\right)\right]^2\right\},$$

$$-\infty<x<\infty$$
$$-\infty<y<\infty$$

Similarly, we can compute the conditional probability density of Y, given that $X=x$:

$$h_2(y|x)=\frac{1}{\sqrt{2\pi}\sigma_Y\sqrt{1-\rho^2}}$$

$$\cdot\exp\left\{-\frac{1}{2\sigma_Y^2(1-\rho^2)}\left[(Y-\mu_Y)-\rho\sigma_Y\left(\frac{x-\mu_X}{\sigma_X}\right)\right]^2\right\}, \quad -\infty<x<\infty,-\infty<y<\infty.$$

Thus, the conditional density $h_1(x|y)$ is a univariate normal probability density function with a mean of $\mu_X+\rho\sigma_Y(y-\mu_X)/\sigma_X$ and a variance of $\sigma_X^2(1-\rho^2)$. That is,

$$h_1(x|y)=\frac{1}{\sqrt{2\pi}\sqrt{\sigma_X^2 1-\rho^2}}$$

$$\cdot\exp\left\{-\frac{1}{2}\left[\frac{x-\left[\mu_X-\rho\sigma_X\left(\frac{y-\mu_Y}{\sigma_Y}\right)\right]}{\sqrt{\sigma_X^2(1-\rho^2)}}\right]^2\right\}, \quad -\infty<x<\infty,-\infty<y<\infty.$$

An analogous interpretation can be given to the conditional probability density $h_2(y|x)$.

4. Special Cases

   a. If the coefficient of correlation is zero, $\rho_{XY}=0$, then X and Y are independently distributed. That is,

   $$f(x,y;\mu_X,\mu_Y,\sigma_X,\sigma_Y)=\frac{1}{2\pi\sigma_X\sigma_Y}\exp\left\{-\frac{1}{2}\left[\frac{x-\mu_X}{\sigma_X}\right]^2+\left[\frac{y-\mu_Y}{\sigma_Y}\right]^2\right\}$$

   $$=\frac{1}{2\pi\sigma_X}\exp\left\{-\frac{1}{2}\left[\frac{x-\mu_X}{\sigma_X}\right]^2\right\}\cdot\frac{1}{\sqrt{2\pi}\sigma_Y}\cdot\exp\left\{-\frac{1}{2}\left[\frac{y-\mu_Y}{\sigma_Y}\right]^2\right\}$$

   $$=f_1(x;\mu_X,\sigma_X)f_2(y;\mu_Y,\sigma_Y).$$

   b. If $\rho_{XY}\neq 0$, it is clear from part 3 that $h_1(x|y)$ and $h_2(y|x)$ depend on y and x, respectively.

   Thus, in the bivariate normal distribution, we have independence if and only if the coefficient of correlation is zero.

5. In this part, we show that, if $f(x,y;\mu_X,\mu_Y,\sigma_X,\sigma_Y)$ equals a constant, then the contour lines for this density are ellipses. It has been shown that $f(x,y;\mu_X,\mu_Y,\sigma_X,\sigma_Y)$ can be reduced to

$$z(u,v,\rho)=\frac{1}{2\pi\sqrt{1-\rho^2}}\cdot\exp\left\{-\frac{1}{2(1-\rho^2)}(u-\rho v)^2-\frac{v^2}{2}\right\},$$

$$-\infty<v<\infty,\ -\infty<u<\infty,\ -1<\rho<1.$$

It is clear that the exponent of this density is always negative and that its constant is always positive. In fact, $z(u,v;\rho)$ attains a maximum when $u=0=v$. Thus, if we let

$$\frac{1}{2\pi\sqrt{1-\rho^2}}\cdot\exp\left\{-\frac{1}{2(1-\rho^2)}(u-\rho v)^2-\frac{v^2}{2}\right\}=c,$$

or equivalently

$$\frac{(u-\rho v)^2}{2(1-\rho^2)}+\frac{v^2}{2}=-\ln 2\pi-\frac{1}{2}\ln(-\rho^2)-\ln c,$$

we have

$$\frac{(u-\rho v)^2}{-(1-\rho^2)2\ln\left[2\pi c\sqrt{1-\rho^2}\right]}+\frac{v^2}{-2\ln\left[2\pi c\sqrt{1-\rho^2}\right]}=1.$$

Then,

$$\frac{(u-\rho v)^2}{\left[\ln\left[\frac{1}{2\pi c\sqrt{1-\rho^2}}\right]\right]^{2(1-\rho^2)}}+\frac{v^2}{\left[\ln\left[\frac{1}{2\pi c\sqrt{1-\rho^2}}\right]\right]^2}=1,$$

which is in the form of the equation of an ellipse. If $\rho=0$, the preceding expression reduces to the circle

$$u^2+v^2 = \ln\left[\frac{1}{2\pi c}\right]^2,$$

which has its center at the origin and a radius of $\sqrt{2\ln[1/2\pi c]}$. Furthermore, it is clear that, if $\rho=1$, the bivariate normal density function is indeterminate; however, if ρ approaches ± 1, the ellipses degenerate into a line, as shown later. This is consistent with earlier remarks regarding the interpretation of the coefficient of correlation: namely, the random variables are linearly related. This relation is illustrated as follows. The expression

$$-\frac{1}{2(1-\rho^2)}(u^2-2\rho uv+v^2)-\ln(2\pi\sqrt{1-\rho^2})-\ln c=0$$

can be written as

$$(u-\rho v)^2 = (\rho^2-1)\{v^2+\ln[4\pi^2 c^2(1-\rho^2)]\}$$

or

$$u-\rho v = \pm\sqrt{\rho^2-1}\{v^2+\ln[4\pi^2 c^2(1-\rho^2)]\}^{1/2}.$$

Taking the limit of both sides of the preceding equation as $\rho \to \pm 1$, we have the desired result:

$$u-v=0$$

and

$$u+v=0.$$

6. Let X and Y be independent normally distributed random variables with means of μ_X and μ_Y and variances of σ_X^2 and σ_Y^2, respectively. We show that $U=X+Y$ is also normally distributed with a mean of $\mu_X+\mu_Y$ and a variance of $\sigma_X^2+\sigma_Y^2$. Let $U=X+Y$ and $V=Y$. Thus, $x=u-v$, $v=y$, and $|J|=1$. Because X and Y are independent, we can use part 4 and write

$$z_1(u) = \int_{-\infty}^{\infty} f_1(u-v) f_2(v) |J| dv, \quad -\infty < u < \infty,$$

$$= \int_{-\infty}^{\infty} \frac{1}{\sigma_X\sqrt{2\pi}} e^{-(1/2\sigma_X^2)[(u-v)-\mu_X]^2} \cdot \frac{1}{\sigma_Y\sqrt{2\pi}} e^{-(1/2\sigma_Y^2)(v-\mu_Y)^2} dv, \quad -\infty < u < \infty,$$

$$= \frac{1}{2\pi\sigma_X\sigma_Y}\int_{-\infty}^{\infty} \exp\left\{-\frac{1}{2}\left[\frac{(u-v)-\mu_X}{\sigma_X}\right]^2 - \frac{1}{2}\left(\frac{v-\mu_Y}{\sigma_Y}\right)^2\right\} dv, \quad -\infty < u < \infty.$$

The exponent can be written as follows:

$$-\frac{1}{2}\left[\frac{(u-v)-\mu_X}{\sigma_X}\right]^2 - \frac{1}{2}\left(\frac{v-u_y}{\sigma_Y}\right)^2$$

$$= -\frac{1}{2}\left[\frac{u^2-2uv+v^2+2\mu_X v+\mu_X^2}{\sigma_X^2}\right.$$

$$\left.+\frac{v^2-2v\mu_Y+\mu_Y^2}{\sigma_Y^2}-\frac{2\mu_X u}{\sigma_X^2}\right]$$

$$= -\frac{1}{2}\left\{\left[v^2\left(\frac{1}{\sigma_X^2}+\frac{1}{\sigma_Y^2}\right)-2v\left(\frac{u-\mu_X}{\sigma_X^2}+\frac{\mu_Y}{\sigma_Y^2}\right)\right]\right.$$

$$\left.+\frac{u^2-2\mu_X u+\mu_X^2}{\sigma_X^2}+\frac{\mu_Y^2}{\sigma_Y^2}\right\}$$

$$= -\frac{1}{2}\left\{v^2\left(\frac{\sigma_Y^2+\sigma_X^2}{\sigma_X^2\sigma_Y^2}\right)-2v\left(\frac{\sigma_Y^2 u-\sigma_Y^2\mu_X+\sigma_X^2\mu_Y}{\sigma_X^2\sigma_Y^2}\right)\right\}$$

$$-\frac{1}{2}\left[\frac{(u-\mu_X)^2}{\sigma_X^2}+\frac{\mu_Y^2}{\sigma_Y^2}\right]$$

$$= \frac{\sigma_X^2+\sigma_Y^2}{2\sigma_X^2\sigma_Y^2}\left\{v^2-2\left[\frac{\sigma_Y^2(u-\mu_X)+\sigma_X^2\mu_Y}{\sigma_X^2+\sigma_Y^2}\right]v\right\}$$

$$-\frac{1}{2}\left[\frac{(u-\mu_X)^2}{\sigma_X^2}+\frac{\mu_Y^2}{\sigma_Y^2}\right].$$

Completing the square on v, the exponent can be written as

$$-\frac{\sigma_X^2+\sigma_Y^2}{2\sigma_X^2\sigma_Y^2}\left\{\left[v-\frac{\sigma_Y^2(u-\mu_X)+\sigma_X^2\mu_Y}{\sigma_X^2+\sigma_Y^2}\right]^2\right.$$

$$\left.-\left[\frac{\sigma_Y^2(u-\mu_X)+\sigma_X^2\mu_Y}{\sigma_X^2+\sigma_Y^2}\right]^2\right\}-\frac{1}{2}\left[\frac{(u-\mu_X)^2}{\sigma_X^2}+\frac{\mu_Y^2}{\sigma_Y^2}\right].$$

Thus,

$$z_1(u) = \frac{1}{2\pi\sigma_X\sigma_Y} \exp\left\{-\frac{1}{2}\left[\frac{(u-\mu_X)^2}{\sigma_X^2} + \frac{\mu_Y^2}{\sigma_Y^2}\right.\right.$$

$$\left.\left. - \left[\frac{\sigma_Y^2(u-\mu_X) + \sigma_X^2\mu_Y}{\sigma_X\sigma_Y\sqrt{\sigma_X^2+\sigma_Y^2}}\right]^2\right]\right\}$$

$$\int_{-\infty}^{\infty} \exp\left\{-\frac{\sigma_X^2+\sigma_Y^2}{2\sigma_X^2\sigma_Y^2}\left[v - \frac{\sigma_Y^2(u-\mu_X)+\sigma_X^2\mu_Y}{\sigma_X^2+\sigma_Y^2}\right]^2\right\} dv, \quad -\infty < u < \infty.$$

Let

$$h = \left[v - \frac{\sigma_Y^2(u-\mu_X)+\sigma_X^2\mu_Y}{\sigma_X^2+\sigma_Y^2}\right] \frac{\sqrt{\sigma_X^2+\sigma_Y^2}}{\sigma_X\sigma_Y};$$

then,

$$\frac{\sigma_X\sigma_Y}{\sqrt{\sigma_X^2+\sigma_Y^2}} dh = dv.$$

Hence,

$$z_1(u) = \frac{1}{\sqrt{2\pi}\sqrt{\sigma_X^2+\sigma_Y^2}} \exp\left\{-\frac{1}{2}\left[\frac{(u-\mu_X)^2}{\sigma_X^2} + \frac{\mu_Y^2}{\sigma_Y^2}\right.\right.$$

$$\left.\left. - \left[\frac{\sigma_Y^2(u-\mu_X)+\sigma_X^2\mu_Y}{\sigma_X\sigma_Y\sqrt{\sigma_X^2+\sigma_Y^2}}\right]^2\right]\right\}$$

$$\cdot \frac{1}{2\pi}\int_{-\infty}^{\infty} e^{-(1/2)h^2} dh, \quad -\infty < u < \infty.$$

The exponent of e can be simplified as follows:

$$-\frac{1}{2}\left[\frac{(u-\mu_X)^2}{\sigma_X^2}+\frac{\mu_Y^2}{\sigma_Y^2}\right]-\left[\frac{\sigma_Y^2(u-\mu_X)+\sigma_X^2\mu_Y}{\sigma_X\sigma_Y\sqrt{\sigma_X^2+\sigma_Y^2}}\right]^2$$

$$=-\frac{1}{2}\left\{\frac{(u-\mu_X)^2}{\sigma_X^2}+\frac{\mu_Y^2}{\sigma_Y^2}-\frac{\sigma_Y^2(u-\mu_X)^2}{\sigma_X^2(\sigma_X^2+\sigma_Y^2)}\right.$$

$$\left.-\frac{\sigma_X^2\mu_Y^2}{\sigma_Y^2(\sigma_X^2+\sigma_Y^2)}-\frac{2(u-\mu_X)\mu_Y}{\sigma_X^2+\sigma_Y^2}\right\}$$

$$=-\frac{1}{2}\left\{\{u-\mu_X^2\}\left[\frac{1}{\sigma_X^2}-\frac{\sigma_Y^2}{\sigma_X^2(\sigma_X^2+\sigma_Y^2)}\right]\right.$$

$$\left.+\mu_Y^2\left[\frac{1}{\sigma_Y^2}-\frac{\sigma_X^2}{\sigma_Y^2(\sigma_X^2+\sigma_Y^2)}\right]-\frac{2\mu_Y(u-\mu_X)}{\sigma_X^2+\sigma_Y^2}\right\}$$

$$=-\frac{1}{2}\left[\frac{1}{\sigma_X^2+\sigma_Y^2}(u-\mu_X)^2-\frac{2\mu_Y(u-\mu_X)}{\sigma_X^2+\sigma_Y^2}+\frac{\mu_Y^2}{\sigma_X^2+\sigma_Y^2}\right]$$

$$=-\frac{1}{2(\sigma_X^2+\sigma_Y^2)}\left[(u-\mu_X)^2-2\mu_Y(u-\mu_X)+\mu_Y^2\right]$$

$$=-\frac{1}{2(\sigma_X^2+\sigma_Y^2)}\left[(u-\mu_X)-\mu_Y\right]^2=-\frac{1}{2}\frac{\left[u-(\mu_X+\mu_Y)\right]^2}{\sigma_X^2+\sigma_Y^2}.$$

Therefore,

$$z_1(u)=\frac{1}{\sqrt{2\pi}\sqrt{\sigma_X^2+\sigma_Y^2}}$$

$$\cdot\exp\left\{-\frac{1}{2}\left[\frac{u-(\mu_X+\mu_Y)}{\sqrt{\sigma_X^2+\sigma_Y^2}}\right]^2\right\},\quad -\infty<u<\infty.$$

Thus, the random variable U is normally distributed with the expected values $\mu_X+\mu_Y$ and $\sigma_X^2+\sigma_Y^2$. Also, the covariance of the bivariate normal distribution is given by

$$\text{Cov}(X,Y)=E\left[(X-\mu_{10})(Y-\mu_{01})\right]=\mu_{11}-\mu_{10}\mu_{01}$$

or

$$\text{Cov}(X,Y)=\mu_X\mu_Y+\rho\sigma_X\sigma_Y-\mu_X\mu_Y=\rho\sigma_X\sigma_Y.$$

6.9 Bivariate Normal Distribution

Therefore,

$$\rho_{X,Y} = \frac{\text{Cov}(X,Y)}{\sigma_X \sigma_Y}.$$

Thus, the coefficient of correlation between X and Y is $\rho_{XY} = \rho$.

7. The conditional expectation of X, given $Y = y$, is computed as follows:

$$E(E|Y=y) = \int_{-\infty}^{\infty} x h_1(x|y) dx = \int_{-\infty}^{\infty} \frac{x}{\sqrt{2\pi}\sigma_X \sqrt{1-\rho^2}}$$

$$\cdot e^{-(1/2)\{x - [\mu_X + \rho\sigma_X(y-\mu_Y/\sigma_Y)]\}^2 / \sqrt{\sigma_{x^2}(1-\rho^2)}\}^2} dx.$$

Let

$$w = \frac{x - \left[\mu_X + \rho\sigma_X \left(\frac{y-\mu_Y}{\sigma_Y}\right)\right]}{\sqrt{\sigma_X^2 (1-\rho^2)}},$$

or

$$x = \sqrt{\sigma_X^2 (1-\rho^2)}\, w + \mu_X + \rho\sigma_X \left(\frac{y-\mu_Y}{\sigma_Y}\right),$$

and

$$\sqrt{\sigma_X^2 (1-\rho^2)}\, dw = dx.$$

Thus,

$$E(X|Y=y) = \sqrt{\sigma_X^2 (1-\rho^2)} \int_{-\infty}^{\infty} \frac{w}{\sqrt{2\pi}} e^{-(1/2)w^2} dw$$

$$+ \mu_X + \rho\sigma_X \left(\frac{y-\mu_Y}{\sigma_Y}\right) \int_{-\infty}^{\infty} \left(\frac{2}{\sqrt{2\pi}}\right) e^{(1/2)w^2} dw$$

$$= \mu_X \rho\sigma_X \left(\frac{y-\mu_Y}{\sigma_Y}\right).$$

In view of the comments in part 3, this result was expected. Similarly, we can obtain

$$E[Y|X=x] = \mu_Y + \rho\sigma_Y \left(\frac{x-\mu_X}{\sigma_X}\right).$$

8. The regression curves of X on Y and Y on X for the bivariate normal density function are given by

$$\xi_1(y) = E(X|Y=y) = \mu_X + \rho\sigma_X \left(\frac{y-\mu_Y}{\sigma_Y}\right)$$

and

$$\xi_2(x) = E(Y|X=x) = \mu_Y + \rho\sigma_Y \left(\frac{x-\mu_X}{\sigma_X}\right),$$

respectively. Thus, for the bivariate normal distribution, the regression curves are straight lines.

Summary

Let S be a sample space; let $X(s)$ and $Y(s)$ be two real valued functions (defined on S) that assign a real number to each element of the sample space. We call the pair $(X(s)=X, Y(s)=Y)$ a *two-dimensional random variable*.

Let (X,Y) be a two-dimensional random variable. If the number of possible values of (X,Y) is finite or countably infinite, then (X,Y) is called a *two-dimensional discrete random variable*. If (X,Y) can assume all values in some uncountable (nondenumerable) subset of the Euclidean plane, then (X,Y) is called a *two-dimensional continuous random variable*.

Let (X,Y) be a two-dimensional discrete random variable. A function $f(x_i, y_j)$ is called the *joint probability density function* of the random variable (X,Y) if

1. $f(x_i, y_j) \geq 0$, for all $(x_i, y_j) \in R_{X,Y}$

2. $\sum\limits_{x_i, y_j) \in R(X,Y)} \sum f(x_i, y_j) = 1$

for $i=1, 2,\ldots n,\ldots$, and $j=1, 2,\ldots k,\ldots$. If (X,Y) is a two-dimensional continuous random variable, then a function $f(x,y)$ is the *joint probability density function* of the random variable (X,Y) if

1. $f(x,y) \geq 0$, for all $(x,y) \in R_{X,Y}$

2. $\iint_{R_{X,Y}} f(x,y) dx\, dy = 1$

The *cumulative distribution function* $F(x,y)$ of a two-dimensional random variable (X,Y) is defined by

$$F(x,y) = Pr(X \leq x, Y \leq y).$$

If (X,Y) is of the discrete type, then

$$F(x,y) = Pr(X \leq x, Y \leq y) = \sum_{y_j \leq y} \sum_{x_i \leq x} f(x_i, y_j);$$

if (X,Y) is of the continuous type, then

$$F(x,y) = Pr(X \leq x, Y \leq y) = \int_{-\infty}^{y}\left\{\int_{-\infty}^{x} f(t,s)\,dt\right\}ds.$$

$F(x,y)$ possesses the following properties:

1. $F(\infty, \infty) = 1$.
2. $F(-\infty, y) = F(x, -\infty) = 0$.
3. $F(x,y)$ is a monotone nondecreasing function in each variable.
4. $F(x,y)$ is continuous, at least from the right, in each variable.
5. For all real values, $x_1, x_2 \, (x_1 < x_2)$ and $y_1, y_2 \, (y_1 < y_2)$, the relation $F(x_2, y_2) - F(x_1, y_2) - F(x_2, y_1) + F(x_1, y_1) \geq 0$ is satisfied.
6. If the first and second partial derivatives of $F(x,y)$ exist, then

$$\frac{\partial^2 F(x,y)}{\partial x \partial y} = f(x,y).$$

Let $f(x_i, y_j)$ be the probability density function of the discrete random variable (X,Y). The *marginal probability density functions* of the variates X and Y are given by

$$f_1(x_i) = \sum_{j=1}^{\infty} f(x_i, y_j)$$

and

$$f_2(y_j) = \sum_{i=1}^{\infty} f(x_i, y_j),$$

respectively. If $f(x,y)$ is the probability density function of the continuous random variable (X,Y), then the *marginal densities* of the variates X and Y are given by

$$f_1(x) = \int_{-\infty}^{\infty} f(x,y)\,dy$$

and

$$f_2(y) = \int_{-\infty}^{\infty} f(x,y)\,dx,$$

respectively.

For every i and j, we define the probabilities

$$h_1(x_i | y_j) = \frac{f(x_i, y_j)}{f_2(y_j)}$$

and

$$h_2(y_j | x_i) = \frac{f(x_i, y_j)}{f_1(x_i)},$$

where $f_2(y_j)$ and $f_1(x_i)$ are greater than zero. If y_j is fixed and x_i varies over all possible values, then $h_1(x_i|y_j)$ is called the *conditional probability density function of the discrete variable* X, under the condition that $Y = y_j$. Similarly, if x_i is fixed and y_j varies over all possible values, then $h_2(y_j|x_i)$ is called the *conditional probability density function of the discrete variate* Y, given that $X = x_i$. Conditional probability density functions of continuous variates can be defined similarly.

The *conditional cumulative distribution functions* of the continuous random variable X given Y and Y given X are

$$F_X(x|y) = Pr(X \le x|Y \le y) = \frac{Pr(X \le x, Y \le y)}{Pr(Y \le y)}$$

and

$$F_Y(y|x) = Pr(Y \le y|X \le x) = \frac{Pr(X \le x, Y \le y)}{Pr(X \le x)},$$

respectively, provided that $Pr(Y \le y) > 0$ and $Pr(X \le x) > 0$, respectively.

Two random variables X and Y are said to be *independent (mutually independent)* if

$$F(x,y) = F_1(x)F_2(y)$$

is satisfied for every real pair of numbers (x, y).

Let $f(x, y)$ be the joint probability density function of the random variable (X, Y). If (X, Y) is of the continuous type, then X and Y are independent random variables if and only if

$$f(x,y) = f_1(x)f_2(x)$$

for all real (x, y).

One function of two discrete random variables was studied, as were two functions of two continuous random variables. More precisely, let $f(x, y)$ be the joint probability density function of the continuous random variable (X, Y). Suppose that

$$U = g_1(X, Y)$$

and

$$V = g_2(X, Y),$$

such that g_1 and g_2 define a one-to-one transformation of the random variable (X, Y) and that the transformation has continuous first partial derivatives with respect to x and y. The joint probability density function of the variate (U, V) is given by

$$z(u,v) = f\left[z_1^{-1}(u,v), z_2^{-1}(u,v)\right]|J|,$$

where $x = z_1^{-1}(u,v)$, $y = z_2^{-1}(u,v)$, and the Jacobian $|J| \ne 0$.

The joint probability density function of the discrete type was also studied. Let $f(x,y)$ be the probability density function of the continuous random variable (X,Y). The probability densities of $U = X+Y$, $U = XY$, and $U = X/Y$ are

$$g(u) = \int_{-\infty}^{\infty} f(u-v, v) \, dv,$$

$$h(u) = \int_{-\infty}^{\infty} f\left(\frac{u}{v}, v\right) \left|\frac{1}{v}\right| dv,$$

and

$$k(u) = \int_{-\infty}^{\infty} f(uv, v) |v| \, dv,$$

respectively.

Let (X,Y) be a two-dimensional random variable with joint probability density function $f(x,y)$. The *moment of order* $k+m$ of the distribution of the variate (X,Y) is given by

$$\mu_{km} = E(X^k Y^m) = \sum_i \sum_j x_i^k y_j^m f(x_i, y_j)$$

if (X,Y) is discrete and by

$$\mu_{km} = E(X^k Y^m) = \int_{-\infty}^{\infty} \int_{-\infty}^{\infty} x^k y^m f(x, y) \, dx \, dy$$

if (X,Y) is continuous. The *central moment of order* $k+m$ of the distribution of the variate (X,Y) is

$$\eta_{km} = E\left[(X - \mu_{10})^k (Y - \mu_{01})^m\right].$$

The second central moment $(k=1, m=1)$ of the probability distribution of the variate (X,Y) is the *covariance of* X *and* Y; that is,

$$\eta_{11} = E\left[(X - \mu_{10})(Y - \mu_{01})\right].$$

The coefficient of correlation ρ_{XY} between X and Y is given by

$$\rho_{XY} = \frac{\eta_{11}}{\sigma_X \sigma_Y}.$$

The coefficient of correlation, if it exists, assumes a value between -1 and $+1$.

Let (X,Y) be a two-dimensional random variable. The *conditional expected value* of the variate X given $Y = y_j$ is

$$E(X | Y = y_j) = \sum_i x_i h_1(x_i | y_j)$$

if (X,Y) is discrete and

$$E(X | Y = y) = \int_{-\infty}^{\infty} x h_1(x | y_j) \, dx$$

if (X,Y) is of the continuous type.

If $E(X|Y=y)$ and $E(Y|X=x)$ exist, then

$$E\big[E(X|Y=y)\big]=E(X)$$

and

$$E\big[E(Y|X=x)\big]=E(Y).$$

The geometric interpretation of conditional expectation is illustrated by a *regression curve*. Let the expected value of the variates X given $Y=y_j$ and Y given $X=x_i$ be denoted using

$$\xi_1(y_j)=E(X|Y=y_j)$$

and

$$\xi_2(x_i)=E(Y|X=x_i).$$

The graph of the collection of points $(\xi_1(y_j),y_j)$ is called the regression curve of *the random variable X on the random variable Y*. Similarly, the graph of the set of points $(x_i,\xi_2(x_i))$ is called the regression curve of *the random variable Y on the random variable X*. We can obtain similar results if the variate (X,Y) is of the continuous type.

The bivariate normal distribution was studied in depth.

In this chapter, we were concerned with a precise and comprehensive study of two random variables and their probability distributions. However, in many cases, we have to deal with random variables of more than two dimensions. Chapter 7 aims to extend the concept of a two-dimensional random variable and its probability distribution function to n-dimensional random variables.

Theoretical Exercises

6.1 Let

$$f(x,y)=\begin{cases}\dfrac{1}{50}(x^2+2y), & x=0,1,2,3, \text{ and } y=x+3,\\ 0, & \text{elsewhere.}\end{cases}$$

a. Show that $f(x,y)$ satisfies the conditions of a probability density function.
b. Sketch $f(x,y)$.

Theoretical Exercises

6.2 Show that the bivariate function given by

$$f(x,y) = \begin{cases} \left(\dfrac{2}{n(n+1)}\right)^2 xy, & x,y = 1,2,\ldots,n, \\ 0, & \text{elsewhere}, \end{cases}$$

is the joint probability density function of the random variable (X,Y).

Hint: $\sum_{i=1}^{n} i = [n(n+1)/2]$.

6.3 Let

$$f(x,y) = c(1-x)(1-y), \quad -1 \le x \le 1,\ -1 \le y \le 1.$$

Find c so that $f(x,y)$ is the joint probability density function of the random variable (X,Y).

6.4 Find c so that

$$f(x,y) = cx(y-x),\ 0 \le x \le 4,\ 4 \le y \le 8$$

is a bivariate probability density function.

6.5 Let

$$f(x,y) = xe^{-xy}, \quad x \ge 0,\ y \ge 1.$$

Is $f(x,y)$ a probability density function? If not, find the proper constant to multiply with $f(x,y)$ so that it is a probability density.

6.6 Find c so that

$$f(x,y) = cxe^{-(x^2+y)}, \quad x \ge 0,\ y \ge 0,$$

is the joint probability density function of the random variable (X,Y).

6.7 Prove Theorem 6.2.1.

6.8 The joint probability density function of the random variable (X,Y) is given by

$$f(x,y) = \begin{cases} \dfrac{1}{5}(3x-y), & 1 \le x \le 2,\ 1 \le y \le 3, \\ 0, & \text{elsewhere}. \end{cases}$$

Find the cumulative distribution function of the random variable (X,Y).

6.9 Find $F(x,y)$ of the random variable (X,Y) whose probability density function is given by

$$f(x,y) = \begin{cases} 2, & 0 \le y \le x \le 1, \\ 0, & \text{elsewhere}. \end{cases}$$

6.10 In Exercises 6.1 through 6.4, find the cumulative distribution function of the random variable (X,Y).

6.11 Find the joint probability density function of the random variable (X,Y) if its cumulative distribution function is given by

$$F(x,y) = \begin{cases} 0, & x \leq 1 \text{ or } y \leq 1, \\ (x-1)(2y-x-1), & 1 < x < y < 2, \\ (x-1)(3-x), & 1 < x < 2, y \geq 2, \\ (y-1)^2, & x \geq y, 1 < y < 2, \\ 1, & x \geq 2, y \geq 2. \end{cases}$$

6.12 The cumulative distribution function of the random variable (X,Y) is given by

$$F(x,y) = \begin{cases} 0, & x \leq 1 \text{ or } y \leq 2, \\ \frac{1}{6}(x-1)(y-2)(3x-y+1), & 1 < x < 2, 2 < y < 4, \\ \frac{1}{6}(y-2)(7-y), & x \geq 2, 2 < y < 4, \\ (x-1)^2, & 1 < x < 2, y \geq 4, \\ 1, & x \geq 2, y \geq 4. \end{cases}$$

Find the joint probability density function of the random variable (X,Y).

6.13 Find the marginal probability density function of the random variables X and Y if their joint probability density function is given in Exercise 6.8.

6.14 The joint probability density function of the random variable (X,Y) is given by

$$f(x,y) = \begin{cases} \frac{1}{9}(2x-y), & x, y = 0,1,2, \text{ such that } y \leq 2x, \\ 0, & \text{elsewhere.} \end{cases}$$

Find

a. $f_1(x)$
b. $f_2(y)$
c. $F_1(x)$
d. $F_2(y)$
e. Sketch the graphs of Exercise 6.14(a) through (d).

6.15 Let

$$f(x,y) = \begin{cases} \binom{y}{x} p^x (1-p)^{y-x} \frac{e^{-\lambda} \lambda^y}{y!}, & \begin{array}{l} x = 0,1,\ldots,y, \\ y = 0,1,2,\ldots, \\ \text{with } y \geq x, \end{array} \\ 0, & \text{elsewhere,} \end{cases}$$

where $0 \leq p \leq 1$. Find

a. $f_1(x)$
b. $f_2(y)$

What are

c. $\Pr(X \leq r), r \leq y$
d. $\Pr(Y \geq k)$

6.16 The bivariate probability density function of the random variable (X,Y) is given by

$$f(x,y)=\begin{cases}\dfrac{2}{19}xy^2, & 1\le x\le 2,\ 2\le y\le 3,\\ 0, & \text{elsewhere.}\end{cases}$$

Find
a. $F(x,y)$
b. $f_1(x)$
c. $f_2(y)$
d. $F_1(x)$
e. $F_2(y)$

6.17 The joint probability density function of the random variable (X,Y) is given by

$$f(x,y)=\begin{cases}e^{-(x+y)}, & x,y>0,\\ 0, & \text{elsewhere.}\end{cases}$$

a. Find $F(x,y)$.
b. Find $f_1(x)$.
c. Find $f_2(y)$.
d. Find $F_1(x)$.
e. Find $F_2(y)$.
f. Find $h_1(x|y)$.
g. Find $h_2(y|x)$.
h. Find $F_X(x|y)$.
i. Find $F_Y(y|x)$.
j. Are the random variables X and Y independent?

6.18 Prove Theorem 6.5.2.

6.19 In Exercise 6.8, are the random variables X and Y independent?

6.20 In Exercise 6.8, find
a. $f_1(x)$
b. $f_2(y)$
c. $h_1(x|y=1)\ h_1(x|y=2)\ h_1(x|y=3)\ h_1(x|y=4)\ h_1(x|y=5)$
d. $h_2(y|x=1)\ h_2(y|x=2)\ h_2(y|x=3)\ h_2(y|x=4)\ h_2(y|x=5)$

6.21 In Exercise 6.15, find
a. $h_1(x|y)$
b. $h_2(y|x)$
c. Determine whether the random variables X and Y are independent.

6.22 Let the random variable (X,Y) have the following joint probability density functions:

a. $f(x,y) = \begin{cases} \frac{1}{14}xy(x-y), & x=2,3, y=1,2, \\ 0, & \text{elsewhere} \end{cases}$

b. $f(x,y) = \begin{cases} \frac{1}{15}(x-3)(x-y), & x=3,4,5, y=2,3,4, \\ 0, & \text{elsewhere} \end{cases}$

c. $f(x,y) = \begin{cases} \frac{1}{11}(x-y), & x=2,3, y=1,2, \\ 0, & \text{elsewhere} \end{cases}$

d. $f(x,y) = \begin{cases} \frac{y}{21}(2x-y), & x=1,2,3, y=0,1,2, \\ 0, & \text{elsewhere} \end{cases}$

e. $f(x,y) = \begin{cases} \frac{3}{2ab}\left(\frac{x^2}{a^2}+\frac{y^2}{b^2}\right), & 0 \leq x \leq a, 0 \leq y \leq b, \\ 0, & \text{elsewhere} \end{cases}$

f. $f(x,y) = \begin{cases} 3x(1-xy), & 0 \leq x \leq 1, 0 \leq y \leq 1, \\ 0, & \text{elsewhere} \end{cases}$

g. $f(x,y) = \begin{cases} \frac{2}{a^2}, & 0 \leq x \leq y \leq a, \\ 0, & \text{elsewhere} \end{cases}$

h. $f(x,y) = \begin{cases} 2xe^{-(x^2+y)}, & x,y \geq 0, \\ 0, & \text{elsewhere} \end{cases}$

i. $f(x,y) = \begin{cases} 3, & 0 \leq y \leq x^2 \leq 1, 0 \leq x \leq 1, \\ 0, & \text{elsewhere} \end{cases}$

j. $f(x,y) = \begin{cases} \frac{3}{4}, & 0 \leq x^2 \leq y \leq 1, \\ 0, & \text{elsewhere} \end{cases}$

k. $f(x,y) = \begin{cases} \frac{3}{2}, & 0 \leq y \leq x^2 \leq 1, \\ 0, & \text{elsewhere} \end{cases}$

l. $f(x,y) = \begin{cases} y, & 0 \leq x, y \leq 1, \\ \frac{1}{4}(2-y), & 0 \leq x \leq 1, 0 \leq y \leq 2, \\ 0, & \text{elsewhere} \end{cases}$

Find the following for each probability density function:
a. $f_1(x)$
b. $f_2(y)$
c. $F(x,y)$
d. $F_1(x)$
e. $F_2(y)$
f. $h_1(x|y)h_1(x|y_0)$
g. $h_2(y|x)h_2(y|x_0)$
h. $F_X(x|y)$
i. $F_Y(y|x)$
j. Determine whether the random variables X and Y are independent.

6.23 If the probability density function of the random variable X and $Y|X$ is given by

$$f(x) = \begin{cases} e^{-x}, & x \geq 0, \\ 0, & \text{elsewhere,} \end{cases}$$

and

$$h_2(y|x) = \begin{cases} \dfrac{e^{-x} x^y}{y!}, & y > 0, \\ 0, & \text{elsewhere,} \end{cases}$$

respectively, find the probability density function of the random variable Y.

6.24 Let

$$f(x,y) = \dfrac{c}{(1+x^2)\sqrt{1-y^2}}, \quad -\infty < x < \infty,\ -1 < y < 1.$$

a. Find c so that $f(x,y)$ has the probability density function of the random variable (X,Y).
b. Determine whether X and Y are independent.

6.25 Let X and Y be two independent random variables, each normally distributed, with the parameters μ_1, σ_1^2 and μ_2, σ_2^2, respectively. Show that the probability density function of $Z = X/Y$ is the *Cauchy distribution;* that is, show that

$$f(z) = \dfrac{\sigma_1 \sigma_2}{\pi(\sigma_1^2 + \sigma_2^2 z^2)}, \quad -\infty < z < \infty.$$

6.26 Suppose that X_1 and X_2 are independent variates with the joint probability density function

$$f(x_1, x_2) = \begin{cases} \dfrac{x_1^{\alpha-1} x_2^{\beta-1} e^{-(x_1+x_2)}}{\Gamma(\alpha)\Gamma(\beta)}, & \begin{array}{l} 0 < x_1 < \infty, \\ 0 < x_2 < \infty, \\ \alpha, \beta > 0, \end{array} \\ 0, & \text{elsewhere.} \end{cases}$$

a. Find the joint probability density function of (Z_1, Z_2), where

$$Z_1 = \frac{X_1}{X_1 + X_2}, \quad Z_2 = X_1 + X_2.$$

b. Find the marginal probability density functions of the random variables Z_1 and Z_2.
c. Determine whether the variates Z_1 and Z_2 are independent.
d. If $\alpha = \beta = 1$, what is the probability that $Z_2 < Z_1$?

6.27 The bivariate probability density function of the random variable (X, Y) is given by

$$f(x, y) = \begin{cases} \dfrac{1}{9} x(y - x), & 0 \le x \le 3,\ 2 \le y \le 4, \\ 0, & \text{elsewhere.} \end{cases}$$

a. Find the probability density function of the variable $Z = X + Y$.
b. What is the probability that Z will assume a value between 2.5 and 5.5?

6.28 The joint probability density function of the random variable (X, Y) is given by

$$f(x, y) = \begin{cases} \dfrac{2x}{y^2}, & 0 \le x \le 1,\ y \ge 1, \\ 0, & \text{elsewhere.} \end{cases}$$

Find the probability density function of the following random variables:
a. $Z_1 = X + Y$
b. $Z_2 = X - Y$
c. $Z_3 = X/Y$
d. $Z_4 = XY$

6.29 Show that if the random variables X and Y are independent and follow the gamma distribution with the parameters α_1 and α_2, respectively, then the functional form of the variates,

$$Z = \frac{X}{X + Y},$$

has a *beta distribution* with the parameters α_1 and α_2.

6.30 Let two independent random variables X_1 and X_2, follow the *Rayleigh distribution*; that is, for $i = 1, 2$, let

$$f(x_i) = \begin{cases} \dfrac{2}{\alpha} x_i e^{-(x_i^2/\alpha)}, & x_i \ge 0,\ \alpha > 0, \\ 0, & \text{elsewhere.} \end{cases}$$

Show that the probability density function of the ratio V of these variables is given by

$$f(v) = \begin{cases} \dfrac{2v}{(1+v^2)^2}, & v \geq 0, \\ 0, & \text{elsewhere.} \end{cases}$$

6.31 Let the variate X have a uniform distribution on $(0, 1)$, and let Y have a symmetrical triangular distribution on $(0, 2)$. That is, let

$$f(y) = \begin{cases} y, & 0 \leq y \leq 1, \\ \dfrac{1}{4}(2-y), & 1 \leq y \leq 2, \\ 0, & \text{elsewhere.} \end{cases}$$

Assume that X and Y are independent random variables, and find the probability density function of $Z = X + Y$.

6.32 Let the bivariate probability density function of the random variable (X,Y) be

$$f(x,y) = \dfrac{1}{2\pi} e^{-(1/2)(x^2+y^2)}, \quad -\infty < x, y < \infty.$$

a. Are the random variables X and Y independent?
b. Are X and Y identically distributed?
c. Find $Pr(X^2 + Y^2 \leq 4)$.
d. Find the probability density functions of the random variables X^2 and Y^2.

Applied Problems

6.1 An experiment consists of drawing six balls from an urn that contains nine red, eight green, and three yellow balls. Let X and Y be, respectively, the number of red and green balls drawn.
  a. Determine the joint probability density function of the random variable (X,Y).
  b. What is $Pr(X=3, Y=3)$?
  c. What is $Pr(X=1, Y \leq 4)$?

6.2 Let X and Y be the number of jacks and queens appearing in a bridge hand of 13 cards.
  a. Obtain the joint probability density function of the variable (X,Y).
  b. What is $Pr(X=4, Y \leq 2)$?
  c. What is $Pr(X=4, Y=4)$?
  d. What is $Pr(X \leq 4, Y \geq 2)$?

6.3 From a standard deck of cards, 20 cards are drawn. Find the joint probability density function of X, the number of hearts drawn, and Y, the number of black cards drawn
  a. with replacement
  b. without replacement

6.4 In Exercise 6.3, find the following probabilities:

a. $Pr\left(X > \dfrac{1}{2}, Y \le \dfrac{3}{4}\right)$

b. $Pr\left(0 < X < \dfrac{3}{4}, -\dfrac{1}{2} \le Y \le \dfrac{1}{2}\right)$

c. $Pr\left(X \ge \dfrac{1}{2} \Big| Y \le -\dfrac{1}{2}\right) - Pr\left(Y \ge \dfrac{1}{2} \Big| X \le -\dfrac{1}{2}\right)$

6.5 In Exercise 6.4, find the following probabilities:

a. $Pr(X \le 1, Y \ge 1)$
b. $Pr(2 \le X \le 6, Y \le 6)$
c. $Pr(X \le 3 | Y \le 3) - Pr(Y \le 3 | X \le 3)$

6.6 In Exercise 6.5, evaluate the following probabilities:

a. $Pr(X \ge 1, Y \ge 3)$
b. $Pr(4 \le X \le 6, Y \le 3)$
c. $Pr(X \le 2, Y \ge 3)$

6.7 In Exercise 6.6, obtain the following probabilities:

a. $Pr\left(X > \dfrac{1}{2}, Y \ge 1\right)$

b. $Pr(X \le 3, Y \ge 3)$

c. $Pr\left(X < \dfrac{1}{2} \Big| Y \le 1\right)$

6.8 The joint probability density function of the random variable (X, Y) is given by Table 6.P.1. Find the cumulative distribution function of the random variable (X, Y).

Table 6.P.1

			x		
y	1	2	3	4	5
1	$\dfrac{1}{12}$	$\dfrac{1}{24}$	0	$\dfrac{1}{24}$	$\dfrac{1}{30}$
2	$\dfrac{1}{24}$	$\dfrac{1}{24}$	$\dfrac{1}{24}$	$\dfrac{1}{24}$	$\dfrac{1}{30}$
3	$\dfrac{1}{12}$	$\dfrac{1}{24}$	$\dfrac{1}{24}$	0	$\dfrac{1}{30}$
4	$\dfrac{1}{12}$	0	$\dfrac{1}{24}$	$\dfrac{1}{24}$	$\dfrac{1}{30}$
5	$\dfrac{1}{24}$	$\dfrac{1}{24}$	$\dfrac{1}{24}$	$\dfrac{1}{24}$	$\dfrac{1}{30}$

Applied Problems

6.9 An urn contains 14 balls: four red, five white, and five blue. Six balls are drawn at random from the urn, without replacement. The joint probability density of X, the number of red balls drawn, and Y, the number of white balls drawn, is given by

$$f(x,y) = \begin{cases} \dfrac{\binom{4}{x}\binom{5}{y}\binom{5}{6-x-y}}{\binom{14}{6}}, & \begin{array}{l} x=0,1,\ldots,4, \\ y=0,1,\ldots,5, \\ 1 \leq x+y \leq 6, \end{array} \\ 0, & \text{elsewhere.} \end{cases}$$

a. Find the marginal probability density function of the random variables X and Y.
b. What is $Pr(X \leq 3)$?
c. What is $Pr(1 < Y \leq 4)$?

6.10 In Exercise 6.16, find the following probabilities:

a. $Pr\left(\dfrac{1}{2} \leq X \leq 1\right)$
b. $Pr(Y \geq 2.5)$

6.11 A container contains 30 electrical components, 9 of which are operable, 10 of which are defective, and 11 of which are semidefective. We draw 15 components from the container, without replacement. The joint probability density function of X, the number of operable components drawn, and Y, the number of defective components drawn, is given by

$$f(x,y) = \begin{cases} \dfrac{\binom{9}{x}\binom{10}{y}\binom{11}{15-x-y}}{\binom{30}{15}}, & \begin{array}{l} x=0,1,\ldots,9, \\ y=0,1,\ldots,10 \\ 4 \leq x+y \leq 15. \end{array} \end{cases}$$

a. Find $h_1(x|y_0)$.
b. Find $h_2(y|x_0)$.
c. Find $h_1(x|y=5)$.
d. Find $h_2(y|x=5)$.
e. Are the random variables X and Y independent?

6.12 The probability densities of the random variables X and Y are given by

$$f_1(x) = \begin{cases} 2, & -\dfrac{1}{4} \leq x \leq \dfrac{1}{4}, \\ 0, & \text{elsewhere,} \end{cases}$$

and

$$f_2(y) = \begin{cases} \dfrac{1}{2}, & 0 \leq y \leq 2, \\ 0, & \text{elsewhere,} \end{cases}$$

respectively.

a. Find the probability density function of the random variable $Z = X - Y$.
b. What is the probability that Z will assume a value greater than zero?

6.13 Let the two-dimensional random variable (X,Y) be uniformly distributed over a parallelogram with the vertices (0, 0), (1, 0), (1, 1), and (2, 1).

a. What is $Pr\left(X < \frac{1}{2}, \ 0 \leq Y \leq \frac{3}{4}\right)$?

b. What is $Pr\left(X \geq 1, \ Y \leq \frac{1}{2}\right)$?

c. What is $Pr\left(Y \leq \frac{1}{2} \mid X \leq 1\right)$?

d. What is the marginal probability density function of the random variable X, defined for $0 \leq x \leq 2$?

6.14 In Problem 6.13, let $Z = X + Y$.

a. Find the probability density function of Z.
b. Find the cumulative distribution function of Z.
c. Sketch $f(z)$.
d. What is the probability that Z will assume a value less than or equal to $3/4$?

6.15 In Exercise 6.1, find the expected value of the random variables:

a. X
b. Y
c. (XY)

6.16 In Exercises 6.4 through 6.6, find the expected value of the random variables:

a. X
b. Y
c. (XY)

6.17 In Exercises 6.8 and 6.9, find the expected value of the random variables:

a. X
b. Y
c. (XY)

6.18 The joint probability density of the discrete random variable (X,Y) is given by

Table 6.P.2

y	1	2	3
2	$\frac{1}{12}$	$\frac{1}{6}$	$\frac{1}{12}$
3	$\frac{1}{6}$	0	$\frac{1}{6}$
4	0	$\frac{1}{3}$	0

a. Determine the marginal probability density functions of X and Y.
b. Calculate the expected value of $X+Y$ and $X-Y$.
c. Calculate the second central moment of (X,Y).
d. Are X and Y independent?

6.19 In Exercises 6.1, 6.2, 6.6, and 6.8, find
   a. $\text{Var}(X)$
   b. $\text{Var}(Y)$
   c. $\text{Cov}(X,Y)$
   d. Coefficient of correlation ρ_{XY}.

6.20 In Exercise 6.22, find
   a. the expected value of X
   b. the expected value of Y
   c. the expected value of (X,Y)
   d. the variance of X
   e. the variance of Y
   f. the variance of (X,Y)
   g. $\text{Cov}(X,Y)$.

6.21 In Exercise 6.25, find the following for the variate Z:
   a. expected value
   b. variance

6.22 In Exercise 6.26, find the expected value and variance of
   a. Z_1
   b. Z_2
   c. (Z_1, Z_2)
   d. $\text{Cov}(Z_1, Z_2)$

6.23 In Exercise 6.27, find the following for the random variable Z:
   a. the mean
   b. the variance

6.24 Using the bivariate probability density of the discrete random variable (X,Y) given in Problem 6.8, find the following for $Z = (1/2)X - 3Y^2$:
   a. expected value
   b. variance

6.25 Using the joint probability density functions given in Exercises 6.4, 6.8, 6.14, and 6.16, find the expected value and variance of
   a. $g(X,Y) = X^2 - 3XY + Y^2$
   b. $V = X - XY + Y^2$
   c. $Z = e^{X+Y}$

6.26 In Exercises 6.3 and 6.4, calculate the ordinary moment of order $k+m$.

6.27 If the random variables X and Y are independent and have equal variances, what is the coefficient of correlation between X and $\alpha X + Y$, where α is some constant?

6.28 In Problem 6.8, calculate
   a. $E(X|Y=3)$
   b. $E(X|Y=5)$
   c. $E(Y|X=3)$
   d. $E(Y|X=5)$

6.29 In Exercise 6.22, calculate
   a. $E(X|Y=y_0)$
   b. $E(Y|X=x_0)$
   c. $E\{E(X|Y=y_0)\}$

6.30 Refer to Problem 6.18.
   a. Calculate the regression curve of the random variable X on the random variable Y.
   b. Calculate the regression curve of the random variable Y on the random variable X.
   c. Sketch both parts.

6.31 Find and sketch the regression curves of X on Y and Y on X for the continuous bivariate distribution given in Exercises 6.3, 6.4, and 6.22(k) and (l).

6.32 The joint density of the random variable is given by

$$f(x,y) = \begin{cases} 2xe^{-(x^2+y)}, & x,y \geq 0, \\ 0, & \text{elsewhere.} \end{cases}$$

Find and sketch the regression curves of X on Y and Y on X.

> "Probability is the very guide of life."
> —Cicero

CHAPTER 7

Sequence of Random Variables

OBJECTIVES:

7.1 Multivariate Probability Density Functions
7.2 Multivariate Cumulative Distribution Functions
7.3 Marginal Probability Distributions
7.4 Conditional Probability Density and Cumulative Distribution Functions
7.5 Sequence of Independent Random Variables
7.6 Functions of Random Variables
7.7 Expected Value and Moments
7.8 Conditional Expectations

John Wilder Tukey (1915–2000)

> "When you have eliminated the impossible, what ever remains, however improbable, must be the truth."
> —Sir Arthur Conan Doyle

Chapter 7 Sequence of Random Variables

INTRODUCTION

In Chapter 6, we extended the study of the one-dimensional random variable to the two-dimensional random variable and the joint probability density function. Many practical problems involving a natural phenomenon may be characterized by more than two attributes. Thus, it is often necessary to consider n-dimensional random variables $(X_1, X_2, \ldots, X_n)$ and their multivariate distributions. Once the transition from the univariate to the bivariate case is understood, the similar generalization of the concept to higher dimensions is readily grasped.

In this chapter, we extend the concept of a random variable and its probability density function to the n-dimensional case. For brevity, we restrict our study to the continuous case; however, many generalizations can be illustrated with a discrete probability density function.

7.1 Multivariate Probability Density Functions

We begin our discussion with the definition of an n-dimensional continuous random variable.

Definition 7.1.1

Let S be a sample space and $X_1(s), X_2(s), \ldots, X_n(s)$ real valued functions (defined on S), each assigning a number to each element of the sample space. Then, we call $(X_1(s) = X_1, X_2(s) = X_2, \ldots, X_n(s) = X_n)$ an ***n*-dimensional discrete random variable**.

Thus, the range space $R_{X_1, X_2, \ldots, X_n}$ is the set of all possible values of $(X_1, X_2, \ldots, X_n)$, which is a subset of n-dimensional Euclidean space.

Definition 7.1.2

Let $(X_1, X_2, \ldots, X_n)$ be an n-dimensional random variable. If $(X_1, X_2, \ldots, X_n)$ can assume all values in some noncountable (nondenumerable) set of the n-dimensional Euclidean space, then we call $(X_1, X_2, \ldots, X_n)$ an ***n*-dimensional continuous random variable**.

The range space of the n-dimensional variate $(X_1, X_2, \ldots, X_n)$ is the set of all possible values of $(X_1, X_2, \ldots, X_n)$, which is a subset of the n-dimensional Euclidean space. For example, if (X, Y, Z) assumes values from the subset $E_1 \subset E$, where $E_1 = \{(x, y, z): x_1 \leq x \leq x_2, y_1 \leq y \leq y_2, z_1 \leq z \leq z_2\}$, we would say that (X, Y, Z) is a three-dimensional continuous random variable, the range space of which is a rectangular parallelepiped, a subset of the three-dimensional Euclidean space. As in the bivariate case, our objective is to obtain probabilities that the n-dimensional random variable will assume certain values. For example, if E_1 is the set above $Pr[(X,Y,Z) \text{ in } E_1] = Pr[x_1 \leq X \leq x_2, y_1 \leq Y \leq y_2, z_1 \leq Z \leq z_2]$, then the probability that X, Y, and Z assume the values simultaneously lies in the set $E_1 \subset E$. The mathematical function that is needed to calculate such a probability is defined in Definition 7.1.3.

■ Definition 7.1.3

Let $(X_1, X_2, \ldots, X_n)$ be an n-dimensional continuous random variable. A function $f(x_1, x_2, \ldots, x_n)$ is called a **multivariate probability density function** or *joint probability density function* of the random variable $(X_1, X_2, \ldots, X_n)$ if the following conditions are satisfied:

1. $f(x_1, x_2, \ldots, x_n) \geq 0$, for all $(x_1, x_2, \ldots, x_n) \in R_{X_1, X_2, \ldots, X_n}$
2. $\int \int \cdots \int_{R_{X_1, X_2, \ldots, X_n}} f(x_1, x_2, \ldots, x_n) dx_1 dx_2 \ldots dx_n = 1$

Here, the integration is performed over $R_{X_1, X_2, \ldots, X_n}$, in which $(X_1, X_2, \ldots, X_n)$ assumes all values.

Furthermore, for all real valued pairs (a_i, b_i), $i = 1, 2, \ldots, n$, such that $a_i < b_i$, we have
$$Pr(a_1 \leq X_1 \leq b_1, a_2 \leq X_2 \leq b_2, \ldots a_n \leq X_n \leq b_n)$$
$$= \int_{a_n}^{b_n} \cdots \int_{a_2}^{b_2} \int_{a_1}^{b_1} f(x_1, x_2, \ldots, x_n) dx_1 dx_2 \ldots dx_n.$$

Thus, with the aid of the multivariate density function, we can calculate various probabilities, such as
$$Pr[(X, Y, Z) \in E_1] = \int \int_{E_1} \int f(x, y, z) dz \, dy \, dx$$
$$= \int_{x_1}^{x_2} \int_{y_1}^{y_2} \int_{z_1}^{z_2} f(x, y, z) dz \, dy \, dx$$
$$= Pr[x_1 \leq X \leq x_2, y_1 \leq Y \leq y_2, z_1 \leq Z \leq z_2].$$

Similar definitions can be given for the discrete case.

Example 7.1.1

The function defined by
$$f(x_1, x_2, x_3, \beta) = \begin{cases} \dfrac{1}{\beta^3} e^{-(1/\beta)(x_1 + x_2 + x_3)}, & x_1, x_2, x_3 > 0, \beta > 0, \\ 0, & \text{elsewhere,} \end{cases}$$

is a probability density function of the three-dimensional variate (X_1, X_2, X_3) because it satisfies the preceding requirements. Thus,

a. $Pr(X_1 \leq 2, X_2 \geq 3, X_3 < 1) = \int_0^2 \int_3^\infty \int_0^1 f(x_1, x_2, x_3) dx_3 dx_2 dx_1$

$= \int_0^2 \int_3^\infty \int_0^1 \frac{1}{\beta^3} e^{-(1/\beta)(x_1+x_2+x_3)} dx_3 dx_2 dx_1$

$= e^{-(6/\beta)} - e^{-(5/\beta)} - e^{-(4/\beta)} + e^{-(3/\beta)}, \beta > 0$.

b. $Pr\left(X_1 + X_2 < 3, X_3 \leq \frac{1}{2}\right) = \int_0^{1/2} \int_0^3 \int_0^{3-x_2} f(x_1, x_2, x_3) dx_1 dx_2 dx_3$

$= \int_0^{1/2} \int_0^3 \int_0^{3-x_2} \frac{1}{\beta^3} e^{-(1/\beta)(x_1+x_2+x_3)} dx_1 dx_2 dx_3$

$= \frac{1}{\beta^2} \left\{ \int_0^{1/2} e^{-(x_3/\beta)} \int_0^3 \left(e^{-(x_2/\beta)} - e^{-(3/\beta)}\right) dx_1 dx_3 \right\}$

$= \frac{1}{\beta^2} \left\{ \left(\beta - \beta e^{-(3/\beta)} - 3e^{-(3/\beta)}\right) \int_0^{1/2} e^{-(x_3/\beta)} dx_3 \right\}$

$= \left(1 - e^{-(1/2\beta)}\right) + \frac{1}{\beta}(3+\beta)\left(e^{-(7/2\beta)} - e^{-(3/\beta)}\right)$

$= \left(1 - e^{-(1/2\beta)}\right)\left(1 - \frac{\beta+3}{\beta} e^{-(3/\beta)}\right), \beta > 0$. ◀

7.1.1 Multinomial Distribution A discrete distribution of importance is a generalization of the Bernoulli distribution, which was discussed in Section 2.4. Consider an experiment consisting of n independent trials, where each trial can have one of several outcomes. Let $x_1, x_2, \ldots, x_k$ denote the possible outcomes of each trial, with $p_1, p_2, \ldots, p_k$ being their corresponding probabilities of occurrence, subject to the conditions

$$\sum_{i=1}^{k} p_i = 1, \ p_i \geq 0 \ (i=1,2,\ldots,k).$$

As a result of n independent trials of the experiment, what is the total probability of obtaining exactly x_1 results of the first type, x_2 of the second type, $\ldots, x_k$ of the kth type, where $x_1 + x_2 + \ldots + x_k = n$? Because the trials are independent, the probability of obtaining any specific type $x_i, i=1,2,\ldots,k$, on a given trial is not affected by the outcome of the other trials. Thus, the product probability of any sequence of outcomes is equal to the product of their separate probabilities:

$$p_1^{x_1} p_2^{x_2} \cdots p_k^{x_k}. \tag{7.1.1}$$

The total number of distinct sequences yielding the stated number of outcomes of each kind is given by

$$\binom{n}{x_1, x_2, \ldots, x_k} = \frac{n!}{x_1! x_2! \ldots x_k!}. \tag{7.1.2}$$

Therefore, the total probability is the product of the expressions in Equations (7.1.1) and (7.1.2):

$$\frac{n!}{x_1! x_2! \ldots x_k!} p_1^{x_1} p_2^{x_2} \cdots p_k^{x_k}.$$

We formally define the multinomial distribution in Definition 7.1.4.

Definition 7.1.4

The k-dimensional discrete random variable $(X_1, X_2, \ldots, X_k)$ is distributed as a **multinomial distribution** if its joint probability density function is given by

$$f(x_1, x_2, \ldots, x_k; p_1, p_2, \ldots, p_k, n) = \begin{cases} \dfrac{n!}{x_1! x_2! \ldots x_k!} p_1^{x_1} p_2^{x_2} \cdots p_k^{x_k}, & \begin{array}{l} x_i = 0, 1, \ldots, n, \\ 0 \leq p_i \leq 1, \\ i = 1, 2, \ldots, k, \end{array} \\ 0, & \text{elsewhere,} \end{cases} \quad (7.1.3)$$

where

$$\sum_{i=1}^{k} p_i = 1$$

and

$$\sum_{i=1}^{k} x_i = n.$$

We can replace x_k with

$$n - \sum_{i=1}^{k-1} x_i$$

and

$$p_k = 1 - \sum_{i=1}^{k-1} p_i.$$

Thus, the multinomial probability density function involves $k-1$ of the x_i's because $k-1$ of them are independent and x_k is exactly determined.

Clearly, $f(x_1, x_2, \ldots, x_k; p_1, p_2, \ldots, p_k, n) \geq 0$. Applying the multinomial theorem

$$(y_1 + y_2 + \ldots + y_k)^n = \sum_{r_1, r_2, \ldots, r_k \geq 0} \binom{n}{r_1, r_2, \ldots, r_k} y_1^{r_1} y_2^{r_2} \cdots y_k^{r_k},$$

where the summation is over all nonnegative integral k-tuples such that $r_1 + r_2 + \ldots + r_k = n$, we can show that

$$\sum_{x_1, x_2, \ldots, x_k \geq 0} f(x_1, x_2, \ldots, x_k; p_1, p_2, \ldots, p_k, n) = 1.$$

Thus, if $k = 2$, the multinomial distribution reduces to the binomial distribution $p_1 = p$, $p_2 = 1 - p$, $x_1 = x$, and $x_2 = n - x$.

Example 7.1.2

Consider the experiment of rolling 10 fair dice in which we desire to obtain the probability of rolling four aces, two deuces, one three, two fours, and one six. Here $p_i = 1/6, i = 1, 2, \ldots, 6$. In view of the multinomial density function, we have

$$f(4,2,1,2,0,1) = \frac{10!}{4!2!1!2!0!1!}\left(\frac{1}{6}\right)^{10}$$

$$= 0.0075.$$

◀

Example 7.1.3

A type of electrical system that consists of five major components, which we denote using $X_1, X_2, \ldots, X_5$, will fail to operate if any one of the five components fails. If simultaneous defects are negligible and the probabilities of failure of components 1 through 5 are 0.3, 0.2, 0.1, 0.3, and 0.1, respectively, what is the probability that a random sample of eight defective systems would consist of three failures due to component X_1, no failure due to X_2 and X_3, four failures due to X_4, and one failure due to X_5?

Solution: Assuming independent trials, we can obtain the necessary probability by applying the multinomial distribution law. Thus,

$$Pr(X_1 = 3, X_2 = 0, X_3 = 0, X_4 = 4, X_5 = 1) = \frac{8!}{3!0!0!4!1!}(0.3)^3(0.3)^4(0.1)$$

$$= 0.0061.$$

◀

7.1.2 Multivariate Normal Distribution In Chapter 3, we discussed and emphasized the importance of the univariate normal density function. The normal distribution in the multivariate case is of equal importance. A comprehensive presentation of this distribution is beyond the scope of this book; however, we define it here and refer you to the book of T. W. Anderson,[1] which gives an extensive study of the multivariate normal distribution.

■ Definition 7.1.5

The n-dimensional continuous random variable $(X_1, X_2, \ldots, X_n)$ is distributed as **multivariate normal** if its joint probability density function is given by

$$f(X, P, \mu) = \frac{\sqrt{|P|}}{(2\pi)^{n/2}} e^{-(1/2)(X-\mu)^T P(X-\mu)}, \quad -\infty < x_i < \infty, \; i = 1, 2, \ldots, n,$$

where

1. X is an $n \times 1$ vector, the elements of which are random variables $X = (x_1, x_2, \ldots, x_n)^T$, $-\infty < x_i < \infty$, $i = 1, 2, \ldots, n$.
2. μ is an $n \times 1$ vector, the elements of which are constants.
3. P is a positive definite symmetrical matrix whose elements p_{ij} are constants: $|P|$ denotes the determinant of the matrix P.

The product $(X-\mu)^T P(X-\mu)$ is called a *quadratic form* in the elements $x_i - \mu_i$, $i = 1, 2, \ldots, n$, and P is called the *matrix of the quadratic form*. If the quadratic form $(X-\mu)^T P(X-\mu)$ is positive for every nonzero vector $X - \mu$, then $(X-\mu)^T P(X-\mu)$ is called a *positive definite form* and P is called a *positive definite symmetrical matrix*. The matrix P is the inverse of the covariance matrix C of the sequence of random variables $(X_1, X_2, \ldots, X_n)$, given by

$$C = \begin{pmatrix} \sigma_{11} & \sigma_{12} & \cdots & \sigma_{1n} \\ \sigma_{21} & \sigma_{22} & \cdots & \sigma_{2n} \\ \vdots & \vdots & & \vdots \\ \sigma_{n1} & \sigma_{n2} & \cdots & \sigma_{nn} \end{pmatrix},$$

where $\sigma_{ij} = E\left[(X_i - \mu_i)(X_j - \mu_j)\right]$ and $E[X_i] = \mu_i$. If $n = 2$, the multivariate normal density function reduces to the bivariate normal density function, which was defined in Chapter 6.

7.2 Multivariate Cumulative Distribution Functions

In this section, we extend the study of cumulative distribution functions to the n-dimensional random variable.

Definition 7.2.1

The function defined by

$$F(x_1, x_2, \ldots, x_n) = Pr(X_1 \leq x_1, X_2 \leq x_2, \ldots X_n \leq x_n)$$

$$= \int_{-\infty}^{x_n} \int_{-\infty}^{x_{n-1}} \cdots \int_{-\infty}^{x_1} f(y_1, y_2, \ldots, y_n) dy_1 \ldots dy_{n-1} dy_n$$

is called the **multivariate cumulative distribution function** or **joint cumulative distribution function** of the continuous n-dimensional random variable $(X_1, X_2, \ldots, X_n)$, which has the joint probability density function $f(x_1, x_2, \ldots, x_n)$.

Theorem 7.2.1 states the properties of the multivariate cumulative distribution function, which are analogous to the univariate and bivariate cases.

Theorem 7.2.1

If $F(x_1, x_2, \ldots, x_n)$ is the joint cumulative distribution function of the random variable $(X_1, X_2, \ldots, X_n)$, then $F(x_1, x_2, \ldots, x_n)$ has the following properties:

1. $F(\infty, \infty, \ldots, \infty) = 1$.
2. $F(x_1, x_2, \ldots, x_n) = 0$ if at least one of the x_i's is $-\infty$.
3. $F(x_1, x_2, \ldots, x_n)$ is monotone nondecreasing in each variate separately.
4. $F(x_1, x_2, \ldots, x_n)$ is continuous from the right in each variate separately.

It is also possible to obtain the joint probability density function of $F(x_1, x_2, \ldots, x_n)$ of the continuous n-dimensional random variable $(X_1, X_2, \ldots, X_n)$. That is, if $F(x_1, x_2, \ldots, x_n)$ is continuous and if its partial derivatives exist for all x_i's, then

$$\frac{\partial^n}{\partial x_1, \partial x_2, \ldots \partial x_n} F(x_1, x_2, \ldots, x_n) = f(x_1, x_2, \ldots, x_n)$$

exists and satisfies the conditions of a multivariate probability density function.

In Section 6.2, we showed that, for the two-dimensional random variable (X, Y), we can write

$$Pr(x_1 \leq X, x_2, y_1 \leq Y_n \leq y_2)$$
$$= F(x_2, y_2) - F(x_2, y_1) - F(x_1, y_2) + F(x_1, y_1). \quad (7.2.1)$$

A generalization of Equation (7.2.1) to the n-dimensional random variable $(X_1, X_2, \ldots, X_n)$ is given by the following expression:

$$Pr(x_1 \leq X_1 \leq x_1 + \Delta x_1, x_2 \leq X_2 \leq x_2 + \Delta x_2, \ldots, x_n \leq X_n \leq x_n + \Delta x_n)$$
$$= F(x_1 + \Delta x_1, x_2 + \Delta x_2, \ldots, x_n + \Delta x_n)$$
$$- \sum_{i=1}^{n} F(x_1 + \Delta x_1, \ldots, x_{i-1} + \Delta x_{i-1}, x_i, x_{i-1} + \Delta x_{i+1}, \ldots, x_n + \Delta x_n)$$
$$+ \sum_{\substack{i,j=1 \\ i<j}}^{n} F(x_1 + \Delta x_1, \ldots, x_{i-1} + \Delta x_{i-1}, x_i, x_{i-1} + \Delta x_{i+1}, \ldots)$$
$$+ \ldots + (-1)^n F(x_1, x_2, \ldots, x_n),$$

for $\Delta x_i > 0, i = 1, 2, \ldots, n$.

Example 7.2.1

The cumulative distribution function of the three-dimensional random variable (X_1, X_2, X_3), the joint probability density function of which is given in Example 7.1.1, is

$$F(x_1, x_2, x_3) = Pr(X_1 \leq x_1, X_2 \leq x_2, X_3 \leq x_3)$$
$$= \int_{-\infty}^{x_3} \int_{-\infty}^{x_2} \int_{-\infty}^{x_1} f(y_1, y_2, y_3) dy_1 dy_2 dy_3$$
$$= \int_{0}^{x_3} \int_{0}^{x_2} \left\{ \int_{-0}^{x_1} \frac{1}{\beta^3} e^{-(1/\beta)(y_1+y_2+y_3)} dy_1 \right\} dy_2 dy_3$$
$$= e^{-(1/\beta)(x_1+x_2)} + e^{-(1/\beta)(x_2+x_3)} + e^{-(1/\beta)(x_1+x_3)} - e^{-(x_1/\beta)} - e^{-(x_2/\beta)}$$
$$- e^{-(x_3/\beta)} - e^{-(1/\beta)(x_1+x_2+x_3)} + 1.$$

Thus,

$$F(x_1,x_2,x_3) = \begin{cases} 1 - e^{-(x_1/\beta)} - e^{-(x_2/\beta)} - e^{-(x_3/\beta)} + e^{-(1/\beta)(x_1+x_2)} \\ \quad + e^{-(1/\beta)(x_1+x_3)} + e^{-(1/\beta)(x_2+x_3)} - e^{-(1/\beta)(x_1+x_2+x_3)}, & x_1,x_2,x_3 \geq 0, \beta > 0, \\ 0, & \text{elsewhere.} \end{cases}$$

It is clear that we can obtain the joint probability density function from $F(x_1,x_2,x_3)$:

$$\frac{\partial^n}{\partial x_1, \partial x_2, \ldots \partial x_n} F(x_1,x_2,x_3) = f(x_1,x_2,x_3).$$

◀

7.3 | Marginal Probability Distributions

Consider the sequence of random variables $(X_1, X_2, \ldots, X_n)$ with the multivariate probability density function $f(x_1, x_2, \ldots, x_n)$. At times, we must obtain the probability density function of a selected subset of the given sequence. For example, we might be interested in the trivariate probability distribution of the variate (X_2, X_3, X_5). As in the bivariate case, this probability density function can be obtained simply by integrating over the entire range of the variate $(X_1, X_4, X_6, \ldots, X_n)$ if the n-dimensional random variable is of the continuous type. Similarly, we sum over the entire range of the random variable $(X_1, X_4, X_6, \ldots, X_n)$ if it is of the discrete type. We call the joint probability density function $h_3(x_2, x_3, x_5)$ the marginal probability density function of the random variable (X_2, X_3, X_5).

Definition 7.3.1

Let $f(x_1, x_2, \ldots, x_n)$ be the multivariate probability density function of the continuous random variable $(X_1, X_2, \ldots, X_n)$. The *marginal probability density function* of the variate $(X_1, X_2, \ldots, X_r)$, $r < n$, is given by

$$h_r(x_1, x_2, \ldots, x_r) = \int_{-\infty}^{\infty} \cdots \int_{-\infty}^{\infty} f(x_1, x_2, \ldots, x_n) \, dx_{r+1}, \ldots, dx_n.$$

The marginal probability density function of any other subset of the sequence $(X_1, X_2, \ldots, X_n)$ is obtained in similar fashion.

Example 7.3.1

Let the k-dimensional discrete random variable $(X_1, X_2, \ldots, X_k)$ be multinormally distributed as

$$f(x_1, x_2, \ldots x_k; p_1, p_2, \ldots p_k, n)$$

$$= \begin{cases} \dfrac{n!}{x_1! x_2! \ldots x_k!} p_1^{x_1} p_2^{x_2} \ldots p_k^{x_k}, & 0 \leq p_i \leq 1, \; x_i = 0, 1, \ldots, k, \\ 0, & \text{elsewhere,} \end{cases}$$

where $\sum_{i=1}^{k} p_i = 1$ and $\sum_{i=1}^{k} x_i = n$. The marginal probability mass function of $(X_1, X_2, \ldots, X_r)$, $r < k$, is obtained as follows:

$$h_r(x_1, x_2, \ldots x_r; p_1, p_2, \ldots p_r, n)$$

$$= \sum_{x_{r+1}, x_{r+2}, \ldots, x_k} \frac{n!}{x_1! x_2! \ldots x_k!} p_1^{x_1} p_2^{x_2} \ldots p_k^{x_k}$$

$$= \frac{n!}{x_1! x_2! \ldots x_r! \left(n - \sum_{i=1}^{r} x_i\right)!} p_1^{x_1} p_2^{x_2} \ldots p_r^{x_r} \left(1 - \sum_{i=1}^{r} p_i\right)^{\left(n - \sum_{i=1}^{r} x_i\right)}$$

$$\sum_{x_{r+1}, x_{r+2}, \ldots, x_k} \frac{\left(n - \sum_{i=1}^{r} x_i\right)!}{x_{r+1}! x_{r+2}! \ldots x_k!} \left[\frac{P_{r+1}}{1 - \sum_{i=1}^{r} p_i}\right]^{x_{r+1}}$$

$$\left[\frac{P_{r+2}}{1 - \sum_{i=1}^{r} p_i}\right]^{x_{r+2}} \ldots \left[\frac{P_k}{1 - \sum_{i=1}^{r} p_i}\right]^{x_k}.$$

In the preceding expression, we have

$$\frac{P_{r+1}}{1 - \sum_{i=1}^{r} p_j} \geq 0, \; i = 1, 2, \ldots k - r,$$

and

$$\sum_{i=1}^{k-r} \frac{P_{r+1}}{1 - \sum_{i=1}^{r} p_j} = 1$$

with the assumption that

$$\left(1 - \sum_{j=1}^{r} p_j\right) > 0.$$

Thus, we have adjusted the sum in $h_r(x_1, x_2, \ldots x_r; p_1, p_2, \ldots p_r, n)$ to be unity because we are simply summing the multinomial probability density function. Hence, the marginal probability density function of the r-dimensional random variable is

$$h_r(x_1, x_2, \ldots x_r, p_1, p_2, \ldots p_r, n)$$

$$= \begin{cases} \dfrac{n!}{x_1! x_2! \ldots x_r! \left(n - \sum_{i=1}^{r} x_i\right)!} p_1^{x_1} p_2^{x_2} \ldots p_r^{x_r} \left(1 - \sum_{j=1}^{r} p_j\right)^{\left(n - \sum_{i=1}^{r} x_i\right)}, & x_i = 0, 1, \ldots, r, \\ 0, & \text{elsewhere.} \end{cases}$$

Given the multivariate cumulative distribution function $F(x_1, x_2, \ldots x_n)$ of the random variable $(X_1, X_2, \ldots, X_n)$, we can obtain various marginal cumulative distribution functions. For example, the marginal cumulative distribution function of $(X_1, X_2, \ldots, X_k)$, $k < n$, is $G_k(x_1, x_2, \ldots x_k)$. That is,

$$Pr(X_1 \leq x_1, X_2 \leq x_2, \ldots, X_k \leq x_k)$$

$$= Pr(X_1 \leq x_1, X_2 \leq x_2, \ldots, X_k \leq x_k, X_{k+1} \leq \infty, \ldots X_n \leq \infty)$$

$$= F(x_1, x_2, \ldots, x_k, \infty, \infty \ldots \infty)$$

$$= G_k(x_1, x_2, \ldots x_k).$$

The marginal cumulative distribution function of any subset of the variates $X_1, X_2, \ldots, X_n$ is obtained in a similar fashion.

◀

Example 7.3.2

Using the cumulative distribution function of the trivariate random variable (X_1, X_2, X_3) given in Example 7.2.1, we obtain the following marginal cumulative distribution functions:

a. $G_1(x_1) = Pr(X_1 \leq x_1, X_2 \leq \infty, X_3 \leq \infty)$

$= F(x_1, \infty, \infty)$

$= 1 - e^{(x_1/\beta)}$

b. $G_2(x_1, x_3) = Pr(X_1 \leq x_1, X_2 \leq \infty, X_3 \leq x_3)$

$= F(x_1, \infty, x_3)$

$= 1 - e^{(x_1/\beta)} - e^{(x_3/\beta)} + e^{(1/\beta)(x_1 + x_3)}$

Thus,

$$G_1(x_1) = \begin{cases} 1 - e^{-(x_1/\beta)}, & x_1 \geq 0, \beta > 0, \\ 0, & \text{elsewhere}, \end{cases}$$

and

$$G_2(x_1, x_3) = \begin{cases} 1 - e^{-(x_1/\beta)} - e^{(x_3/\beta)} + e^{-(1/\beta)(x_1+x_3)}, & x_1, x_3 \geq 0, \beta > 0, \\ 0, & \text{elsewhere}. \end{cases}$$

◀

7.4 | Conditional Probability Density and Cumulative Distribution Functions

We begin by generalizing the basic definitions introduced in Section 6.4 for the bivariate random variable to the n-dimensional case.

■ Definition 7.4.1

Let $f(x_1, x_2, \ldots, x_n)$ be the multivariate density function of the continuous random variable $(X_1, X_2, \ldots, X_n)$. The *conditional probability density function* of the random variable

$$(X_1, X_2, \ldots, X_r), r < n,$$

given $(X_{r+1}, X_{r+2}, \ldots, X_n)$, is defined by

$$g_r(x_1, x_2, \ldots, x_r \mid x_{r+1}, \ldots, x_n) = \frac{f(x_1, x_2, \ldots, x_n)}{h_{n-r}(x_{r+1}, x_{r+2}, \ldots, x_n)}. \quad (7.4.1)$$

Of course, $h_{n-r}(x_{r+1}, x_{r+2}, \ldots, x_n)$ is the marginal probability density function, which must be greater than zero, as discussed in the previous section. Also, the conditional density in Equation (7.4.1) is nonnegative, and

$$\int_{-\infty}^{\infty} \cdots \int_{-\infty}^{\infty} g_r(x_1, x_2, \ldots, x_r \mid x_{r+1}, \ldots, x_n) dx_1, \ldots, dx_r$$

$$= \int_{-\infty}^{\infty} \cdots \int_{-\infty}^{\infty} \frac{f(x_1, x_2, \ldots, x_r, \ldots, x_n)}{h_{n-r}(x_{r+1}, x_{r+2}, \ldots, x_n)} dx_1, \ldots, dx_r$$

$$= \frac{h_{n-r}(x_{r+1}, x_{r+2}, \ldots, x_n)}{h_{n-r}(x_{r+1}, x_{r+2}, \ldots, x_n)}$$

$$= 1.$$

The conditional probability density function of any other collection of the variates, given the others, is defined similarly.

Example 7.4.1

Using the results of Example 7.3.1, we calculate the conditional probability density function $g_r(x_{r+1},\ldots,x_k | x_1,x_2,\ldots,x_r)$ of the multinomial distribution law.

Solution: Here,

$$g_{k-r}(x_{r+1},\ldots,x_k | x_1,x_2,\ldots,x_r) = \frac{f(x_1,x_2,\ldots,x_k;p_1,p_2,\ldots,p_k,n)}{h_r(x_1,x_2,\ldots,x_k;p_1,p_2,\ldots,p_k,n)}$$

$$= \frac{\dfrac{n!}{x_1!x_2!\ldots x_k!} p_1^{x_1} p_2^{x_2} \cdots p_k^{x_k}}{\dfrac{n!}{x_1!x_2!\ldots x_r!\left[n-\sum_{i=1}^{r} x_i\right]!} p_1^{x_1} p_2^{x_2} \cdots p_r^{x_r} \left[1-\sum_{i=1}^{r} p_i\right]^{\left[n-\sum_{i=1}^{r} x_i\right]}}$$

$$= \frac{\left[n-\sum_{i=1}^{r} x_i\right]!}{x_{r+1}!x_{r+2}!\ldots x_k!} \left[\frac{p_{r+1}}{1-\sum_{i=1}^{r} p_i}\right]^{x_{r+1}}$$

$$\cdot \left[\frac{p_{r+2}}{1-\sum_{i=1}^{r} p_i}\right]^{x_{r+2}} \cdots \left[\frac{p_k}{1-\sum_{i=1}^{r} p_i}\right]^{x_k},$$

where

$$\sum_{i=1}^{k} x_i = n,$$
$$p_i \geq 0,$$

and

$$\sum_{i=1}^{k} p_k = 1.$$

The conditional cumulative distribution function of the continuous $(k-r)$-dimensional random variable $(X_{r+1}, X_{r+2},\ldots, X_k)$, given $(X_1, X_2,\ldots, X_r)$, $r < k$, is given by

$$F(x_{r+1}, x_{r+2}, \ldots, x_k | x_1, x_2, \ldots, x_r)$$

$$= Pr\left(X_{r+1} \leq x_{r+1}, X_{r+2} \leq x_{r+2}, \ldots, \times X_k \leq x_k | X_1 \leq x_1, X_2 \leq x_2, \ldots, X_r \leq x_r\right)$$

$$= \frac{Pr(X_1 \leq x_1, X_2 \leq x_2, \ldots, X_k \leq x_k)}{Pr(X_1 \leq x_1, X_2 \leq x_2, \ldots, X_r \leq x_r)}$$

$$= \frac{F(x_1, x_2, \ldots x_k)}{F_r(x_1, x_2, \ldots x_r)}. \qquad (7.4.2)$$

Extending the mathematical reasoning used in Section 6.4 for the two-dimensional random variable to show that

$$F(x|y) = \frac{\int_{-\infty}^{x} \int_{-\infty}^{x} f(t,s) \, ds \, dt}{F(y)}, \quad F(y) > 0,$$

we can write the expression in Equation (7.4.2) as follows:

$$F(x_{r+1}, x_{r+2}, \ldots x_k | x_1, x_2, \ldots x_r)$$

$$= \frac{\int_{-\infty}^{x_1} \int_{-\infty}^{x_2} \cdots \int_{-\infty}^{x_k} f(s_1, s_2, \ldots, s_k) \, ds_1 \, ds_2 \ldots ds_k}{F_r(x_1, x_2, \ldots x_r)}, \quad (7.4.3)$$

where $f(x_1, x_2, \ldots, x_k)$ is the multivariate probability density function of the k-dimensional continuous random variable $(X_1, X_2, \ldots, X_k)$ and $F_r(x_1, x_2, \ldots, x_r)$, $r < k$, is the marginal cumulative distribution function of the random variable $(X_1, X_2, \ldots, X_r)$. It is assumed that the denominator of Equation (7.4.3) is positive.

We saw in Section 6.4 that we can express marginal cumulative distribution functions in terms of their conditional cumulative distributions. This result can be generalized to apply to more than two random variables. For simplicity in notation, we illustrate this extension using a four-dimensional random variable. Let the joint probability density function of the continuous random variable (X_1, X_2, X_3, X_4) be $f(x_1, x_2, x_3, x_4)$. The conditional distribution of the random variable (X_3, X_4), given (X_1, X_2), is

$$F(x_3, x_4 | x_1, x_2) = \int_{-\infty}^{x_3} \int_{-\infty}^{x_4} h(s_3, s_4 | x_1, x_2) \, ds_4 \, ds_3$$

$$= \int_{-\infty}^{x_3} \int_{-\infty}^{x_4} \frac{f(x_1, x_2, s_3, s_4)}{h_2(x_1, x_2)} \, ds_4 \, ds_3$$

or

$$\int_{-\infty}^{x_3} \int_{-\infty}^{x_4} f(x_1, x_2, s_3, s_4) \, ds_4 \, ds_3 = h_2(x_1, x_2) F(x_3, x_4 | x_1, x_2).$$

It follows that

$$\int_{-\infty}^{x_3} \int_{-\infty}^{x_4} \left\{ \int_{-\infty}^{\infty} \int_{-\infty}^{\infty} f(x_1, x_2, s_3, s_4) \, dx_1 \, dx_2 \right\} ds_3 \, ds_4$$

$$= \int_{-\infty}^{\infty} \int_{-\infty}^{\infty} h_2(x_1, x_2) F(x_3, x_4 | x_1, x_2) \, dx_1 \, dx_2$$

or

$$\int_{-\infty}^{x_3} \int_{-\infty}^{x_4} f_2(s_3, s_4) \, ds_3 \, ds_4$$

$$= \int_{-\infty}^{\infty} \int_{-\infty}^{\infty} h_2(x_1, x_2) F(x_3, x_4 | x_1, x_2) \, dx_1 \, dx_2.$$

Thus, we can express the marginal cumulative distribution function of the random variable (X_3, X_4) as a function of the conditional cumulative distribution function $F(x_3, x_4 | x_1, x_2)$:

$$F_2(x_3, x_4) = \int_{-\infty}^{\infty} \int_{-\infty}^{\infty} h_2(x_1, x_2) F(x_3, x_4 | x_1, x_2) dx_1 \, dx_2.$$

Similarly,

$$F_1(x_4) = \int_{-\infty}^{\infty} \int_{-\infty}^{\infty} \int_{-\infty}^{\infty} h_2(x_1, x_2, x_3) F(x_4 | x_1, x_2, x_3) dx_1 \, dx_2 \, dx_3$$

and

$$F_3(x_2, x_3, x_4) = \int_{-\infty}^{\infty} h_1(x_1) F(x_2, x_3, x_4 | x_1) dx_1.$$

The conditional distribution function in the multivariate case satisfies the properties of the ordinary cumulative distribution function, as shown in Section 6.4 for the two-dimensional random variable.

◀

7.5 Sequence of Independent Random Variables

Let $F(x_1, x_2, \ldots, x_n)$, $F_1(x_1), F_2(x_2), \ldots, F_n(x_n)$, represent the cumulative distribution function of the n-dimensional random variable $(X_1, X_2, \ldots, X_n)$ and the marginal cumulative distribution functions of the random variables $X_1, X_2, \ldots, X_n$, respectively.

■ **Definition 7.5.1**

The random variables in the sequence $(X_1, X_2, \ldots, X_n)$ are said to be *independent (mutually independent)* if

$$F(x_1, x_2, \ldots, x_n) = F_1(x_1) \, F_2(x_2) \ldots F_n(x_n) \quad (7.5.1)$$

for all real n-tuples, $(x_1, x_2, \ldots, x_n)$.

Thus, the random variables in the sequence $(X_1, X_2, \ldots, X_n)$ are independent if the events of the sequence $\{X_1 \leq x_1\}, \{X_2 \leq x_2\}, \ldots, \{X_n \leq x_n\}$ are independent for every real number $x_1, x_2, \ldots, x_n$ or

$$Pr(X_1 \leq x_1, X_2 \leq x_2, \ldots, X_n \leq x_n) = Pr(X_1 \leq x_1) Pr(X_2 \leq x_2) \ldots Pr(X_n \leq x_n).$$

Theorem 7.5.1

Let $f(x_1, x_2, \ldots, x_n)$ be the multivariate probability density function of the n-dimensional random variable $(X_1, X_2, \ldots, X_n)$ of the continuous type. The random variables $X_1, X_2, \ldots, X_n$ are mutually independent if and only if

$$f(x_1, x_2, \ldots, x_n) = f_1(x_1) f_2(x_2) \cdots f_n(x_n),$$

where $f_1(x_1) f_2(x_2), \ldots, f_n(x_n)$ are the marginal probability density functions of the variates $X_1, X_2, \ldots, X_n$, respectively.

Proof Applying the relation between probability density and cumulative distribution functions, as stated in Section 7.2, we have

$$\frac{\partial^n F(x_1, x_2, \ldots, x_n)}{\partial x_1 \partial x_2 \cdots \partial x_n} = f(x_1, x_2, \ldots, x_n)$$

$$= \frac{dF_1(x_1)}{dx_1} \cdot \frac{dF_2(x_2)}{dx_2} \cdots \frac{dF_n(x_n)}{dx_n}$$

$$= f_1(x_1) f_2(x_2) \cdots f_n(x_n).$$

Conversely,

$$F(x_1, x_2, \ldots, x_n) = \int_{-\infty}^{x_n} \int_{-\infty}^{x_{n-1}} \cdots \int_{-\infty}^{x_1} f(s_1, s_2, \ldots, s_n) \, ds_1 \, ds_2 \cdots ds_n$$

$$= \int_{-\infty}^{x_n} \int_{-\infty}^{x_{n-1}} \cdots \int_{-\infty}^{x_1} f_1(s_1) f_2(s_2) \cdots f_n(s_n) \, ds_1 \, ds_2 \cdots ds_n$$

$$= \int_{-\infty}^{x_1} f_1(s_1) ds_1 \cdots \int_{-\infty}^{x_{n-1}} f_{n-1}(s_{n-1}) ds_{n-1} \int_{-\infty}^{x_n} f_n(s_n) ds_n$$

$$= F_1(x_1) \cdots F_{n-1}(x_{n-1}) F_n(x_n).$$

A similar theorem can be stated for a sequence of discrete random variables.

Example 7.5.1

Using the cumulative distribution function of the three-dimensional random variable (X_1, X_2, X_3) of the continuous type, as given in Example 7.2.1, we can obtain the marginal cumulative distributions of the variates X_1, X_2, and X_3:

$$F_1(x_1) = \begin{cases} 1 - e^{-(x_1/\beta)}, & x_1 > 0, \beta > 0, \\ 0, & \text{elsewhere,} \end{cases}$$

$$F_2(x_2) = \begin{cases} 1 - e^{-(x_2/\beta)}, & x_2 > 0, \beta > 0, \\ 0, & \text{elsewhere,} \end{cases}$$

and

$$F_3(x_3) = \begin{cases} 1-e^{-(x_3/\beta)}, & x_3 > 0, \beta > 0, \\ 0, & \text{elsewhere.} \end{cases}$$

In view of Equation (7.5.1), we have

$$F_1(x_1)F_2(x_2)F_3(x_3)$$

$$= \left\{1-e^{-(x_1/\beta)}\right\}\left\{1-e^{-(x_2/\beta)}\right\}\left\{1-e^{-(x_3/\beta)}\right\}$$

$$= 1-e^{-(x_1/\beta)}-e^{-(x_2/\beta)}-e^{-(x_3/\beta)}+e^{-(1/\beta)(x_1+x_2)}+e^{-(1/\beta)(x_1+x_3)}+e^{-(1/\beta)(x_2+x_3)}-e^{-(1/\beta)(x_1+x_2+x_3)}$$

$$= F(x_1, x_2, x_3).$$

Hence, the random variables X_1, X_2, and X_3 are independent. This independence can also be shown using Theorem 7.5.1.

Mutual independence of a sequence of random variables can equivalently be defined using the notion of the conditional probability density function, as illustrated for the two-dimensional case in Section 6.5. We can show that any subsequence of random variables of a sequence of independent random variables is also independent. Let $(X_1, X_2, \ldots, X_n)$ be a sequence of n independent variates. For every $r \leq n$, the random variables $X_{i_1}, X_{i_2}, \ldots, X_{i_r}, 1 \leq i_1 < i_2 < \ldots < i_r \leq n$, are also independent. For simplicity, we show the independence for $i_1 = 1, i_2 = 2, \ldots, i_r = r$. It follows from Equation 7.5.1 that

$$F_r(x_1, x_2, \ldots, x_r) = F(x_1, x_2, \ldots, x_r, \infty, \infty, \ldots, \infty)$$

$$= \lim_{x_{r+1} \to \infty} \cdots \lim_{x_n \to \infty}\left[F(x_1, x_2, \ldots, x_r, x_{r+1}, \ldots, x_n)\right]$$

$$= \lim_{x_{r+1} \to \infty} \cdots \lim_{x_n \to \infty}\left[F_1(x_1)F_2(x_2)\ldots F_r(x_r)F(x_{r+1})\ldots F(x_n)\right]$$

$$= F_1(x_1)F_2(x_2)\ldots F_r(x_r)\lim_{x_{r+1} \to \infty}F(x_{r+1})\ldots \lim_{x_n \to \infty}F(x_n)$$

$$= F_1(x_1)F_2(x_2)\ldots F_r(x_r)F_{r+1}(\infty)\ldots F_n(\infty)$$

$$= F_1(x_1)F_2(x_2)\ldots F_r(x_r).$$

Thus, if the random variables $X_1, X_2, \ldots, X_n$ are mutually independent, then the random variables $X_1, X_2, \ldots, X_r, r \leq n$, are also independent.

7.6 Functions of Random Variables

In Section 6.6, we discussed in detail various aspects of deriving distributions of functions of two-dimensional random variables. In this section, we extend the discussion to the more complicated derivation of distributions of a sequence of functions of the n-dimensional random variable.

Suppose that $f(x_1, x_2, \ldots, x_n)$ is the multivariate probability density function of the n-dimensional random variable $(X_1, X_2, \ldots, X_n)$ of the continuous type; we are interested in obtaining the joint probability density function of the random variable $(Y_1, Y_2, \ldots, Y_n)$, where

$$Y_1 = g_1(X_1, X_2, \ldots, X_n)$$
$$Y_2 = g_2(X_1, X_2, \ldots, X_n)$$
$$\vdots \qquad \vdots$$
$$Y_n = g_n(X_1, X_2, \ldots, X_n). \qquad (7.6.1)$$

To obtain the new distribution, as in the two-dimensional case, we must assume that the functions $g_1, g_2, \ldots, g_n$ define a one-to-one transformation that maps an n-dimensional set E_n of the Euclidean n-space, $x_1, x_2, \ldots, x_n$, onto an n-dimensional set E^* in the $y_1, y_2, \ldots, y_n$ space. We also must have continuous first partial derivatives with respect to $x_1, x_2, \ldots, x_n$. These conditions imply that we can solve for the inverses of $g_1, g_2, \ldots, g_n$:

$$x_1 = z_1^{-1}(y_1, y_2, \ldots, y_n)$$
$$x_2 = z_2^{-1}(y_1, y_2, \ldots, y_n)$$
$$\vdots \qquad \vdots$$
$$x_n = z_n^{-1}(y_1, y_2, \ldots, y_n).$$

Thus, the functions $z_1^{-1}, z_2^{-1}, \ldots, z_n^{-1}$ are one to one and have continuous first partial derivatives with respect to $y_1, y_2, \ldots, y_n$. Having made the preceding assumptions, we can obtain the multivariate probability density function as follows:

$$z(y_1, y_2, \ldots, y_n) = f\left[z_1^{-1}(y_1, y_2, \ldots, y_n),\right.$$
$$\left. z_2^{-1}(y_1, y_2, \ldots, y_n), \ldots, x_n = z_n^{-1}(y_1, y_2, \ldots, y_n)\right]|J|,$$

where $|J|$ denotes the absolute value of the Jacobian of the transformation, defined by the n by n determinant

$$\begin{vmatrix} \dfrac{\partial x_1}{\partial y_1} & \dfrac{\partial x_1}{\partial y_2} & \cdots & \dfrac{\partial x_1}{\partial y_n} \\ \dfrac{\partial x_2}{\partial y_1} & \dfrac{\partial x_2}{\partial y_2} & \cdots & \dfrac{\partial x_2}{\partial y_n} \\ \vdots & & & \\ \dfrac{\partial x_n}{\partial y_1} & \dfrac{\partial x_n}{\partial y_2} & \cdots & \dfrac{\partial x_n}{\partial y_n} \end{vmatrix},$$

which cannot be equal to zero.

The preceding discussion can be intuitively presented as follows: Let $E_n \subset E$ be an n-dimensional subset (an event of E that consists of n-tuples) in the $x_1, x_2, \ldots, x_n$ space, and let $E_n^* \subset E^*$ be a subset (an event of E^* that consists of n-tuples) in the $y_1, y_2, \ldots, y_n$ space. Thus, under the condition of one-to-one transformation, the events

$$(X_1, X_2, \ldots, X_n) \in E_n$$

and

$$(Y_1, Y_2, \ldots, Y_n) \in E_n^*$$

are equivalent, and

$$Pr\left[(X_1, X_2, \ldots, X_n) \in E_n\right]$$

$$= \int\int_{E_n} \cdots \int f(x_1, x_2, \ldots, x_n) dx_1 \, dx_2 \ldots dx_n$$

$$Pr\left[(Y_1, Y_2, \ldots, Y_n) \in E_n^*\right]$$

$$= \int\int_{E_n^*} \cdots \int f\left[z_1^{-1}(y_1, y_2, \ldots, y_n),\right.$$

$$\left. z_2^{-1}(y_1, y_2, \ldots, y_n), z_n^{-1}(y_1, y_2, \ldots, y_n)\right]$$

$$\times |J| \, dy_1 \, dy_2 \ldots dy_n.$$

Hence, we summarize the preceding discussion by generalizing Theorem 6.6.1 to the n-dimensional case.

Theorem 7.6.1

Let $f(x_1, x_2, \ldots, x_n)$ be the multivariate probability density function of the continuous random variable $(X_1, X_2, \ldots, X_n)$. Suppose that

$$Y_1 = g_1(X_1, X_2, \ldots, X_n)$$
$$Y_2 = g_2(X_1, X_2, \ldots, X_n)$$
$$\vdots$$
$$Y_n = g_n(X_1, X_2, \ldots, X_n)$$

are such that $g_1, g_2, \ldots, g_n$ define a one-to-one transformation of the random variable $(X_1, X_2, \ldots, X_n)$ and that the variable has continuous first partial derivatives with respect to $x_1, x_2, \ldots, x_n$. The joint probability density function of the random variable $(Y_1, Y_2, \ldots, Y_n)$ is given by

$$z(y_1, y_2, \ldots, y_n) = f\left[z_1^{-1}(y_1, y_2, \ldots, y_n),\right.$$

$$\left. z_2^{-1}(y_1, y_2, \ldots, y_n), z_n^{-1}(y_1, y_2, \ldots, y_n)\right]|J|,$$

$$(y_1, y_2, \ldots, y_n) \in S_n^*,$$

and zero elsewhere.

Example 7.6.1 (Dirichlet distribution)

Let the multivariate probability density function of the $n+1$-dimensional random variable $(X_1, X_2, \ldots, X_n, X_{n+1})$ be given by

$$(x_1, x_2, \ldots, x_n, x_{n+1}, \alpha_1, \alpha_2, \ldots, \alpha_n, \alpha_{n+1})$$

$$= \begin{cases} \prod_{i=1}^{n+1} \dfrac{1}{\Gamma(\alpha_i)} x_i^{\alpha_i - 1} e^{-x_i}, & x_i \geq 0, i = 1, 2, \ldots, n, n+1, \\ 0, & \text{elsewhere.} \end{cases} \quad (7.6.2)$$

Suppose that

$$Y_1 = \frac{X_1}{X_1 + X_2 + \ldots + X_{n+1}}$$

$$Y_2 = \frac{X_2}{X_1 + X_2 + \ldots + X_{n+1}}$$

$$\vdots$$

$$Y_n = \frac{X_n}{X_1 + X_2 + \ldots + X_{n+1}}$$

$$Y_{n+1} = X_1 + X_2 + \ldots + X_{n+1},$$

and we wish to obtain the probability density function of the random variable $(Y_1, Y_2, \ldots, Y_n, Y_{n+1})$.

Solution: The preceding transformation maps the set

$$E_n = \{(x_1, x_2, \ldots, x_{n+1}), x_i \geq 0, i = 1, 2, \ldots, n+1\}$$

onto the set

$$E_n^* = \{(y_1, y_2, \ldots, y_n, y_{n+1}),$$

$$y_i \geq 0, i = 1, 2, \ldots, n+1, y_1 + y_2 + \ldots + y_n < 1\}.$$

The single-valued inverse functions are

$$x_1 = y_1 y_{n+1}, x_2 = y_2 y_{n+1}, \ldots, x_n = y_n y_{n+1},$$

$$x_{n+1} = y_{n+1}(1 - y_1 - y_2 - \ldots - y_n),$$

and the Jacobian of transformations is

$$J = \begin{vmatrix} y_{n+1} & 0 & 0 & \cdots & 0 & y_1 \\ 0 & y_{n+1} & 0 & \cdots & 0 & y_2 \\ 0 & 0 & y_{n+1} & \cdots & 0 & y_3 \\ \vdots & \vdots & \vdots & \vdots & \vdots & \vdots \\ 0 & 0 & 0 & \cdots & y_{n+1} & y_n \\ -y_{n+1} & -y_{n+1} & -y_{n+1} & \cdots & -y_{n+1} & (1 - y_1 - y_2 - \ldots - y_n) \end{vmatrix} = y_{n+1}^n.$$

Thus,

$$z(y_1, y_2, \ldots y_n, y_{n+1})$$
$$= f\left[y_1 y_{n+1}, y_2 y_{n+1}, \ldots, y_n y_{n+1}, y_{n+1}(1 - y_1 - y_2 - \ldots - y_n)\right] |J|$$
$$= \frac{1}{\Gamma(\alpha_1)} (y_1 y_{n+1})^{\alpha_1 - 1} e^{-y_1 y_{n+1}} \frac{1}{\Gamma(\alpha_2)} (y_2 y_{n+1})^{\alpha_2 - 1} e^{-y_2 y_{n+1}} \ldots$$
$$\frac{1}{\Gamma(\alpha_n)} (y_n y_{n+1})^{\alpha_n - 1} e^{-y_n y_{n+1}}$$
$$\frac{1}{\Gamma(\alpha_{n+1})} \left[y_{n+1}(1 - y_1 - y_2 - \ldots - y_n)^{\alpha_{n+1}-1}\right]$$
$$e^{-[y_{n+1}(1 - y_1 - y_2 - \ldots - y_n)]} y_{n+1}^n$$

or

$$= \begin{cases} \dfrac{y_1^{\alpha_1 - 1} y_2^{\alpha_2 - 1} \ldots y_n^{\alpha_n - 1} y_{n+1}^{\sum_{i=1}^{n+1} \alpha_i - 1} (1 - y_1 - y_2 - \ldots - y_n)^{\alpha_{n+1}-1} e^{-y_{n+1}}}{\Pi_{i=1}^{n+1} \Gamma(\alpha_i)}, & (y_1, y_2, \ldots, y_n, y_{n+1}) \in S_n^* \subset S^*, \\ 0, & \text{elsewhere.} \end{cases}$$

The marginal probability density function of the n-dimensional random variable $(Y_1, Y_2, \ldots, Y_n)$ is obtained by

$$z_n(y_1, y_2, \ldots y_n) = \int_{-\infty}^{\infty} z(y_1, y_2, \ldots y_n, y_{n+1}) dy_{n+1}$$
$$= \frac{y_1^{\alpha_1 - 1} y_2^{\alpha_2 - 1} \ldots y_n^{\alpha_n - 1} (1 - y_1 - y_2 - \ldots - y_n)^{\alpha_{n+1}-1}}{\Pi_{i=1}^{n+1} \Gamma(\alpha_i)}$$
$$= \int_0^{\infty} \sum_{i=1}^{n+1} \alpha_i^{-1} e^{-y_{n+1}} dy_{n+1}.$$

Thus, because

$$= \left\{\frac{1}{\Gamma\left(\sum_{i=1}^{n+1}\right)} \int_0^{\infty} \sum_{y_{n+1}}^{n+1} \alpha_i^{-1} e^{-y_{n+1}} dy_{n+1}\right\} = 1,$$

we have

$$z(y_1, y_2, \ldots y_n)$$
$$= \begin{cases} \dfrac{\Gamma\left(\sum_{i=1}^{n+1} \alpha_i\right)}{\Pi_{i=1}^{n+1} \Gamma(\alpha_i)} y_1^{\alpha_1 - 1} y_2^{\alpha_2 - 1} \ldots y_n^{\alpha_n - 1} (1 - y_1 - y_2 - \ldots - y_n)^{\alpha_{n+1}-1}, & (y_1, y_2, \ldots, y_n) \in S_n^*, \\ 0, & \text{elsewhere.} \end{cases} \quad (7.6.3)$$

The probability density function $z_n(y_1, y_2, \ldots, y_n)$ is called the *Dirichlet distribution*, with the parameters $\alpha_i > 0, i = 1, 2, \ldots, n+1$. This distribution has been used extensively in mathematical biology. The multivariate probability density in Equation (7.6.2) is simply the product of $n+1$ independent random variables having a gamma distribution with $\beta_i = 1$ and $\alpha_i > 0, i = 1, 2, \ldots, n+1$. Also, if we let $n = 1$ in Equation (7.6.3), the Dirichlet density reduces to the beta distribution, with the parameters $\alpha > 0$ and $\beta = 1$.

If the n-dimensional random variable $(X_1, X_2, \ldots, X_n)$ is discrete, then the multivariate probability density function of the sequence of variates $(Y_1, Y_2, \ldots, Y_n)$, as defined in Equation (7.6.1), is

$$p(y_1, y_2, \ldots y_n) = Pr(Y = y_1, Y = y_2, \ldots, Y = y_n)$$

$$= Pr\left[X_1 = z_1^{-1}(y_1, y_2, \ldots y_n), X_2 = z_2^{-1}(y_1, y_2, \ldots y_n), \right.$$

$$\left. \ldots, X_n = z_n^{-1}(y_1, y_2, \ldots y_n)\right].$$

In the preceding discussion, we derived the joint probability density function of a sequence of n functional forms $g_1, g_2, \ldots, g_n$ of the n-dimensional random variable $(X_1, X_2, \ldots, X_n)$. However, in many problems, the number of functional forms is less than the dimension of the given sequence of random variables. In such cases, to be able to apply Theorem 7.6.1, we must introduce *auxiliary* variables, as illustrated for the two-dimensional case.

A problem often arises in deriving distributions when the transformation involved is not one to one. For example, suppose that the transformation in Equation (7.6.1), considered earlier, which maps E_n in the $x_1, x_2, \ldots, x_n$ space into E_n^* in the $y_1, y_2, \ldots, y_n$ space, is not one to one; that is, to each point in E_n there corresponds one point in E_n^*, but to some point in E_n^* there may correspond more than one point of E_n. In such a situation, we partition E_n into a finite number of mutually exclusive sets $E_{n1}, E_{n2}, \ldots, E_{nr}$, with $\bigcup_{i=1}^{r} E_{n_i} = E_n$ such that the transformation in Equation (7.6.1) defines a one-to-one transformation of each E_{n_i}, $i = 1, 2, \ldots, n$, onto E_n^*. That is, for each point (n-tuple) in E_n^*, there corresponds at most one point in each E_{n_i}. Thus, we have r sets of n inverse functions, with one set for each of the transformations given by

$$x_1 = z_{1_i}^{-1}(y_1, y_2, \ldots, y_n)$$
$$x_2 = z_{2_i}^{-1}(y_1, y_2, \ldots, y_n),$$
$$\vdots$$
$$x_n = z_{n_i}^{-1}(y_1, y_2, \ldots, y_n)$$

$i=1,2,\ldots,r$. We assume that the transformations have continuous first partial derivatives and the Jacobian for each $i, i=1,2,\ldots,r$, is

$$J = \begin{vmatrix} \dfrac{\partial z^{-1}_{1_i}}{\partial y_1} & \dfrac{\partial z^{-1}_{1_i}}{\partial y_2} & \cdots & \dfrac{\partial z^{-1}_{1_i}}{\partial y_n} \\ \dfrac{\partial z^{-1}_{2_i}}{\partial y_1} & \dfrac{\partial z^{-1}_{2_i}}{\partial y_2} & \cdots & \dfrac{\partial z^{-1}_{2_i}}{\partial y_n} \\ \vdots & & & \\ \dfrac{\partial z^{-1}_{n_i}}{\partial y_1} & \dfrac{\partial z^{-1}_{n_i}}{\partial y_2} & \cdots & \dfrac{\partial z^{-1}_{n_i}}{\partial y_n} \end{vmatrix},$$

which cannot be equal to zero. Thus, we can obtain the joint probability density function of the n-dimensional random variable $(Y_1, Y_2, \ldots, Y_n)$, as defined by Equation (7.6.1), for each partition of the set E_n. In addition, because the r partitions are mutually exclusive, the sum of the probability density functions defined in each partition is the desired joint probability density function:

$$z(y_1, y_2, \ldots y_n) = \begin{cases} \sum_{i=1}^{n+1} f\left[z_1^{-1}(y_1, y_2, \ldots, y_n), z_1^{-1}(y_1, y_2, \ldots, y_n), \\ \ldots, z_{n_i}^{-1}(y_1, y_2, \ldots, y_n)\right] |J_i| (y_1, y_2, \ldots, y_n) \in S_n^*, \\ 0, \quad \text{elsewhere.} \end{cases}$$

When the given sequence of random variables $(Y_1, Y_2, \ldots, Y_n)$ is discrete and the n-dimensional random variable $(X_1, X_2, \ldots, X_n)$, as given in Equation (7.6.1), is continuous, its discrete probability mass function is

$$p(y_1, y_2, \ldots y_n) = Pr(Y = y_1, Y = y_2, \ldots, Y = y_n)$$

$$= \sum_{i=1}^{r} Pr\left[X_1 = z_{1_i}^{-1}(y_1, y_2, \ldots y_n), X_2\right.$$

$$= z_{2_i}^{-1}(y_1, y_2, \ldots y_n) \times \ldots, X_n$$

$$= z_{n_i}^{-1}(y_1, y_2, \ldots y_n)\bigg].$$

7.7 Expected Value and Moments

The extension of Section 6.7 to the n-dimensional case is straightforward; thus, our discussion is brief.

> **Definition 7.7.1**
>
> Let $(X_1, X_2, \ldots, X_n)$ be an n-dimensional continuous random variable with the joint probability density function $f(x_1, x_2, \ldots, x_n)$. The multiple integral
>
> $$\int_{-\infty}^{\infty} \cdots \int_{-\infty}^{\infty} \int_{-\infty}^{\infty} x_1, x_2, \ldots, x_n f(x_1, x_2, \ldots, x_n) \, dx_1 \, dx_2 \ldots dx_n$$
>
> is called the **expected value of the product of the variates** $(X_1, X_2, \ldots, X_n)$ if the following inequality is satisfied:
>
> $$\int_{-\infty}^{\infty} \cdots \int_{-\infty}^{\infty} \int_{-\infty}^{\infty} |x_1, x_2, \ldots, x_n| f(x_1, x_2, \ldots, x_n) \, dx_1 \, dx_2 \ldots dx_n < \infty.$$

A generalized version of the properties of expectation, discussed in Section 5.2, is given here.

> **Theorem 7.7.1**
>
> Let $(X_1, X_2, \ldots, X_n)$ be an n-dimensional random variable of the continuous type with the joint probability density function $f(x_1, x_2, \ldots, x_n)$. If $U = g(X_1, X_2, \ldots, X_n)$, then the expected value of U is
>
> $$E(U) = E[g(X_1, X_2, \ldots, X_n)]$$
>
> $$\int_{-\infty}^{\infty} \cdots \int_{-\infty}^{\infty} \int_{-\infty}^{\infty} g(x_1, x_2, \ldots, x_n) f(x_1, x_2, \ldots, x_n) \, dx_1 \, dx_2 \ldots dx_n.$$

We omit the proof because it is similar to that of Theorem 5.2.2. However, understand that the multiple integral must be absolutely convergent for $E(U)$ to exist.

> **Theorem 7.7.2**
>
> Let $(X_1, X_2, \ldots, X_n)$ be an n-dimensional random variable with the probability density function $f(x_1, x_2, \ldots, x_n)$. If
>
> $$Y_1 = g_1(X_i), Y_2 = g_2(X_i), \ldots, Y_n = g_n(X_i), i = 1, 2, \ldots, n,$$

then
$$E(Y_1, Y_2, \ldots, Y_n)$$
$$= E\left[g_1(x_i) g_2(x_i) \cdots g_n(x_i)\right]$$
$$= \int_{-\infty}^{\infty} \cdots \int_{-\infty}^{\infty} \int_{-\infty}^{\infty} g_1(x_i) g_2(x_i) \cdots g_n(x_i) f(x_1, x_2, \ldots, x_n) dx_1 dx_2 \ldots dx_n$$

if the variate $(X_1, X_2, \ldots, X_n)$ is continuous. We can also obtain $E(Y_1, Y_2, \ldots, Y_n)$ by first finding, if possible, the multivariate distribution of $(Y_1, Y_2, \ldots, Y_n)$, $z(y_1, y_2, \ldots, y_n)$, as discussed in the previous section, and then applying Definition 7.7.1 to obtain

$$E(Y_1, Y_2, \ldots, Y_n)$$
$$= \int_{-\infty}^{\infty} \cdots \int_{-\infty}^{\infty} \int_{-\infty}^{\infty} y_1 y_2 \cdots y_n z(y_1, y_2, \ldots, y_n) dy_1 dy_2 \ldots dy_n.$$

■ **Theorem 7.7.3**

Let $(X_1, X_2, \ldots, X_n)$ be an n-dimensional random variable. If the variates in the sequence $(X_1, X_2, \ldots, X_n)$ are mutually independent, then

1. $E(X_1, X_2, \ldots, X_n) = \prod_{i=1}^{n} E(X_i)$
2. $\text{Var}(X_1 + X_2 + \ldots + X_n) = \sum_{i=1}^{n} \text{Var}(X_i)$

■ **Definition 7.7.2**

Let $(X_1, X_2, \ldots, X_n)$ be an n-dimensional random variable of the continuous type with the probability density function $f(x_1, x_2, \ldots, x_n)$. The **moment of order** $k_1 + k_2 + \ldots + k_n$ of the multivariate distribution of the variate $(X_1, X_2, \ldots, X_n)$ is

$$E\left(X_1^{k_1} X_2^{k_2} \ldots X_n^{k_n}\right)$$
$$= \int_{-\infty}^{\infty} \cdots \int_{-\infty}^{\infty} \int_{-\infty}^{\infty} x_1^{k_1} x_2^{k_2} \ldots x_n^{k_n} f(x_1, x_2, \ldots, x_n) dx_1 dx_2 \ldots dx_n$$
$$= \mu_{k_1, k_2, \ldots, k_n}.$$

The **central moment of order** $k_1 + k_2 + \ldots + k_n$ is

$$E\left[(X_1 - \mu_1)^{k_1} (X_2 - \mu_2)^{k_2} \ldots (X_n - \mu_n)^{k_n}\right]$$
$$= \int_{-\infty}^{\infty} \cdots \int_{-\infty}^{\infty} \int_{-\infty}^{\infty} (x_1 - \mu_1)^{k_1} (x_2 - \mu_2)^{k_2} \ldots (x_n - \mu_n)^{k_n} f(x_1, x_2, \ldots, x_n) dx_1 dx_2 \ldots dx_n$$
$$= \eta_{k_1, k_2, \ldots, k_n},$$

where $E(X_i) = \mu_i, i = 1, 2, \ldots, n$.

Thus, if $k_i = k_j = 1$, $1 \leq i < j \leq n$, with the remaining k's equal to zero, then

$$\eta_{0,0\ldots,0,1,0,\ldots,0,1,0,\ldots,0} = E\left[(X_i - \mu_i)(X_j - \mu_j)\right]$$

$$= \text{Cov}(X_i, X_j).$$

A random variable X is *chi-square distributed*, $\chi^2(k)$, with k degrees of freedom if its probability density function is

$$f(x;k) = \begin{cases} \dfrac{1}{\Gamma\left(\dfrac{k}{2}\right)2^{k/2}} x^{k/2-1} e^{-x/2}, & x \geq 0, \\ 0, & \text{elsewhere.} \end{cases}$$

Thus, the $\chi^2(k)$ distribution is simply the gamma distribution with the parameters $\alpha = k/2$ and $\beta = 2$.

Theorem 7.7.4

Let the sequence of random variables $X_1, X_2, \ldots, X_n$ be mutually independent, and let each variable be chi-square distributed, $\chi^2(k_1), \chi^2(k_2), \ldots, \chi^2(k_n)$, with $k_1, k_2, \ldots, k_n$ degrees of freedom, respectively. If $U = X_1 + X_2 + \ldots + X_n$, then U is chi-square distributed with r degrees of freedom, $\chi^2(r), r = k_1 + k_2 + \ldots + k_n$.

Proof The proof of this theorem is straightforward; however, we can obtain a similar result as follows:

Let the sequence of random variables $X_1, X_2, \ldots, X_n$ be mutually independent, and let each one be normally distributed with a mean of μ and a variance of σ^2. Then, the random variable

$$U = \sum_{j=1}^{n} \left(\frac{X_j - \mu}{\sigma}\right)^2$$

in $\chi^2(n)$.

Example 7.7.1

Let the sequence of random variables $X_1, X_2, \ldots, X_n$ be mutually independent, and let each variable be normally distributed with the parameters μ and σ^2. The distribution of the random variable $U = \dfrac{1}{n}\sum_{j=1}^{n} X_j$ is also normally distributed with the parameters

$$\mu = \sum_{j=1}^{n}\left(\frac{1}{n}\mu\right)$$

and

$$\frac{\sigma^2}{n} = \sum_{j=1}^{n}\left(\frac{1}{n}\sigma\right)^2.$$

The random variable U is the *sample mean*, and it is usually denoted by $\bar{X}$. Thus,

$$f\left(\bar{x};\mu,\frac{\sigma^2}{n}\right) = \frac{\sqrt{n}}{\sqrt{2\pi}\sigma}\exp\left\{-\frac{1}{2}\left(\frac{\bar{x}-\mu}{\sigma/\sqrt{n}}\right)^2\right\}, -\infty<\bar{x}<\infty, n=1,2,\ldots,\sigma>0.$$

Furthermore, if $\bar{X} = (1/n)\sum_{i=1}^{n} X_i$, then the distribution of $\{(n-1)S^2/\sigma^2\}$,

$$S^2 = \frac{1}{n-1}\sum_{i=1}^{n}(X_i - \bar{X})^2,$$

is $\chi^2(n-1)$. The variate S^2 is called the *sample variance*, and it is used to estimate the variance of a given distribution function.

7.8 | Conditional Expectations

We saw in the study of two-dimensional random variables that conditional expectation is simply the expected value of a variate computed with respect to a conditional distribution and that its geometric interpretation is a regression curve. The extension of conditional expectation to the multivariate case is briefly discussed here.

■ Definition 7.8.1

Let $f(x_1, x_2, \ldots, x_n)$ be the joint probability density function of the continuous n-dimensional random variable $(X_1, X_2, \ldots, X_n)$. The **conditional expected value of the variate** $(X_1, X_2, \ldots, X_n)$, $r < n$, given that $X_{r+1} = x_{r+1}, X_{r+2} = x_{r+2}, \ldots, X_n = x_n$, is

$$E(X_1 X_2 \ldots X_r | X_{r+1} = x_{r+1}, X_{r+2} = x_{r+2}, \ldots, X_n = x_n)$$

$$= \int_{-\infty}^{\infty}\int_{-\infty}^{\infty}\cdots\int_{-\infty}^{\infty} x_1, x_2, \ldots, x_r h(x_1, x_2, \ldots, x_r | x_{r+1}, x_{r+2}, \ldots, x_n) dx_1 dx_2 \ldots dx_r. \quad (7.8.1)$$

The preceding definition is interpreted as the average value of $(X_1, X_2, \ldots, X_n)$ when it is known that $X_{r+1} = x_{r+1}, X_{r+2} = x_{r+2}, \ldots, X_n = x_n$. For the conditional expectation to exist, the right-hand side of Equation (7.8.1) must be absolutely convergent.

The conditional expectation of $(X_1, X_2, \ldots, X_n)$, given $X_{r+1} = x_{r+1}, X_{r+2} = x_{r+2}, \ldots, X_n = x_n, r < n$, is constant if $x_{r+1}, x_{r+2}, \ldots, x_n$ are fixed values. However, in general, the $E(X_1 X_2 \ldots X_r | X_{r+1} = x_{r+1}, X_{r+2} = x_{r+2}, \ldots, X_n = x_n)$ is a function of $x_{r+1}, x_{r+2}, \ldots, x_n$ values of the $(n-r)$-dimensional random variable $(X_{r+1}, X_{r+2}, \ldots, X_n)$; thus, it is a random variable.

If the random variables in the sequence $(X_1, X_2, \ldots, X_n)$ are mutually independent, then

$$E(X_1 X_2 \ldots X_r | X_{r+1} = x_{r+1}, X_{r+2} = x_{r+2}, \ldots, X_n = x_n)$$

$$= \int_{-\infty}^{\infty} \int_{-\infty}^{\infty} \ldots \int_{-\infty}^{\infty} x_1 x_2 \ldots x_r \frac{f(x_1, x_2, \ldots, x_n)}{f_{n-r}(x_{r+1}, x_{r+2}, \ldots, x_n)} dx_1 dx_2 \ldots dx_r$$

$$= \int_{-\infty}^{\infty} \int_{-\infty}^{\infty} \ldots \int_{-\infty}^{\infty} x_1 x_2 \ldots x_r \frac{f_r(x_1, x_2, \ldots, x_r) f_{n-r}(x_{r+1}, x_{r+2}, \ldots, x_n)}{f_{n-r}(x_{r+1}, x_{r+2}, \ldots, x_n)} dx_1 dx_2 \ldots dx_r$$

$$= \int_{-\infty}^{\infty} \int_{-\infty}^{\infty} \ldots \int_{-\infty}^{\infty} x_1 x_2 \ldots x_r f_r(x_1, x_2, \ldots, x_r) dx_1 dx_2 \ldots dx_r$$

$$= \int_{-\infty}^{\infty} x_1 f_1(x_1) dx_1 \int_{-\infty}^{\infty} x_2 f_2(x_2) dx_2 \ldots \int_{-\infty}^{\infty} x_r f(x_r) dx_r$$

$$= E(X_1) E(X_2) \ldots E(X_r).$$

■ **Theorem 7.8.1**

Let $(X_1, X_2, \ldots, X_n)$ be an n-dimensional random variable. If the conditional expectation

$$E(X_1 X_2 \ldots X_r | X_{r+1} = x_{r+1}, X_{r+2} = x_{r+2}, \ldots, X_n = x_n), r < n,$$

exists, then

$$E\{E(X_1 X_2 \ldots X_r | X_{r+1} = x_{r+1}, X_{r+2} = x_{r+2}, \ldots, X_n = x_n)\} = E(X_1 X_2 \ldots X_r).$$

Proof Let $f(x_1, x_2, \ldots, x_r) h_{n-r}(x_{r+1}, x_{r+2}, \ldots, x_n)$ and $h(x_1, x_2, \ldots, x_r | x_{r+1}, x_{r+2}, \ldots, x_n)$ be the probability density function of the continuous random variable $(X_1, X_2, \ldots, X_n)$, the marginal probability density of the variate $(X_{r+1}, X_{r+2}, \ldots, X_n)$,

7.8 Conditional Expectations

and the conditional probability density function of the random variable $(X_1 X_2 \ldots X_r | X_{r+1} = x_{r+1}, X_{r+2} = x_{r+2}, \ldots, X_n = x_n)$, respectively. Thus,

$$E\Big[E(X_1 X_2 \ldots X_r | X_{r+1} = x_{r+1}, X_{r+2} = x_{r+2}, \ldots, X_n = x_n)\Big]$$

$$= \int_{-\infty}^{\infty} \int_{-\infty}^{\infty} \ldots \int_{-\infty}^{\infty} E(X_1 X_2 \ldots X_r$$

$$| X_{r+1} = x_{r+1}, X_{r+2} = x_{r+2}, \ldots, X_n = x_n)$$

$$h_{n-r}(x_{r+1}, x_{r+2}, \ldots, x_n) dx_{r+1} dx_{r+2} \ldots dx_n$$

$$= \int_{-\infty}^{\infty} \int_{-\infty}^{\infty} \ldots \int_{-\infty}^{\infty} \Big\{ \int_{-\infty}^{\infty} \int_{-\infty}^{\infty} \ldots \int_{-\infty}^{\infty} x_1 x_2 \ldots x_r h$$

$$\times (x_1, x_2, \ldots, x_r | x_{r+1}, x_{r+2}, \ldots, x_n) dx_1 dx_2 \ldots dx_r \Big\}$$

$$h_{n-r}(x_{r+1}, x_{r+2}, \ldots, x_n) dx_{r+1} dx_{r+2} \ldots dx_n$$

$$= \int_{-\infty}^{\infty} \int_{-\infty}^{\infty} \ldots \int_{-\infty}^{\infty} \Big\{ \int_{-\infty}^{\infty} \int_{-\infty}^{\infty} \ldots \int_{-\infty}^{\infty} x_1 x_2 \ldots x_r$$

$$\frac{f(x_1, x_2, \ldots, x_n)}{h_{n-r}(x_{r+1}, x_{r+2}, \ldots, x_n)} dx_1 dx_2 \ldots dx_r \Big\}$$

$$h_{n-r}(x_{r+1}, x_{r+2}, \ldots, x_n) dx_{r+1} dx_{r+2} \ldots dx_n.$$

Because the expected value exists, we can change the order of integration and have

$$E\Big[E(X_1 X_2 \ldots X_r | X_{r+1} = x_{r+1}, X_{r+2} = x_{r+2}, \ldots, X_n = x_n)\Big]$$

$$= \int_{-\infty}^{\infty} \int_{-\infty}^{\infty} \ldots \int_{-\infty}^{\infty} x_1 x_2 \ldots x_r \Big\{ \int_{-\infty}^{\infty} \int_{-\infty}^{\infty} \ldots \int_{-\infty}^{\infty} \frac{f(x_1, x_2, \ldots, x_n)}{h_{n-r}(x_{r+1}, x_{r+2}, \ldots, x_n)}$$

$$h_{n-r}(x_{r+1}, x_{r+2}, \ldots, x_n) dx_{r+1} dx_{r+2} \ldots dx_n \Big\} dx_1 dx_2 \ldots dx_r$$

$$= \int_{-\infty}^{\infty} \int_{-\infty}^{\infty} \ldots \int_{-\infty}^{\infty} x_1 x_2 \ldots x_r h_r(x_1, x_2, \ldots, x_r) dx_1 dx_2 \ldots dx_r$$

$$= E(X_1 X_2 \ldots X_r).$$

Thus, the expected value of the conditional expected value is equal to the unconditional expectation.

If $g(X_1, X_2, \ldots, X_n)$ is some functional form of the n-dimensional random variable $(X_1, X_2, \ldots, X_n)$, the joint probability density of which is $f(x_1, x_2, \ldots, x_n)$, then

$$E\left[g(X_1, X_2, \ldots, X_n) \middle| x_{11} \leq X_1 \leq x_{12}, x_{21} \leq X_2 \leq x_{22}, \ldots, x_{r1} \leq X_r \leq x_{r2}\right]$$

$$= \int_{-\infty}^{\infty}\int_{-\infty}^{\infty} \cdots \int_{-\infty}^{\infty} g(x_1, x_2, \ldots, x_n)$$

$$\times \frac{f(x_1, x_2, \ldots, x_n)}{\Pr(x_{11} \leq X_1 \leq x_{12}, x_{21} \leq X_2 \leq x_{22}, \ldots, x_{r1} \leq X_r \leq x_{r2})} dx_1 \, dx_2 \ldots dx_n$$

provided that $\Pr(x_{11} \leq X_1 \leq x_{12}, x_{21} \leq X_2 \leq x_{22}, \ldots, x_{r1} \leq X_r \leq x_{r2}) \neq 0$ and the multiple integral is absolutely convergent. Applying the mathematical method used in Section 6.8, the preceding expression reduces to

$$E\left[g(X_1, X_2, \ldots, X_n) \middle| x_{11} \leq X_1 \leq x_{12}, x_{21} \leq X_2 \leq x_{22}, \ldots, x_{r1} \leq X_r \leq x_{r2}\right]$$

$$= \int_{-\infty}^{\infty}\int_{-\infty}^{\infty} \cdots \int_{-\infty}^{\infty} g(x_1, x_2, \ldots, x_n) \frac{f(x_1, x_2, \ldots, x_n)}{h_r(x_1, x_2, \ldots, x_r)}$$

$$\times dx_{r+1} dx_{r+2} \ldots dx_n,$$

with

$$h_r(x_1, x_2, \ldots, x_r) > 0.$$

Summary

Let S be a sample space and $X_1(s), X_2(s), \ldots, X_n(s)$ real valued functions (defined on S), each assigning a number to each element of the sample space. Then, we call $(X_1(s) = X_1, X_2(s) = X_2, \ldots, X_n(s) = X_n)$ an n-*dimensional random variable*.

Let $(X_1, X_2, \ldots, X_n)$ be an n-dimensional random variable. If $(X_1, X_2, \ldots, X_n)$ can assume all values in some noncountable (nondenumerable) set of the n-dimensional Euclidean space, then we call $(X_1, X_2, \ldots, X_n)$ an n-*dimensional continuous random variable*.

Let $(X_1, X_2, \ldots, X_n)$ be an *n*-dimensional continuous random variable. A function $f(x_1, x_2, \ldots, x_n)$ is called a *multivariate or joint density function* of the random variable $(X_1, X_2, \ldots, X_n)$ if the following conditions are satisfied:

1. $f(x_1, x_2, \ldots, x_n) \geq 0$, for all $x_1, x_2, \ldots, x_n \in R_{X_1, X_2, \ldots, X_n}$
2. $\int \int_{R_{X_1, X_2, \ldots, X_n}} \cdots \int f(x_1, x_2, \ldots, x_n) dx_1 dx_2 \ldots dx_n = 1$

Here, the integration is performed over $R_{X_1, X_2, \ldots, X_n}$ in which $(X_1, X_2, \ldots, X_n)$ assumes all values.

The *multinomial* and *multivariate normal* probability density functions were discussed.

The *multivariate cumulative distribution function* of the continuous *n*-dimensional random variable $(X_1, X_2, \ldots, X_n)$, which has the joint probability density function $f(x_1, x_2, \ldots, x_n)$, is given by

$$F(x_1, x_2, \ldots, x_n) = Pr(X_1 \leq x_1, X_2 \leq x_2, \ldots, X_n \leq x_n)$$

$$= \int_{-\infty}^{x_n} \int_{-\infty}^{x_{n-1}} \cdots \int_{-\infty}^{x_1} f(y_1, y_2, \ldots y_n) dy_1 \ldots dy_{n-1} dy_n.$$

It possesses the following properties:

1. $F(\infty, \infty, \ldots, \infty) = 1$.
2. $F(x_1, x_2, \ldots, x_n) = 0$, if at least one of the x_i's is $-\infty$.
3. $F(x_1, x_2, \ldots, x_n)$ is monotone nondecreasing in each variate separately.
4. $F(x_1, x_2, \ldots, x_n)$ is continuous from the right in each variate separately.
5. If $F(x_1, x_2, \ldots, x_n)$ is continuous and if its partial derivatives exist for all x_i's, then

$$\frac{\partial^n}{\partial x_1 \partial x_2 \ldots \partial x_n} F(x_1, x_2, \ldots, x_n) = f(x_1, x_2, \ldots, x_n).$$

Let $f(x_1, x_2, \ldots, x_n)$ be the joint probability density function of the variate $(X_1, X_2, \ldots, X_n)$. The *marginal probability density function* of $(X_1, X_2, \ldots, X_r), r < n$, is

$$h_r(x_1, x_2, \ldots, x_r) = \int_{-\infty}^{\infty} \cdots \int_{-\infty}^{\infty} f(x_1, x_2, \ldots, x_n) dx_{r+1} \ldots dx_n.$$

The *conditional probability density function* of the random variable $(X_1, X_2, \ldots, X_r), r < n$, given

$$(X_{r+1}, X_{r+2}, \ldots, X_n),$$

is

$$g_r(x_1, x_2, \ldots, x_r | x_{r+1}, \ldots, x_n)$$

$$= \frac{f(x_1, x_2, \ldots, x_n)}{h_{n-r}(x_{r+1}, x_{r+2}, \ldots, x_n)}, h_{n-r}(x_{r+1}, x_{r+2}, \ldots, x_n) > 0.$$

The random variables of the sequence $(X_1, X_2, \ldots, X_n)$ are said to be *mutually independent* if

$$F(x_1, x_2, \ldots, x_n) = F_1(x_1) F_2(x_2) \ldots F_n(x_n)$$

for all real n-tuples $(x_1, x_2, \ldots, x_n)$.

The method of deriving probability distributions of a sequence of functions of the n-dimensional random variable is given by Theorem 7.6.1.

The study of *expected value and moments* is discussed for n random variables.

The *conditional expected value* of the variate $(X_1, X_2, \ldots, X_r)$, $r < n$, given $X_{r+1} = x_{r+1}, X_{r+2} = x_{r+2}, \ldots, X_n = x_n$, is given by

$$E(X_1 X_2 \ldots X_r | X_{r+1} = x_{r+1}, X_{r+2} = x_{r+2}, \ldots, X_n = x_n)$$

$$= \int_{-\infty}^{\infty} \cdots \int_{-\infty}^{\infty} x_1, x_2, \ldots, x_r h(x_1, x_2, \ldots, x_r | x_{r+1}, x_{r+2}, \ldots, x_n) dx_1 dx_2 \ldots dx_r.$$

Applied Problems

7.1 A die is rolled 10 times. If X, Y, and Z are random variables representing the number of sixes, fives, and fours, respectively, that turn up, derive the joint probability density function $f(x, y, z)$.

7.2 Fifteen cards are drawn from an ordinary deck of cards. Let X, Y, and Z be random variables representing the number of spades, hearts, and diamonds, respectively, that are drawn. Exhibit the joint probability density function $f(x, y, z)$ if
  a. there is replacement
  b. there is no replacement

7.3 Five coins are tossed n times. Write the probability density of W, the number of times no heads appear; X, the number of times one head appears; Y, the number of times two heads appear; and Z, the number of times three heads appear.

7.4 Five cards are drawn n times from a deck of cards containing five aces, four kings, and three queens. Find the probability density of W, the number of times no aces appear; X, the number of times one ace appears; Y, the number of times three aces appear; and Z, the number of times four aces appear if
  a. there is replacement
  b. there is no replacement

7.5 The probability density function of X, Y, and Z is uniform throughout a cube that has vertices of (3,2,0), (3,2,1), (3,3,0), (3,3,1), (4,2,0), (4,2,1), (4,3,0), and (4,3,1). Obtain the joint probability density function of the random variable (X, Y, Z).

7.6 If the probability density of X, Y, and Z is uniform throughout $0 \le x \le y \le z \le 1$, find the joint probability density function $f(x, y, z)$.

7.7 If $f(x, y, z) = k$ for $0 \le x^2 \le y \le z \le 2$ and equals zero elsewhere, find k so that the function $f(x, y, z)$ is the joint probability density function of the random variable (X, Y, Z).

7.8 If

$$f(w, x, y, z) = \begin{cases} kwxyz, & \text{for } 1 \le w \le 7, \ 2 \le x \le 4, \\ & 3 \le y \le 5, \ 6 \le z \le 8, \\ 0, & \text{elsewhere,} \end{cases}$$

find k so that the function $f(w, x, y, z)$ is a probability density function of the four-dimensional random variable (W, X, Y, Z).

7.9 Write the joint probability function for W, X, Y, Z if it is uniform throughout a four-dimensional figure that has the following 16 vertices: (a, b, c, d), where $a = 0, 1$, $b = 1, 3/2$, $c = 1, 2$, and $d = 2, 5/2$.

7.10 If

$$f(w, x, y, z) = \begin{cases} kwxyz(z-y)(y-x)(x-w), & 0 \le w \le x \le y \le z \le 1, \\ 0, & \text{elsewhere,} \end{cases}$$

find k so that the function $f(w, x, y, z)$ is the joint probability density function of the random variable (W, X, Y, Z).

7.11 Find the cumulative distributions of the probability densities in Problems 7.5 through 7.7.

7.12 For Problems 7.8 through 7.10, find the cumulative distributions of the joint probability density functions.

7.13 For Problems 7.1, 7.2, and 7.5 through 7.7, find the marginal probability densities of
   a. X
   b. Y
   c. X, Z
   d. Y, Z

7.14 For Problems 7.3, 7.4, and 7.8 through 7.10, find the marginal probability densities of
   a. W
   b. (W, Z)
   c. (X, Y)
   d. (X, Z)
   e. (W, X, Y)
   f. (W, X, Z)
   g. (W, X, Y)
   h. (X, Y, Z)

7.15 For Problems 7.1, 7.2, and 7.5 through 7.7, find the following conditional probability density functions:
- a. $f(x|y)$
- b. $g(y|z)$
- c. $k_1(z|x,y)$
- d. $k_2(z|x,y)$
- e. $l(x,y|z)$

7.16 For Problems 7.1, 7.2, and 7.5 through 7.7, find the following conditional cumulative distribution functions:
- a. $F(x|y,z)$
- b. $G(y|x,z)$
- c. $H(z|x)$
- d. $L(x,z|y)$
- e. $M(y,z|x)$

7.17 For Problems 7.3, 7.4, and 7.8 through 7.10, find the following conditional cumulative distribution functions:
- a. $h(w|x)$
- b. $f(x|w,y)$
- c. $g(y|w,x,z)$
- d. $l(w,z|y)$
- e. $m(w,x,y|z)$

7.18 For Problems 7.3, 7.4, and 7.8 through 7.10, find the following conditional cumulative distribution functions:
- a. $G(w|x,y,z)$
- b. $F(x|y,z)$
- c. $H(z|x)$
- d. $L(x,z|w,y)$
- e. $M(x,y,z|w)$

7.19 Determine whether the random variables involved in Problems 7.1 through 7.5 are mutually independent.

7.20 Determine whether the random variables involved in Problems 7.6 through 7.10 are mutually independent.

7.21 Let X_1 and X_2 be two independent random variables that have the same probability distribution. For $i = 1, 2$,

$$Pr(X_i = -1) = \frac{1}{2}$$

and

$$Pr(X_i = 1) = \frac{1}{2}.$$

Further, let $X_3 = X_1 X_2$. Show that (X_1, X_3) and (X_2, X_3) are independent but that (X_1, X_2, X_3) are not independent.

7.22 Show that, for $n = 2$, $P^{-1} = \begin{bmatrix} \sigma_{11}^2 & \sigma_{12}^2 \\ \sigma_{21}^2 & \sigma_{22}^2 \end{bmatrix}$, and $\rho = \sigma_{12}/\sigma_1\sigma_2$, the multivariate normal probability density function is given by

$$f(x) = \frac{1}{(2\pi)^{n/2}\sqrt{|P^{-1}|}} e^{-(1/2)(x-\mu)^T P(x-\mu)},$$

which reduces to

$$f(x_1, x_2) = \frac{1}{(2\pi)\sigma_1\sigma_2\sqrt{1-\rho^2}} e^{-[1/2(1-\rho^2)]\left[\left(\frac{x_1-\mu_1}{\sigma_1}\right)^2 \right.}$$

$$\left. +2\rho\left(\frac{x_1-\mu_1}{\sigma_1}\right)\left(\frac{x_2-\mu_2}{\sigma_2}\right)+\left(\frac{x_2-\mu_2}{\sigma_2}\right)^2\right].$$

7.23 Let $Y_j = \sum_{i=1}^{j} X_i$ for $j = 1, 2, \ldots, n$. Find the joint probability density of $Y_1, Y_2, \ldots, Y_n$ if the joint probability density function of the n-dimensional random variable $(X_1, X_2, \ldots, X_n)$ is given by

a. $f(x_1, x_2, \ldots, x_n) = \frac{1}{(2\pi)^{n/2}\sigma^n} e^{-(1/2\sigma^2)\sum_{i=1}^{n}(x_i - \mu)^2}$, $-\infty < x_i < \infty$

b. $f(x_1, x_2, \ldots, x_n) = \frac{1}{\beta^{n\alpha}[\Gamma(\alpha)]^n} \Pi_{i=1}^{n} x_i^{\alpha-1} e^{-(1/\beta)\sum_{i=1}^{n} x_i}$, $x_i > 0$

c. $f(x_1, x_2, \ldots, x_n) = \frac{\lambda^{\sum_{i=1}^{n} x_i} e^{-\lambda}}{\Pi_{i=1}^{n} x_i!}$, $x_i = 0, 1, 2, \ldots$

7.24 Find Problem 7.23(a) through (c) if

a. $Y_j = j \sum_{i=1}^{j} X_i$

b. $Y_j = j \sum_{i=1}^{n} X_i$

c. $Y_j = \dfrac{X_j}{\sum_{i=1}^{n} X_i}$

7.25 For the probability densities of Problems 7.1, 7.2, and 7.5 through 7.7, find
   a. $E(X Y Z)$
   b. $E(X^2 Y^3 Z^4)$
   c. $E(3X - Y^2 + 4Z)$
   d. $E(X+Y+Z)^2$

7.26 For the probability densities of Problems 7.3, 7.4, and 7.8 through 7.10, find
   a. $E(W X Y Z)$
   b. $E(W^2 X^3 Y^4 Z^5)$
   c. $E(W + 2X^2 - 3Y + 4Z^2)$
   d. $E(W+X+Y+Z)^2$

7.27 Let X_1, X_2, and X_3 be independent normally distributed random variables, with means of 1, 2, and 5 and standard deviations of 2, 2, and 4, respectively.
   a. Find the moment-generating function $Y = (X_1 - 2X_2 + X_3)$.
   b. Determine the probability that Y exceeds 8.

7.28 For the probability densities of Problems 7.5 through 7.7, find the moment of order
   a. $1+2+3$
   b. $3+3+3$
   c. $n_1 + n_2 + n_3$

7.29 Do Problem 7.28 for central moments.

7.30 For the probability densities of Problems 7.8 through 7.10, find the moment of order
   a. $1+2+3+4$
   b. $3+3+3+3$
   c. $n_1 + n_2 + n_3 + n_4$

7.31 Do Problem 7.30 for central moments.

7.32 Let $X_1, X_2, \ldots, X_n$ be a sequence of random variables that have the same variance, σ^2. If $\bar{X} = (1/n) \sum_{i=1}^{n} X_i$, show that
   a. $Var(\bar{X}) = \dfrac{\sigma^2}{n}$
   b. $E\left[\sum_{i=1}^{n} (X_i - \bar{X})^2 \right] = (n-1)\sigma^2$

7.33 Let $X_1, X_2, \ldots, X_n$ be a sequence of random variables that have a common standard deviation, σ. The coefficient of correlation between any two of them is ρ. Show that
   a. the variance of their mean is given by

   $$\frac{\sigma^2}{n} + \left(1 - \frac{1}{n}\right)\sigma^2$$

   b. $E\left[\sum_{i=1}^{n} (X_i - \bar{X})^2 \right] = (n-1)(1-\rho)\sigma^2$

7.34 Find the probability density function of

$$\frac{1}{n}\sum_{i=1}^{n} X_i$$

if, for $i = 1, 2, \ldots, n$,
- **a.** X_i is binomially distributed with the parameters m and p.
- **b.** X_i has the Poisson distribution with the parameter λ.
- **c.** X_i has the geometric probability density function with the parameter p.
- **d.** X_i has the negative binomial distribution with the parameters k and p.
- **e.** X_i has the gamma probability density with the parameters α and β.
- **f.** $f(x_i) = \frac{1}{2} e^{-|x_i|}, \quad -\infty < x_i < \infty$.
- **g.** $f(x_i) = \begin{cases} x_i, & 0 \le x_i \le 1, \\ 2 - x_i, & 1 < x \le 2, \\ 0, & \text{elsewhere.} \end{cases}$

7.35 For the probability densities of Problems 7.1, 7.2, and 7.5 through 7.7, find the following conditional expectations:
- **a.** $E(X|Y=y)$
- **b.** $E(X|Y,z)$
- **c.** $E(XY|Z=z)$

7.36 For the probability densities of Problems 7.3, 7.4, and 7.8, 7.9 through 7.10, find
- **a.** $E(W|X=x)$
- **b.** $E(W|X=x, Y=y)$
- **c.** $E(W|x,y,z)$
- **d.** $E(WX|Y=y)$
- **e.** $E(WX|y,z)$
- **f.** $E(WXY|Z=z)$

7.37 Find
- **a.** $E(X+Y+Z|X,Y,Z>2)$ for the probability densities of Problems 7.1 and 7.2
- **b.** $E\left(X+Y+Z \Big| X > \frac{7}{2}, Y > \frac{5}{2}, Z > \frac{1}{2}\right)$ for Problem 7.5
- **c.** $E(X+Y+Z|X,Y,Z>2)$ for the probability densities of Problems 7.6 and 7.7

7.38 Find
- **a.** $E(X+Y+Z|W,X,Y,Z>k)$ for the probability densities of Problems 7.3 and 7.4
- **b.** $E(W+X+Y+Z|W,X>3,Y>4,Z>7)$ for the probability density of Problem 7.8

7.39 For Problem 7.10, find

a. $E\left(W+X+Y+Z \mid W > \dfrac{1}{2}, X > \dfrac{5}{4}, Y > \dfrac{3}{2}, Z > \dfrac{9}{4}\right)$

b. $E\left(W+X+Y+Z \mid W, X, Y, Z > \dfrac{1}{2}\right)$

7.40 Let $X_1, X_2, \ldots, X_n$ be a sequence of mutually independent random variables, each of which is normally distributed with a mean of μ and a variance of σ^2. Show that

$$U = \sum_{i=1}^{n} \left(\dfrac{x_i - \mu}{\sigma}\right)^2$$

is chi-square distributed with n degrees of freedom.

References

[1] Anderson, T. W. *An Introduction to Multivariate Statistical Analysis.* New York: John Wiley & Sons, Inc., 1958.

"It is a truth very certain that when it is not in our power to determine what is true we ought to follow what is most probable."
—Descartes

CHAPTER 8

Limit Theorems

OBJECTIVES:

8.1 Chebyshev's Inequality
8.2 Bernoulli's Law of Large Numbers
8.3 Weak and Strong Laws of Large Numbers
8.4 Central Limit Theorem
8.5 DeMoivre-Laplace Theorem
8.6 Normal Approximation to the Poisson Distribution
8.7 Normal Approximation to the Gamma Distribution

Abraham De Moivre (1667–1754)

"The ways in which we think, marshal our evidence and formulate our arguments in every field today is influenced by techniques first applied in science."
—Central Advisory Council for Education

Abraham De Moivre was a French mathematician known his work on the normal distribution and probability theory. is famous for DeMoivre's theorem, which links complex numbers and trigonometry. He fled France and went to England to escape the persecution of Protestants. In England, he wrote a book on probability theory titled *The Doctrine of Chances*. This book was popular among gamblers. The normal distribution was first introduced by De Moivre in an article in 1733 in the context of approximating certain binomial distributions for large *n* and is now called the DeMoivre-Laplace theorem.

■ INTRODUCTION

We begin this chapter with the classical inequality, derived by Pafnuty Lvovich Chebyshev, which justifies the role that the standard deviation plays as a measure of variability. Following that discussion, we expound on and clarify, in Sections 8.2 and 8.3, the relative frequency interpretation of probability that was introduced in Chapter 1. Beginning our study is an introduction to the law of large numbers, determined by Bernoulli; we continue our discussion into more general forms of this law. Specifically, we distinguish between the weak and the strong laws of large numbers and illustrate their meanings with various examples.

The remainder of the chapter is devoted to the central limit theorem, which is one of the most important theorems in the whole of mathematics. In a sequence of mutually independent random variables $(X_1, X_2, \ldots, X_n)$ their sum, for sufficiently large n, is *approximately normally distributed* under certain conditions. To be practical, we confine our study to two versions of the central limit theorem. To conclude, we prove the asymptotic normality of the binomial, Poisson, gamma, and $\chi^2(n)$ probability distributions and illustrate their usefulness in several problems.

8.1 Chebyshev's Inequality

In this section, we discuss a well-known inequality that was derived by the great Russian probabilist Pafnuty Lvovich Chebyshev (Tschebysheff) (1821–1894). Chebyshev's inequality justifies our previous statement that the variance measures variability about the expected value of a random variable.

Theorem 8.1.1 (Chebyshev's inequality)

Let X be a discrete or continuous random variable, with $E(X) = \mu$ and $\text{Var}(X) = \sigma^2$. Then, for any positive number k, we have

$$Pr\left(|X - \mu| \geq k\sigma\right) \leq \frac{1}{k^2}. \tag{8.1.1}$$

An equivalent of the expression in Equation (8.1.1) is

$$Pr\left(|X - \mu| < k\sigma\right) \geq 1 - \frac{1}{k^2}. \tag{8.1.2}$$

Proof Let $f(x)$ be the probability density function of the continuous random variable X. We know that the definition of variance is

$$\text{Var}(X) = \sigma^2 = \int_{-\infty}^{\infty} (x - \mu)^2 f(x) dx. \tag{8.1.3}$$

For some $c > 0$, we can break the range of integration in Equation (8.1.3) into three parts, as shown in Figure 8.1.1. Accordingly,

$$\sigma^2 = \int_{-\infty}^{\mu - \sqrt{c}} (x - \mu)^2 f(x) dx + \int_{\mu - \sqrt{c}}^{\mu + \sqrt{c}} (x - \mu)^2 f(x) dx + \int_{\mu + \sqrt{c}}^{\infty} (x - \mu)^2 f(x) dx$$

$$\geq \int_{-\infty}^{\mu - \sqrt{c}} (x - \mu)^2 f(x) dx + \int_{\mu + \sqrt{c}}^{\infty} (x - \mu)^2 f(x) dx \tag{8.1.3}$$

because the middle term is nonnegative. We replace the factor $(x-\mu)^2$ with c. Note that $(x-\mu)^2 \geq c$ if and only if $x \geq (x-\mu)^2$ or $x \leq \mu - \sqrt{c}$. Thus, the inequality in Equation (8.1.2) still holds and

$$\sigma^2 \geq \int_{-\infty}^{\mu-\sqrt{c}} cf(x)dx + \int_{\mu+\sqrt{c}}^{\infty} cf(x)dx = c\{Pr(X \leq \mu - \sqrt{c}) + Pr(X \geq \mu + \sqrt{c})\}$$

or

$$Pr\{(X-\mu) \leq -\sqrt{c}\} + Pr\{(X-\mu) \geq \sqrt{c}\} \leq \frac{\sigma^2}{c}.$$

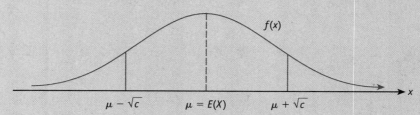

Figure 8.1.1

It follows that

$$Pr(|X-\mu| \geq \sqrt{c}) \leq \frac{\sigma^2}{c}. \quad (8.1.4)$$

If we let $c = k^2\sigma^2$ in Equation (8.1.4), we have the desired result:

$$Pr(|X-\mu| \geq k\sigma) \leq \frac{1}{k^2}.$$

The proof is analogous when the random variable is discrete.

What is important about Chebyshev's inequality is that we can compute upper (or lower) bounds for quantities such as $Pr(|X-\mu| \geq k)$ from our knowledge of σ^2 and μ without knowing the probability density function of the random variable X. If the probability density function of the variate X is known, we can compute $E(X)$ and $Var(X)$ and thus be able to compute the preceding probabilities exactly. However, in many instances involving physical phenomena, the distribution is not known; in such cases, the inequality in Equation (8.1.1) gives us important information about the random variable.

To justify our initial remark that the variance measures (or controls) the spread or dispersion of the distribution of the variate X, with respect to its $E(X)$, we can write the inequality in Equation (8.1.2) as

$$Pr(\mu - k\sigma \leq X \leq \mu + k\sigma) \geq 1 - \frac{1}{k^2}.$$

Thus, the values of X falling in the interval from $\mu - k\sigma$ to $\mu + k\sigma$ account for at least $1 - 1/k^2$ of the total probability. That is, the probability that the variate assumes a value within two standard deviations of its expected value is at least 3/4, the probability that it assumes a value within three standard deviations of μ is at least 8/9, the probability that it assumes a value within four standard deviations of the μ is at least 15/16, and so on. Therefore, if the variance is small, most of the probability distribution of X is concentrated near $E(X)$. Note that $\text{Var}(X) = 0$ (the random variable is *degenerate*) implies that all probability is concentrated at the point $E(X)$. This condition is obvious in the inequality in Equation (8.1.4) for $c > 0$:

$$Pr\left(|X - \mu| \geq \sqrt{c}\right) = 0$$

or

$$Pr\left(|X - \mu| < \sqrt{c}\right) = 1.$$

A generalization of Chebyshev's inequality is the important inequality developed by Kolmogorov.

Theorem 8.1.2 (Kolmogorov's inequality)

Let $X_1, X_2, \ldots, X_n$ be a sequence of independent random variables (discrete or continuous), with $E(X_i^2) < \infty$, $i = 1, 2, \ldots n$, and $Y_r = X_1 + X_2 + \ldots + X_r$, $r \leq n$. Then, for any positive number α, we have

$$Pr\left(\max_{1 \leq i \leq r} |Y_i - E(Y_i)| \geq \alpha\right) \leq \frac{1}{\alpha^2}[\text{Var}(Y_n)], \quad \text{Var}(Y_n) = \text{Var}\left(\sum_{i=1}^{n} X_i\right).$$

When $n = 1$ and $k = \alpha/\sqrt{\text{Var}(Y_n)}$, we obtain Chebyshev's inequality. For the proof of the theorem, we suggest Bernard Harris's book.

8.2 Bernoulli's Law of Large Numbers

In Section 1.1, we defined the relative frequency interpretation of probability as

$$\lim_{n \to \infty} \frac{S_n^*}{n} = p,$$

where n is the total number of trials in a given problem, S_n^* is the number of occurrences of the event S^*, and p is the probability of the event. The limit in the definition suggests that its applicability to various scientific problems where n is almost always finite is not extensive. However, despite this limitation, the frequency interpretation of probability is one of the most popular definitions among scientists. From a practical point of view, the limitation affects the accuracy of measuring p; the accuracy increases as the number of trials becomes very large. This convergence is stated more precisely as the *law of large numbers,* which was first discovered by Bernoulli.

Theorem 8.2.1 (Bernoulli's law of large numbers).

Let S_n^* be the observed number of successes in n independent repeated trials; p is the probability of success, and $1-p$ is the probability of failure of each trial. Let

$$f_n = \frac{S_n^*}{n}$$

be the relative frequency of the observed number of successes in n trials. Then, for every $\epsilon > 0$, we have

$$Pr\left(|f_n - p| < \epsilon\right) \to 1 \text{ as } n \to \infty \qquad (8.2.1)$$

or, equivalently,

$$Pr\left(|f_n - p| \geq \epsilon\right) \to 0 \text{ as } n \to \infty. \qquad (8.2.2)$$

Proof The number of times the event S^* occurs, S_n^*, is a random variable binomially distributed, with $E(S_n^*) = np$ and $\text{Var}(S_n^*) = np(1-p)$. It follows that

$$E(f_n) = E\left(\frac{S_n^*}{n}\right) = \frac{1}{n} E(S_n^*) = p$$

and

$$\text{Var}(f_n) = \text{Var}\left(\frac{S_n^*}{n}\right)$$

$$= \frac{1}{n^2} \text{Var}(S_n^*) = \frac{p(1-p)}{n}.$$

The conditions of Chebyshev's inequality are satisfied. Thus, applying Equation (8.1.1), we have

$$Pr\left(|f_n - p| \geq k\sqrt{\frac{p(1-p)}{n}}\right) \leq \frac{1}{k^2}. \qquad (8.2.3)$$

Choosing

$$\epsilon = k\sqrt{\frac{p(1-p)}{n}},$$

which implies

$$k^2 = \frac{n\epsilon^2}{p(1-p)},$$

continued

the inequality in Equation (8.2.3) can be written as

$$Pr\left(|f_n - p| \geq \epsilon\right) \leq \frac{p(1-p)}{n\epsilon^2}. \qquad (8.2.4)$$

Thus, for any $\epsilon > 0$, no matter how small,

$$Pr\left(|f_n - p| \geq \epsilon\right) \leq \frac{p(1-p)}{n\epsilon^2} \to 0 \text{ as } n \to \infty$$

or

$$Pr\left(|f_n - p| < \epsilon\right) = 1 - Pr\left(|f_n - p| \geq \epsilon\right)$$

$$= 1 - \frac{p(1-p)}{n\epsilon^2} \to 1 \text{ as } n \to \infty.$$

Therefore, the expressions in Equations (8.2.1) and (8.2.2) simply state that as the number of trials n tends to infinity, the relative frequency of successes in the n trials converges to the true value of p, the probability of successes on any individual trial. However, from the practical point of view, n is finite, and we are interested in determining the number of trials needed to place the observed relative frequency within a specified distance ϵ from the true value of p with a high probability. Thus, for a preassigned $\epsilon > 0$, we want to determine n so that

$$Pr\left(|f_n - p| < \epsilon\right) \geq 1 - \alpha, \, 0 < \alpha < 1. \qquad (8.2.5)$$

The value of n that satisfies the preceding inequality is obtained as follows: The inequality in Equation (8.2.4) can be written as

$$Pr\left(|f_n - p| < \epsilon\right) \geq 1 - \frac{p(1-p)}{n\epsilon^2}. \qquad (8.2.6)$$

Because $p(1-p) \leq 1/4$ for all p, $0 \leq p \leq 1$, the inequality in Equation (8.2.6) becomes

$$Pr\left(|f_n - p| < \epsilon\right) \geq 1 - \frac{1}{4n\epsilon^2}. \qquad (8.2.7)$$

Choosing $n \geq 1/4\alpha\epsilon^2$, $0 < \alpha < 1$, the inequality in Equation (8.2.5) is satisfied. Thus, with this choice of n, we can conclude that, with a probability at least $1 - \alpha$, the relative frequency is within ϵ units of the true value of p. The validity of $p(1-p) \leq 1/4$ follows because

$$y = p(1-p), \, 0 \leq p \leq 1$$

attains its maximum when $p = 1/2$, which implies that $y = 1/4$ and thus that $p(1-p) \leq 1/4$.

Example 8.2.1

Suppose that the failure of a certain component follows the binomial probability distribution with an unknown mean value p. That is, $f(x) = p^x(1-p)^{1-x}$ for $x = 0, 1$ and zero elsewhere. How many components must we test so that the sample mean, $\bar{X}$, lies within 0.4 of the true state with a probability at least as great as 0.95?

Solution: Here $\epsilon = 0.4$; using the inequality in Equation (8.2.7), we have

$$Pr\left(|\bar{X} - p| < 0.4\right) \geq 1 - \frac{1}{4n(0.4)^2} = 0.95$$

or

$$n \geq \frac{1}{4(0.05)(0.4)^2} = \frac{1}{0.032} = 31.25.$$

Thus, with a sample of $n \geq 32$ components, we can conclude that

$$Pr\left(-0.4 < (\bar{X} - p) < 0.4\right) \geq 0.95.$$

8.3 Weak and Strong Laws of Large Numbers

In this section, we continue our discussion into more general forms of the law of large numbers.

Let $X_1, X_2, \ldots, X_n$ be a sequence of mutually independent random variables with the common probability density function $E(X_i) = \mu$ and

$$S_n = X_1 + X_2 + \ldots + X_n.$$

Assume that the random variable $X_i = 1$ or 0, with the respective probability p or $1 - p$, depending on whether the ith trial results in a success or a failure. Then, S_n is binomially distributed, with $E(S_n) = np$ and $Var(S_n) = np(1-p)$. Applying the results of the previous section, we can write

$$Pr\left(\left|\frac{S_n}{n} - p\right| < \epsilon\right) \to 1 \text{ as } n \to \infty.$$

That is, for large n, the average portion of successes S_n/n is likely to lie near p with high probability.

The preceding connection of Bernoulli trials and the theory of random variables is a special case of the *weak law of large numbers (WLLN)*.

Theorem 8.3.1 (WLLN)

Let $X_1, X_2, \ldots, X_n$ be a sequence of mutually independent random variables with a common distribution. If $\mu = E(X_i)$ exists, then for every $\epsilon > 0$,

$$Pr\left(\left|\frac{S_n}{n} - \mu\right| < \epsilon\right) \to 1 \text{ as } n \to \infty. \tag{8.3.1}$$

Thus, without any knowledge of the probability distribution function of the variate S_n, the WLLN states that the sample mean $\bar{X} = S_n/n = (1/n)\sum_1^n x_i$ will differ from the population mean by less than an arbitrary constant, $\epsilon > 0$, with a probability that tends to 1 as n tends to ∞. If we assume the existence of the common variance $\sigma^2 = \text{Var}(X_i)$, applying Chebyshev's inequality to the proof of Theorem 8.3.1 is trivial.

The theorem of the WLLN was first proved in 1929 by the Russian mathematician Aleksandr Yakovlevich Khintchin. See William Feller's book for a further discussion. However, here we prove the WLLN for discrete random variables.

Proof For $i = 1, 2, \ldots, n$ and a fixed $\delta > 0$, we define two new sequences of random variables, which are functions of X_i, as follows:

$$Y_i = \begin{cases} X_i, & \text{if } |X_i| \leq \delta n, \\ 0, & \text{if } |X_i| > \delta n, \end{cases} \text{ and } Z_i = \begin{cases} 0, & \text{if } |X_i| \leq \delta n, \\ X_i, & \text{if } |X_i| > \delta n. \end{cases} \tag{8.3.2}$$

Let $p(x_j)$ be the common probability density function of $X_i, i = 1, 2, \ldots, n$. Because $E(X_i) = \mu$ is assumed to exist,

$$\sum_j |x_j| p(x_j) = C < \infty.$$

The expected value of Y_i is

$$\mu_n = E(Y_i) = \sum_{|x_j| \leq \delta_n} x_j p(x_j),$$

where the summation is over those j's for which $|x_j| \leq \delta_n$. Then, $n \to \infty |x_j| \leq \delta_n$ for all j so that $\mu_n \to \mu$. This condition implies that for all $\gamma > 0$ there exists a k_1 such that, for all $n \geq k_1$, the variance of Y is

$$\text{Var}(Y_i) = E(Y_i^2) - [E(Y_i)]^2 \leq E(Y_i^2)$$

$$= \sum_{|x_j| \leq \delta_n} x_j^2 p(x_j)$$

$$= \sum_{|x_j| \leq \delta_n} |x_j||x_j| p(x_j) \leq \sum_{|x_j| \leq \delta_n} n|x_j| p(x_j).$$

But

$$n\delta \sum_{|x_j| \leq \delta_n} |x_j| p(x_j) \leq n\delta \sum_{\text{all } j} |x_j| p(x_j) = n\delta C.$$

Thus,

$$\text{Var}(Y_i) \leq n\delta C.$$

The Y_is are mutually independent random variables because the X_is were defined as such. Therefore,

$$\text{Var}(Y_1, Y_2, \ldots, Y_n) = \sum_{i=1}^{n} \text{Var}(Y_i) \leq \sum_{i=1}^{n} n\delta C$$

$$= n^2 \delta C < \infty.$$

Because the variance exists and is finite, we apply Chebyshev's inequality and have

$$Pr\left(\left|\frac{Y_1 + Y_2 + \ldots + Y_n}{n} - \mu_n\right| \geq \gamma\right)$$

$$\leq \frac{\text{Var}\left(\frac{Y_1 + Y_2 + \ldots + Y_n}{n}\right)}{\gamma^2} \leq \frac{n^2 \delta C}{n^2 \gamma^2} = \frac{\delta C}{\gamma^2}. \tag{8.3.4}$$

The inequality in Equation (8.3.4) can be written as

$$Pr\left(\left|\left(\frac{Y_1 + Y_2 + \ldots + Y_n}{n} - \mu\right) + (\mu - \mu_n)\right| < \gamma\right) \geq 1 - \frac{\delta C}{\gamma^2}. \tag{8.3.5}$$

Because $|a+b| \leq |a| + |b|$, $|a-b| \geq |a| - |b|$, and $|\mu_n - \mu| < \gamma$ for $n \geq k_1$, we can write

$$\gamma > \left|\frac{Y_1 + Y_2 + \ldots + Y_n}{n} - \mu - (\mu_n - \mu)\right|$$

$$\geq \left|\frac{Y_1 + Y_2 + \ldots + Y_n}{n} - \mu\right| - |\mu - \mu_n|$$

$$> \left|\frac{Y_1 + Y_2 + \ldots + Y_n}{n} - \mu\right| - \gamma.$$

Then,

$$2\gamma > \left|\frac{Y_1 + Y_2 + \ldots + Y_n}{n} - \mu\right|$$

continued

if
$$\gamma < \left| \frac{Y_1 + Y_2 + \ldots + Y_n}{n} - \mu_n \right|,$$

and the inequality in Equation (8.3.5) becomes

$$Pr\left(\left| \frac{Y_1 + Y_2 + \ldots + Y_n}{n} - \mu \right| < 2\gamma \right)$$

$$= Pr\left(\left| \frac{Y_1 + Y_2 + \ldots + Y_n}{n} - \mu_n \right| < \gamma \right) \geq 1 - \frac{\delta C}{\gamma^2}$$

or

$$Pr\left(\left| \frac{Y_1 + Y_2 + \ldots + Y_n}{n} - \mu \right| \geq 2\gamma \right) \leq \frac{\delta C}{\gamma^2}. \tag{8.3.6}$$

Also, $Z_i = 0$ if $Y_i = X_i$, and

$$Pr(Z_i \neq 0) = Pr(Y_i = 0) = Pr(|X_i| > \delta n)$$

$$= \sum_{|x_j| > \delta_n} p(x_j)$$

$$= \sum_{|x_j| > \delta_n} \left| \frac{x_j}{x_j} \right| p(x_j)$$

$$= \frac{1}{\delta n} \sum_{|x_j| \leq \delta_n} |x_j| p(x_j) \to 0 \text{ as } n \to \infty.$$

Hence, for an arbitrary $\delta_1 > 0$, there exists a k_2 such that, for $n \geq k_2$, $Pr(Z_i \neq 0) \leq \delta_1$. Thus,

$$Pr(Z_1 + Z_2 + \ldots + Z_n \neq 0)$$

$$\leq Pr(Z_1 \neq 0 \text{ or } Z_2 \neq 0 \text{ or } \ldots \text{ or } Z_n \neq 0)$$

$$\leq \sum_{i=1}^{n} Pr(Z_i \neq 0)$$

$$\leq \sum_{i=1}^{n} \delta_1 = n\delta_1.$$

However,
$$S_n = X_1 + X_2 + \ldots + X_n$$
$$= (Y_1 + Z_1) + (Y_2 + Z_2) + \ldots + (Y_n + Z_n)$$
$$= (Y_1 + Y_2 + \ldots + Y_n) + (Z_1 + Z_2 + \ldots + Z_n)$$

and

$$Pr\left(\left|\frac{S_n}{n} - \mu\right| \geq 2\gamma\right)$$
$$= Pr\left(\left|\frac{Y_1 + Y_2 + \ldots + Y_n}{n} + \frac{Z_1 + Z_2 + \ldots + Z_n}{n} - \mu\right| \geq 2\gamma\right)$$
$$\leq Pr\left(\left|\frac{Y_1 + Y_2 + \ldots + Y_n}{n} - \mu\right| \geq 2\gamma\right)$$
$$+ Pr\left(\left|\frac{Z_1 + Z_2 + \ldots + Z_n}{n}\right| \neq 0\right)$$
$$= Pr\left(\left|\frac{Y_1 + Y_2 + \ldots + Y_n}{n} - \mu\right| \geq 2\gamma\right)$$
$$+ Pr(Z_1 + Z_2 + \ldots + Z_n \neq 0).$$

For $k = \max(k_1, k_2)$, $n > k$, the preceding inequality becomes

$$Pr\left(\left|\frac{S_n}{n} - \mu\right| > 2\gamma\right) \leq \frac{\delta C}{\gamma^2} + n\delta_1 = \gamma_1. \tag{8.3.7}$$

Because δ, δ_1, and γ are arbitrary, the right-hand side of Equation (8.3.7) can be made as small as we wish. Therefore,

$$Pr\left(\left|\frac{S_n}{n} - \mu\right| \geq 2\gamma\right) \to 0 \text{ as } n \to \infty.$$

The assumption of a common probability distribution for the X_is in the preceding discussion can be eliminated. Let the sequence $X_1, X_2, \ldots, X_n$ of random variables be mutually independent but not necessarily identically distributed, with a finite mean of $\mu_i = E(X_i)$ and a variance of $\sigma_i^2 = \text{Var}(X_i)$. If

$$m_n = E(S_n) = E(X_1 + X_2 + \ldots + X_n)$$
$$= \mu_1 + \mu_2 + \ldots + \mu_n$$

and

$$s_n^2 = \text{Var}(S_n) = \text{Var}(X_1 + X_2 + \ldots + X_n)$$
$$= \text{Var}(X_1) + \text{Var}(X_2) + \ldots + \text{Var}(X_n)$$
$$= \sigma_1^2 + \sigma_2^2 + \ldots + \sigma_n^2,$$

then the WLLN holds for the sequence of variates $X_1, X_2, \ldots, X_n$. That is,

$$Pr\left(\left|\frac{S_n - m_n}{n}\right| < \epsilon\right) \to 1 \text{ as } n \to \infty.$$

> **Theorem 8.3.2**
>
> A sufficient condition for the sequence of random variables $X_1, X_2, \ldots, X_n$ to satisfy the WLLN is
>
> $$\frac{s_n}{n} \to 0 \text{ as } n \to \infty.$$
>
> **Proof** Here,
>
> $$E\left(\frac{S_n}{n}\right) = \frac{m_n}{n},$$
>
> and
>
> $$\text{Var}\left(\frac{S_n}{n}\right) = \frac{1}{n^2}\text{Var}(S_n) = \frac{s_n^2}{n^2}.$$
>
> Applying Chebyshev's inequality to S_n/n, we have
>
> $$Pr\left(\left|\frac{S_n}{n} - \frac{m_n}{n}\right| \geq k\frac{s_n}{n}\right) \leq \frac{1}{k^2}$$
>
> or
>
> $$Pr\left(\left|\frac{S_n}{n} - \frac{m_n}{n}\right| \geq \epsilon\right) \leq \frac{s_n^2}{n^2\epsilon^2}. \qquad (8.3.8)$$
>
> If $s_n/n \to 0$ as $n \to \infty$, then, for $\epsilon > 0$, $s_n^2/n^2\epsilon^2 \to 0$ as $n \to \infty$, and we have
>
> $$Pr\left(\left|\frac{S_n - m_n}{n}\right| < e\right) \to 1 \text{ as } n \to \infty.$$

Example 8.3.1

Let $X_1, X_2, \ldots, X_n$ represent the sales of a particular product per operating day. Assume that the random variables are independent and, for various seasonal variations, they have different probability distributions. Past experience suggests that a good approximation for the expected sales m_n is $2000; a good approximation for the variance s_n^2 is $1600. If we assume that $n = 320$ and $\epsilon = 10$, we have, using the inequality in Equation (8.3.8):

$$Pr\left(\left|\frac{S_n}{n} - 2000\right| \geq 10\right) \leq \frac{1600}{(10)^2 \, 320^2}$$

or

$$Pr\left(1900 < \frac{S_n}{n} < 2010\right) \geq 0.95.$$

Example 8.3.2 (Feller)

Let $X_1, X_2, \ldots, X_n$ be a sequence of mutually independent random variables, with probability distributions defined as follows:

a. $Pr\left(X_j = \sqrt{j}\right) = \frac{1}{2}$ and $Pr\left(X_j = -\sqrt{j}\right) = \frac{1}{2}$

b. $Pr\left(X_j = 2^j\right) = 2^{-(2j+1)}$, $Pr\left(X_j = -2^j\right) = 2^{-(2j+1)}$, and $Pr\left(X_j = 0\right) = 1 - 2^{-2j}$

c. $Pr\left(X_j = 2^j\right) = \frac{1}{2}$ and $Pr\left(X_j = -2^j\right) = \frac{1}{2}$

We can show that the sequence of variates with the probability distribution given by Example 8.3.2(b) satisfies the WLLN, but the sequence with the probability distributions given by Example 8.3.2(a) and (c) does not satisfy the condition for the WLLN.

Solution:

a. $E(X_j) = \frac{1}{2}\sqrt{j} + \frac{1}{2}\left(-\sqrt{j}\right) = 0$

$\operatorname{Var}(X_j) = E(X_j^2) - \left[E(X_j)\right]^2$

$= \frac{1}{2}\left(\sqrt{j}\right)^2 + \frac{1}{2}\left(-\sqrt{j}\right)^2$

$= j$

Thus,

$$s^2 = \sum_{j=1}^{n} j = \frac{n(n+1)}{2}$$

or

$$\frac{s_n^2}{n^2} = \frac{(n+1)}{2n} \to \frac{1}{2} \text{ as } n \to \infty,$$

and the sufficient condition, stated in Theorem 8.3.2, for the WLLN is not satisfied.

b. $E(X_j) = 2^{-(2j+1)}(2^j) + 2^{-(2j+1)}(-2^j) = 0$

$\text{Var}(X_j) = E(X_j^2) = 2^{-(2j+1)}(2^j)^2 + 2^{-(2j+1)}(2^j)^2 = 1.$

Thus,

$$s_n^2 = \sum_{j=1}^{n} \text{Var}(X_j) = n.$$

Therefore,

$$\frac{s_n^2}{n^2} = \frac{1}{n} \to 0 \text{ as } n \to \infty,$$

which implies that the sequence of variates obeys the WLLN.

c. $E(X_j) = \frac{1}{2}(2^j) + \frac{1}{2}(-2^j) = 0$

$\text{Var}(X_j) = E(X_j^2) = \frac{1}{2}(2^j)^2 + \frac{1}{2}(-2^j)^2 = 2^{2j}.$

Now

$$s_n^2 = \sum_{j=1}^{n} \text{Var}(X_j) = 2^2 + 2^4 + \ldots + 2^{2n}$$

$$= \frac{4(4^n - 1)}{3}$$

or

$$\frac{s_n^2}{n^2} = \frac{4(4^n - 1)}{3n^2},$$

which diverges as $n \to \infty$; again, the condition for the WLLN is not satisfied. ◀

Example 8.3.3

Let $X_1, X_2, \ldots, X_n$ be a sequence of mutually independent random variables, with probability distributions given by

a. $\Pr(X_i = i) = \frac{1}{2}$ and $\Pr(X_i = -i) = \frac{1}{2}$

b. $\Pr\left(X_i = \sqrt{\ln(i+\alpha)}\right) = \frac{1}{2}$ and $\Pr\left(X_i = -\sqrt{\ln(i+\alpha)}\right) = \frac{1}{2}$, α being a positive integer

The sequence of variates $X_1, X_2, \ldots, X_n$, the probability distribution of which is given by Example 8.3.3(a), does not satisfy the condition for the WLLN:

$$E(X_i) = 0, \quad \text{Var}(X_i) = \frac{1}{2}i^2 + \frac{1}{2}i^2 = i^2,$$

and

$$\frac{s_n^2}{n^2} = \frac{\text{Var}(S_n)}{n^2} = \frac{n(n+1)(2n+1)}{6n^2},$$

which diverges as $n \to \infty$. However, in Example 8.3.3(b), we have

$$E(X_i) = \frac{1}{2}\sqrt{\ln(i+\alpha)} - \frac{1}{2}\sqrt{\ln(i+\alpha)} = 0$$

and

$$\text{Var}(X_i) = \frac{1}{2}\ln(i+\alpha) + \frac{1}{2}\ln(i+\alpha) = \ln(i+\alpha).$$

Also,

$$\text{Var}(S_n) = \sum_{i=1}^{n} \text{Var}(X_i) = \sum_{i=1}^{n} \ln(i+\alpha) = \ln \Pi_{i=1}^{n}(i+\alpha)$$

$$= \ln\left[(1+\alpha)(2+\alpha)\ldots(n+\alpha)\right]$$

$$= \ln \frac{(n+\alpha)!}{\alpha!}$$

$$= \ln(n+\alpha)! - \ln \alpha!.$$

Applying Stirling's formula, we can write $\text{Var}(S_n)$ as

$$\text{Var}(S_n) \approx \ln\left[\sqrt{2\pi}\, e^{-(n+\alpha)} (n+\alpha)^{n+\alpha+1/2}\right] - \ln \alpha!$$

or

$$\frac{\text{Var}(S_n)}{n^2} \approx \frac{\ln \sqrt{2\pi}}{n^2} - \frac{(n+\alpha)}{n^2} - \frac{\ln \alpha!}{n^2} + \frac{(n+\alpha+1/2)\ln(n+\alpha)}{n^2}.$$

Thus, $s_n/n \to 0$ as $n \to \infty$, and the sequence of variates $X_1, X_2, \ldots, X_n$ obeys the WLLN.

An equivalent way of stating the WLLN from Theorem 8.3.1 is

$$\lim_{n \to \infty} Pr\left(\left|\frac{S_n}{n} - \mu\right| < \epsilon\right) = 1$$

for $\epsilon > 0$, which means that s_n/n converges to μ in *probability* or *stochastically* as n increases indefinitely.

There are instances in which the WLLN does not convey any information concerning the probability of convergence of s_n/n to μ. For example, the inequality in Equation (8.2.7) can be written as

$$Pr\left(\left|\frac{S_n^*}{n}-p\right|\geq\epsilon\right)\leq\frac{1}{4\epsilon^2 n},$$

which tends to zero as $n\to\infty$. However, we can choose an r such that

$$Pr\left(\left|\frac{S_r^*}{r}-p\right|\geq\epsilon+\left|\frac{S_{r+1}^*}{r+1}-p\right|\geq\epsilon+\ldots+\right)$$

$$\leq Pr\left(\left|\frac{S_r^*}{r}-p\right|\geq\epsilon\right)+Pr\left(\left|\frac{S_r^*}{r}-p\right|\geq\epsilon\right)+\ldots+$$

$$\leq\frac{1}{4\epsilon^2 r}+\frac{1}{4\epsilon^2(r+1)}+\ldots+$$

$$=\frac{1}{4\epsilon^2}\left(\frac{1}{r}+\frac{1}{r+1}+\ldots+\right),$$

which diverges and tells nothing about the probability. To overcome this difficulty, we have a stronger form of the WLLN, which is called the *strong law of large numbers (SLLN)* and is defined as follows:

Definition 8.3.1

The sequence $X_1, X_2, \ldots, X_m, \ldots, X_n$ of mutually independent and identically distributed random variables with finite second moments, $E(X_i^2)<\infty$, is said to obey the SLLN if, for $\epsilon>0$, $E(X_i)=\mu$, $\text{Var}(X_i)=\sigma^2$, and $S_r = X_1+X_2+\ldots+X_r$, we have

$$Pr\left(\left|\frac{S_r}{r}-\mu\right|\geq\epsilon\right)\leq\frac{2\sigma^2}{\epsilon^2 m} \qquad (8.3.9)$$

for at least one r, $m\leq r\leq n$.

The bound of the SLLN is twice that of the bound we can obtain using the WLLN. Also, the inequality in Equation (8.3.9) is independent of n, even though we think of it as a very large number. The condition in which the random variables have a common probability density function can be eliminated, and the SLLN still holds.

An equivalent form of the SLLN is

$$Pr\left(\lim_{n\to\infty} S_n/n = \mu\right) = 1, \qquad (8.3.10)$$

which states that, in almost every infinite sequence of variates, $S_1, S_2, \ldots, S_r, \ldots$, where $S_n = X_1+X_2+\ldots+X_n$, the sequence tends to the limit μ. More precisely, the sequence S_n/n converges to μ, with a probability of 1, if Equation (8.3.10) holds.

A sufficient condition for the SLLN to apply to the preceding sequence of mutually independent random variables that are not necessarily identically distributed is given by the *Kolmogorov criterion*.

Theorem 8.3.3 (Kolmogorov criterion)

The convergence of the series

$$\sum_{r=1}^{\infty} \frac{\sigma_r^2}{r^2}$$

is a sufficient condition for the SLLN to apply to a sequence of mutually independent random variables, with $\text{Var}(X_r) = \sigma_r^2$.

The SLLN for Bernoulli trials can be stated as follows: Let $X_1, X_2, \ldots, X_m, \ldots, X_n$ be n Bernoulli trials, with a probability p of success on each trial. If $S_r = X_1 + X_2 + \ldots + X_r$ and $\epsilon > 0$, we have

$$Pr\left(\left|\frac{S_r}{r} - p\right| < \epsilon\right) \geq 1 - \frac{2p(1-p)}{\epsilon^2 m} \quad (8.3.11)$$

whenever $m \leq r \leq n$. Because $p(1-p) \leq 1/4$, the inequality in Equation (8.3.11) becomes

$$Pr\left(\left|\frac{S_r}{r} - p\right| < \epsilon\right) \geq 1 - \frac{1}{2\epsilon^2 m}.$$

Example 8.3.4

Let $X_1, X_2, \ldots, X_n$ be a sequence of mutually independent random variables whose probability distribution is defined by

$$Pr(X_n = 0) = 1 - \frac{1}{n}$$

and

$$Pr(X_n = 1) = \frac{1}{n}.$$

Here,

$$E(X_r) = \frac{1}{r}$$

and

$$\text{Var}(X_r) = \frac{r-1}{r^2}.$$

Thus, the series

$$\sum_{r=1}^{\infty} \frac{\text{Var}(X_r)}{r^2} = \sum_{r=1}^{\infty} \frac{r-1}{r^4}$$

converges, and the preceding sequence of random variables obeys the SLLN.

8.4 Central Limit Theorem

In Section 4.3, we mentioned that few random phenomena precisely obey a normal probability law; however, the laws that they do follow can often, under certain conditions, be closely approximated by the normal probability law. In this sense, the normal distribution is of paramount importance. In particular, if a random variable X is represented by the sum of n independent identically distributed random variables $X_1, X_2, \ldots, X_n$,

$$S_n = X_1 + X_2 + \ldots + X_n,$$

then S_n, for sufficiently large n, is *approximately normally distributed,* provided that the random variables possess finite means and variances. This remarkable result is known as the (classical) *central limit theorem.* The central limit theorem is one of the most important theorems in the whole of mathematics; the efforts that scientists have expended on it, especially in the area of statistics, are certainly justified.

There are two main versions of the central limit theorem that you should be familiar with at this level of mathematical maturity: one is when the random variables in the sequence $X_1, X_2, \ldots, X_n$ are independent and identically distributed; the other is when the random variables in the sequence $X_1, X_2, \ldots, X_n, \ldots$ are independent but not identically distributed. We include some additional conditions in the theorem; however, from the practical point of view, the preceding distinction is significant. The original formulation of the central limit theorem was the work, in the early nineteenth century, of Laplace and Gauss. However, no formal presentation was made until 1901, when Aleksandr Mikhailovich Lyapunov gave a rigorous mathematical proof of the theorem.

There are a number of different versions and extensions of the central limit theorem, but as we indicated earlier, we restrict ourselves to the versions represented by Theorems 8.4.1 and 8.4.2.

Theorem 8.4.1

Let $X_1, X_2, \ldots, X_n$ be a sequence of n independent and identically distributed random variables. Let $\mu = E(X_i)$ and $\sigma^2 = \text{Var}(X_i)$ be the common expectation and variance. Let

$$S_n = X_1 + X_2 + \ldots + X_n.$$

Then, $E(S_n) = n\mu$ and $\text{Var}(S_n) = n\sigma^2$. Let the random variable Z_n be defined by

$$Z_n = \frac{S_n - n\mu}{\sqrt{n}\,\sigma}.$$

The variate Z_n is approximately normally distributed, with a of mean 0 and a variance of 1 as $n \to \infty$. That is,

$$\lim_{n \to \infty} Pr(Z_n \leq x) = \frac{1}{\sqrt{2\pi}} \int_{-\infty}^{z} e^{-(1/2)t^2} dt = F(z),$$

or equivalently, the distribution of the random variable $\bar{X}$ may be approximated as

$$\bar{X} : N\left(\mu, \frac{\sigma^2}{n}\right) \text{ as } n \to \infty.$$

The rigorous proof of this theorem requires advanced mathematical techniques and is beyond the scope of this book.

However, an astonishing concept related to the central limit theorem is that nothing is said about the common distribution of the random variables except that they have finite variance. The distribution of Z_n, $f(z)$, may be approximated by the $N(0,1)$ as $n \to \infty$. From an applied point of view, the assumption of finite variance is not a critical restriction, because in almost all practical situations, the range of the variate will be finite, which makes the variance finite.

The central limit theorem also justifies a result important for its practical applications: by using the theorem, we can approximate the probability distribution of the sample mean,

$$\bar{X}_n = \frac{1}{n}\sum_{i=1}^{n} X_i,$$

with a normal distribution having a mean of μ and a variance of σ^2/n when n is sufficiently large. It is not correct to state that the distribution of $\bar{X}_n$ approaches that of a normal distribution as $n \to \infty$, because under the condition of the central limit theorem, the variance of $\bar{X}_n$ approaches 0 as $n \to \infty$. Therefore, if

$$\bar{X}_n = \frac{1}{n}\sum_{i=1}^{n} X_i,$$

then the variate Y_n,

$$Y_n = \frac{\bar{X}_n - E(\bar{X}_n)}{\sqrt{\operatorname{Var}\bar{X}_n}}, \tag{8.4.1}$$

is approximately normally distributed with a mean of 0 and a variance of 1 when n is sufficiently large. Here, $E(\bar{X}_n) = \mu$ and $\operatorname{Var}(\bar{X}_n) = \sigma^2/n$.

Example 8.4.1

Let $X_1, X_2, \ldots, X_n$ be independent discrete random variables identically distributed as

$$f(x_i) = \begin{cases} 0.2, & x_i = 0, 1, \ldots, 4, \\ 0, & \text{elsewhere for } i = 1, 2, \ldots, n. \end{cases}$$

We have

$$\mu = E(X_n) = \sum_{r=0}^{4} r = \frac{1}{5}(0+1+2+3+4) = 2.0$$

and

$$\sigma^2 = \operatorname{Var}(X_n) = \frac{1}{5}\sum_{r=0}^{4}(r-\mu)^2 = \frac{1}{5}\sum_{r=0}^{4} r^2 - \mu^2$$

$$= \frac{1}{5}(30) - 4 = 6 - 4 = 2.0.$$

Let the random variable $\bar{X}_{100}$ be defined by

$$\bar{X}_{100} = \frac{1}{100}\sum_{i=1}^{100} X_i.$$

What is the probability that $\bar{X}_{100}$ will exceed 2?

Solution: Applying Equation (8.4.1), we have

$$Pr(\bar{X}_{100} > 2) = 1 - Pr(\bar{X}_{100} \leq 2)$$

$$= 1 - Pr\left(\frac{\bar{X}_{100} - 2.0}{1.414/10} \leq \frac{2.0 - 2.0}{1.414/10}\right) = 1 - Pr(Z \leq 0),$$

where $Z: N(0,1)$. Thus, we get

$$Pr(\bar{X}_{100} > 2) \approx 1 - \frac{2}{\sqrt{2\pi}}\int_{-\infty}^{0} e^{-z^2/2}\, dz$$

$$= 1 - F(0)$$

$$= 0.5.$$

◀

Theorem 8.4.2

Let $X_1, X_2, \ldots, X_n$ be a sequence of n independent random variables with a finite mean of $E(X_i)$; for some $\beta > 0$, let

$$E\left[\left|X_r - E(X_r)\right|\right]^{2+\beta}$$

be finite. If

$$S_n = X_1 + X_2 + \ldots + X_n,$$

then

$$Z_n = \frac{S_n - n\mu}{\sqrt{n}\sigma}$$

may be approximately distributed as an $N(0,1)$ variate as $n \to \infty$, provided that

$$\lim_{n \to \infty}\left[\mathrm{Var}(S_n)\right]^{-(1+\beta/2)} \sum_{r=1}^{n} E\left[\left|X_r - E(X_r)\right|^{2+\beta}\right] = 0.$$

Theorem 8.4.2 is due to Lyapunov, and we will not prove it. It is important because the random variables need not be identically distributed.

Example 8.4.2

Let $X_1, X_2, \ldots, X_n$ be a sequence of independent random variables identically distributed as

$$f(x_i) = \begin{cases} \dfrac{1}{\alpha - \beta}, & \alpha \leq x_i \leq \beta, \\ 0, & \text{elsewhere} \end{cases} \text{ for } i = 1, 2, \ldots, n.$$

Then,

$$E(X_i) = \frac{\alpha + \beta}{2}$$

and

$$\operatorname{Var}(X_i) = \frac{(\beta - \alpha)^2}{12}, \quad i = 1, 2, \ldots, n.$$

Let $Y_n = X_1 + X + \ldots + X_n$. Applying the central limit theorem, we obtain

$$\Pr(y_1 \leq Y_n \leq y_2)$$

$$= \left[\frac{y_1 - n\left(\frac{\alpha+\beta}{2}\right)}{\sqrt{n}\,\frac{\beta-\alpha}{2\sqrt{3}}} \leq \frac{Y_n - n\left(\frac{\alpha+\beta}{2}\right)}{\sqrt{n}\,\frac{\beta-\alpha}{2\sqrt{3}}} \leq \frac{y_2 - n\left(\frac{\alpha+\beta}{2}\right)}{\sqrt{n}\,\frac{\beta-\alpha}{2\sqrt{3}}} \right]$$

$$\approx \frac{1}{\sqrt{2\pi}} \int_a^b e^{-(1/2)y_n^2} \, dy_n$$

$$= F(b) - F(a),$$

where

$$a = \frac{y_1 - n\left(\frac{\alpha+\beta}{2}\right)}{\sqrt{n}\,\frac{\beta-\alpha}{2\sqrt{3}}}, \quad b = \frac{y_2 - n\left(\frac{\alpha+\beta}{2}\right)}{\sqrt{n}\,\frac{\beta-\alpha}{2\sqrt{3}}}, \quad \alpha < y_1 < y_2 < \beta.$$

◀

8.5 DeMoivre-Laplace Theorem

Let $X_1, X_2, \ldots, X_n$ be a sequence of independent random variables. Assume that the random variable X_i equals 1 if the ith trial results in a success, with the probability p, and 0 otherwise, with the probability $1 - p$; that is, for every $i, i = 1, 2, \ldots, n$,

$$\Pr(X_i = 1) = p$$

and

$$Pr(X_i=0)=1-p, 0<p<1.$$

Then, the sum $Y_n = X_1 + X + \ldots + X_n$ is binomially distributed with the mean np and the variance $np(1-p)$. Applying Theorem 8.4.1, the variate Z_n, defined by

$$Z_n = \frac{Y_n - np}{\sqrt{np(1-p)}},$$

is asymptotically normally distributed, with a mean of 0 and a variance of 1. Thus, for n that is sufficiently large, we can approximate the binomial probability distribution with the normal probability law. This special case of the central limit theorem is known as the *DeMoivre-Laplace theorem,* which was first formulated in 1733. This theorem is of considerable theoretical and practical importance. We state it and prove it, using an approach different from that in our introductory discussion.

Theorem 8.5.1

Let $X_1, X_2, \ldots, X_n$ be a sequence of independent random variables with the distribution

$$f(x_i) = \begin{cases} p, & x_i = 1, \\ 1-p, & x_i = 0, \end{cases} \text{ for } i = 1, 2, \ldots, n,$$

so that $Y_n = X_1 + X_2 + \ldots + X_n$ is binomially distributed with the parameters n and p. Then, for sufficiently large n,

$$\binom{n}{y_n} p^{y_n}(1-p)^{n-y_n} \approx \frac{1}{\sqrt{2\pi}\sqrt{np(1-p)}} \exp\left\{-\frac{1}{2}\left(\frac{y_n - np}{\sqrt{np(1-p)}}\right)^2\right\}. \quad (8.5.1)$$

Proof Recall that Stirling's approximation is given by

$$n! \approx n^{n+1/2} e^{-n} \sqrt{2\pi}. \quad (8.5.2)$$

We can then write $\binom{n}{y_n}$ of the expression in Equation (8.5.1) as follows:

$$\binom{n}{y_n} = \frac{n!}{y_n!(n-y_n)!}$$

$$\approx \frac{n^{n+1/2} e^{-n} \sqrt{2\pi}}{y_n^{y_n+1/2} e^{-y_n} \sqrt{2\pi} (n-y_n)^{n-y_n+1/2} e^{-(n-y_n)} \sqrt{2\pi}}$$

$$\approx \frac{1}{\sqrt{2\pi}} \frac{n^{n+1/2} e^{-n}}{y_n^{y_n+1/2} (n-y_n)^{n-y_n+1/2} e^{-(y_n-n+y_n)}}. \quad (8.5.3)$$

8.5 DeMoivre-Laplace Theorem

Multiplying the denominator of Equation (8.5.3) by $n^{1/2}$ and the numerator by $n^{1/2+y_n-y_n}$, we have

$$\binom{n}{y_n} \approx \frac{1}{\sqrt{2\pi}} \frac{n^{y_n+1/2} n^{n-y_n+1/2}}{n^{1/2} y_n^{y_n+1/2} (n-y_n)^{n-y_n+1/2}}. \tag{8.5.4}$$

Equation (8.5.4) becomes

$$\binom{n}{y_n} \approx \frac{1}{\sqrt{2\pi}\sqrt{n}} \left(\frac{n}{y_n}\right)^{y_n+1/2} \left(\frac{n}{n-y_n}\right)^{n-y_n+1/2}. \tag{8.5.5}$$

The quantity $p^{y_n}(1-p)^{n-y_n}$ in the expression in Equation (8.5.1) can be written as

$$p^{y_n}(1-p)^{n-y_n} = \frac{p^{y_n+1/2}(1-p)^{n-y_n+1/2}}{p^{1/2}(1-p)^{1/2}}. \tag{8.5.6}$$

Substituting Equations (8.5.5) and (8.5.6) into the expression in Equation (8.5.1), we have

$$\binom{n}{y_n} p^{y_n}(1-p)^{n-y_n}$$

$$\approx \frac{1}{\sqrt{2\pi}\sqrt{np(1-p)}} \frac{p^{y_n+1/2}}{\left(\frac{y_n}{n}\right)^{y_n+1/2}} \frac{(1-p)^{n-y_n+1/2}}{\left[\frac{(n-y_n)}{n}\right]^{n-y_n+1/2}}$$

$$= \frac{1}{\sqrt{2\pi}\sqrt{np(1-p)}} \left[\left(\frac{y_n}{np}\right)^{y_n+1/2}\right]^{-1} \left\{\left[\frac{(n-y_n)}{n(1-p)}\right]^{n-y_n+1/2}\right\}^{-1}. \tag{8.5.7}$$

Letting

$$f(y_n; p, n) = \binom{n}{y_n} p^{y_n}(1-p)^{n-y_n}$$

and taking the ln of both sides of Equation (8.5.7) results in

$$\ln f(y_n; p, n)$$

$$\approx \ln\left[\frac{1}{\sqrt{2\pi}\sqrt{np(1-p)}}\right] - \ln\left(\frac{y_n}{n}\right)^{y_n+1/2} - \ln\left[\frac{(n-y_n)}{n(1-p)}\right]^{n-y_n+1/2}$$

$$= -\ln\left(\sqrt{2\pi}\sqrt{np(1-p)}\right) - \left(y_n + \frac{1}{2}\right)\ln\left(\frac{y_n}{np}\right)$$

$$- \left(n - y_n + \frac{1}{2}\right)\ln\left[\frac{(n-y_n)}{n(1-p)}\right]. \tag{8.5.8}$$

Let

$$t = \frac{y_n - np}{\sqrt{np(1-p)}},$$

which implies that

$$\frac{y_n}{np} = 1 + t\frac{\sqrt{1-p}}{\sqrt{np}}$$

and

$$\frac{n - y_n}{n - (1-p)} = 1 - t\frac{\sqrt{p}}{\sqrt{n(1-p)}}.$$

Thus, Equation (8.5.8) becomes

$$\ln f(y_n; p, n)$$

$$\approx -\ln\left[\sqrt{2\pi}\sqrt{np(1-p)}\right] - \left[np + t\sqrt{np(1-p)} + \frac{1}{2}\right]$$

$$\approx \ln\left[1 + t\frac{\sqrt{1-p}}{\sqrt{np}}\right] - \left[n + \frac{1}{2}\left(np + t\sqrt{np(1-p)}\right)\right]$$

$$\approx \ln\left[1 - t\frac{\sqrt{p}}{\sqrt{n(1-p)}}\right]. \tag{8.5.9}$$

We use the expansion

$$\ln(1+z) = z - \frac{z^2}{2} + \frac{z^3}{3} - \cdots,$$

which is valid for $|z| < 1$. Such a power series is permissible to use because $\left(t\sqrt{1-p}\right)/\sqrt{np} \to 0$ for large values of n. Also, the manner in which t is defined suggests that large values occur with very small probabilities. Thus, two of the terms in Equation (8.5.9) can be expanded:

$$\ln\left[1 + t\sqrt{\frac{1-p}{np}}\right]$$

$$\approx t\sqrt{\frac{1-p}{np}} - \frac{t^2(1-p)}{2np} + \frac{t^3(1-p)\sqrt{1-p}}{3np\sqrt{np}} - \cdots. \tag{8.5.10}$$

and

$$\ln\left[1-t\sqrt{\frac{1-p}{n(1-p)}}\right]$$

$$=-t\sqrt{\frac{p}{n(1-p)}}-\frac{t^2 p}{2n(1-p)}-\frac{t^3 p\sqrt{p}}{3n(1-p)\sqrt{n(1-p)}}-\cdots. \qquad (8.5.11)$$

Substituting Equations (8.5.10) and (8.5.11) into Equation (8.5.9), we have

$$\ln f(y_n;n,p)$$

$$\approx -\ln\left[\sqrt{2\pi}\sqrt{np(1-p)}\right]-\left[np+t\sqrt{np(1-p)}+\frac{1}{2}\right]$$

$$\times\left[t\frac{\sqrt{1-p}}{\sqrt{np}}-\frac{t^2(1-p)}{2np}+\frac{t^3(1-p)\sqrt{1-p}}{3np\sqrt{np}}-\cdots\right]$$

$$-\left[n(1-p)-t\sqrt{np(1-p)}+\frac{1}{2}\right]\left[-t\sqrt{\frac{p}{n(1-p)}}-\frac{t^2 p}{2n(1-p)}\right.$$

$$\left.-\frac{t^3 p\sqrt{p}}{3n(1-p)\sqrt{n(1-p)}}-\cdots\right]. \qquad (8.5.12)$$

Collecting the powers of t in Equation (8.5.12), we have

$$\ln f(y_n;n,p)$$

$$\approx -\ln\left[\sqrt{2\pi}\sqrt{np(1-p)}\right]$$

$$-\left[np\frac{\sqrt{1-p}}{\sqrt{np}}+\frac{1}{2}\sqrt{\frac{(1-p)}{np}}-n(1-p)\sqrt{\frac{p}{n(1-p)}}-\frac{1}{2}\sqrt{\frac{p}{n(1-p)}}\right]t$$

$$-\left[-\frac{(1-p)}{2}+(1-p)-\frac{(1-p)}{4np}-\frac{p}{2}+p-\frac{p}{4n(1-p)}\right]t^2-\cdots. \qquad (8.5.13)$$

Equation (8.5.13) reduces to

$$\ln f(y_n;n,p)\approx -\ln\left[\sqrt{2\pi}\sqrt{np(1-p)}\right]+\frac{1}{2\sqrt{n}}\left[\sqrt{\frac{p}{1-p}}-\sqrt{\frac{1-p}{p}}\right]t$$

$$-\frac{1}{2}t^2+\frac{1}{4n}\left[\sqrt{\frac{p}{1-p}}+\sqrt{\frac{1-p}{p}}\right]t^2+\cdots, \qquad (8.5.14)$$

where, as $n \to \infty$, the coefficients of t and t^2 tend to zero. Also, coefficients of higher powers of t tend to zero so that

$$\ln f(y_n; n, p) \approx -\ln\left[\sqrt{2\pi}\sqrt{np(1-p)}\right] - \frac{1}{2}t^2$$

or

$$f(y_n; n, p) \approx \frac{1}{\sqrt{2\pi}\sqrt{np(1-p)}} e^{-t^2/2}. \tag{8.5.15}$$

Therefore, substituting $t = y_n - np / \sqrt{np(1-p)}$ into Equation (8.5.15), we have

$$f(y_n; n, p) = \binom{n}{y_n} p^{y_n} (1-p)^{n-y_n}$$

$$\approx \frac{1}{\sqrt{2\pi}\sqrt{np(1-p)}} \exp\left\{-\frac{1}{2}\left(\frac{y_n - np}{\sqrt{np(1-p)}}\right)^2\right\}.$$

Thus, the probability that a random phenomenon obeying the binomial probability distribution with the parameters n and p will have an observed value between two arbitrary integers y_1, y_2 $(y_1 < y_2)$ is given by applying the DeMoivre-Laplace theorem as follows:

$$Pr(y_1 < Y_n < y_2) = \sum_{r=y_1}^{y_2} \binom{n}{r} p^r (1-p)^{n-r}$$

$$= Pr\left(\frac{y_1 - np}{\sqrt{np(1-p)}} < \frac{Y_n - np}{\sqrt{np(1-p)}} < \frac{y_2 - np}{\sqrt{np(1-p)}}\right)$$

$$\approx \frac{1}{\sqrt{2\pi}} \int_{\frac{y_1 - np}{\sqrt{np(1-p)}}}^{\frac{y_2 - np}{\sqrt{np(1-p)}}} e^{-(1/2)z^2} dz$$

$$= F\left(\frac{y_2 - np}{\sqrt{np(1-p)}}\right) - F\left(\frac{y_1 - np}{\sqrt{np(1-p)}}\right). \tag{8.5.16}$$

However, in the preceding discussion, we are approximating probabilities of a *discrete* density with a *continuous* probability distribution, and we need to introduce the conventional *correction for continuity*. In the discrete case, the probability is concentrated at the integers, but when we approximate such probabilities with a continuous distribution, the corresponding probability is spread over a rectangle, with the base stretching from $-(1/2)$ to $1/2$ on either side of the integer involved (Figure 8.5.1).

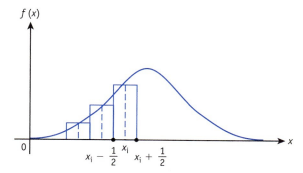

Figure 8.5.1

That is, if we were to approximate the probability up to point x_i, we would integrate $f(x)$ from $-\infty$ to x_i, in which case we would omit the area from x_i to $x_i + \frac{1}{2}$. Thus, we need

$$Pr(X \leq x_i) \approx \int_{-\infty}^{x_i+1/2} f(x)dx. \tag{8.5.17}$$

However, if we were to approximate the probability that $X < x_i$, we would integrate $f(x)$ from $-\infty$ to x_i and we would improperly include the area from $x_i - 1/2$ to x_i. Thus, we must change our upper limit:

$$Pr(X \leq x_i) \approx \int_{-\infty}^{x_i-1/2} f(x)dx. \tag{8.5.18}$$

Combining Equations (8.5.17) and (8.5.18), we obtain

$$Pr(X = x_i) = Pr(X \leq x_i) - Pr(X < x_i)$$

$$= \int_{x_i-1/2}^{x_i+1/2} f(x)dx.$$

Therefore, to improve our approximation of the binomial distribution, we include the *correction for continuity* in Equation (8.5.16):

$$Pr(y_1 < Y_n < y_2) \approx \frac{1}{\sqrt{2\pi}} \int_{\frac{y_1-np-1/2}{\sqrt{np(1-p)}}}^{\frac{y_2-np+1/2}{\sqrt{np(1-p)}}} e^{-(1/2)y_n^2} dy_n$$

$$= F\left(\frac{y_2 - np + 1/2}{\sqrt{np(1-p)}}\right) - F\left(\frac{y_1 - np - 1/2}{\sqrt{np(1-p)}}\right). \tag{8.5.19}$$

For example, in a random phenomenon where the binomial law is applicable with $p=1/2, n=10$, the normal approximation with the correction factor is significantly closer to the exact probabilities, as shown in Table 8.5.1.

Table 8.5.1 Comparison of Binomial Probabilities with Normal Approximation

		$p = 1/2, n = 10$	
y	Exact Binomial Probabilities: Eq. (2.4.1)	Normal Approximation: Eq. (8.5.16)	Normal Approximation: Eq. (8.5.19)
0	0.00098	0.00079	0.00219
1	0.00977	0.00491	0.01136
2	0.04395	0.02302	0.04350
3	0.11719	0.07508	0.11405
4	0.20508	0.16050	0.20340
5	0.24609	0.23570	0.25100
6	0.20508	0.23570	0.20340
7	0.11719	0.16050	0.11405
8	0.04395	0.07508	0.04350
9	0.00977	0.02302	0.01136
10	0.00098	0.00491	0.00219

A geometric illustration of the correctional factor for the preceding example is shown in Figure 8.5.2.

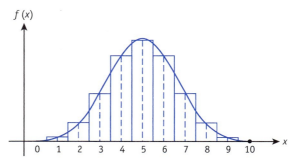

Figure 8.5.2

Example 8.5.1 illustrates the fruitfulness of the normal approximation from the computational point of view.

Example 8.5.1

Suppose that a "loaded" coin is tossed 2000 times. If the probability of a head occurring at any given toss is 0.75, what is the probability that the number of tosses on which heads will occur is between 1475 and 1535?

Solution: An exact probability to this question is given by

$$Pr\left(1475 \leq y_n \leq 1535\right) = \sum_{y=1475}^{1535} \binom{2000}{y} \left(\frac{3}{4}\right)^y \left(\frac{1}{4}\right)^{2000-y}.$$

The calculation of the preceding probability is quite laborious, but an acceptable approximation to the sum is given by Equation (8.5.19):

$$Pr\left(1475 \leq Y_n \leq 1535\right)$$

$$= Pr\left(\frac{1475 - 1500 - 1/2}{\sqrt{375}} \leq \frac{Y_n - 1500}{\sqrt{375}} \leq \frac{1535 - 1500 + 1/2}{\sqrt{375}}\right)$$

$$\approx \frac{1}{\sqrt{2\pi}} \int_{-\frac{51\sqrt{15}}{150}}^{\frac{71\sqrt{15}}{150}} e^{-(1/2)z^2} \, dz = \frac{1}{\sqrt{2\pi}} \int_{-1.32}^{1.83} e^{-(1/2)z^2} \, dz$$

$$= F(1.83) - F(-1.27)$$

$$= 0.8730$$

Example 8.5.2

In a certain industrial complex, we inspected 100 components of an electrical system. Each component is classified as being either operable or defective. We assign the value 1 if the component is operable and the value 0 if the component is defective. From previous experience, we know that the probability of finding an operable component is $p = 0.8$. What is the probability that the number of operable components will be more than 70 and less than 80?

Solution: The variate Y_n in Theorem 8.5.1 can assume values from 0 to 100. Also,

$$E[Y_n] = np = 80, \quad \text{Var}(Y_n) = np(1-p) = 16,$$

and

$$Pr(70 < Y_n < 80) = Pr\left(\frac{70 - 80 + 1/2}{4} < \frac{Y_n - 80}{4} < \frac{80 - 80 - 1/2}{4}\right)$$

$$= Pr\left(-2.38 < \frac{Y_n - 80}{4} < -0.13\right)$$

$$\approx \frac{1}{\sqrt{2\pi}} \int_{-2.38}^{-0.13} e^{-(1/2)z^2} \, dz$$

$$= F(-0.13) - F(-2.38).$$

From tables of the standardized normal distribution, we find the preceding probability to be 0.4396.

Example 8.5.3

A shirt manufacturer knows that, on the average, 2% of his product will not meet quality specifications. Find the greatest number of shirts constituting a lot that will have, with a probability of 0.95, fewer than five defectives.

Solution: If we let the random variable X be the number of defectives in the lot, then $E(Y_n) = 0.02n$ and $\text{Var}(Y_n) = 0.196n$, where

$$Y_n = \sum_{i=1}^{n} X_i.$$

Using Theorem 8.5.1, we obtain

$$0.95 \leq Pr(X<5) = Pr\left(\frac{Y_n - 0.02n}{\sqrt{0.0196n}} < \frac{5 - 1/2 - 0.02n}{\sqrt{0.0196n}}\right)$$

$$\approx \frac{1}{\sqrt{2\pi}} \int_{-\infty}^{\frac{5-1/2-0.02n}{\sqrt{0.0196n}}} e^{-(1/2)} \, dz$$

$$= F\left(\frac{5 - 1/2 - 0.02n}{\sqrt{0.0196n}}\right),$$

which implies that

$$\left(5 - \frac{1}{2} - 0.02n\right) \Big/ \sqrt{0.0196n} \geq 1.645$$

or

$$0.0004n^2 - 0.23271616n + 20.25 \geq 0.$$

Two solutions result:

$$n \leq 106.5176$$

and

$$n \geq 475.2728.$$

The second solution, however, does not satisfy the original relationship:

$$\frac{5 - 1/2 - 0.02n}{\sqrt{0.0196n}} \geq 1.645.$$

Therefore, we disregard it and accept the first one. Our solution, then, is $n = 106$.

8.6 Normal Approximation to the Poisson Distribution

In Chapter 3, we derived the Poisson distribution as a limit of a sequence of binomial distributions: many of the properties of the binomial probability density also apply to the Poisson probability distribution, and we can approximate the Poisson probability density function with the normal distribution. One can illustrate this approximation by showing that if $X_1, X_2, \ldots, X_n$ is a sequence of independent identically Poisson-distributed random variables with the parameter λ, then $Y_n = X_1 + X_2 + \ldots + X_n$ has the Poisson distribution with a mean and a variance of $n\lambda$. In view of the central limit theorem, we obtain the desired result.

Theorem 8.6.1

If $X_1, X_2, \ldots, X_n$ is a sequence of independent identically Poisson-distributed random variables with the parameter λ, then the distribution of the variate Z_n,

$$Z_n = \frac{Y_n - E(Y_n)}{\sqrt{\text{Var}(Y_n)}} = \frac{Y_n - n\lambda}{\sqrt{n\lambda}},$$

may be approximated by a normal distribution with a mean of 0 and a variance of 1 as $n \to \infty$, where

$$Y_n = X_1 + X_2 + \ldots + X_n.$$

In this case, we also employ the conventional correction factor of 1/2 for continuity, because we are approximating probabilities of a discrete distribution with a continuous probability density function, as discussed in the previous section. Hence, for a random phenomenon obeying the Poisson law with the parameter λ, the probability that we will have an observed value less than or equal to an arbitrary integer y is given by

$$Pr(Y_n \leq y) = \sum_{r=0}^{y} \frac{e^{-\lambda} \lambda^r}{r!}$$

$$= Pr\left(\frac{Y_n - n\lambda}{\sqrt{n\lambda}} \leq \frac{y - n\lambda + 1/2}{\sqrt{n\lambda}}\right)$$

$$\approx \frac{1}{\sqrt{2\pi}} \int_{-\infty}^{\frac{y-n\lambda+1/2}{\sqrt{n\lambda}}} e^{-(1/2)t^2} \, dt$$

$$= F\left(\frac{y - n\lambda + 1/2}{\sqrt{n\lambda}}\right).$$

The preceding discussion indicates that any Poisson-distributed random variable with the parameter λ is asymptotically normally distributed as $\lambda \to \infty$ because of the additive property of the independent identically Poisson-distributed random variables. The preceding approximation can be also shown by using Stirling's formula.

Example 8.6.1

A drug manufacturer receives a shipment of 10,000 calibrated "eyedroppers" for administering H1N1 vaccine. If the calibration mark is missing on 500 droppers, which are scattered randomly throughout the shipment, what is the probability that, at most, 2 defective droppers will be detected in a random sample of 125?

Solution: The population of 10,000 droppers is very large, and the Poisson law seems to be applicable. The probability that a dropper is defective is $500/10{,}000 = 0.05$, and the probability of finding at most 2 defectives is

$$Pr(Y_n \leq 2) = \sum_{y_n=0}^{2} \frac{e^{-6.25}(3.25)^{Y_n}}{Y_n!}$$

$$\approx Pr\left(\frac{0-1/2-6.25}{2.5} \leq \frac{Y_n-6.25}{2.5} \leq \frac{2+1/2-6.25}{2.5}\right)$$

$$= \int_{\frac{.5+6.25}{2.5}}^{\frac{2.5-6.25}{2.5}} e^{-z^2/2} dz$$

$$= F(-1.5) - F(-2.70)$$

$$= F(2.70) - F(1.5)$$

$$= 0.9965 - 0.9332$$

$$= 0.0633.$$

The probability that the number of defectives falls between one and three inclusively is

$$Pr(1 \leq Y_n \leq 3) = \sum_{y_n=1}^{2} \frac{e^{-6.25}(6.25)^{Y_n}}{y_n!}$$

$$\approx Pr\left(\frac{0-1/2-6.25}{2.5} \leq \frac{Y_n-6.25}{2.5} \leq \frac{3+1/2-6.25}{2.5}\right)$$

$$= \frac{1}{\sqrt{2\pi}2.5} \int_{1-1/2}^{3+1/2} \exp\left\{-\frac{1}{2}\left(\frac{y_n-6.25}{2.5}\right)^2\right\} dy_n$$

$$= \frac{1}{\sqrt{2\pi}} \int_{\frac{1-1/2-6.25}{2.5}}^{\frac{3+1/2-6.25}{2.5}} e^{-(1/2)z^2} dz$$

$$= F(-1.1) - F(-2.30)$$

$$= F(2.3) - F(1.1)$$

$$= 0.9893 - 0.8643$$

$$= 0.1250.$$

8.7 Normal Approximation to the Gamma Distribution

In Section 3.3, we introduced the gamma probability density function and illustrated its usefulness. We now show that this distribution can be approximated by the normal probability density for a sufficiently large n. As one might expect, as a consequence of this approximation, the $\chi^2(n)$ distribution is asymptotically normally distributed with a mean of n and a variance of $2n$. This last result is significant in the study of statistics.

Theorem 8.7.1

Let $X_1, X_2, \ldots, X_n$ be a sequence of independent identically gamma-distributed random variables with the parameters α, β. If $Y_n = X_1 + X_2 + \ldots + X_n$, then the random variable Z_n,

$$Z_n = \frac{Y_n - E(Y_n)}{\sqrt{\text{Var}(Y_n)}} = \frac{Y_n - n\alpha\beta}{\beta\sqrt{\alpha n}},$$

is normally distributed, with a mean of 0 and a variance of 1 as $n \to \infty$. Note that $E(X_i) = \alpha\beta$ and $\text{Var}(X_i) = \alpha\beta^2$, $i = 1, 2, \ldots, n$, and that

$$E(Y_n) = E\left(\sum_{i=1}^{n} X_i\right) = \sum_{i=1}^{n} E(X_i) = n\alpha\beta$$

and

$$\text{Var}(Y_n) = \text{Var}\left(\sum_{i=1}^{n} X_i\right) = \sum_{i=1}^{n} \text{Var}(X_i) = n\alpha\beta^2.$$

Thus, the probability that a gamma-distributed random variable Y_n will assume a value between y_1 and y_2, $0 \leq Y_1 < Y_2$, is

$$Pr(y_1 \leq Y_n \leq y_2) = \int_{y_1}^{y_2} \frac{1}{\Gamma(\alpha)\beta^2} y_n^{\alpha-1} e^{-(y_n/\beta)} dy_n$$

$$= Pr\left(\frac{y_1 - \alpha\beta n}{\beta\sqrt{\alpha n}} \leq \frac{y_1 - \alpha\beta n}{\beta\sqrt{\alpha\beta}} \leq \frac{y_2 - \alpha\beta n}{\beta\sqrt{\alpha n}}\right)$$

$$\approx \frac{1}{\sqrt{2\pi}} \int_{\frac{y_1 - \alpha\beta n}{\beta\sqrt{\alpha n}}}^{\frac{y_2 - \alpha\beta n}{\beta\sqrt{\alpha n}}} e^{-(1/2)z^2} dz$$

Example 8.7.1

Suppose that the life of a certain electrical component follows the gamma distribution with the parameters $\alpha = 1$ and $\beta = 10$. A company purchases 100 such components. What is the probability that the total life of these components will be between 850 and 1090 units?

Solution: Because the additive property of the gamma distribution holds, applying Theorem 8.7.1 with

$$Y_{100} = X_1 + X_2 + \ldots + X_{100},$$

$$E(Y_{100}) = n\alpha\beta = (100)(1)(10) = 1000,$$

and

$$\text{Var}(Y_{100}) = (100)(1)(10) = 10{,}000,$$

we obtain

$$Pr(850 \leq Y_{100} \leq 1090) = Pr\left(\frac{850-1000}{100} \leq \frac{Y_{100}-1000}{100} \leq \frac{1090-1000}{100}\right)$$

$$= Pr\left(-1.5 \leq \frac{Y_{100}-1000}{100} \leq 0.9\right)$$

$$\approx \frac{1}{\sqrt{2\pi}} \int_{-1.5}^{0.9} e^{-z^2/2} dz$$

$$= F(0.9) - F(-1.5)$$

$$= 0.7491.$$

The exact probability is given by

$$Pr(850 \leq Y_{100} \leq 1090) = \frac{1}{10} \int_{850}^{1090} y^{99} e^{-(y/10)} dy$$

$$= 0.7561.$$

Example 8.7.2

Let $X_1, X_2, \ldots, X_n$ be a sequence of independent identically *chi-square–distributed* random variables, each with 1 degree of freedom. It was shown by Theorem 7.7.7 that the sum $Y_n = X_1 + X_2 + \ldots + X_n$ is $\chi^2(n)$ distributed with n degrees of freedom; that is, the additive property of the central limit theorem holds. The expected value and variance of $\chi^2(n)$ are n and $2n$, respectively. Thus in view of the central limit theorem, the distribution of the random variable

$$Z_n = \frac{Y_n - n}{\sqrt{2n}}$$

may be approximated by a normal distribution with a mean of 0 and a variance of 1 for an effectively large n. Also, because $\chi^2(n)$ is a special form of the gamma distribution

$(\alpha = n/2, \beta = 2)$, the normal approximation is evident from Theorem 8.7.1. Thus, for $0 \leq y_1 \leq y_2$,

$$Pr(y_1 \leq Y_n \leq y_2) = \int_{y_1}^{y_2} \frac{1}{\Gamma\left(\frac{n}{2}\right) 2^{n/2}} y_n^{(n/2)-1} e^{-[y_n/2]} dy_n$$

$$= Pr\left(\frac{y_1 - n}{\sqrt{2n}} \leq \frac{Y_n - n}{\sqrt{2n}} \leq \frac{y_2 - n}{\sqrt{2n}}\right)$$

$$\approx \frac{1}{\sqrt{2\pi}} \int_{\frac{y_1-n}{\sqrt{2n}}}^{\frac{y_2-n}{\sqrt{2n}}} e^{-(1/2)z^2} dz$$

$$= F\left(\frac{y_2 - n}{\sqrt{2n}}\right) - F\left(\frac{y_1 - n}{\sqrt{2n}}\right).$$

Summary

The important *Chebyshev inequality,* which justifies the variance measuring variability about the expected value of a random variable, states that for any variate X with $E(X) = \mu$ and $\text{Var}(X) = \sigma^2$, we have

$$Pr\left(|X - \mu| \geq k\sigma\right) \leq \frac{1}{k^2},$$

where k is a positive number. A generalization of Chebyshev's inequality is given by *Kolmogorov's inequality.*

Bernoulli's law of large numbers was discussed. Let S_n^* be the observed number of successes in n independent repeated trials, with a probability p of success and a probability $1 - p$ of failure of each trial. Let

$$f_n = \frac{S_n^*}{n}$$

be the relative frequency of the observed number of successes in n trials. Then, for every $\epsilon > 0$, we have

$$Pr\left(|f_n - p| < \epsilon\right) \to 1 \text{ as } n \to \infty.$$

A more general law is given by the *weak law of large numbers (WLLN).* Let $X_1, X_2, \ldots, X_n$ be a sequence of mutually independent random variables with a common distribution. Let $E(X_i) = \mu$ exist. Then, for every $\epsilon > 0$,

$$Pr\left(\left|\frac{S_n}{n} - \mu\right| < \epsilon\right) \to 1 \text{ as } n \to \infty,$$

where S_n is the number of successes in n trials. A stronger form of the WLLN is the *strong law of large numbers (SLLN)*. A sufficient condition for the SLLN to apply to a sequence of mutually independent random variables not necessarily identically distributed is given by the *Kolmogorov criterion*.

There are two main versions of the *central limit theorem:* one in which the sequence of random variables $X_1, X_2, \ldots, X_n$ is independent and has the same probability distribution and one in which the random variables $X_1, X_2, \ldots, X_n, \ldots$ are independent but not identically distributed.

The *DeMoivre-Laplace theorem* was discussed. For n that is sufficiently large, we can approximate the binomial probability distribution with the normal probability law. Also, the normal approximation to the Poisson and gamma distributions was studied.

Applied Problems

8.1 Suppose that the number of cars arriving at a busy intersection in a large city has a Poisson distribution with a mean of 120. Determine a lower bound for the probability that the number of cars arriving in a given 20-minute period will be between 100 and 140, using *Chebyshev's inequality.*

8.2 A random variable X has a mean of 24 and a variance of 9. Obtain a bound on the probability that the random variable X assumes a value between 16 and 28.

8.3 Find the smallest value of n in a binomial distribution for which we can assert that

$$Pr\left(\left|\frac{X}{n} - p\right| < 0.1\right) \geq 0.90.$$

8.4 How large should the size of a random sample be so that we can be 90% certain that the sample mean $\bar{X}$ will not deviate from the true mean by more than $\sigma/2$?

8.5 Prove Chebyshev's inequality for the discrete case.

8.6 Consider a random sample of size 100, taken from a normal population with a mean of 0.5 and a variance of 0.04; that is, $X_1, X_2, \ldots, X_{100}$ are independent and identically distributed as $N(0.5, 0.4)$. Prove without the use of tables that

$$Pr\left(0.42 \leq \bar{X} \leq 0.58\right) \geq 0.9375,$$

where

$$\bar{x} = \frac{1}{100}\sum_{i=1}^{100} x_i.$$

8.7 The failure of a certain type of lightbulb is characterized by the Poisson probability density function with the parameter $\lambda = 0.08n$. That is,

$$f(x) = \begin{cases} \dfrac{e^{-0.08n}(0.08n)^x}{x!}, & x = 0, 1, 2, \ldots, \\ 0, & \text{elsewhere.} \end{cases}$$

How many lightbulbs must one test so that the sample mean will lie within 0.6 of the true mean, with probability at least as great as 0.90?

8.8 A random sample of size n is to be taken to determine the true proportion of cancer victims in a certain country. Determine the sample size n required to have probability of at least 0.95 that the observed proportion of cancer victims will differ from the true proportion by more than

   a. 1%
   b. 5%

8.9 Color blindness appears in 2% of the people of a certain population. How large must a random sample be to obtain a probability of 0.99 or more that a color-blind person will be in the sample?

8.10 Let $Pr(X \leq 0) = 0$. Show that if $\mu = E(X)$ exists, then

$$Pr(X \geq 2\mu) \leq 1/2.$$

8.11 Let $X_1, X_2, \ldots, X_n$ be a sequence of mutually independent random variables with the probability distribution given by

$$Pr\left(X_j = -\frac{1}{j}\right) = \frac{j^2}{1+j^2}$$

and

$$Pr(X_j = j) = \frac{1}{1+j^2}, \; j = 1, 2, \ldots, n.$$

Show that the sequence of random variables satisfies Theorem 8.3.2.

8.12 Let $X_1, X_2, \ldots, X_n$ be a sequence of mutually independent random variables with the probability defined function given by

$$Pr\left(X_r = \frac{1}{\sqrt{r}}\right) = \frac{1}{2}$$

and

$$Pr\left(X_r = -\frac{1}{\sqrt{r}}\right) = \frac{1}{2}, \; r = 1, 2, \ldots, n.$$

Does the sequence of random variables obey the *weak law of large numbers (WLLN)*? If so, according to which theorem?

8.13 Let $X_1, X_2, \ldots, X_n$ be a sequence of mutually independent random variables with the probability density function given by

$$Pr\left(X_k = \frac{1}{2^k}\right) = \frac{1}{2}$$

and

$$Pr\left(X_k = \frac{1}{2^k}\right) = \frac{1}{2}, \; k = 1, 2, \ldots, n.$$

Does the sequence of random variables obey the WLLN? If so, according to which theorem?

8.14 Let $Y_1, Y_2, \ldots, Y_n$ be a sequence of mutually independent random variables whose probability density is defined by

$$Pr(X_r = 0) = \frac{1}{r}$$

and

$$Pr(X_r = 1) = \frac{1}{r}, \quad r = 1, 2, \ldots, n.$$

Show that the sequence of random variables obeys the *strong law of large numbers* (*SLLN*).

8.15 Prove or disprove that the sequence of mutually independent random variables $X_1, X_2, \ldots, X_n$ satisfies the WLLN, where

$$Pr(X_j = c^{-j}) = \frac{1}{2},$$

$$Pr(X_i = -c^{-j}) = \frac{1}{2}, \quad j = 1, 2, \ldots, n,$$

and $c > 1$ is a constant.

8.16 A random sample of the size 144 is taken from an infinite population with a mean of $\mu = 53$ and a variance of $\sigma^2 = 324$. What is the probability that $\bar{X}_{144}$ is between 51 and 55?

8.17 From a distribution, the probability density function is given by

$$f(x) = \begin{cases} 3x^2, & 0 \leq x \leq 1, \\ 0, & \text{elsewhere.} \end{cases}$$

a. Find the approximate probability that the mean of a random sample of the size 25, taken from this distribution, lies between 2/5 and 3/5.
b. Repeat the problem for a sample of the size 64.

8.18 Suppose that the lifetime X in hours of a certain type of electrical component has the exponential distribution

$$f(x) = \begin{cases} \frac{1}{3} e^{-(1/3)x}, & x > 0, \\ 0, & \text{elsewhere.} \end{cases}$$

If a random sample of the size 36 is taken from these components, what is the probability that the mean $\bar{X}_{36}$ is less than 2 hours?

8.19 Consider a game of chance in which a person may win \$5 or lose \$1, \$2, or \$3 with a probability of 0.25 each. After 36 plays of the game, what is the probability that the person will have an average gain or loss between –\$2 and +\$2 dollars?

8.20 During tests of reliability of a certain item, the probability of failure is 0.1. Find the probability that, during tests of 100 items, the number of failures will be
a. less than 3 hours
b. between 3 and 5 hours
c. more than 4 hours

8.21 A condenser fails during a time T, with a probability of 0.2. Find the probability that, among 144 condensers during time T
  a. at least 15 will fail
  b. fewer than 10 will fail
  c. between 16 and 24 will fail

8.22 Given that an event occurs with a probability of 0.4, how many trials are necessary for at least 25 occurrences with a probability ≥ 0.90?

8.23 (Feller) Let $X_1, X_2, \ldots, X_n$ be a sequence of mutually independent random variables such that

$$Pr(X_n = -1) = Pr(X_n + 1) = \frac{(1 - 2^{-n})}{2}$$

and

$$Pr(X_n = -2n) = Pr(X_n + 2^n) = 2^{-(n+1)}.$$

Show that both the WLLN and the SLLN apply to the sequence.

8.24 Suppose that a "loaded" coin is tossed a large number of times. If the probability of a head occurring at any given toss is 0.60, find the number of tosses required so that the experiment will reveal fewer than 20 tails, with a probability of at least 0.95.

8.25 The probability that a gunner will hit a target is $1/30$. How many times must he fire at the target so that he will be 95% certain of hitting the target at least once? Solve the problem using the Poisson distribution and the normal approximation to the Poisson distribution, and compare your answers.

8.26 Suppose that 2500 customers subscribe to a telephone exchange. There are 80 trunk lines available. Any one customer has a probability of 0.03 of needing a trunk line on a given call. Considering the situation as 2500 trials with a probability of "success" of $p = 0.03$, what is the approximate probability that the 2500 customers will "tie up" the 80 trunk lines at any given time?

8.27 The exponential distribution is a special case of the gamma distribution when $\alpha = 1$ and $\beta > 0$. Suppose that the lifetime of a certain brand of lightbulb is exponentially distributed, with a mean of 100 hours. If a person buys 50 bulbs for her apartment complex, what is the approximate probability that the total of the lifetimes is greater than 4000 hours?

8.28 Let $X_1, X_2, \ldots, X_n$ be a sequence of independent identically distributed random variables with a mean of μ and a variance of σ^2. Show that

$$Z_n = \frac{\bar{X}_n - E(\bar{X})}{\sqrt{Var(\bar{X})}}$$

is approximately normally distributed with a mean of 0 and a variance of 1 for a sufficiently large n.

8.29 Let $X_1, X_2, \ldots, X_n$ be a sequence of independent identically chi-square–distributed random variables, each with 1 degree of freedom. Show using the technique employed in Theorem 8.4.1 that

$$Z_n = \frac{Y_n - E(Y_n)}{\sqrt{\operatorname{Var}(Y_n)}},$$

where $Y_n = X_1 + X_2 + \ldots + X_n$ is approximately normally distributed with a mean of 0 and a variance of 1 for a sufficiently large n.

8.30 Let $X_1, X_2, \ldots, X_n$ be a sequence of independent identically, normally distributed random variables with a mean of 0 and a variance of 1.

   a. Show that

$$S_n = \sum_{i=1}^{n} X_i^2$$

   has the chi-square distribution, with n degrees of freedom.

   b. Prove that

$$Z_n = \frac{S_n - E(S_n)}{\sqrt{\operatorname{Var}(S_n)}}$$

   is approximately normally distributed with a mean of 0 and a variance of 1 as $n \to \infty$.

8.31 Let $X_1, X_2, \ldots, X_n$ be a sequence of independent identically, normally distributed random variables with a mean of μ and a variance of σ^2.

   a. Show that

$$D_n = \frac{(n-1)S^2}{\sigma^2},$$

   where $S^2 = [1/(n-1)] \sum_{i=1}^{n} (x_i - \bar{x})^2$ has a chi-square distribution with $n-1$ degrees of freedom.

   b. Prove that

$$Z_n = \frac{D_n - E(D_n)}{\sqrt{\operatorname{Var}(D_n)}}$$

   is approximately normally distributed with a mean of 0 and a variance of 1 for a sufficiently large n.

8.32 Let $X_1, X_2, \ldots, X_n$ be a sequence of independent identically, normally distributed random variables with the parameter μ and the variance σ^2.

  a. Show that the probability density of

$$t_n = \frac{(\bar{X} - \mu)\sqrt{n}}{\sqrt{s}}$$

  is given by

$$f(t_n) = \frac{\Gamma\left(\frac{n}{2}\right)}{\sqrt{n-1}\sqrt{\pi}\,\Gamma\left(\frac{n-1}{2}\right)} \left(1 + \frac{t_n^2}{n-1}\right)^{-n/2}, \quad -\infty < t < \infty$$

  (*Student-t distribution* with $n-1$ degrees of freedom).

  b. Prove that $Z_n = \dfrac{t_n - E(t_n)}{\sqrt{\text{Var}(t_n)}}$ is approximately normally distributed, with a mean of 0 and a variance of 1 for a sufficiently large n.

8.33 Let $X_1, X_2, \ldots, X_n$ be a sequence of random variables that are characterized by the binomial probability density

$$Pr(X_n = k) = \binom{n}{k} p^k (1-p)^{n-k}, \quad k = 0, 1, 2, \ldots, n, \; 0 < p < 1.$$

Use the *DeMoivre-Laplace theorem* to obtain an analogous result for the sequence of random variables defined by $V_n = \dfrac{X_n}{n}$.

8.34 Of a large lot a certain type of electrical component, 18% are defective. A component is selected at random and is marked defective or nondefective. Before choosing the next component, we return the first one to the lot so that the probability of selecting a defective component remains 0.18. In this manner, we inspect n components. What should the value of n be so that the probability will be 0.90 that the frequency of the defective components will lie between 0.16 and 0.20?

8.35 Let $X_1, X_2, \ldots, X_n$ be a sequence of mutually independent random variables that are uniformly distributed; that is, let

$$(x_i) = \begin{cases} \dfrac{1}{\beta - \alpha}, & \alpha \leq x_i \leq \beta, \; i = 1, 2, \ldots, n. \\ 0, & \text{elsewhere,} \end{cases}$$

Show, using the moment-generating function, that

$$Z_n = \frac{Y_n - E(Y_n)}{\sqrt{\text{Var}(Y_n)}},$$

where $Y_n = (X_1 + X_2 + \ldots + X_n)/n$ is approximately normally distributed with a mean of 0 and a variance of 1 as $n \to \infty$.

8.36 Let $X_1, X_2, \ldots, X_n$ be a sequence of mutually independent Poisson-distributed random variables with the common parameter λ. Using the technique employed in Theorem 8.5.1, show that

$$Z_n = \frac{Y_n - E(Y_n)}{\sqrt{\text{Var}(Y_n)}},$$

where $Y_n = (X_1 + X_2 + \ldots + X_n)$ is approximately normally distributed with a mean of 0 and a variance of 1 as $n \to \infty$.

"A random walk down Wall Street."
—*Burton G. Malkiel*

CHAPTER 9

Finite Markov Chains

OBJECTIVES:

9.1 Basic Concepts in Markov Chains
9.2 n-Step Transition Probabilities
9.3 Evaluation of P^n
9.4 Classification of States

Andrei Andreevich Markov
(1856–1922)

"Statistical analysis and statistical reasoning have proved to be powerful weapons in the armory of rational enquiry in sciences and the arts."
—*Sir Francis Galton*

The work of Russian mathematician Andrei Andreevich Markov is one of the starting points of the modern theory of stochastic processes. In his early works, Markov concentrated on number theory and analysis, with particular interest in continued fractions, approximation theory, convergence of sets, and limits of integrals. He was the first to present a complete and strict proof of the central limit theorem—of great importance to statistics. Markov later became occupied with probability theory and began his study of mutually dependent variables, resulting in the notion of chained events (Markov chains).

Markov attended the Russian University of St. Petersburg and was appointed professor there in 1886. Ten years later, he became a member of the Academy of Sciences. A selection of his work in number theory and probability was published in Moscow in 1951.

■ INTRODUCTION

One important concept in probability theory is known as the stochastic process. A *stochastic* or *random process* is defined as a family of random variables $\{X(t)\}$ describing an empirical process, the development in time of which is governed by probabilistic laws. The parameter t is often interpreted as time and may be either discrete or continuous. Some examples of stochastic processes are provided by the growth of populations, such as bacteria colonies; the paths traced by moving particles in Brownian notion; and the fluctuating particles emitted by a radioactive source. Applications of stochastic processes occur in many fields; in particular, they appear in agriculture, biology, economics, engineering, medicine, oceanography, and psychology. A complete treatment of the subject may be found in the Suggested Supplementary Reading section at the end of the chapter.

In this chapter, we are concerned with a special class of stochastic processes, termed *Markov processes* or *chains*, which has been investigated quite extensively. The basic concepts of Markov chains were introduced in 1907 by the Russian mathematician Markov. A Markov chain may be defined as a random process whose development is treated as a series of transitions between certain values, called the *states*, of the process, which are finite or countably infinite and which possess the property that the future probabilistic behavior of the process depends only on the present (given) state, not on the method by which the process arrived in that state.

Markov chain theory is of great importance in many branches of science and engineering. In this chapter, we give a brief introduction to the basic concepts of a special class of Markov chains, namely, finite Markov chains that have a discrete parameter. There are three other basic classifications of Markov processes:

1. discrete state space and continuous parameter
2. continuous state space and discrete parameter
3. continuous state space and continuous parameter

A complete treatment of these can be found in Parzen.[1]

9.1 Basic Concepts in Markov Chains

Consider a random process that is represented by a sequence of random variables $X_1, X_2, X_3, \ldots$ defined on the space T of all possible values that the variates can assume. The space T is called the *state space* of the sequence, and the different values that the variates can assume are called the *states*. Thus, if we denote the states of the space by $i_1, i_2, i_3, \ldots, i_{n-1}, i_n$, one can pose the following question: What is the probability that the random variable $X_n = i_{k_n}$, given that $X_1 = i_1, X_2 = i_2, \ldots X_{n-1} = i_{n-1}$? If the structure of the random process is such that the conditional probabilities of the question depend only on the value that the random variable X_{n-1} assumes and are independent of all previous values, then we call such a process a *Markov chain*.

Definition 9.1.1

Let $i_{k_1}, i_{k_2}, i_{k_3}, \ldots, i_{k_n}$ be any sequence of states. The sequence of variates $X_1, X_2, X_3, \ldots$ is called a Markov chain if

$$Pr\left(X_n = i_{k_n} \mid X_i = i_{k_1}, \ldots X_{n-1} = i_{k_{n-1}}\right) = Pr\left(X_n = i_{k_n} \mid X_{n-1} = i_{k_{n-1}}\right). \tag{9.1.1}$$

Thus, an intuitive interpretation of a Markov chain is simply that the probability of going from the k_{n-1} st state, $i_{k_{n-1}}$, to the k_n th state, i_{k_n}, does *not* depend on how we got to the k_{n-1} st state. That is, only the present state of the process determines its future.

We can interpret the state space as the set of outcomes of a given process. If the outcome on the $(n-1)$ st trial is $i_{k_{n-1}}$, we may use the terminology that the process is in the state $i_{k_{n-1}}$ at the $(n-1)$ st step or at time $(n-1)$. Thus, in a Markovian sense, the outcome of any trial depends on the outcome of the immediately preceding trial.

With each Markov chain, we associate a set of *transition probabilities*.

■ Definition 9.1.2

The conditional probabilities that the process moves to state j at time n, given that it was in state i at time $n-1$, are called transition probabilities and are denoted by p_{ij},

$$p_{ij} = Pr(X_n = j | X_{n-1} = i), \tag{9.1.2}$$

with the subscript of p indicating the direction of transition $i \to j$. The transition probabilities satisfy the conditions

$$p_{ij} \geq 0 \text{ for all } i \text{ and } j \tag{9.1.3}$$

and

$$\sum_{j=1}^{n} P_{ij} = 1 \text{ for every } i. \tag{9.1.4}$$

■ Definition 9.1.3

If the transition probabilities p_{ij} depend only on the states i and j and not on the time $n-1$, then the conditional probabilities are stationary or constant.

Markov chains with stationary transition probabilities are called *homogeneous Markov chains*. Stationary transition probabilities mean that if the process is in a state i_k at time 0 and in state i_r at time t_2, and if the process starts in state i_k at time t_1, then the probability of being in state i_r at time $t_1 + t_2$ is the same as the probability if the process had started in state i_k at time 0. That is, no matter when the process arrived at state i_k, the probability of transition to state i_r remains the same. In this chapter, we are concerned only with Markov chains that have stationary transition probabilities.

The behavior of such Markov chains is described by the transition or stochastic matrices of the processes.

Definition 9.1.4

The **transition matrix** or **stochastic matrix** of a process with the states $i_1, i_2, i_3, \ldots, i_n$ and transition probabilities p_{ij}, $i, j = 1, 2, \ldots, n$, is

$$P = \begin{pmatrix} p_{11} & p_{12} & p_{13} & \cdots & p_{1j} & \cdots & p_{1n} \\ p_{21} & p_{22} & p_{23} & \cdots & p_{2j} & \cdots & p_{2n} \\ \vdots & \vdots & \vdots & & \vdots & & \vdots \\ p_{i1} & p_{i2} & p_{i3} & \cdots & p_{ij} & \cdots & p_{in} \\ \vdots & \vdots & \vdots & & \vdots & & \vdots \\ p_{n1} & p_{n2} & p_{n3} & \cdots & p_{nj} & \cdots & p_{nn} \end{pmatrix}. \tag{9.1.5}$$

Clearly, P is a square matrix. Each entry of the transition matrix P, p_{ij}, $i, j = 1, 2, \ldots, n$, must be nonnegative, and each row must sum to 1; that is, the conditions in Equations (9.1.3) and (9.1.4),

$$p_{ij} \geq 0, \, i, j = 1, 2, \ldots, n,$$

and

$$\sum_{j=1}^{n} P_{ij} = 1, \text{ for all } i,$$

hold. The ith row of the stochastic matrix, P,

$$\left(p_{i1}, p_{i2}, p_{i3}, \ldots, p_{ij}, \ldots p_{in}\right),$$

is called a *probability vector*. It represents the probabilities of all possible outcomes of the next trial. If, in addition to the rows, the columns of P sum to unity, the matrix is called *doubly stochastic*.

State or transition diagrams are quite helpful in gaining insight into the behavior of Markov chains that have stationary transition probabilities. For example, let the transition matrix of a certain process that has Markovian states be given by

$$P = \begin{array}{c} \\ i_1 \\ i_2 \\ i_3 \\ i_4 \end{array} \begin{array}{c} \begin{array}{cccc} i_1 & i_2 & i_3 & i_4 \end{array} \\ \begin{bmatrix} \frac{1}{3} & \frac{1}{3} & 0 & \frac{1}{3} \\ 0 & \frac{1}{2} & 0 & \frac{1}{2} \\ \frac{1}{4} & \frac{1}{2} & 0 & \frac{1}{4} \\ \frac{1}{4} & \frac{1}{6} & \frac{1}{3} & \frac{1}{4} \end{bmatrix} \end{array}$$

The state diagram of such a transition matrix is shown in Figure 9.1.1, in which circles denote the states of the process and arrows denote the transition probabilities from state i to state j in *single steps*.

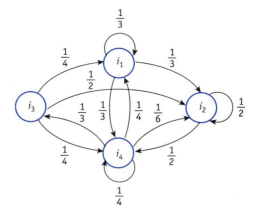

Figure 9.1.1

Conversely, looking at a state diagram of the form in Figure 9.1.2, we can write its transition probability matrix as

$$P = \begin{array}{c} \\ i_1 \\ i_2 \\ i_3 \\ i_4 \\ i_5 \end{array} \begin{array}{c} \begin{array}{ccccc} i_1 & i_2 & i_3 & i_4 & i_5 \end{array} \\ \left[\begin{array}{ccccc} \frac{1}{2} & \frac{1}{4} & 0 & 0 & \frac{1}{4} \\ 0 & \frac{1}{6} & \frac{1}{2} & 0 & \frac{1}{3} \\ 0 & 0 & \frac{1}{3} & \frac{1}{3} & \frac{1}{3} \\ \frac{1}{2} & 0 & 0 & \frac{3}{8} & \frac{1}{8} \\ \frac{1}{4} & \frac{1}{6} & \frac{1}{3} & \frac{1}{4} & 0 \end{array} \right] \end{array}.$$

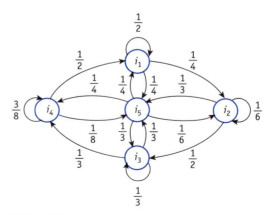

Figure 9.1.2

Example 9.1.1 (Random-walk problem)

A woman is standing at one of six states arranged in a straight path,

$$i_1 - i_2 - i_3 - i_4 - i_5 - i_6,$$

where i_1 and i_6 are referred to as the boundary states. Assume that she moves only in steps of one full unit and moves to the right with the probability p or to the left with the probability $q = 1 - p$. She moves until she reaches one of the two boundary states. The transition matrix for this random walk is

$$P = \begin{array}{c} \\ i_1 \\ i_2 \\ i_3 \\ i_4 \\ i_5 \\ i_6 \end{array} \begin{array}{cccccc} i_1 & i_2 & i_3 & i_4 & i_5 & i_6 \end{array} \\ \left[\begin{array}{cccccc} 1 & 0 & 0 & 0 & 0 & 0 \\ q & 0 & p & 0 & 0 & 0 \\ 0 & q & 0 & p & 0 & 0 \\ 0 & 0 & q & 0 & p & 0 \\ 0 & 0 & 0 & q & 0 & p \\ 0 & 0 & 0 & 0 & 0 & 1 \end{array} \right].$$

The first and last rows of the matrix correspond to the boundary states. The second row corresponds to the woman moving from state i_2 to state i_3 with the probability p or to state i_1 with the probability q, the third row corresponds to the woman moving from state i_3 to state i_4 with the probability p or to state i_2 with the probability q, and so on.

Example 9.1.2

Four quarterbacks are warming up by throwing a football to one another. Let i_1 through i_4 denote the four quarterbacks. It has been observed that i_1 is as likely to throw the ball to i_2 as to i_3 and i_4. Player i_2 never throws to i_3 but splits his throws between i_1 and i_4. Quarterback i_3 throws twice as many passes to i_1 as to i_4 and never throws to i_2, but i_4 throws only to i_1. This process forms a Markov chain because the player who is about to throw the ball is not influenced by the player who had the ball before him. The one-step transition matrix is

$$P = \begin{array}{c} \\ i_1 \\ i_2 \\ i_3 \\ i_4 \end{array} \begin{array}{cccc} i_1 & i_2 & i_3 & i_4 \end{array} \\ \left[\begin{array}{cccc} 0 & \frac{1}{3} & \frac{1}{3} & \frac{1}{3} \\ \frac{1}{2} & 0 & 0 & \frac{1}{2} \\ \frac{2}{3} & 0 & 0 & \frac{1}{3} \\ 1 & 0 & 0 & 0 \end{array} \right].$$

The state diagram of the process is shown in Figure 9.1.3.

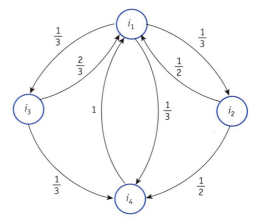

Figure 9.1.3

Example 9.1.3 (Gambler's ruin problem)

Suppose that an individual named Nicole is playing a game of chance against an adversary. Let each player begin the game with $3. On each play, it is possible for Nicole to win $1 from her opponent or to lose $1 to her opponent. The game is stopped either when Nicole loses all of her money or when she wins all of her opponent's money—that is, when Nicole's fortune reaches either $0 or $6. Assume that the probability of Nicole's winning on each play of the game is equal to p and that the probability of her losing on each play is equal to $q=1-p$. Thus, the probability of her fortune remaining the same is 0. When $p>q$, the game is advantageous to Nicole; when $p=q$, the game is fair; and when $p<q$, the game is advantageous to Nicole's opponent.

If $i_1, i_2, \ldots, i_7$ represent the possible states of Nicole's fortune, where $i_1 = 0, i_2 = 1, \ldots, i_7 = 6$, then her fortune may be represented by a random walk among these possible states. The probability of a step to the right (Nicole winning $1) is p, and the probability of a step to the left (Nicole losing $1) is q, for each play. The transition probability matrix for this Markov chain is given by

$$P = \begin{array}{c} \\ i_1 \\ i_2 \\ i_3 \\ i_4 \\ i_5 \\ i_6 \\ i_7 \end{array} \begin{array}{c} \begin{array}{ccccccc} i_1 & i_2 & i_3 & i_4 & i_5 & i_6 & i_7 \end{array} \\ \left[\begin{array}{ccccccc} 1 & 0 & 0 & 0 & 0 & 0 & 0 \\ q & 0 & p & 0 & 0 & 0 & 0 \\ 0 & q & 0 & p & 0 & 0 & 0 \\ 0 & 0 & q & 0 & p & 0 & 0 \\ 0 & 0 & 0 & q & 0 & p & 0 \\ 0 & 0 & 0 & 0 & q & 0 & p \\ 0 & 0 & 0 & 0 & 0 & 0 & 1 \end{array} \right] \end{array}.$$

Example 9.1.4 (Biological population problem)

Suppose that a finite biological population of N (fixed) individuals consists of individuals of either type A or type a. The population undergoes a birth and death process such that N remains fixed: At the discrete instant of time $t_1 < t_2 \ldots < t_n < \ldots$, one individual dies and is replaced by another individual of type A or type a. If, just before time instant t_k (when an individual is replaced) there are m individuals of type A and $N-m$ of type a present in the population, then we assume that the probability that an A individual dies (or will be replaced) is mp_1/Q_m and the probability that an a individual dies is $[(N-m)p_2]/Q_m$, where $Q_m = p_1 m + p_2(N-m)$. This assumption is based on the premise that a single type A individual dies at each time instant with the probability $p_1/(p_1+p_2)$ and a single type a individual dies at each time instant with the probability $p_2/(p_1+p_2)$ so that, for m of type A and $N-m$ of type a at each instant, the preceding probabilities apply. No difference in the birth pattern of the types is assumed; thus, the new individual has the probability m/N of being type A and the probability $(N-m)/N$ of being type a. The Markov chain $X_1, X_2, \ldots, X_n, \ldots$, where X_n is the number of individuals of type A in the population at time instant t_n, $n = 1, 2, 3, \ldots$, describes the population growth process. The possible states for the Markov chain are $i_1 = 0, i_2 = 1, \ldots, i_{N+1} = N$ (that is, $i_{m+1} = m$), with transition probabilities given by

$$P_{m,m-1} = \frac{p_1 m(N-m)}{Q_m N},$$

$$P_{m,m+1} = \frac{p_2 m(N-m)}{Q_m N},$$

$$P_{mm} = 1 - P_{m,m-1} - P_{m,m+1},$$

and

$$P_{mk} = 0 \text{ if } |k - m| > 1.$$

Therefore, the one-step transition probability matrix is given by

$$P = \begin{pmatrix} 1 & 0 & 0 & 0 & 0 & \cdots & 0 & 0 \\ \frac{p_1(N-1)}{Q_1 N} & 1 - \left((p_1+p_2)\frac{N-1}{Q_1 N}\right) & \frac{p_1(N-1)}{Q_1 N} & 0 & 0 & \cdots & 0 & 0 \\ 0 & \frac{2p_1(N-2)}{Q_2 N} & 1 - \left((p_1+p_2)\frac{2N-2}{Q_2 N}\right) & \frac{2p_2(N-2)}{Q_2 N} & 0 & \cdots & 0 & 0 \\ \vdots & \vdots & \vdots & \vdots & \vdots & \vdots & \vdots & \vdots \\ 0 & 0 & 0 & 0 & 0 & 0 & 0 & 0 \\ 0 & 0 & 0 & 0 & 0 & 0 & 0 & 1 \end{pmatrix}.$$

Example 9.1.5

Suppose that a signal is received that consists of only the digits 0, 1, 7, and 9 such that the next digit in the signal depends on the previous one received, one digit being received at each of the time instants t_n, $n = 1, 2, 3, \ldots$. Assume that if a 0 is received at time instant t_k, then only a 1 may be received at time t_{k+1}. If a 1 is received at time t_k, then at t_{k+1}, a 0 is received with a probability of $1/2$ and a 9 is received with a probability of $1/2$. If a 7 is received at the kth instant, then at the $(k+1)$ th instant, a 0 is received with a probability of $1/2$ and a 1 is received with a probability of $1/2$ and if a 9 is received, then, with a probability of 1, a 7 follows. This process constitutes a Markov chain, $X_1, X_2, \ldots, X_n, \ldots$, where X_n, $n = 1, 2, 3, \ldots$, takes the value of the digit received at time t_n. The states of the process are $i_1 = 0$, $i_2 = 1$, $i_3 = 7$, and $i_4 = 9$. The one-step transition probability matrix is given by

$$P = \begin{array}{c} \\ i_1 \\ i_2 \\ i_3 \\ i_4 \end{array} \begin{array}{cccc} i_1 & i_2 & i_3 & i_4 \\ \begin{bmatrix} 0 & 1 & 0 & 0 \\ \frac{1}{2} & 0 & 0 & \frac{1}{2} \\ \frac{1}{2} & \frac{1}{2} & 0 & 0 \\ 0 & 0 & 1 & 0 \end{bmatrix} \end{array}.$$

That is,

$$p_{12} = 1, \ p_{21} = \frac{1}{2}, \ p_{24} = \frac{1}{2}, \ p_{31} = \frac{1}{2}, \ p_{32} = \frac{1}{2}, \ p_{43} = 1,$$

and all other p_{ij}s are 0.

It is of considerable importance in a random process that has Markovian states to obtain the probability that the process will move through a given sequence of states. In general, this probability,

$$i_1 \to i_2 \to i_3 \to \ldots \to i_{n-1} \to i_n,$$

is the product of the probability of starting at state i_1 and the one-step transition probabilities up to the boundary state i_n. Applying the general law of compound probability from Theorem 1.3.1, we can verify the preceding statement as follows: We can write

$$Pr(X_1 = i_1, X_2 = i_2, X_3 = i_3, \ldots, X_{n-1} = i_{n-1}, X_n = i_n)$$

$$= Pr(X_1 = i_1) Pr(X_2 = i_2 | X_1 = i_1) Pr(X_3 = i_3 | X_1 = i_1, X_2 = i_2) \cdots$$

$$Pr(X_n = i_n | X_1 = i_1, X_2 = i_2, \ldots, X_{n-1} = i_{n-1}). \tag{9.1.6}$$

Because the sequence of states is Markovian, Equation (9.1.6) becomes

$$Pr(X_1 = i_1, X_2 = i_2, X_3 = i_3, \ldots, X_{n-1} = i_{n-1}, X_n = i_n)$$

$$= Pr(X_1 = i_1) Pr(X_2 = i_2 | X_1 = i_1) Pr(X_3 = i_3 | X_2 = i_2) \cdots Pr(X_n = i_n | X_{n-1} = i_{n-1})$$

$$= Pr(X_1 = i_1) p_{i_1 i_2} p_{i_2 i_3} \cdots p_{i_{n-1} i_n}, \tag{9.1.7}$$

where $Pr(X_i = i_1)$ is the probability that the process is initially in state i_1. In a given Markov chain, the probability given by the expression in Equation (9.1.7) is called the *initial probability density* because we are concerned with the probability distribution at the starting set of observations. The process does not necessarily have to start at state i_1, but it must begin at one of the states $i_1, i_2, \ldots, i_n$ so that

$$p_{i_k} \geq 0, \text{ for all } k,$$

and

$$\sum_{k=1}^{n} p_{i_k} = 1.$$

Thus, the initial probability distribution generates a sequence of initial probabilities

$$\pi_0 = \left(p_{i_1}, p_{i_2}, p_{i_3} \cdots p_{i_{n-1}} p_{i_n} \right),$$

which we refer to as the *initial probability vector* of the process and denote using π_0.

In many applications of random Markovian processes, we are given the initial probability distribution and the transition matrix. The question that we would like to answer with this information is the following: What is the probability that, at a particular time, the process is at a certain state? Before giving the theoretical formulation of the preceding question, we illustrate the method with an example that uses the concept of a *tree diagram*.

Example 9.1.6

Consider the random-walk problem formulated in Example 9.1.1. Assuming $p = q = 1/2$, the transition matrix becomes

$$P = \begin{array}{c} \\ i_1 \\ i_2 \\ i_3 \\ i_4 \\ i_5 \\ i_6 \end{array} \begin{array}{c} \begin{matrix} i_1 & i_2 & i_3 & i_4 & i_5 & i_6 \end{matrix} \\ \begin{bmatrix} 1 & 0 & 0 & 0 & 0 & 0 \\ \frac{1}{2} & 0 & \frac{1}{2} & 0 & 0 & 0 \\ 0 & \frac{1}{2} & 0 & \frac{1}{2} & 0 & 0 \\ 0 & 0 & \frac{1}{2} & 0 & \frac{1}{2} & 0 \\ 0 & 0 & 0 & \frac{1}{2} & 0 & \frac{1}{2} \\ 0 & 0 & 0 & 0 & 0 & 1 \end{bmatrix} \end{array}.$$

A tree diagram of the transition matrix is shown in Figure 9.1.4. The starting point of the process is assumed to be in state i_4.

9.1 Basic Concepts in Markov Chains

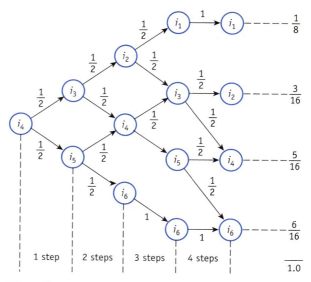

Figure 9.1.4

Path Probabilities

Suppose that we want to obtain the probability that the process at time $n=3$, or at the third step, is in state i_3. This probability is determined by adding all path probabilities that lead to state i_3: $1/8+1/8=1/4$. Similarly, the probability that we are in state i_1 at the fourth step is $1/2+1/8=5/8$. Starting at state i_4, the probability of being in state i_5 after two steps is 0.

One can also obtain the various steps of probability distribution vectors from a tree diagram. For the preceding example, the probability distribution vectors are

$$
\begin{array}{cccccccc}
& & i_1 & i_2 & i_3 & i_4 & i_5 & i_6 \\
\pi_0 = & (& 0 & 0 & 0 & 1 & 0 & 0) \\
\pi_1 = & (& 0 & 0 & \frac{1}{2} & 0 & \frac{1}{2} & 0) \\
\pi_2 = & (& 0 & \frac{1}{4} & 0 & \frac{1}{2} & 0 & \frac{1}{4}) \\
\pi_3 = & (& \frac{1}{8} & 0 & \frac{3}{8} & 0 & \frac{1}{4} & \frac{1}{4}) \\
\pi_4 = & (& \frac{1}{8} & \frac{3}{16} & 0 & \frac{5}{16} & 0 & \frac{3}{8}).
\end{array}
$$

Thus, from π_3 we can conclude that, starting from state i_4, the process will be at states i_1, i_2, i_3, i_4, i_5, and i_6 at the third step, with probabilities $1/8, 0, 3/8, 0, 1/4$, and $1/4$, respectively.

The mathematical formulation of the earlier question is given by

$$Pr(X_k = j) = \sum_i Pr(X_{k-1} = i) Pr(X_k = j | X_{k-1} = i)$$

$$= \sum_i p_j^{(k-1)} p_{ij}$$

$$= p_j^{(k)}; \qquad (9.1.8)$$

that is, the probability that the process is in state j at the kth step is equal to the sum over all probabilities that the process is in state i at the $(k-1)$ st step; on the next step, the process moves from state i to state j.

Note that $p_j^{(k-1)}$ is the $(k-1)$ st step probability. Thus, for $k=1$, we get

$$p_j^{(1)} = \sum_i p_j^{(0)} p_{ij},$$

where $p_j^{(0)}$ is the initial $(n=0 \text{ steps})$ probability, which constitutes the probability vector

$$\pi_0 = \left(p_1^{(0)} p_2^{(0)} \cdots p_m^{(0)} \right).$$

For $k=2$,

$$p_j^{(2)} = \sum_i p_j^{(1)} p_{ij},$$

where $p_j^{(1)}$ is the jth element of the first-step probability, which constitutes the probability vector

$$\pi_1 = \left(p_1^{(1)} p_2^{(2)} \cdots p_m^{(1)} \right).$$

Equation (9.1.8), which can easily be shown to be true by induction on k, is precisely the product of a row vector (probability distribution vector) and a matrix (transition matrix). Thus, Equation (9.1.8) can be written in the form

$$\pi_k = \pi_{k-1} P, \text{ for } k \geq i. \qquad (9.1.9)$$

This situation can be seen as follows:

$$\left(p_1^{(k-1)} p_2^{(k-1)} \cdots p_i^{(k-1)} \cdots p_m^{(k-1)} \right) \begin{pmatrix} p_{11} & p_{12} & \cdots & p_{1j} & \cdots & p_{1m} \\ p_{21} & p_{22} & \cdots & p_{2j} & \cdots & p_{2m} \\ \vdots & \vdots & & \vdots & & \vdots \\ p_{i1} & p_{i2} & \cdots & p_{ij} & \cdots & p_{im} \\ \vdots & \vdots & \cdots & \vdots & \cdots & \vdots \\ p_{m1} & p_{m2} & \cdots & p_{mj} & \cdots & p_{mm} \end{pmatrix}$$

$$= \left(\sum_{i=1}^{m} p_i^{(k-1)} p_{i1}, \sum_{i=1}^{m} p_i^{(k-1)} p_{i2}, \ldots, \sum_{i=1}^{m} p_i^{(k-1)} p_{ij}, \ldots, \sum_{i=1}^{m} p_i^{(k-1)} p_{im}, \right) = \pi_k$$

By successive iteration of Equation (9.1.9), the kth step probability distribution vector π_k results in

$$\pi_k = \pi_{k-1} P$$
$$= \pi_{k-2} P^2 = \pi_{k-3} P^3 = \cdots = \pi_1 P^{k-1} = \pi_0 P^k. \quad (9.1.10)$$

Thus, to obtain the kth step probability distribution vector, we multiply the initial probability distribution vector by the one-step transition probability matrix raised to the kth power. Therefore, given the initial probability distribution and the transition matrix of a Markov chain and using Equation (9.1.9) or (9.1.10), we can obtain the probability distribution vector at any given time.

Example 9.1.7

We can use Equation (9.1.9) to calculate the one-, two-, three-, and four-step probability distribution vectors for the transition matrix given in Example 9.1.6.

a.

$$\pi_1 = \pi_0 P = \begin{pmatrix} 0 & 0 & 0 & 1 & 0 & 0 \end{pmatrix} \begin{pmatrix} 1 & 0 & 0 & 0 & 0 & 0 \\ \frac{1}{2} & 0 & \frac{1}{2} & 0 & 0 & 0 \\ 0 & \frac{1}{2} & 0 & \frac{1}{2} & 0 & 0 \\ 0 & 0 & \frac{1}{2} & 0 & \frac{1}{2} & 0 \\ 0 & 0 & 0 & \frac{1}{2} & 0 & \frac{1}{2} \\ 0 & 0 & 0 & 0 & 0 & 1 \end{pmatrix} = \begin{pmatrix} 0 & 0 & \frac{1}{2} & 0 & \frac{1}{2} & 0 \end{pmatrix}$$

b.

$$\pi_2 = \pi_1 P = \begin{pmatrix} 0 & 0 & \frac{1}{2} & 0 & \frac{1}{2} & 0 \end{pmatrix} \begin{pmatrix} 1 & 0 & 0 & 0 & 0 & 0 \\ \frac{1}{2} & 0 & \frac{1}{2} & 0 & 0 & 0 \\ 0 & \frac{1}{2} & 0 & \frac{1}{2} & 0 & 0 \\ 0 & 0 & \frac{1}{2} & 0 & \frac{1}{2} & 0 \\ 0 & 0 & 0 & \frac{1}{2} & 0 & \frac{1}{2} \\ 0 & 0 & 0 & 0 & 0 & 1 \end{pmatrix} = \begin{pmatrix} 0 & \frac{1}{4} & 0 & \frac{1}{2} & 0 & \frac{1}{4} \end{pmatrix}$$

c.

$$\pi_3 = \pi_2 P = \begin{pmatrix} 0 & \frac{1}{4} & 0 & \frac{1}{2} & 0 & \frac{1}{4} \end{pmatrix} \begin{pmatrix} 1 & 0 & 0 & 0 & 0 & 0 \\ \frac{1}{2} & 0 & \frac{1}{2} & 0 & 0 & 0 \\ 0 & \frac{1}{2} & 0 & \frac{1}{2} & 0 & 0 \\ 0 & 0 & \frac{1}{2} & 0 & \frac{1}{2} & 0 \\ 0 & 0 & 0 & \frac{1}{2} & 0 & \frac{1}{2} \\ 0 & 0 & 0 & 0 & 0 & 1 \end{pmatrix} = \begin{pmatrix} \frac{1}{8} & 0 & \frac{3}{8} & 0 & \frac{1}{4} & \frac{1}{4} \end{pmatrix}$$

d.

$$\pi_4 = \pi_3 P = \begin{pmatrix} \frac{1}{8} & 0 & \frac{3}{8} & 0 & \frac{1}{4} & \frac{1}{4} \end{pmatrix} \begin{pmatrix} 1 & 0 & 0 & 0 & 0 & 0 \\ \frac{1}{2} & 0 & \frac{1}{2} & 0 & 0 & 0 \\ 0 & \frac{1}{2} & 0 & \frac{1}{2} & 0 & 0 \\ 0 & 0 & \frac{1}{2} & 0 & \frac{1}{2} & 0 \\ 0 & 0 & 0 & \frac{1}{2} & 0 & \frac{1}{2} \\ 0 & 0 & 0 & 0 & 0 & 1 \end{pmatrix} = \begin{pmatrix} \frac{1}{8} & \frac{3}{16} & 0 & \frac{5}{16} & 0 & \frac{3}{8} \end{pmatrix}$$

Hence, the probability distribution vectors that we calculate using Equation (9.1.9) are the same as those obtained from the tree diagram.

Thus, a random process that results in a sequence of states that are Markovian, with stationary transition probabilities, is completely specified (defined) when we know the initial probability density and the transition probabilities.

9.2 n-Step Transition Probabilities

The entry p_{ij} in the transition matrix P of a Markov chain is the probability that the process changes from state i to state j in a single step. In many processes, however, it is impossible to change from state i to state j in one step. For instance, in the state diagram shown by Figure 9.1.1, it is impossible for the process to change from i_2 to i_1 in one step, but it can make the change in two steps via state i_4.

We know that the one-step transition probability $i \to j$ is given by

$$p_{ij} = Pr\left[X_n = j \mid X_{n-1} = i\right].$$

For a two-step transition $i \to r \to j$, the transition probabilities that the process changes from state $i \to r$ and from state $r \to j$ are independent. Thus,

$$Pr\left[X_{n-1} = r \mid X_{n-2} = i\right] \cdot Pr\left[X_n = j \mid X_{n-1} = r\right] = p_{ir} p_{rj}.$$

Summing over all possible intermediate states, we obtain the two-step transition probability

$$p_{ij}^{(2)} = \sum_r p_{ir} p_{rj}.$$

Similarly, the three-step transition probability is given recursively by

$$p_{ij}^{(3)} = \sum_r p_{ir}^{(2)} p_{rj};$$

in general, the $(n+1)$-step transition probabilities are given by

$$p_{ij}^{(n+1)} = \sum_r p_{ir}^{(n)} p_{rj}. \qquad (9.2.1)$$

9.2 n-Step Transition Probabilities

We proceed to show that, for all r,

$$Pr(X_{m+r} = j | X_m = i) = p_{ij}^{(r)}. \tag{9.2.2}$$

For $r = 1$, Equation (9.2.2) is true by the definition of a Markov chain.

Assume that Equation (9.2.2) is true for $r = n$. We have

$$Pr(X_{m+n+1} = j | X_m = i) = \sum_k Pr(X_{m+n} = k, X_{m+n+1} = j | X_m = i). \tag{9.2.3}$$

Recall that

$$Pr(AB|C) = Pr(A|C) Pr(B|AC).$$

Thus, Equation (9.2.3) can be written as

$$Pr(X_{m+n+1} = j | X_m = i) = \sum_k Pr(X_{m+n} = k | X_m = i) Pr(X_{m+n+1} = j | X_m = i, X_{m+n} = k).$$

Applying the Markov property, the preceding equation becomes

$$Pr(X_{m+n+1} = j | X_m = i) = \sum_k Pr(X_{m+n} = k | X_m = i) Pr(X_{m+n+1} = j | X_{m+n} = k)$$

$$= \sum_k p_{ik}^{(n)} p_{kj}$$

$$= p_{ij}^{(n+1)}.$$

Therefore, by induction, we have shown that Equation (9.2.2) is true for all r.

A generalization of Equation (9.2.1) can be formulated as follows: Suppose that a process is to move from state i into state j in $m+n$ steps and we wish to obtain the transition probability $p_{ij}^{(m+n)}$. The first m steps take the process from state i into some intermediate state k, and the second n steps take the process from state k to state j. The product of these two transition probabilities $p_{ik}^{(m)} p_{kj}^{(n)}$, summed over all possible intermediate states that the process may visit, gives the desired $(m+n)$-step transition probability,

$$p_{ij}^{(m+n)} = \sum_k p_{ik}^{(m)} p_{kj}^{(n)}, \tag{9.2.4}$$

where

$$p_{ij}^{(0)} = \begin{cases} 1 & \text{for } i = j, \\ 0 & \text{for } i \neq j. \end{cases}$$

For $m = 1$, Equation (9.2.4) reduces to Equation (9.2.1). It can be shown by induction that Equation (9.2.4) is true for all m.

In Section 9.1, we defined a one-step transition matrix, which describes the behavior of a Markov chain. Similarly, for n-step transition probabilities, we can define the n-*step transition matrix*.

Definition 9.2.1

The n-step transition matrix of a process, with states $i_1, i_2, \ldots, i_k$ and n-step transition probabilities $p_{ij}^{(n)}$, $i,j = 1,2,\ldots,k$, is defined as

$$p^{(n)} = \begin{pmatrix} p_{11}^{(n)} & p_{12}^{(n)} & p_{13}^{(n)} & \cdots & p_{1j}^{(n)} & \cdots & p_{1k}^{(n)} \\ p_{21}^{(n)} & p_{22}^{(n)} & p_{23}^{(n)} & \cdots & p_{2j}^{(n)} & \cdots & p_{2k}^{(n)} \\ \vdots & \vdots & \vdots & & \vdots & & \vdots \\ p_{i1}^{(n)} & p_{i2}^{(n)} & p_{i3}^{(n)} & \cdots & p_{ij}^{(n)} & \cdots & p_{ik}^{(n)} \\ \vdots & \vdots & \vdots & & \vdots & & \vdots \\ p_{k1}^{(n)} & p_{k2}^{(n)} & p_{k3}^{(n)} & \cdots & p_{kj}^{(n)} & \cdots & p_{kk}^{(n)} \end{pmatrix},$$

where

$$p_{ij} \geq 0, \quad i,j = 1,2,\ldots,k,$$

and

$$\sum_j p_{ij}^{(n)} = 1, \text{ for all } i.$$

Clearly, $p^{(n)}$ is a square matrix, and $p_{ij}^{(n)}$ denotes the transition probability that the process moves from state i to state j in n steps.

Equation (9.2.4) is, by definition (see Section 0.8),

$$p^{(m+n)} = p^{(m)} p^{(n)}. \tag{9.2.5}$$

Equation (9.2.5) is the matrix form of the well-known *equation of Chapman-Kolmogorov*, which is significant in the theory of Markov chains. Also, from the application point of view, Equation (9.2.5) is a useful result. We should mention that matrices do not commute, in general, but that transition matrices do satisfy the condition

$$p^{(m+n)} = p^{(m)} p^{(n)} = p^{(m)} p^{(n)}.$$

A fundamental property of the transition matrix P is that the n-step transition matrix is equal to the nth power of P:

$$p^{(n)} = p^n. \tag{9.2.6}$$

To verify Equation (9.2.6), consider the process to be in state i at time k; we are interested in obtaining the probability $p_{ij}^{(n)}$ that the process is in state j at time $k+n$. Because the process is in state i at time k, the initial probability distribution vector $\pi_i = (0 \ 0 \ 0 \ \ldots \ 1 \ \ldots \ 0 \ 0 \ 0)$ has a 1 at the ith state and a 0 everywhere else. As a consequence of Equation (9.1.9), we have

$$\pi_n = \pi_0 P^n = \begin{pmatrix} 0 & 0 & 0 & \ldots & 1 & \ldots & 0 & 0 & 0 \end{pmatrix} \begin{pmatrix} p_{11} & p_{12} & p_{13} & \cdots & p_{1j} & \cdots & p_{1k} \\ p_{21} & p_{22} & p_{23} & \cdots & p_{2j} & \cdots & p_{2k} \\ \vdots & \vdots & \vdots & & \vdots & & \vdots \\ p_{i1} & p_{i2} & p_{i3} & \cdots & p_{ij} & \cdots & p_{ik} \\ \vdots & \vdots & \vdots & & \vdots & & \vdots \\ p_{k1} & p_{k2} & p_{k3} & \cdots & p_{kj} & \cdots & p_{kk} \end{pmatrix}$$

$$= \begin{pmatrix} p_{i1} & p_{i2} & \cdots & p_{ij} & \cdots & p_{ik} \end{pmatrix}^n$$

$$= \begin{pmatrix} p_{i1}^{(n)} & p_{i2}^{(n)} & \cdots & p_{ij}^{(n)} & \cdots & p_{ik}^{(n)} \end{pmatrix},$$

that is, the ith row of the transition matrix P^n. Thus, $p_{ij}^{(n)}$ is the jth component of the ith row of P^n, and $P^{(n)} = P^n$.

Example 9.2.1

Consider the one-step transition matrix given in Example 9.1.2:

$$P = \begin{matrix} & \begin{matrix} i_1 & i_2 & i_3 & i_4 \end{matrix} \\ \begin{matrix} i_1 \\ i_2 \\ i_3 \\ i_4 \end{matrix} & \begin{bmatrix} 0 & \frac{1}{3} & \frac{1}{3} & \frac{1}{3} \\ \frac{1}{2} & 0 & 0 & \frac{1}{2} \\ \frac{2}{3} & 0 & 0 & \frac{1}{3} \\ 1 & 0 & 0 & 0 \end{bmatrix} \end{matrix}.$$

The two-step transition matrix P^2 is

$$P^2 = P \cdot P = \begin{pmatrix} 0 & \frac{1}{3} & \frac{1}{3} & \frac{1}{3} \\ \frac{1}{2} & 0 & 0 & \frac{1}{2} \\ \frac{2}{3} & 0 & 0 & \frac{1}{3} \\ 1 & 0 & 0 & 0 \end{pmatrix} \begin{pmatrix} 0 & \frac{1}{3} & \frac{1}{3} & \frac{1}{3} \\ \frac{1}{2} & 0 & 0 & \frac{1}{2} \\ \frac{2}{3} & 0 & 0 & \frac{1}{3} \\ 1 & 0 & 0 & 0 \end{pmatrix} = \begin{pmatrix} \frac{13}{18} & 0 & 0 & \frac{5}{18} \\ \frac{1}{2} & \frac{1}{6} & \frac{1}{6} & \frac{1}{6} \\ \frac{1}{3} & \frac{2}{9} & \frac{2}{9} & \frac{2}{9} \\ 0 & \frac{1}{3} & \frac{1}{3} & \frac{1}{3} \end{pmatrix}.$$

The three-step transition matrix P^3 is

$$P^3 = P^2 \cdot P = \begin{pmatrix} \frac{13}{18} & 0 & 0 & \frac{5}{18} \\ \frac{1}{2} & \frac{1}{6} & \frac{1}{6} & \frac{1}{6} \\ \frac{1}{3} & \frac{2}{9} & \frac{2}{9} & \frac{2}{9} \\ 0 & \frac{1}{3} & \frac{1}{3} & \frac{1}{3} \end{pmatrix} \begin{pmatrix} 0 & \frac{1}{3} & \frac{1}{3} & \frac{1}{3} \\ \frac{1}{2} & 0 & 0 & \frac{1}{2} \\ \frac{2}{3} & 0 & 0 & \frac{1}{3} \\ 1 & 0 & 0 & 0 \end{pmatrix} = \begin{pmatrix} \frac{5}{18} & \frac{13}{54} & \frac{13}{54} & \frac{13}{54} \\ \frac{13}{36} & \frac{1}{6} & \frac{1}{6} & \frac{11}{36} \\ \frac{13}{27} & \frac{1}{9} & \frac{1}{9} & \frac{8}{27} \\ \frac{13}{18} & 0 & 0 & \frac{5}{18} \end{pmatrix}.$$

the third row of P^3,

$$\left(\frac{13}{27} \quad \frac{1}{9} \quad \frac{1}{9} \quad \frac{8}{27}\right),$$

denotes that, after three throws, the ball is in the hands of players i_1, i_2, i_3, and i_4 with respective probabilities 13/27, 1/9, 1/9, and 8/27.

Example 9.2.2

In the gambler's ruin problem in Section 9.1, the entries of the three-step transition probability matrix are $p_{jk}^{(3)}$. The probability of being in state i_k after three plays, given that the fortune was in state i_j, is given by

$$P^3 = \begin{pmatrix} 1 & 0 & 0 & 0 & 0 & 0 & 0 \\ q+pq^2 & 0 & 2p^2q & 0 & p^3 & 0 & 0 \\ q^2 & 2pq^2 & 0 & 3p^2q & 0 & p^3 & 0 \\ q^3 & 0 & 3pq^2 & 0 & 3p^2q & 0 & p^3 \\ 0 & q^3 & 0 & 3p^2q & 0 & 2pq^2 & p^2 \\ 0 & 0 & q^3 & 0 & 2pq^2 & 0 & p+p^2q \\ 0 & 0 & 0 & 0 & 0 & 0 & 1 \end{pmatrix}.$$

Example 9.2.3

In Example 9.1.5, we may find the probabilities of receiving certain combinations of n digits by considering the n-step transition probabilities and the n-step transition probability matrix P^n. For example, because

$$p_{22}^{(3)} = \sum_{j=1}^{4} \sum_{k=1}^{4} p_{2k} p_{kj} p_{j2} = p_{24} p_{43} p_{32},$$

from the preceding values of p_{jk}, $j,k = 1,2,3,4$, we may find the probability that the combination 1971 will be received in the next three transitions, given that the first 1 was received at the last time instant—that is, that the process is in state $i_2 = 1$:

$$P^3 = \begin{pmatrix} 0 & \frac{1}{2} & \frac{1}{2} & 0 \\ \frac{1}{2} & \frac{1}{4} & 0 & \frac{1}{4} \\ \frac{1}{4} & \frac{1}{4} & \frac{1}{4} & \frac{1}{4} \\ \frac{1}{4} & \frac{1}{2} & 0 & \frac{1}{4} \end{pmatrix}$$

so that $p_{22}^{(3)} = 1/4$.

Thus, the probability for a transition in n steps may simply be obtained by raising the one-step transition matrix to the nth power.

When n is fairly large, computing p^n is laborious; but, in the next section, we present a shorter method for accomplishing such calculations.

9.3 Evaluation of P^n

Before we proceed with this section, we discuss some preliminary concepts concerning matrices. Those notions are essential to the understanding of what follows.

> **9.3.1 Characteristic Equation** The characteristic equation of any square $k \times k$ matrix A is
> $$|A - \lambda I| = 0,$$
> and its k solutions are called the *eigenvalues* of matrix A. For every distinct eigenvalue of matrix A, there exists an eigenvector $\mathbf{x}$ such that $|A - \lambda I|\mathbf{x} = 0$.

We now give some illustrative examples.

Example 9.3.1

Let matrix A be given by
$$A = \begin{bmatrix} 0 & I \\ 1-d & d \end{bmatrix}.$$

The characteristic equation is
$$0 = |A - \lambda I| = \left| \begin{bmatrix} 0 & I \\ 1-d & d \end{bmatrix} - \begin{bmatrix} \lambda & 0 \\ 0 & \lambda \end{bmatrix} \right| = \begin{vmatrix} -\lambda & 1 \\ 1-d & d-\lambda \end{vmatrix}$$
$$= -\lambda(d-\lambda) - (1-d) = \lambda^2 - d\lambda - 1 + d.$$

Thus, the characteristic equation of matrix A is
$$\lambda^2 - d\lambda - 1 + d = 0.$$

To solve the characteristic equation, we write
$$\lambda = \frac{d \pm \sqrt{d^2 + 4(1-d)}}{2} = \frac{d \pm \sqrt{(d-2)^2}}{2} = \frac{d \pm (d-2)2}{2} = d-1, 1.$$

Therefore, the eigenvalues of matrix A are $\lambda_1 = d-1$ and $\lambda_2 = 1$. The eigenvector corresponding to the eigenvalue λ_1 is $\mathbf{x}_1$ such that
$$0 = (A - \lambda_1 I)\mathbf{x}_1 = \left(\begin{bmatrix} 0 & 1 \\ 1-d & d \end{bmatrix} - \begin{bmatrix} d-1 & 0 \\ 0 & d-1 \end{bmatrix} \right) \mathbf{x}_1 = \begin{bmatrix} 1-d & 1 \\ 1-d & 1 \end{bmatrix} \mathbf{x}_1$$

At this point, we express the eigenvector $\mathbf{x}_1$ as $\begin{bmatrix} a \\ b \end{bmatrix}$ and solve for the components of $\mathbf{x}_1$, namely, a and b:

$$\begin{bmatrix} 1-d & 1 \\ 1-d & 1 \end{bmatrix} \begin{bmatrix} a \\ b \end{bmatrix} = \begin{bmatrix} (1-d)a+b \\ (1-d)a+b \end{bmatrix} = \begin{bmatrix} 0 \\ 0 \end{bmatrix}.$$

This implies that

$$b = (d-1)a.$$

Thus,

$$\mathbf{x}_1 = \begin{bmatrix} a \\ (d-1)a \end{bmatrix}.$$

We let a equal any convenient number, say, 1, and we determine that the eigenvector corresponding to $\lambda_1 = d-1$ is

$$\mathbf{x}_1 = \begin{bmatrix} 1 \\ d-1 \end{bmatrix}.$$

We can easily check to see that $(A - \lambda_1 I)\mathbf{x}_1 = 0$. The eigenvector corresponding to λ_2 is $\mathbf{x}_2$ such that

$$\mathbf{0} = (A - \lambda_2 I)\mathbf{x}_2 = \left(\begin{bmatrix} 0 & 1 \\ 1-d & d \end{bmatrix} - \begin{bmatrix} 1 & 0 \\ 0 & 1 \end{bmatrix} \right) \mathbf{x}_2 = \begin{bmatrix} -1 & 1 \\ 1-d & d-1 \end{bmatrix} \mathbf{x}_2.$$

Expressing $\mathbf{x}_2$ as $\begin{bmatrix} a \\ b \end{bmatrix}$ and solving for a and b gives

$$\begin{bmatrix} -1 & 1 \\ 1-d & d-1 \end{bmatrix} \begin{bmatrix} a \\ b \end{bmatrix} = \begin{bmatrix} -(a-b) \\ (1-d)(a-b) \end{bmatrix} = \begin{bmatrix} 0 \\ 0 \end{bmatrix}.$$

This implies that $a = b$, so $\mathbf{x}_2 = \begin{bmatrix} b \\ b \end{bmatrix}$. Letting b equal any convenient number, say, 1, we see that the eigenvector corresponding to $\lambda_2 = 1$ is

$$\mathbf{x}_2 = \begin{bmatrix} 1 \\ 1 \end{bmatrix}.$$

If we construct a matrix B, the columns of which are the eigenvectors of A (A having distinct eigenvalues), then $B^{-1}AB = \Lambda$, where Λ is a diagonal matrix of the eigenvalues of A. (A diagonal matrix is a square matrix, the nonzero elements of which appear only on the upper left to lower right main diagonal.) To illustrate, we immediately can construct B from $\mathbf{x}_1$ and $\mathbf{x}_2$, getting

$$B = \begin{bmatrix} 1 & 1 \\ d-1 & 1 \end{bmatrix},$$

$$B^{-1} = \frac{1}{2-d} \begin{bmatrix} 1 & -1 \\ 1-d & 1 \end{bmatrix},$$

and

$$B^{-1}AB = \frac{1}{2-d}\begin{bmatrix} 1 & -1 \\ 1-d & 1 \end{bmatrix}\begin{bmatrix} 0 & 1 \\ 1-d & d \end{bmatrix}\begin{bmatrix} 1 & 1 \\ d-1 & 1 \end{bmatrix}$$

$$= \frac{1}{2-d}\begin{bmatrix} d-1 & 1-d \\ 1-d & 1 \end{bmatrix}\begin{bmatrix} 1 & 1 \\ d-1 & 1 \end{bmatrix}$$

$$= \frac{1}{2-d}\begin{bmatrix} (d-1)(2-d) & 0 \\ 0 & 2-d \end{bmatrix} = \begin{bmatrix} d-1 & 0 \\ 0 & 1 \end{bmatrix}.$$

Thus, we have that

$$B^{-1}AB = \Lambda$$

or

$$\frac{1}{2-d}\begin{bmatrix} 1 & -1 \\ 1-d & 1 \end{bmatrix}\begin{bmatrix} 0 & 1 \\ 1-d & d \end{bmatrix}\begin{bmatrix} 1 & 1 \\ d-1 & 1 \end{bmatrix} = \begin{bmatrix} d-1 & 0 \\ 0 & 1 \end{bmatrix}.$$

Example 9.3.2

Let matrix A be given by

$$A = \begin{bmatrix} 1 & 0 & 0 \\ \frac{3}{4} & \frac{1}{4} & 0 \\ \frac{1}{4} & \frac{1}{4} & \frac{1}{2} \end{bmatrix}.$$

The characteristic equation is

$$\begin{vmatrix} 1-\lambda & 0 & 0 \\ \frac{3}{4} & \frac{1}{4}-\lambda & 0 \\ \frac{1}{4} & \frac{1}{4} & \frac{1}{2}-\lambda \end{vmatrix} = 0.$$

Solving it gives $\lambda_1 = 1$, $\lambda_2 = 1/4$, and $\lambda_3 = 1/2$. To obtain the eigenvector for $\lambda_1 = 1$, we write

$$\begin{bmatrix} 0 & 0 & 0 \\ \frac{3}{4} & -\frac{3}{4} & 0 \\ \frac{1}{4} & \frac{1}{4} & -\frac{1}{2} \end{bmatrix}\begin{bmatrix} a \\ b \\ c \end{bmatrix} = \begin{bmatrix} 0 \\ \frac{3}{4}(a-b) \\ \frac{1}{4}(a+b-2c) \end{bmatrix} = \begin{bmatrix} 0 \\ 0 \\ 0 \end{bmatrix}.$$

which implies that $a=b=c$; so, we get

$$\mathbf{x}_1 = \begin{bmatrix} 1 \\ 1 \\ 1 \end{bmatrix}.$$

For $\lambda_2 = 1/4$, we have

$$\begin{bmatrix} \frac{3}{4} & 0 & 0 \\ \frac{3}{4} & 0 & 0 \\ \frac{1}{4} & \frac{1}{4} & \frac{1}{4} \end{bmatrix} \begin{bmatrix} a \\ b \\ c \end{bmatrix} = \begin{bmatrix} \frac{3}{4}a \\ \frac{3}{4}a \\ \frac{1}{4}(a+b+c) \end{bmatrix} = \begin{bmatrix} 0 \\ 0 \\ 0 \end{bmatrix},$$

which implies that $a=0$ and $b=-c$. Therefore,

$$\mathbf{x}_2 = \begin{bmatrix} 0 \\ -1 \\ 1 \end{bmatrix}.$$

Finally, for $\lambda_3 = 1/2$, we have

$$\begin{bmatrix} \frac{1}{2} & 0 & 0 \\ \frac{3}{4} & -\frac{1}{4} & 0 \\ \frac{1}{4} & \frac{1}{4} & 0 \end{bmatrix} \begin{bmatrix} a \\ b \\ c \end{bmatrix} = \begin{bmatrix} \frac{1}{2}a \\ \frac{1}{4}(3a-b) \\ \frac{1}{4}(a+b) \end{bmatrix} = \begin{bmatrix} 0 \\ 0 \\ 0 \end{bmatrix}.$$

which implies that $a=b=0$. Thus,

$$\mathbf{x}_3 = \begin{bmatrix} 0 \\ 0 \\ 1 \end{bmatrix}.$$

Constructing matrix B gives us

$$B = \begin{bmatrix} 1 & 0 & 0 \\ 1 & -1 & 0 \\ 1 & 1 & 1 \end{bmatrix}$$

and

$$B^{-1} = \begin{bmatrix} 1 & 0 & 0 \\ 1 & -1 & 0 \\ -2 & 1 & 1 \end{bmatrix}.$$

Thus,

$$B^{-1}AB = \begin{bmatrix} 1 & 0 & 0 \\ 1 & -1 & 0 \\ -2 & 1 & 1 \end{bmatrix} \begin{bmatrix} 1 & 0 & 0 \\ \frac{3}{4} & \frac{1}{4} & 0 \\ \frac{1}{4} & \frac{1}{4} & \frac{1}{2} \end{bmatrix} \begin{bmatrix} 1 & 0 & 0 \\ 1 & -1 & 0 \\ 1 & 1 & 1 \end{bmatrix}$$

$$= \begin{bmatrix} 1 & 0 & 0 \\ \frac{1}{4} & -\frac{1}{4} & 0 \\ -1 & \frac{1}{2} & \frac{1}{2} \end{bmatrix} \begin{bmatrix} 1 & 0 & 0 \\ 1 & -1 & 0 \\ 1 & 1 & 1 \end{bmatrix} = \begin{bmatrix} 1 & 0 & 0 \\ 0 & \frac{1}{4} & 0 \\ 0 & 0 & \frac{1}{2} \end{bmatrix}.$$

In the previous section, we showed how to calculate the probability distribution vectors at any stage n of a Markov process by using the relationship $\pi_n = \pi_0 P^n$. But the calculation of P^n could be prohibitive if the order of P, n, or both were large. So, instead of multiplying P by itself n times, we describe an alternative method for computing P^n.

Suppose that P is an $m \times m$ transition matrix. Further suppose that all of its eigenvalues are distinct. One can then theoretically calculate m eigenvectors of P. If we let C be the matrix that has columns that are the m eigenvectors, then $C^{-1}PC = \Lambda$, where Λ is a matrix on whose main diagonal appear the m eigenvalues; everywhere else are zeros.

Thus, $P = C\Lambda C^{-1}$ and $P^n = C\Lambda^n C^{-1}$. We state this result in Theorem 9.3.1.

■ **Theorem 9.3.1**

Let P be an $m \times m$ matrix with distinct eigenvalues, and let P equal $C\Lambda C^{-1}$, where C is a matrix whose columns are the eigenvectors of P and Λ is a diagonal matrix of the eigenvalues of P. Then, $P^n = C\Lambda^n C^{-1}$.

Proof The proof is induction. For $n=1$, we have $P = C\Lambda C^{-1}$, which is given. If we assume $P^n = C\Lambda^n C^{-1}$ to be true for $n = k$, then $P^{k+1} = P^k$. $P = (C\Lambda^k C^{-1})(C\Lambda C^{-1}) = C\Lambda^{k+1}C^{-1}$, and the theorem is proved.

Example 9.3.3

Let

$$P = \begin{bmatrix} \alpha & 1-\alpha \\ 1-\beta & \beta \end{bmatrix}.$$

We wish to find P^n. In the first step, we find the eigenvalues of P:

$$0 = |P - \lambda I| = \begin{bmatrix} \alpha - \lambda & 1-\alpha \\ 1-\beta & \beta - \lambda \end{bmatrix} = (\alpha - \lambda)(\beta - \lambda) - (1-\alpha)(1-\beta)$$

$$= \alpha\beta - (\alpha + \beta)\lambda + \lambda^2 - 1 + (\alpha + \beta) - \alpha\beta$$

$$= \lambda^2 - (\alpha + \beta)\lambda - (1 - \alpha - \beta).$$

Solving for λ, we get

$$\lambda = \frac{\alpha+\beta \pm \sqrt{(\alpha+\beta)^2 + 4(1-\alpha-\beta)}}{2} = \frac{\alpha+\beta \pm \sqrt{(\alpha+\beta)^2 - 4(\alpha+\beta) + 4}}{2}$$

$$= \frac{\alpha+\beta \pm \sqrt{(\alpha+\beta-2)^2}}{2}$$

$$= \frac{\alpha+\beta \pm (\alpha+\beta-2)}{2} = 1, \alpha+\beta-1.$$

Thus, the eigenvalues of P are 1 and $\alpha+\beta-1$. The second step is to find the eigenvectors of P: For $\lambda = 1$,

$$0 = (P - \lambda I)\mathbf{x} = \begin{bmatrix} \alpha-1 & 1-\alpha \\ 1-\beta & \beta-1 \end{bmatrix} \begin{bmatrix} x_1 \\ x_2 \end{bmatrix} = \begin{bmatrix} (\alpha-1)x_1 + (1-\alpha)x_2 \\ (1-\beta)x_1 + (\beta-1)x_2 \end{bmatrix}$$

$$= \begin{bmatrix} (1-\alpha) & (x_2 - x_1) \\ -(1-\beta) & (x_2 - x_1) \end{bmatrix} = \begin{bmatrix} 0 \\ 0 \end{bmatrix}.$$

Because $1-\alpha$, $1-\beta \neq 0$, we conclude that $x_2 = x_1$; and, therefore, an eigenvector is $\begin{bmatrix} 1 \\ 1 \end{bmatrix}$. For $\lambda = \alpha+\beta-1$,

$$0 = (P - \lambda I)\mathbf{x} = \begin{bmatrix} 1-\beta & 1-\alpha \\ 1-\beta & 1-\alpha \end{bmatrix} \begin{bmatrix} x_1 \\ x_2 \end{bmatrix} = \begin{bmatrix} (1-\beta)x_1 + (1-\alpha)x_2 \\ (1-\beta)x_1 + (1-\alpha)x_2 \end{bmatrix} = \begin{bmatrix} 0 \\ 0 \end{bmatrix}.$$

The solution is that $x_1 = 1-\alpha$ and $x_2 = -1(1-\beta)$. Thus, the other eigenvector is $\begin{bmatrix} 1-\alpha \\ -(1-\beta) \end{bmatrix}$. We can now determine matrix C, the columns of which are the eigenvectors. Thus,

$$C = \begin{bmatrix} 1 & 1-\alpha \\ 1 & -(1-\beta) \end{bmatrix}$$

and

$$C^{-1} = -\frac{1}{2-\alpha-\beta} \begin{bmatrix} -(1-\beta) & -(1-\alpha) \\ -1 & 1 \end{bmatrix} = \frac{1}{2-\alpha-\beta} \begin{bmatrix} 1-\beta & 1-\alpha \\ 1 & -1 \end{bmatrix}.$$

Finally, we can find

$$P^n = C\Lambda^n C^{-1} = \frac{1}{2-\alpha-\beta}\begin{bmatrix} 1 & 1-\alpha \\ 1 & -(1-\beta) \end{bmatrix}\begin{bmatrix} 1 & 0 \\ 0 & (\alpha+\beta-1)^n \end{bmatrix}\begin{bmatrix} 1-\beta & 1-\alpha \\ 1 & -1 \end{bmatrix}$$

$$= \frac{1}{2\alpha-\beta}\begin{bmatrix} 1 & (1-\alpha)(\alpha+\beta-1)^n \\ 1 & -(1-\beta)(\alpha+\beta-1)^n \end{bmatrix}\begin{bmatrix} 1-\beta & 1-\alpha \\ 1 & -1 \end{bmatrix}$$

$$= \frac{1}{2-\alpha-\beta}\begin{bmatrix} 1-\beta+(1-\alpha)(\alpha+\beta-1)^n & (1-\alpha)\left[1-(\alpha+\beta-1)^n\right] \\ (1-\beta)\left[1-(\alpha+\beta-1)^n\right] & 1-\alpha+(1-\beta)(\alpha+\beta-1)^n \end{bmatrix}.$$

Example 9.3.4

Let

$$P = \begin{bmatrix} 1 & 0 & 0 \\ \frac{1}{2} & \frac{1}{2} & 0 \\ \frac{1}{3} & \frac{1}{2} & \frac{1}{6} \end{bmatrix}.$$

To find the eigenvalues, we write

$$0 = |P - \lambda I| = \begin{vmatrix} 1-\lambda & 0 & 0 \\ \frac{1}{2} & \frac{1}{2}-\lambda & 0 \\ \frac{1}{3} & \frac{1}{2} & \frac{1}{6}-\lambda \end{vmatrix} = (1-\lambda)\left(\frac{1}{2}-\lambda\right)\left(\frac{1}{6}-\lambda\right) = 0,$$

which implies that $\lambda = 1, 1/2, 1/6$. To find the eigenvector for $\lambda = 1$, we must solve

$$\mathbf{0} = (P - \lambda I) = \begin{bmatrix} 0 & 0 & 0 \\ \frac{1}{2} & -\frac{1}{2} & 0 \\ \frac{1}{3} & \frac{1}{2} & -\frac{5}{6} \end{bmatrix}\begin{bmatrix} x_1 \\ x_2 \\ x_3 \end{bmatrix} = \begin{bmatrix} 0 \\ \frac{1}{2}x_1 - \frac{1}{2}x_2 \\ \frac{1}{3}x_1 + \frac{1}{2}x_2 - \frac{5}{6}x_3 \end{bmatrix} = \begin{bmatrix} 0 \\ 0 \\ 0 \end{bmatrix},$$

which implies that $x_1 = x_2$ and $x_2 = x_3$. Therefore, an eigenvector corresponding to $\lambda = 1$ is $\begin{bmatrix} 1 \\ 1 \\ 1 \end{bmatrix}$.

For $\lambda = 1/2$, we have

$$\begin{bmatrix} \frac{1}{2} & 0 & 0 \\ \frac{1}{2} & 0 & 0 \\ \frac{1}{3} & \frac{1}{2} & -\frac{1}{3} \end{bmatrix} \begin{bmatrix} x_1 \\ x_2 \\ x_3 \end{bmatrix} = \begin{bmatrix} \frac{1}{2}x_1 \\ \frac{1}{2}x_2 \\ \frac{1}{3}x_1 + \frac{1}{2}x_2 - \frac{1}{3}x_3 \end{bmatrix} = \begin{bmatrix} 0 \\ 0 \\ 0 \end{bmatrix}.$$

This equation implies that $x_1 = 0$ and $3x_2 = 2x_3$. So, if we let $x_3 = 3$, then $x_2 = 2$ and an eigenvector for $\lambda = 1/2$ is $\begin{bmatrix} 0 \\ 2 \\ 3 \end{bmatrix}$. Finally, for $\lambda = 1/6$, we write

$$\begin{bmatrix} \frac{5}{6} & 0 & 0 \\ \frac{1}{2} & \frac{1}{3} & 0 \\ \frac{1}{3} & \frac{1}{2} & 0 \end{bmatrix} \begin{bmatrix} x_1 \\ x_2 \\ x_3 \end{bmatrix} = \begin{bmatrix} \frac{5}{6}x_1 \\ \frac{1}{2}x_1 + \frac{1}{3}x_2 \\ \frac{1}{3}x_1 + \frac{1}{2}x_2 \end{bmatrix} = \begin{bmatrix} 0 \\ 0 \\ 0 \end{bmatrix},$$

which requires that $x_1 = x_2 = 0$. Thus, an eigenvector for $\lambda = 1/6$ is $\begin{bmatrix} 0 \\ 0 \\ 1 \end{bmatrix}$.

Matrix C then is

$$\begin{bmatrix} 1 & 0 & 0 \\ 1 & 2 & 0 \\ 1 & 3 & 1 \end{bmatrix},$$

and

$$C^{-1} = \frac{1}{2} \begin{bmatrix} 2 & 0 & 0 \\ -1 & 1 & 0 \\ 1 & -3 & 2 \end{bmatrix}.$$

To find P^n, we write $P^n = C\Lambda^n C^{-1}$ or

$$P^n = \frac{1}{2} \begin{bmatrix} 1 & 0 & 0 \\ 1 & 2 & 0 \\ 1 & 3 & 1 \end{bmatrix} \begin{bmatrix} 1 & 0 & 0 \\ 0 & \frac{1}{2^n} & 0 \\ 0 & 0 & \frac{1}{6^n} \end{bmatrix} \begin{bmatrix} 2 & 0 & 0 \\ -1 & 1 & 0 \\ 1 & -3 & 1 \end{bmatrix}$$

$$= \frac{1}{2} \begin{bmatrix} 1 & 0 & 0 \\ 1 & \frac{2}{2^n} & 0 \\ 0 & \frac{3}{2^n} & \frac{1}{6^n} \end{bmatrix} \begin{bmatrix} 2 & 0 & 0 \\ -1 & 1 & 0 \\ 1 & -3 & 2 \end{bmatrix}$$

$$= \frac{1}{2} \begin{bmatrix} 2 & 0 & 0 \\ 2 - \frac{2}{2^n} & \frac{2}{2^n} & 0 \\ 2 - \frac{3}{2^n} + \frac{1}{6^n} & \frac{3}{2^n} - \frac{3}{6^n} & \frac{2}{6^n} \end{bmatrix}$$

$$= \begin{bmatrix} 1 & 0 & 0 \\ 1 - \frac{1}{2^n} & \frac{1}{2^n} & 0 \\ 1 - \frac{3}{2^{n+1}} + \frac{3}{6^{n+1}} & \frac{3}{2^{n+1}} - \frac{9}{6^{n+1}} & \frac{1}{6^n} \end{bmatrix}.$$

If the eigenvalues of P are not all distinct, then in general, we cannot find matrix B such that $B^{-1}PB = \Lambda$. Instead, we find matrix C such that $C^{-1}PC = J$, where J is the *Jordan canonical form* of matrix P, which has all eigenvalues of P on its main diagonal, some on its first superdiagonal (the diagonal row of elements just above the main diagonal), and zeros elsewhere. The first superdiagonal of J can be made to consist of ones, located immediately above those eigenvalues on the main diagonal that are repetitions of eigenvalues that have already occurred on the main diagonal (viewed from upper left to lower right), with zeros elsewhere. For example, suppose that the eigenvalues of P are $1/2$, $1/4$, $1/4$, and $1/4$. The Jordan canonical form can be made to look like any of the following four configurations:

1. $\begin{bmatrix} \frac{1}{2} & 0 & 0 & 0 \\ 0 & \frac{1}{4} & 1 & 0 \\ 0 & 0 & \frac{1}{4} & 1 \\ 0 & 0 & 0 & \frac{1}{4} \end{bmatrix}$

2. $\begin{bmatrix} \frac{1}{4} & 0 & 0 & 0 \\ 0 & \frac{1}{2} & 1 & 0 \\ 0 & 0 & \frac{1}{4} & 1 \\ 0 & 0 & 0 & \frac{1}{4} \end{bmatrix}$

3. $\begin{bmatrix} \frac{1}{4} & 1 & 0 & 0 \\ 0 & \frac{1}{4} & 0 & 0 \\ 0 & 0 & \frac{1}{2} & 1 \\ 0 & 0 & 0 & \frac{1}{4} \end{bmatrix}$

4. $\begin{bmatrix} \frac{1}{4} & 1 & 0 & 0 \\ 0 & \frac{1}{4} & 1 & 0 \\ 0 & 0 & \frac{1}{4} & 0 \\ 0 & 0 & 0 & \frac{1}{2} \end{bmatrix}$

We notice that 1 never occurs on the first superdiagonal immediately above $1/2$ because $1/2$ is not repeated on the main diagonal. Thus, 0 always appears on the first superdiagonal immediately above $1/2$. Viewing the main diagonal from upper left to lower right, $1/4$ appears three times. On the first superdiagonal, immediately above the first occurrence of $1/4$, 0 appears; however, on the first superdiagonal, immediately above the second and third occurrences of $1/4$, 1 appears. Henceforth, when we speak of J, we mean the form of J as shown in part 1, which can be obtained by the proper ordering of the columns of C. In general, then, the main diagonal of J, going from upper left to lower right, has as its first elements the unrepeated eigenvalues of A. The repeating eigenvalues of A appear in unbroken strings of length equal to their respective multiplicities. Finally, the first superdiagonal of J consists of ones and zeros, the location of which was described previously.

We denote any square matrix that has ones and zeros on its first superdiagonal and zeros elsewhere as matrix M. Observe that M^2 has ones and zeros only on its second superdiagonal (the diagonal row immediately above the first superdiagonal) and zeros elsewhere. Similarly, M^3 has ones and zeros only on its third superdiagonal and zeros elsewhere. Thus, if M is a $k \times k$ matrix, $M^k = 0$.

According to the preceding paragraph, we can now express J as $J = \Lambda + M$ so that, if $C^{-1}PC = J$, we can write $C^{-1}PC = \Lambda + M$, which may be rewritten as $P = C(\Lambda + M)C^{-1}$. Finally, by Theorem 9.2.1, we have

$$P^n = C(\Lambda + M)^n C^{-1}.$$

9.3 Evaluation of P^n

If P is a $k \times k$ matrix, then so is Λ and so is M. Accordingly,

$$(\Lambda+M)^n = \sum_{i=1}^{n}\binom{h}{i} M^i \Lambda^{n-1} = \Lambda^n + nM\Lambda^{n-1} + \frac{n(n-1)}{2} M^2 \Lambda^{n-2} + \cdots + M^n.$$

But because $M^j = 0$ for $j \geq k$, we have

$$(\Lambda+M)^n = \Lambda^n + nM\Lambda^{n-1} + \frac{n(n-1)}{2} M^2 \Lambda^{n-2} + \cdots$$

$$+ \frac{n(n-1)\cdots(n-k+2)}{(k-1)!} M^{k-1} \Lambda^{n-k+1}$$

or

$$(\Lambda+M)^n = \sum_{i=1}^{k-1}\binom{n}{i} M^i \Lambda^{n-1}.$$

To illustrate, suppose A is a 4×4 matrix with the eigenvalues $1/2$, $1/4$, $1/4$, and $1/4$. Then, $J = \Lambda + M$ may be expressed as

$$\begin{bmatrix} \frac{1}{2} & 0 & 0 & 0 \\ 0 & \frac{1}{4} & 1 & 0 \\ 0 & 0 & \frac{1}{4} & 1 \\ 0 & 0 & 0 & \frac{1}{4} \end{bmatrix} = \begin{bmatrix} \frac{1}{2} & 0 & 0 & 0 \\ 0 & \frac{1}{4} & 0 & 0 \\ 0 & 0 & \frac{1}{4} & 0 \\ 0 & 0 & 0 & \frac{1}{4} \end{bmatrix} + \begin{bmatrix} 0 & 0 & 0 & 0 \\ 0 & 0 & 1 & 0 \\ 0 & 0 & 0 & 1 \\ 0 & 0 & 0 & 0 \end{bmatrix},$$

and $J^n = (\Lambda+M)^n$ may be written as

$$\begin{bmatrix} \frac{1}{2} & 0 & 0 & 0 \\ 0 & \frac{1}{4} & 1 & 0 \\ 0 & 0 & \frac{1}{4} & 1 \\ 0 & 0 & 0 & \frac{1}{4} \end{bmatrix}^n = \left(\begin{bmatrix} \frac{1}{2} & 0 & 0 & 0 \\ 0 & \frac{1}{4} & 0 & 0 \\ 0 & 0 & \frac{1}{4} & 0 \\ 0 & 0 & 0 & \frac{1}{4} \end{bmatrix} + \begin{bmatrix} 0 & 0 & 0 & 0 \\ 0 & 0 & 1 & 0 \\ 0 & 0 & 0 & 1 \\ 0 & 0 & 0 & 0 \end{bmatrix}\right)^n$$

$$= \begin{bmatrix} \frac{1}{2} & 0 & 0 & 0 \\ 0 & \frac{1}{4} & 0 & 0 \\ 0 & 0 & \frac{1}{4} & 0 \\ 0 & 0 & 0 & \frac{1}{4} \end{bmatrix} + n\begin{bmatrix} 0 & 0 & 0 & 0 \\ 0 & 0 & 1 & 0 \\ 0 & 0 & 0 & 1 \\ 0 & 0 & 0 & 0 \end{bmatrix}\begin{bmatrix} \frac{1}{2} & 0 & 0 & 0 \\ 0 & \frac{1}{4} & 0 & 0 \\ 0 & 0 & \frac{1}{4} & 0 \\ 0 & 0 & 0 & \frac{1}{4} \end{bmatrix}^{-1}$$

$$+\frac{n(n-1)}{2}\begin{bmatrix}0&0&0&0\\0&0&1&0\\0&0&0&1\\0&0&0&0\end{bmatrix}\begin{bmatrix}\frac{1}{2}&0&0&0\\0&\frac{1}{4}&0&0\\0&0&\frac{1}{4}&0\\0&0&0&\frac{1}{4}\end{bmatrix}^{n-2}+\frac{m(n-1)(n-2)}{6}\begin{bmatrix}0&0&0&0\\0&0&1&0\\0&0&0&1\\0&0&0&0\end{bmatrix}\begin{bmatrix}\frac{1}{2}&0&0&0\\0&\frac{1}{4}&0&0\\0&0&\frac{1}{4}&0\\0&0&0&\frac{1}{4}\end{bmatrix}^{n-3}$$

$$=\begin{bmatrix}\frac{1}{2^n}&0&0&0\\0&\frac{1}{4^n}&0&0\\0&0&\frac{1}{4^n}&0\\0&0&0&\frac{1}{4^n}\end{bmatrix}+\begin{bmatrix}0&0&0&0\\0&0&n&0\\0&0&0&n\\0&0&0&0\end{bmatrix}\begin{bmatrix}\frac{1}{2^{n-1}}&0&0&0\\0&\frac{1}{4^{n-1}}&0&0\\0&0&\frac{1}{4^{n-1}}&0\\0&0&0&\frac{1}{4^{n-1}}\end{bmatrix}$$

$$+\begin{bmatrix}0&0&0&0\\0&0&1&\frac{n(n-1)}{2}\\0&0&0&0\\0&0&0&0\end{bmatrix}\begin{bmatrix}\frac{1}{2^{n-2}}&0&0&0\\0&\frac{1}{4^{n-2}}&0&0\\0&0&\frac{1}{4^{n-2}}&0\\0&0&0&\frac{1}{4^{n-2}}\end{bmatrix}+\begin{bmatrix}0&0&0&0\\0&0&0&0\\0&0&0&0\\0&0&0&0\end{bmatrix}\begin{bmatrix}\frac{1}{2^{n-3}}&0&0&0\\0&\frac{1}{4^{n-3}}&0&0\\0&0&\frac{1}{4^{n-3}}&0\\0&0&0&\frac{1}{4^{n-3}}\end{bmatrix}$$

$$=\begin{bmatrix}\frac{1}{2^n}&0&0&0\\0&\frac{1}{4^n}&0&0\\0&0&\frac{1}{4^n}&0\\0&0&0&\frac{1}{4^n}\end{bmatrix}+\begin{bmatrix}0&0&0&0\\0&0&\frac{1}{4^{n-1}}&0\\0&0&0&\frac{1}{4^{n-1}}\\0&0&0&0\end{bmatrix}+\begin{bmatrix}0&0&0&0\\0&0&0&\frac{n(n-1)1}{2\cdot 4^{n-2}}\\0&0&0&0\\0&0&0&0\end{bmatrix}+\begin{bmatrix}0&0&0&0\\0&0&0&0\\0&0&0&0\\0&0&0&0\end{bmatrix}$$

$$=\begin{bmatrix}\frac{1}{2^n}&0&0&0\\0&\frac{1}{4^n}&\frac{n}{4^{n-1}}&\frac{n(n-1)}{2\cdot 4^{n-2}}\\0&0&\frac{1}{4^n}&\frac{n}{4^{n-1}}\\0&0&0&\frac{1}{4^n}\end{bmatrix}.$$

Thus, if we premultiply this matrix by C and postmultiply it by C^{-1}, we obtain P^n.

9.3 Evaluation of P^n

We now indicate the procedure for obtaining C such that $C^{-1}PC = J$, where P has eigenvalues of multiplicity greater than or equal to one. Let P be a square $k \times k$ matrix, the ith eigenvalue of which occurs with the multiplicity $m_i, i = 1, 2, \ldots, l$. Thus, P has l different eigenvalues, and $\sum_{i=1}^{l} m_i = k$. We wish to obtain matrix C such that $C^{-1}PC = J$, where J is as defined earlier.

To do so, we proceed as follows:

1. Determine an eigenvector for λ_i; call it $\mathbf{x}_1$.
2. Determine $\mathbf{x}_2$ such that $(A - \lambda_i I)\mathbf{x}_2 = \mathbf{x}_1$.
3. Determine $\mathbf{x}_3$ such that $(A - \lambda_i I)\mathbf{x}_3 = \mathbf{x}_2$.
4. In general, determine $\mathbf{x}_j$ such that $(A - \lambda_i I)\mathbf{x}_j = \mathbf{x}_{j-1}$.
5. Continue in this matter until $j = m_i, i = 1, 2, \ldots, l$.
6. Use the eigenvectors so derived, in the order of derivation, to make up the columns of C.

Examples 9.3.5 and 9.4.6 illustrate this procedure.

Example 9.3.5

Let

$$A = \begin{pmatrix} 1 & 0 & 0 & 0 \\ \frac{3}{4} & \frac{1}{4} & 0 & 0 \\ \frac{1}{2} & \frac{1}{4} & \frac{1}{4} & 0 \\ \frac{1}{4} & \frac{1}{4} & \frac{1}{4} & \frac{1}{4} \end{pmatrix}$$

so that $\lambda_1 = 1$, $\lambda_2 = 1/4$, $\lambda_3 = 1/4$, and $\lambda_4 = 1/4$.

The eigenvector for $\lambda_1 = 1$,

$$\begin{pmatrix} 0 & 0 & 0 & 0 \\ \frac{3}{4} & -\frac{3}{4} & 0 & 0 \\ \frac{1}{2} & \frac{1}{4} & -\frac{3}{4} & 0 \\ \frac{1}{4} & \frac{1}{4} & \frac{1}{4} & -\frac{3}{4} \end{pmatrix} \begin{pmatrix} a \\ b \\ c \\ d \end{pmatrix} = \begin{pmatrix} 0 \\ \frac{3}{4}(a-b) \\ \frac{1}{4}(2a+b-3c) \\ \frac{1}{4}(a+b+c-3d) \end{pmatrix} = \begin{pmatrix} 0 \\ 0 \\ 0 \\ 0 \end{pmatrix},$$

implies that $a = b = c = d$. Therefore,

$$\mathbf{x}_1 = \begin{pmatrix} 1 \\ 1 \\ 1 \\ 1 \end{pmatrix}.$$

The eigenvector for $\lambda_2 = 1/4$,

$$\begin{pmatrix} \frac{3}{4} & 0 & 0 & 0 \\ \frac{3}{4} & 0 & 0 & 0 \\ \frac{1}{2} & \frac{1}{4} & 0 & 0 \\ \frac{1}{4} & \frac{1}{4} & \frac{1}{4} & 0 \end{pmatrix} \begin{pmatrix} a \\ b \\ c \\ d \end{pmatrix} = \begin{pmatrix} \frac{3a}{4} \\ \frac{3a}{4} \\ \frac{1}{4}(2a+b) \\ \frac{1}{4}(a+b+c) \end{pmatrix} = \begin{pmatrix} 0 \\ 0 \\ 0 \\ 0 \end{pmatrix},$$

implies that $a = b = c = 0$. Therefore,

$$\mathbf{x}_2 = \begin{pmatrix} 0 \\ 0 \\ 0 \\ 1 \end{pmatrix}.$$

The eigenvector for $\lambda_3 = 1/4$,

$$\begin{pmatrix} \frac{3}{4} & 0 & 0 & 0 \\ \frac{3}{4} & 0 & 0 & 0 \\ \frac{1}{2} & \frac{1}{4} & 0 & 0 \\ \frac{1}{4} & \frac{1}{4} & \frac{1}{4} & 0 \end{pmatrix} \begin{pmatrix} a \\ b \\ c \\ d \end{pmatrix} = \begin{pmatrix} \frac{3a}{4} \\ \frac{3a}{4} \\ \frac{1}{4}(2a+b) \\ \frac{1}{4}(a+b+c) \end{pmatrix} = \begin{pmatrix} 0 \\ 0 \\ 0 \\ 1 \end{pmatrix},$$

implies that $a = b = 0$ and $c = 4$. Therefore,

$$\mathbf{x}_3 = \begin{pmatrix} 0 \\ 0 \\ 4 \\ 1 \end{pmatrix}.$$

The eigenvector for $\lambda_4 = 1/4$,

$$\begin{pmatrix} \frac{3}{4} & 0 & 0 & 0 \\ \frac{3}{4} & 0 & 0 & 0 \\ \frac{1}{2} & \frac{1}{4} & 0 & 0 \\ \frac{1}{4} & \frac{1}{4} & \frac{1}{4} & 0 \end{pmatrix} \begin{pmatrix} a \\ b \\ c \\ d \end{pmatrix} = \begin{pmatrix} \frac{3a}{4} \\ \frac{3a}{4} \\ \frac{1}{4}(2a+b) \\ \frac{1}{4}(a+b+c) \end{pmatrix} = \begin{pmatrix} 0 \\ 0 \\ 4 \\ 1 \end{pmatrix},$$

implies that $a = b = 0$ and $c = 4$. Therefore,

$$\mathbf{x}_3 = \begin{pmatrix} 0 \\ 0 \\ 4 \\ 1 \end{pmatrix}.$$

The eigenvector for $\lambda_4 = 1/4$,

$$\begin{pmatrix} \frac{3}{4} & 0 & 0 & 0 \\ \frac{3}{4} & 0 & 0 & 0 \\ \frac{1}{2} & \frac{1}{4} & 0 & 0 \\ \frac{1}{4} & \frac{1}{4} & \frac{1}{4} & 0 \end{pmatrix} \begin{pmatrix} a \\ b \\ c \\ d \end{pmatrix} = \begin{pmatrix} \frac{3a}{4} \\ \frac{3a}{4} \\ \frac{1}{4}(2a+b) \\ \frac{1}{4}(a+b+c) \end{pmatrix} = \begin{pmatrix} 0 \\ 0 \\ 4 \\ 1 \end{pmatrix},$$

implies that $a=0$, $b=16$, and $c=-12$. Therefore,

$$\mathbf{x}_4 = \begin{pmatrix} 0 \\ 16 \\ -12 \\ 1 \end{pmatrix}.$$

Thus, we obtain

$$C = \begin{pmatrix} 1 & 0 & 0 & 0 \\ 1 & 0 & 0 & 16 \\ 1 & 0 & 4 & -12 \\ 1 & 1 & 1 & 1 \end{pmatrix}.$$

As an exercise, calculate C^{-1} and show that

$$C^{-1}PC = J = \begin{pmatrix} 1 & 0 & 0 & 0 \\ 0 & \frac{1}{4} & 1 & 0 \\ 0 & 0 & \frac{1}{4} & 0 \\ 0 & 0 & 0 & \frac{1}{4} \end{pmatrix};$$

finally, calculate $P^n = CJ^nC^{-1} = C(\Lambda+M)^n C^{-1}$, where

$$\Lambda = \begin{pmatrix} 1 & 0 & 0 & 0 \\ 0 & \frac{1}{4} & 0 & 0 \\ 0 & 0 & \frac{1}{4} & 0 \\ 0 & 0 & 0 & \frac{1}{4} \end{pmatrix}$$

and
$$M = \begin{pmatrix} 0 & 0 & 0 & 0 \\ 0 & 0 & 1 & 0 \\ 0 & 0 & 0 & 1 \\ 0 & 0 & 0 & 0 \end{pmatrix}.$$

Example 9.3.6

Let
$$P = \begin{pmatrix} \frac{1}{2} & 0 & 0 & \frac{1}{2} \\ 0 & \frac{1}{2} & \frac{1}{2} & 0 \\ \frac{1}{2} & 0 & \frac{1}{2} & 0 \\ 0 & 0 & 0 & 1 \end{pmatrix}.$$

The characteristic equation of P is

$$0 = |P - \lambda I|$$

$$= \begin{vmatrix} \frac{1}{2}-\lambda & 0 & 0 & \frac{1}{2} \\ 0 & \frac{1}{2}-\lambda & \frac{1}{2} & 0 \\ \frac{1}{2} & 0 & \frac{1}{2}-\lambda & 0 \\ 0 & 0 & 0 & 1-\lambda \end{vmatrix}$$

$$= (1-\lambda) \begin{vmatrix} \frac{1}{2}-\lambda & 0 & 0 \\ 0 & \frac{1}{2}-\lambda & \frac{1}{2} \\ \frac{1}{2} & 0 & \frac{1}{2}-\lambda \end{vmatrix}$$

$$= (1-\lambda)\left(\frac{1}{2}-\lambda\right) \begin{vmatrix} \frac{1}{2}-\lambda & \frac{1}{2} \\ 0 & \frac{1}{2}-\lambda \end{vmatrix}$$

$$= (1-\lambda)\left(\frac{1}{2}-\lambda\right)^3,$$

which implies that $\lambda_1 = 1$ and $\lambda_2 = \lambda_3 = \lambda_4 = 1/2$.

The eigenvector for $\lambda_1 = 1$,

$$\begin{pmatrix} -\frac{1}{2} & 0 & 0 & \frac{1}{2} \\ 0 & -\frac{1}{2} & \frac{1}{2} & 0 \\ \frac{1}{2} & 0 & -\frac{1}{2} & 0 \\ 0 & 0 & 0 & 0 \end{pmatrix} \begin{pmatrix} a \\ b \\ c \\ d \end{pmatrix} = \begin{pmatrix} \frac{1}{2}(d-a) \\ \frac{1}{2}(c-b) \\ \frac{1}{2}(a-c) \\ 0 \end{pmatrix} = \begin{pmatrix} 0 \\ 0 \\ 0 \\ 0 \end{pmatrix},$$

implies that $a = b = c = d$. Therefore,

$$\mathbf{x}_1 = \begin{pmatrix} 1 \\ 1 \\ 1 \\ 1 \end{pmatrix}.$$

The eigenvector for $\lambda_2 = 1/2$,

$$\begin{pmatrix} 0 & 0 & 0 & \frac{1}{2} \\ 0 & 0 & \frac{1}{2} & 0 \\ \frac{1}{2} & 0 & 0 & 0 \\ 0 & 0 & 0 & \frac{1}{2} \end{pmatrix} \begin{pmatrix} a \\ b \\ c \\ d \end{pmatrix} = \begin{pmatrix} \frac{d}{2} \\ \frac{c}{2} \\ \frac{a}{2} \\ \frac{d}{2} \end{pmatrix} = \begin{pmatrix} 0 \\ 0 \\ 0 \\ 0 \end{pmatrix},$$

implies that $a = b = c = 0$. Therefore,

$$\mathbf{x}_2 = \begin{pmatrix} 0 \\ 1 \\ 0 \\ 0 \end{pmatrix}.$$

The eigenvector for $\lambda_3 = 1/2$,

$$\begin{pmatrix} \frac{d}{2} \\ \frac{c}{2} \\ \frac{a}{2} \\ \frac{d}{2} \end{pmatrix} = \begin{pmatrix} 0 \\ 1 \\ 0 \\ 0 \end{pmatrix},$$

implies that $a = b = 0$ and $c = 2$. Therefore,

$$\mathbf{x}_3 = \begin{pmatrix} 0 \\ 1 \\ 2 \\ 0 \end{pmatrix}.$$

The eigenvector for $\lambda_4 = 1/2$,

$$\begin{pmatrix} \frac{d}{2} \\ \frac{c}{2} \\ \frac{a}{2} \\ \frac{d}{2} \end{pmatrix} = \begin{pmatrix} 0 \\ 1 \\ 2 \\ 0 \end{pmatrix},$$

implies that $a = 4$, $c = 2$, and $d = 0$. Therefore,

$$\mathbf{x}_4 = \begin{pmatrix} 4 \\ 1 \\ 2 \\ 0 \end{pmatrix}.$$

Thus, we obtain

$$C = \begin{pmatrix} 1 & 0 & 0 & 4 \\ 1 & 1 & 1 & 1 \\ 1 & 0 & 2 & 2 \\ 1 & 0 & 0 & 0 \end{pmatrix}$$

and

$$C^{-1} = \begin{pmatrix} 0 & 0 & 0 & 1 \\ 0 & 1 & -\frac{1}{2} & -\frac{1}{2} \\ -\frac{1}{4} & 0 & \frac{1}{2} & -\frac{1}{4} \\ \frac{1}{4} & 0 & 0 & -\frac{1}{4} \end{pmatrix}.$$

As a check to see that $C^{-1}PC = J$, we write

$$\begin{pmatrix} 0 & 0 & 0 & 1 \\ 0 & 1 & -\frac{1}{2} & -\frac{1}{2} \\ -\frac{1}{4} & 0 & \frac{1}{2} & -\frac{1}{4} \\ \frac{1}{4} & 0 & 0 & -\frac{1}{4} \end{pmatrix} \begin{pmatrix} \frac{1}{2} & 0 & 0 & \frac{1}{2} \\ 0 & \frac{1}{2} & \frac{1}{2} & 0 \\ \frac{1}{2} & 0 & \frac{1}{2} & 0 \\ 0 & 0 & 0 & 1 \end{pmatrix} \begin{pmatrix} 1 & 0 & 0 & 4 \\ 1 & 1 & 1 & 1 \\ 1 & 0 & 2 & 2 \\ 1 & 0 & 0 & 0 \end{pmatrix}$$

9.3 Evaluation of P^n

$$= \begin{pmatrix} 0 & 0 & 0 & 1 \\ -\frac{1}{4} & \frac{1}{2} & \frac{1}{4} & -\frac{1}{2} \\ \frac{1}{8} & 0 & \frac{1}{4} & -\frac{3}{8} \\ \frac{1}{8} & 0 & 0 & -\frac{1}{8} \end{pmatrix} \begin{pmatrix} 1 & 0 & 0 & 4 \\ 1 & 1 & 1 & 1 \\ 1 & 0 & 2 & 2 \\ 1 & 0 & 0 & 0 \end{pmatrix} = \begin{pmatrix} 1 & 0 & 0 & 0 \\ 0 & \frac{1}{2} & 1 & 0 \\ 0 & 0 & \frac{1}{2} & 0 \\ 0 & 0 & 0 & \frac{1}{2} \end{pmatrix}.$$

To calculate P^n, we write $P^n = CJ^n C^{-1}$ or

$$P^n = C(\Lambda + M)^n C^{-1}. \tag{9.3.1}$$

First, we must calculate $(\Lambda + M)^n$:

$$(\Lambda + M)^n = \left(\begin{pmatrix} 1 & 0 & 0 & 0 \\ 0 & \frac{1}{2} & 1 & 0 \\ 0 & 0 & \frac{1}{2} & 0 \\ 0 & 0 & 0 & \frac{1}{2} \end{pmatrix} + \begin{pmatrix} 0 & 0 & 0 & 0 \\ 0 & 0 & 1 & 0 \\ 0 & 0 & 0 & 1 \\ 0 & 0 & 0 & 0 \end{pmatrix} \right)^n$$

$$= \begin{pmatrix} 1 & 0 & 0 & 0 \\ 0 & \frac{1}{2^n} & 0 & 0 \\ 0 & 0 & \frac{1}{2^n} & 0 \\ 0 & 0 & 0 & \frac{1}{2^n} \end{pmatrix} + \begin{pmatrix} 0 & 0 & 0 & 0 \\ 0 & 0 & n & 0 \\ 0 & 0 & 0 & n \\ 0 & 0 & 0 & 0 \end{pmatrix} \begin{pmatrix} 1 & 0 & 0 & 0 \\ 0 & \frac{1}{2^{n-1}} & 0 & 0 \\ 0 & 0 & \frac{1}{2^{n-1}} & 0 \\ 0 & 0 & 0 & \frac{1}{2^{n-1}} \end{pmatrix}$$

$$+ \begin{pmatrix} 0 & 0 & 0 & 0 \\ 0 & 0 & 0 & \frac{n(n-1)}{2} \\ 0 & 0 & 0 & 0 \\ 0 & 0 & 0 & 0 \end{pmatrix} \begin{pmatrix} 1 & 0 & 0 & 0 \\ 0 & \frac{1}{2^{n-2}} & 0 & 0 \\ 0 & 0 & \frac{1}{2^{n-2}} & 0 \\ 0 & 0 & 0 & \frac{1}{2^{n-2}} \end{pmatrix}$$

$$+ \begin{pmatrix} 0 & 0 & 0 & 0 \\ 0 & 0 & 0 & 0 \\ 0 & 0 & 0 & 0 \\ 0 & 0 & 0 & 0 \end{pmatrix} \begin{pmatrix} 1 & 0 & 0 & 0 \\ 0 & \frac{1}{2^{n-3}} & 0 & 0 \\ 0 & 0 & \frac{1}{2^{n-3}} & 0 \\ 0 & 0 & 0 & \frac{1}{2^{n-3}} \end{pmatrix}$$

$$= \begin{pmatrix} 1 & 0 & 0 & 0 \\ 0 & \frac{1}{2^n} & 0 & 0 \\ 0 & 0 & \frac{1}{2^n} & 0 \\ 0 & 0 & 0 & \frac{1}{2^n} \end{pmatrix} + \begin{pmatrix} 0 & 0 & 0 & 0 \\ 0 & 0 & \frac{n}{2^{n-1}} & 0 \\ 0 & 0 & 0 & \frac{n}{2^{n-1}} \\ 0 & 0 & 0 & 0 \end{pmatrix} + \begin{pmatrix} 0 & 0 & 0 & 0 \\ 0 & 0 & 0 & \frac{n(n-1)}{2^{n-1}} \\ 0 & 0 & 0 & 0 \\ 0 & 0 & 0 & 0 \end{pmatrix} + 0$$

$$= \begin{pmatrix} 1 & 0 & 0 & 0 \\ 0 & \frac{1}{2^n} & \frac{n}{2^{n-1}} & \frac{n(n-1)}{2^{n-1}} \\ 0 & 0 & \frac{1}{2^n} & \frac{n}{2^{n-1}} \\ 0 & 0 & 0 & \frac{1}{2^n} \end{pmatrix}.$$

Substituting this value into Equation (9.3.1) gives

$$P^n = \begin{pmatrix} 1 & 0 & 0 & 4 \\ 1 & 1 & 1 & 1 \\ 1 & 0 & 2 & 2 \\ 1 & 0 & 0 & 0 \end{pmatrix} \begin{pmatrix} 1 & 0 & 0 & 0 \\ 0 & \frac{1}{2^n} & \frac{n}{2^{n-1}} & \frac{n(n-1)}{2^{n-1}} \\ 0 & 0 & \frac{1}{2^n} & \frac{n}{2^{n-1}} \\ 0 & 0 & 0 & \frac{1}{2^n} \end{pmatrix} \begin{pmatrix} 0 & 0 & 0 & 1 \\ 0 & 1 & -\frac{1}{2} & -\frac{1}{2} \\ -\frac{1}{4} & 0 & \frac{1}{2} & -\frac{1}{4} \\ \frac{1}{4} & 0 & 0 & -\frac{1}{4} \end{pmatrix}$$

$$= \begin{pmatrix} 1 & 0 & 0 & \frac{4}{2^n} \\ 1 & \frac{1}{2^n} & \frac{2n+1n}{2^n} & \frac{2n^2+1}{2^n} \\ 1 & 0 & \frac{2}{2^n} & \frac{2(2n+1)}{2^n} \\ 1 & 0 & 0 & 0 \end{pmatrix} \begin{pmatrix} 0 & 0 & 0 & 1 \\ 0 & 1 & -\frac{1}{2} & -\frac{1}{2} \\ -\frac{1}{4} & 0 & \frac{1}{2} & -\frac{1}{4} \\ \frac{1}{4} & 0 & 0 & -\frac{1}{4} \end{pmatrix}.$$

$$= \begin{pmatrix} \frac{1}{2^n} & 0 & 0 & 1-\frac{1}{2^n} \\ \frac{n(n-1)}{2^{n+1}} & \frac{1}{2^n} & \frac{n}{2^n} & 1-\frac{n^3+n+2}{2^{n+1}} \\ \frac{n}{2^n} & 0 & \frac{1}{2^n} & 1-\frac{n+1}{2^n} \\ 0 & 0 & 0 & 1 \end{pmatrix}.$$

9.4 Classification of States

In this section, we discuss the classification of states of a given Markov chain. The classification of states is important in the physical interpretation of states and in the study of the asymptotic behavior of the n-step transition probabilities $P_{ij}^{(n)}$.

Definition 9.4.1

A transition matrix P, all entries of which are positive, is called a *positive transition matrix*.

The importance of positive matrices is seen by considering an initial distribution π_0 and a transition matrix P, with P^γ positive for some integer $\gamma \geq 1$. For $\beta > \gamma$, we have

$$\pi_\beta = \pi_0 P^\beta,$$

which has only positive entries. This shows that after the random process has continued for a certain number of steps, the probability that the process is in a particular state, $i_k, k = 1, 2, \ldots, n$, is positive.

Example 9.4.1

In Example 9.2.3, P^5 is a positive transition matrix:

$$P^5 = \begin{pmatrix} \frac{1}{8} & \frac{1}{2} & \frac{1}{4} & \frac{1}{8} \\ \frac{3}{8} & \frac{1}{4} & \frac{1}{8} & \frac{1}{4} \\ \frac{5}{16} & \frac{3}{8} & \frac{1}{8} & \frac{3}{16} \\ \frac{1}{4} & \frac{1}{4} & \frac{1}{4} & \frac{1}{4} \end{pmatrix}.$$

Thus, no matter what the initial distribution is, after five steps there is a positive probability that the process is in each of the four states $i_1, i_2, i_3,$ and i_4.

Definition 9.4.2

A Markov chain is called ergodic if it is possible to move from every state to every other state. Some synonyms of the term *ergodic* are *nonnull* and *nonperiodic*. It is clear that a Markov chain with a positive transition matrix is ergodic.

Definition 9.4.3

A state i of a Markov chain is called an *absorbing state* if for every positive integer, $k > 0$, $p_{ii}^{(k)} = 1$.

Hence, once a random process enters an *absorbing state,* it remains there forever. A Markov chain may have one or more such states, or it may have none. Clearly, if an ergodic set contains only one element, then that element is an absorbing state. In the random-walk example in Section 9.1, the one-step transition matrix of which is given by

$$P = \begin{array}{c} \\ i_1 \\ i_2 \\ i_3 \\ i_4 \\ i_5 \\ i_6 \end{array} \begin{array}{cccccc} i_1 & i_2 & i_3 & i_4 & i_5 & i_6 \\ \left[\begin{array}{cccccc} 1 & 0 & 0 & 0 & 0 & 0 \\ q & 0 & p & 0 & 0 & 0 \\ 0 & q & 0 & p & 0 & 0 \\ 0 & 0 & q & 0 & p & 0 \\ 0 & 0 & 0 & q & 0 & p \\ 0 & 0 & 0 & 0 & 0 & 1 \end{array}\right] \end{array},$$

states i_1 and i_6 are *absorbing states.* The chain is obviously not ergodic, because it is impossible to leave states i_1 and i_6. However, a subset of the states is *ergodic,* namely,

$$P_E = \begin{array}{c} \\ i_2 \\ i_3 \\ i_4 \\ i_5 \end{array} \begin{array}{cccc} i_2 & i_3 & i_4 & i_5 \\ \left(\begin{array}{cccc} 0 & p & 0 & 0 \\ q & 0 & p & 0 \\ 0 & q & 0 & p \\ 0 & 0 & q & 0 \end{array}\right) \end{array}.$$

It is possible to move from every state to every other state in $\{i_2, i_3, i_4, i_5\}$. This can be clearly seen from the state diagram of P_E (Figure 9.4.1). Thus, the set of states $G = \{i_k : k = 1, 2, \ldots, 6\}$ is not ergodic, but the subset $E = \{i_k : k = 2, 3, 4, 5\}$ is ergodic.

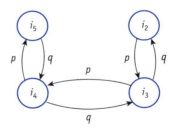

Figure 9.4.1

■ **Definition 9.4.4**

A state i of a Markov chain is called *transient (nonrecurrent)* if there exists a state j and an integer k such that $p_{ij}^{(k)} > 0$ and $n_{ji}^{(k)} = 0, k = 0, 1, 2, \ldots$. Otherwise, state i is called *nontransient.* Thus, a nontransient state is ergodic.

The definition says that i is a transient state if there exists any other state j to which the process, in some number of steps, can get from state i but from which it can never return to state i. Thus, if i is a transient state, $p_{ii} < 1$.

Definition 9.4.5

A Markov chain is called an *absorbing chain* if all states of the chain are non-transient.

Definition 9.4.6

A state i is called *persistent (recurrent)* if, for every state j, the existence of an integer k_i such that $p_{ij}^{(k_i)} > 0$ implies the existence of an integer k_j such that $p_{ji}^{(k_j)} > 0$.

The preceding definition says that no matter what state history may occur, once the process enters a persistent state it will always be able in some number of steps to return to that state. Thus, if i is a persistent state, $p_{ii} = 1$; the probability is one that, having started at i, the process will eventually return to i.

Definition 9.4.7

A persistent state i is called *periodic* if there exists an integer $d > 1$ such that $p_{ii}^{(k)} = 0$ for all values of k other than $d, 2d, 3d, \ldots$. A state that is not periodic is called *aperiodic*.

We illustrate this definition by considering a random process shown by the state diagram in Figure 9.4.2.

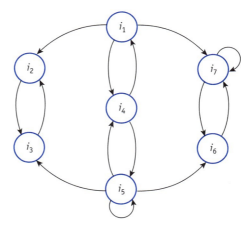

Figure 9.4.2

The only states that the process can leave in some way such that it may never return to them are states i_1, i_4, and i_5. Thus, states i_1, i_4, and i_5 are *transient* states. Clearly, states i_2, i_3, i_6, and i_7 are *persistent*. States i_2 and i_3 are the only *periodic* states.

9.4.1 Absorbing Markov Chains
The following is an interesting question regarding a Markov Chain: If a random process starts at a transient state, what is the mean number of times the process is in a given (transient) state? We answer this question by formulating the *fundamental matrix* or *normal matrix N* for absorbing Markov chains.[2]

Given the transition matrix P for a Markov chain, we rearrange it by row and column operations so that we obtain a partition of P into four submatrices, one of which contains all ergodic sets and another of which contains all transient sets. Matrix P is then of the form

$$P_{s\times s} = \left(\begin{array}{c|c} \overbrace{E}^{s-r} & \overbrace{0}^{r} \\ \hline G & T \end{array} \right) \begin{array}{l} \}s-r \\ \}r \end{array},$$

where

1. E is an $(s-r)\times(s-r)$ matrix, which deals with the process after it has reached an ergodic set.
2. O is an $(s-r)\times r$ zero matrix.
3. G is an $r\times(s-r)$ matrix of transition probabilities from transient states to ergodic states.
4. T is an $r\times r$ matrix of transition probabilities between transient states.

The preceding partition of matrix P for a finite absorbing Markov chain becomes

$$P_{s\times s} = \left(\begin{array}{c|c} \overbrace{I}^{s-r} & \overbrace{0}^{r} \\ \hline G & T \end{array} \right) \begin{array}{l} \}s-r \\ \}r \end{array}, \qquad (9.4.1)$$

where I is the $(s-r)\times(s-r)$ identity matrix.

■ **Definition 9.4.8**

Matrix N, given by

$$N = (I-T)^{-1},$$

is called the *fundamental matrix* or *normal matrix* for an absorbing Markov chain.

Matrix N answers the question posed earlier; that is, it yields the average number of times that a random process is in a given transient state j, given that the process started at state i.

The existence of such a matrix depends on $(I-T)$ possessing an inverse. This dependence can be shown by proving that

$$(I-T)^{-1} = I + T + T^2 + \cdots + T^n + \cdots$$

$$= \sum_{n=0}^{\infty} T^n$$

converges.

We illustrate the preceding remarks in Example 9.4.2.

Example 9.4.2

Consider the random-walk example, the one-step transition probabilities of which are given as in Example 9.1.1. States i_1 and i_6 are absorbing states. Thus, the stochastic matrix P must be rearranged, using Equation (9.4.1), into the form

$$P_{s \times s} = \left(\begin{array}{c|c} E & 0 \\ \hline G & T \end{array} \right) \begin{array}{l} \} s-r \\ \} r \end{array}$$

with column groupings $\overbrace{}^{s-r} \overbrace{}^{r}$

where

$$I = \begin{pmatrix} 1 & 0 \\ 0 & 1 \end{pmatrix},$$

$$O = \begin{pmatrix} 0 & 0 & 0 & 0 \\ 0 & 0 & 0 & 0 \end{pmatrix},$$

$$G = \begin{pmatrix} q & 0 \\ 0 & 0 \\ 0 & 0 \\ 0 & p \end{pmatrix},$$

and

$$T = \begin{pmatrix} 0 & p & 0 & 0 \\ q & 0 & p & 0 \\ 0 & q & 0 & p \\ 0 & 0 & q & 0 \end{pmatrix}.$$

Here,

$$(I - T) = \begin{pmatrix} 1 & -p & 0 & 0 \\ -q & 0 & -p & 0 \\ 0 & -q & 1 & -p \\ 0 & 0 & -q & 1 \end{pmatrix},$$

and the *normal matrix* is given by

$$N = (I - T)^{-1}$$

$$= \frac{1}{1 - 3pq + p^2 q^2} \begin{pmatrix} & i_2 & i_3 & i_4 & i_5 \\ & 1 - 2pq & p(1 - pq) & p^2 & p^3 \\ & q(1 - pq) & 1 - pq & p & p^2 \\ & q^2 & q & 1 - pq & p(1 - pq) \\ & q^2 & q & p(1 - pq) & 1 - 2pq \end{pmatrix}.$$

Thus, the mean of the total number of times the random walk is in a given transient state, i_2, i_3, i_4, or i_5, is given by matrix N. If we assume $p = q = 1/2$, that is, if we assume that it is equally likely that the woman will move to the right or to the left, the fundamental matrix becomes

$$N = (I - T)^{-1} = \frac{1}{5} \begin{pmatrix} i_2 & i_3 & i_4 & i_5 \\ 8 & 6 & 4 & 2 \\ 6 & 12 & 8 & 4 \\ 4 & 8 & 12 & 4 \\ 2 & 4 & 6 & 8 \end{pmatrix}.$$

Hence, the mean of the total number of times the process is in a given transient state is shown by matrix N. For example, if the random walk starts in state i_2, then it will be in state i_3 an average of $6/5$ times and in states i_4 and i_5 an average of $4/5$ and $2/5$ times, respectively. If the process is in state i_5, it will be in state i_2 an average of $2/5$ times, which is the minimum mean of the total number of times the process is in a transient state.

One can use the *normal matrix* N to answer a number of additional questions that concern absorbing Markov chains and that are significant in applications:

1. What is the variance of the number of times that a process is in state j, given that it started in state i?

2. Given that the process started in state i, what is the expected number of transitions before absorption?

3. What is the variance of the number of transitions before absorption, given that the process started in a transient state?

4. What is the probability that the process will be in an absorbing state, given that it started in a transient state?

As we mentioned earlier, we will not give a mathematically rigorous presentation as the answer to these questions. We only indicate how one can use the fundamental matrix to answer them. We refer you to the book by Kemeny and Snell[2] for an elegant presentation of the subject.

■ Theorem 9.4.1

The variance of the number of times m that a process is in a transient state i is given by

$$\text{Var}(m_{j|i}) = V_1 = N(2D - I) - S,$$

where

1. $N_{r \times r} = (I - T)^{-1}$.
2. D is an $r \times r$ diagonal matrix containing the main diagonal elements of the normal matrix.
3. I is an $r \times r$ identity matrix.
4. S is an $r \times r$ matrix whose elements are the squares of the elements of the normal matrix.

Thus, V_1 is an $r \times r$ matrix of variances of the number of times the process is in state j, given that it started in state i.

Example 9.4.3

We compute V_1 for the random-walk problem, for which we obtained the normal matrix in the previous example:

$$D = \begin{pmatrix} \frac{1-2pq}{1-3pq+p^2q^2} & 0 & 0 & 0 \\ 0 & \frac{1-pq}{1-3pq+p^2q^2} & 0 & 0 \\ 0 & 0 & \frac{1-pq}{1-3pq+p^2q^2} & 0 \\ 0 & 0 & 0 & \frac{1-2pq}{1-3pq+p^2q^2} \end{pmatrix}$$

$$S = \frac{1}{(1-3pq+p^2q^2)^2} \times \begin{pmatrix} (1-2pq)^2 & p^2(1-pq)^2 & p^4 & p^6 \\ q^2(1-pq)^2 & (1-pq)^2 & p^2 & p^4 \\ q^4 & q^2 & (1-pq)^2 & p^2(1-pq)^2 \\ q^6 & q^4 & q^2(1-pq)^2 & (1-2pq)^2 \end{pmatrix}$$

$$2D - I = \frac{1}{1-3pq+p^2q^2}$$

$$\times \begin{pmatrix} 1-pq-p^2q^2 & 0 & 0 & 0 \\ 0 & 1+pq-p^3q^3 & 0 & 0 \\ 0 & 0 & 1+pq+p^2q^2 & 0 \\ 0 & 0 & 0 & 1-pq-p^2q^2 \end{pmatrix}$$

$$N(2D - I) = \frac{1}{(1-3pq+p^2q^2)^2}$$

$$\times \begin{pmatrix} 1-3pq+p^2q^2+2p^3q^3 & p-2p^3q^2+p^4q^3 & p^2+p^3q-p^4q^2 & p^3-p^4q-p^5q^2 \\ q-2pq^2+p^3q^4 & 1-2p^2q^2+p^3q^3 & p+p^2q-p^3q^2 & p^2-p^3q-p^4q^2 \\ q^2-pq^3-p^2q^4 & q+pq^2-p^3q^3 & 1-2p^2q^2+p^3q^3 & p-2p^2q+4p^4q^3 \\ q^3-pq^4-p^2q^5 & q^2+pq^3-p^2q^4 & q-2p^2q^3+p^3q^4 & 1-3pq+p^2q^2+2p^3q^3 \end{pmatrix}$$

Hence,

$$V_1 = N(2D-I) - S = \frac{1}{(1-3pq+p^2q^2)^2}$$

$$\times \begin{pmatrix} pq-3p^2q^2+2p^3q^3 & pq+2p^4q-p^5q^2 & p^2q+2p^3q-p^4q^2 & p^3q+p^6q \\ pq-2p^2q^2-p^2q^5 & 2pq-3p^2q^2+p^3q^3 & p-p^3-p^3q^2 & p^2q-p^4q+p^5q \\ q^3-3pq^3-p^2q^4 & q-q^3-p^2q^3 & 2pq-3p^2q^2+p^3q^3 & pq-2p^2q-p^4q^3 \\ pq^3+pq^6 & pq^2+2pq^3-p^2q^4 & pq+2pq^4-p^2q^5 & pq-3p^2q^2+2p^3q^3 \end{pmatrix}$$

with columns labeled i_2, i_3, i_4, i_5.

If $p=q=1/2$, the variance of the number of times the process is in a given transient state is computed as

$$D = \begin{pmatrix} \frac{8}{5} & 0 & 0 & 0 \\ 0 & \frac{12}{5} & 0 & 0 \\ 0 & 0 & \frac{12}{5} & 0 \\ 0 & 0 & 0 & \frac{8}{5} \end{pmatrix},$$

$$S = \frac{1}{25} \begin{pmatrix} 64 & 36 & 16 & 4 \\ 36 & 144 & 64 & 16 \\ 16 & 64 & 144 & 36 \\ 4 & 16 & 36 & 64 \end{pmatrix},$$

and

$$V_1 = N(2D-I) - S = \frac{1}{25} \begin{pmatrix} 24 & 78 & 60 & 18 \\ 30 & 84 & 88 & 28 \\ 28 & 88 & 84 & 30 \\ 18 & 60 & 78 & 24 \end{pmatrix}.$$

with columns labeled i_2, i_3, i_4, i_5.

Thus, if the initial state is i_3, the variances of the number of items for which the process is in states i_2, i_3, i_4, and i_5 are $6/5$, $84/25$, $88/25$, and $28/25$, respectively. No matter the state in which the process starts, the variances are the largest in the middle states, that is, in states i_3 and i_4.

■ Theorem 9.4.2

Let $m(t)$ be the function that gives the number of steps that the process is in a transient state. If the process starts at state i, then

$$E[m(t)] = \psi = N\xi$$

and

$$\text{Var}[m(t)] = \psi_1 = (2N-1)\psi - \psi_{sq},$$

where

1. $N_{r \times r} = (I-T)^{-1}$.
2. ξ is an $r \times 1$ vector of ones.
3. I is an $r \times r$ identity matrix.
4. Ψ_{sq} is an $r \times 1$ vector, the elements of which are the square of the elements of the vector Ψ.

Hence, ψ is an $r \times 1$ vector of the expected number of transitions before absorption, and ψ_1 is an $r \times 1$ vector of the variances of the number of transitions before absorption.

Example 9.4.4

We can obtain the vectors ψ and ψ_1 for the random-walk problem given in Example 9.4.3:

$$\psi = N\xi$$

$$= \frac{1}{1-3pq+p^2q^2} \begin{pmatrix} 1-2pq & p(1-pq) & p^2 & p^3 \\ q(1-pq) & 1-pq & p & p^2 \\ q^2 & q & 1-pq & p(1-pq) \\ q^3 & q^2 & q(1-pq) & 1-2pq \end{pmatrix} \begin{pmatrix} 1 \\ 1 \\ 1 \\ 1 \end{pmatrix}$$

$$= \frac{1}{1-3pq+p^2q^2} \begin{pmatrix} q+2p^2+2p^3 \\ 2q+4p^2-p^3 \\ 3q+p^2+p^3 \\ p+2q^2+2q^3 \end{pmatrix} \begin{matrix} i_2 \\ i_3 \\ i_4 \\ i_5 \end{matrix}.$$

The variance ψ_1 is obtained as follows:

$$2N - I = \frac{1}{1-3pq+p^2q^2} \begin{pmatrix} 1-pq-p^2q^2 & 2p(1-pq) & 2p^2 & 2p^3 \\ 2q(1-pq) & 1+pq-p^2q^2 & 2p & 2p^2 \\ 2q^2 & 2q & 1+pq-p^2q^2 & 2p(1-pq) \\ 2q^3 & 2q^2 & 2q(1-pq) & 1-pq-p^2q^2 \end{pmatrix}$$

and

$$\psi_1 = (2N-I)\psi - \psi_{sq} \frac{1}{(1-3pq+p^2q^2)^2}$$

$$\times \begin{pmatrix} 2p^2+10p^3-2p^4-6p^5-4p^6+5pq-pq^2 \\ \quad +2p^2q-6p^3q-10p^4q+2p^6q-5p^2q^3-2p^4q^2+4q^3p^3 \\ 4p^2+3p^3-14p^4+8p^5-p^6+8pq-12p^2q \\ \quad +4p^2q^2-2pq^4+5p^4q-8p^4q^2+p^5q^2 \\ 3q-5q^2+2q^3+3p^2-2p^5-p^6+2p^2q+7pq^2+4pq^3 \\ \quad -8p^3q+p^4q+4p^2q^2+4p^3q^2+7p^2q^3-4p^2q^4-p^4q^2-p^5q^2 \\ 8q^2+6q^3-2p^4-8q^5-4q^6+pq-4pq^2-12pq^3-2pq^4-2p^2q^4 \\ \quad -2p^2q^5+p^2q+2p^3q+8p^2q^2+4p^3q^3+4p^2q^3-5p^3q^2-2p^4q^2 \end{pmatrix} \begin{matrix} i_2 \\ i_3 \\ i_4 \\ i_5 \end{matrix}.$$

If $p=q=1/2$, we compute the expected number of transitions before absorption for the random-walk problem:

$$\psi = N\xi = \frac{1}{5}\begin{pmatrix} & i_2 & i_3 & i_4 & i_5 \\ & 8 & 6 & 4 & 2 \\ & 6 & 12 & 8 & 4 \\ & 4 & 8 & 12 & 6 \\ & 2 & 4 & 6 & 8 \end{pmatrix}\begin{pmatrix} 1 \\ 1 \\ 1 \\ 1 \end{pmatrix}$$

$$= \begin{matrix} i_2 \\ i_3 \\ i_4 \\ i_5 \end{matrix}\begin{pmatrix} 4 \\ 6 \\ 6 \\ 4 \end{pmatrix}.$$

Hence, the process reaches an absorbing state more quickly from states i_2 or i_5, which is reasonable because it is easier to reach the boundary states from the outside states of the absorbing Markov chain. The symmetry of the vector ψ occurs because $p=q$.

The variance of the number of steps before absorption is computed as

$$(2N-I) = \begin{pmatrix} \frac{11}{5} & \frac{12}{5} & \frac{8}{5} & \frac{4}{5} \\ \frac{12}{5} & \frac{19}{5} & \frac{16}{5} & \frac{8}{5} \\ \frac{8}{5} & \frac{16}{5} & \frac{19}{5} & \frac{12}{5} \\ \frac{4}{5} & \frac{8}{5} & \frac{12}{5} & \frac{11}{5} \end{pmatrix},$$

$$(2N-I)\psi = \frac{1}{5}\begin{pmatrix} 11 & 12 & 8 & 4 \\ 12 & 19 & 16 & 8 \\ 8 & 16 & 19 & 12 \\ 4 & 8 & 12 & 11 \end{pmatrix}\begin{pmatrix} 4 \\ 6 \\ 6 \\ 4 \end{pmatrix} = \begin{pmatrix} 36 \\ 58 \\ 58 \\ 36 \end{pmatrix},$$

and

$$\psi_1 = (2N-I)\psi - \psi_{sq} = \begin{pmatrix} 36 \\ 58 \\ 58 \\ 36 \end{pmatrix} - \begin{pmatrix} 16 \\ 36 \\ 36 \\ 16 \end{pmatrix} = \begin{matrix} i_2 \\ i_3 \\ i_4 \\ i_5 \end{matrix}\begin{pmatrix} 20 \\ 22 \\ 22 \\ 20 \end{pmatrix}.$$

Thus, the variance of the number of steps before the process enters an absorbing state is smallest for the outside states, as one would expect.

Theorem 9.4.3

Let the process start at transient state i. The probability that the process ends up in an absorbing state j is given by

$$A = NG,$$

where

$$N_{r \times r} = (I - T)^{-1}$$

and G is an $r \times (s-r)$ matrix of transition probabilities from the transient states to absorbing states. Thus, A is an $r \times (s-r)$ matrix of probabilities that the absorbing states will capture the random process.

Example 9.4.5

The absorbing probabilities for the random-walk problem are

$$A = NG = \frac{1}{1 - 3pq + p^2q^2} \begin{pmatrix} (1-2pq) & p(1-pq) & p^2 & p^3 \\ q(1-pq) & (1-pq) & p & p^2 \\ q^2 & q & (1-pq) & p(1-pq) \\ q^3 & q^2 & q(1-pq) & (1-2pq) \end{pmatrix} \begin{pmatrix} q & 0 \\ 0 & 0 \\ 0 & 0 \\ 0 & p \end{pmatrix}$$

$$= \frac{1}{1-3pq+p^2q^2} \begin{pmatrix} q(1-2pq) & p^4 \\ q^2(1-pq) & p^3 \\ q^3 & p^2(1-pq) \\ q^4 & p(1-2pq) \end{pmatrix} \begin{matrix} i_2 \\ i_3 \\ i_4 \\ i_5 \end{matrix}$$

$\quad\quad\quad\quad\quad\quad i_1 \quad\quad\quad\quad i_6$

The probability that the process is completed in the absorbing states i_1 and i_6, given that it started in state i_5, is

$$\frac{q^4}{1-3pq+p^2q^2}$$

and

$$\frac{p(1-2pq)}{1-3pq+p^2q^2},$$

respectively.

Suppose that $p=q=1/2$; then, matrix A becomes

$$A = \frac{1}{2}\begin{pmatrix} \overset{i_2}{8} & \overset{i_3}{6} & \overset{i_4}{4} & \overset{i_5}{2} \\ 6 & 12 & 8 & 4 \\ 4 & 8 & 12 & 6 \\ 2 & 4 & 6 & 8 \end{pmatrix}\begin{pmatrix} \frac{1}{2} & 0 \\ 0 & 0 \\ 0 & 0 \\ 0 & \frac{1}{2} \end{pmatrix}$$

$$= \begin{array}{c} i_2 \\ i_3 \\ i_4 \\ i_5 \end{array}\begin{pmatrix} \overset{i_1}{\frac{4}{5}} & \overset{i_6}{\frac{1}{5}} \\ \frac{3}{5} & \frac{2}{5} \\ \frac{2}{5} & \frac{3}{5} \\ \frac{1}{5} & \frac{4}{5} \end{pmatrix}.$$

Thus, if the random walk starts at state i_3, the probability that it ends up in the absorbing states i_1 and i_6 is 3/5 and 2/5, respectively. Again, the symmetry exhibited by matrix A occurs because $p=q=1/2$.

9.4.2 Regular Markov Chains In this section, we define the concept of a regular Markov chain and its fundamental property, and we illustrate its meaning with an example.

■ **Definition 9.4.9**

A Markov chain that consists of no transient states and has a single ergodic set is called a *regular Markov chain*.

■ **Definition 9.4.10**

A stochastic matrix of a regular Markov chain is called a *regular transition matrix*.

It can be shown that a regular transition matrix is in fact a *positive transition matrix*. That is, a Markov chain is regular if and only if it is possible to be in any state after some number n of steps, regardless of the state in which the process started. However, the Markov chain given by the state diagram (Figure 9.4.3) consists of a transient set, states $\{i_1,i_3,i_6\}$, and an ergodic set, states $\{i_2,i_4,i_5\}$, where $\alpha, \beta > 0$ and $\alpha+\beta=1$.

9.4 Classification of States

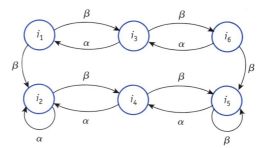

Figure 9.4.3

The single ergodic set constitutes a regular Markov chain, and its stochastic matrix P is regular for $n=2$. That is, every entry of P^2 is positive:

$$P = \begin{matrix} & \begin{matrix} i_2 & i_4 & i_5 \end{matrix} \\ \begin{matrix} i_2 \\ i_4 \\ i_5 \end{matrix} & \begin{pmatrix} \alpha & \beta & 0 \\ \alpha & 0 & \beta \\ 0 & \alpha & \beta \end{pmatrix} \end{matrix}$$

and

$$P^2 = \begin{pmatrix} \alpha(\alpha+\beta) & \alpha\beta & \beta^2 \\ \alpha^2 & 2\alpha\beta & \beta^2 \\ \alpha^2 & \alpha\beta & \beta(\alpha+\beta) \end{pmatrix}.$$

The fundamental property of regular transition matrices is given by Theorem 9.4.4, the proof of which is beyond the scope of this book.

Theorem 9.4.4

Let P be a regular transition matrix. Then,

1. P has a unique fixed probability vector $\gamma = (\gamma_1, \gamma_2, \ldots, \gamma_n)$, the components of which are positive, such that $\gamma P = \gamma$.
2. The sequence $P, P^2, \ldots, P^n$ of powers of P approaches a probability matrix W, each row of which has the same probability vector γ, and $PW = WP = W$.
3. If π is any probability vector, then the sequence of vectors $\pi P, \pi P^2, \ldots, \pi P^n$ approaches the vector γ as n tends to infinity.

By stating that matrix P^n approaches matrix W, we mean that each entry of P^n approaches the corresponding entry of W as n tends to infinity. Similarly, saying that πP^n approaches the vector γ means that each component of πP^n approaches the corresponding component of γ. We refer to matrix W as the limiting matrix of a regular transition matrix. We saw in Section 9.2 that if the random process starts in each of the states with probabilities given by the vector π, then the probabilities of being in each of the states after n steps are given by πP. The preceding theorem states that if P is a regular transition matrix, then for large n, πP^n is approximately equal to the vector γ. The probability vector γ is unique. Thus, for a regular Markov chain, for large values of n, the probabilities of being in each of the states are independent of the initial probability vector.

Example 9.4.6

Consider the Markov chain generated by the four quarterbacks illustrated in Example 9.1.2. The Markov chain is regular, and its regular stochastic matrix is given by

$$P = \begin{array}{c} \\ i_1 \\ i_2 \\ i_3 \\ i_4 \end{array} \begin{array}{c} i_1 \quad i_2 \quad i_3 \quad i_4 \end{array} \\ \left(\begin{array}{cccc} 0 & \frac{1}{3} & \frac{1}{3} & \frac{1}{3} \\ \frac{1}{2} & 0 & 0 & \frac{1}{2} \\ \frac{2}{3} & 0 & 0 & \frac{1}{3} \\ 1 & 0 & 0 & 0 \end{array} \right).$$

We seek a probability vector $\gamma = (\gamma_1, \gamma_2, \gamma_3, \gamma_4)$ such that $\gamma P = \gamma$:

$$(\gamma_1, \gamma_2, \gamma_3, \gamma_4) \left(\begin{array}{cccc} 0 & \frac{1}{3} & \frac{1}{3} & \frac{1}{3} \\ \frac{1}{2} & 0 & 0 & \frac{1}{2} \\ \frac{2}{3} & 0 & 0 & \frac{1}{3} \\ 1 & 0 & 0 & 0 \end{array} \right) = \left(\begin{array}{c} \gamma_1 \\ \gamma_2 \\ \gamma_3 \\ \gamma_4 \end{array} \right).$$

Solving the set of equations,

$$\gamma_1 + \gamma_2 + \gamma_3 + \gamma_4 = 1$$

$$\frac{1}{2}\gamma_2 + \frac{2}{3}\gamma_3 + \gamma_4 = \gamma_1$$

$$\frac{1}{3}\gamma_1 = \gamma_2$$

$$\frac{1}{3}\gamma_1 = \gamma_3$$

$$\frac{1}{3}\gamma_1 + \frac{1}{2}\gamma_2 + \frac{1}{3}\gamma_3 = \gamma_4,$$

where the first equation is possible because γ must be a probability vector, we have the unique solution:

$$\gamma = \left(\frac{18}{41}, \frac{6}{41}, \frac{6}{41}, \frac{11}{41}\right).$$

In view of Theorem 9.4.4, the limiting matrix A is

$$W = \begin{array}{c} \\ i_1 \\ i_2 \\ i_3 \\ i_4 \end{array} \begin{array}{cccc} i_1 & i_2 & i_3 & i_4 \end{array} \\ \left(\begin{array}{cccc} \frac{18}{41} & \frac{6}{41} & \frac{6}{41} & \frac{11}{41} \\ \frac{18}{41} & \frac{6}{41} & \frac{6}{41} & \frac{11}{41} \\ \frac{18}{41} & \frac{6}{41} & \frac{6}{41} & \frac{11}{41} \\ \frac{18}{41} & \frac{6}{41} & \frac{6}{41} & \frac{11}{41} \end{array}\right) = \left(\begin{array}{cccc} 0.4390 & 0.1463 & 0.1463 & 0.2683 \\ 0.4390 & 0.1463 & 0.1463 & 0.2683 \\ 0.4390 & 0.1463 & 0.1463 & 0.2683 \\ 0.4390 & 0.1463 & 0.1463 & 0.2683 \end{array}\right).$$

Each row of the probability matrix W gives the long-range prediction regarding to whom each of the four quarterbacks will throw the football. We also show that the convergence of P^n to matrix W is fairly fast; that is, $n = 15$,

$$P^{15} = \begin{pmatrix} 0.43908 & 0.14629 & 0.14629 & 0.26832 \\ 0.43908 & 0.14630 & 0.14630 & 0.26829 \\ 0.43902 & 0.14634 & 0.14634 & 0.26827 \\ 0.43888 & 0.14642 & 0.14642 & 0.26824 \end{pmatrix}.$$

Thus, the first row gives the probabilities of the ball being thrown to each quarterback 15 throws after it begins with quarterback i_1.

A transition matrix with identical rows is called a *steady-state matrix*. In the previous example, we illustrated the important property that transition matrices asymptotically approach a *steady-state matrix*. Two basic properties of steady-state matrices are given in Corollary 9.4.1.

> **Corollary 9.4.1** Let P_S be an $n \times n$ steady-state matrix with the rows $(s_1, s_2, \ldots s_n)$. Then,
>
> 1. For all $n \times n$ transition matrices P, we have
>
> $$P \times P_S = P_S.$$
>
> 2. For π, a $1 \times n$ probability vector, we have
>
> $$\pi P_S = (s_1, s_2, \ldots s_n).$$

We conclude our brief discussion of finite Markov chains by illustrating the method by which one can obtain the mean and variance of the number of steps necessary to go from state i to state j for the first time (first-passage times) for regular Markov chains. Corresponding to the fundamental matrix N for absorbing Markov chains, we define[2] a similar matrix for regular Markov chains.

Definition 9.4.11

Let P be a regular transition matrix and W its limiting matrix. The matrix defined by

$$L = \left[I - (P - W)\right]^{-1}$$

is called the *fundamental matrix for regular Markov chains.*

Definition 9.4.12

The *first-passage time* t_k is the number of steps necessary before entering state i_k for the first time. The *mean first-passage matrix* is denoted by F, and its entries are denoted by f_{ij}:

$$F = E\left(t_{j|i}\right).$$

The mean first-passage matrix for regular Markov chains is given by Theorem 9.4.5.

Theorem 9.4.5

Let L be the fundamental matrix for a regular Markov chain. The mean first-passage matrix is given by

$$F = \left(I - L + HL_{dg}\right)Q,$$

where

1. $H = WQ$, W is the steady-state matrix.
2. Q is a diagonal matrix with diagonal elements $d_{ii} = 1/\gamma_i$.
3. L_{dg} is a diagonal matrix containing the main diagonal elements of the fundamental matrix L.

Example 9.4.7

We can compute the mean first-passage matrix for the regular stochastic matrix given in Example 9.4.6. We first compute the fundamental matrix:

$$[I - (P - W)] = \begin{pmatrix} 1.4390 & -0.1870 & -0.1870 & -0.0650 \\ -0.0610 & 1.1463 & 0.1463 & -0.2317 \\ -0.2276 & 0.1463 & 1.1463 & -0.0650 \\ -0.5610 & 0.1463 & 0.1463 & 1.2683 \end{pmatrix}$$

and

$$L = [I - (P - W)]^{-1} = \begin{pmatrix} 0.7388 & 0.0999 & 0.0999 & 0.0613 \\ 0.0803 & 0.8804 & -0.1196 & 0.1588 \\ 0.1535 & -0.0952 & 0.9048 & 0.0369 \\ 0.2998 & -0.0464 & -0.0464 & 0.7930 \end{pmatrix}.$$

9.4 Classification of States

Here,

$$H = WQ = \begin{pmatrix} \frac{18}{41} & \frac{6}{41} & \frac{6}{41} & \frac{11}{41} \\ \frac{18}{41} & \frac{6}{41} & \frac{6}{41} & \frac{11}{41} \\ \frac{18}{41} & \frac{6}{41} & \frac{6}{41} & \frac{11}{41} \\ \frac{18}{41} & \frac{6}{41} & \frac{6}{41} & \frac{11}{41} \end{pmatrix} \begin{pmatrix} \frac{41}{18} & 0 & 0 & 0 \\ 0 & \frac{41}{6} & 0 & 0 \\ 0 & 0 & \frac{41}{6} & 0 \\ 0 & 0 & 0 & \frac{41}{11} \end{pmatrix}$$

$$= \begin{pmatrix} 1 & 1 & 1 & 1 \\ 1 & 1 & 1 & 1 \\ 1 & 1 & 1 & 1 \\ 1 & 1 & 1 & 1 \end{pmatrix}$$

and

$$HL_{dg} = \begin{pmatrix} 1 & 1 & 1 & 1 \\ 1 & 1 & 1 & 1 \\ 1 & 1 & 1 & 1 \\ 1 & 1 & 1 & 1 \end{pmatrix} \begin{pmatrix} 0.7388 & 0 & 0 & 0 \\ 0 & 0.8804 & 0 & 0 \\ 0 & 0 & 0.9048 & 0 \\ 0 & 0 & 0 & 0.7930 \end{pmatrix}$$

$$= \begin{pmatrix} 0.7388 & 0.8804 & 0.9048 & 0.7930 \\ 0.7388 & 0.8804 & 0.9048 & 0.7930 \\ 0.7388 & 0.8804 & 0.9048 & 0.7930 \\ 0.7388 & 0.8804 & 0.9048 & 0.7930 \end{pmatrix}.$$

Thus,

$$F = (I - L + HL_{dg})Q$$

$$= \begin{pmatrix} 1.0000 & 0.7805 & 0.8049 & 0.7317 \\ 0.6585 & 1.0000 & 1.0244 & 0.6342 \\ 0.5853 & 0.9756 & 1.0000 & 0.6561 \\ 0.4390 & 0.9268 & 0.9512 & 1.0000 \end{pmatrix} \begin{pmatrix} \frac{41}{18} & 0 & 0 & 0 \\ 0 & \frac{41}{6} & 0 & 0 \\ 0 & 0 & \frac{41}{6} & 0 \\ 0 & 0 & 0 & \frac{41}{11} \end{pmatrix}$$

$$= \begin{pmatrix} & i_1 & i_2 & i_3 & i_4 \\ & 2.2778 & 5.3641 & 5.5000 & 2.7272 \\ & 1.4999 & 6.8333 & 7.0000 & 2.3639 \\ & 1.3332 & 6.6666 & 6.8333 & 2.4455 \\ & 0.9999 & 6.3331 & 6.4999 & 3.7273 \end{pmatrix}.$$

Therefore, we conclude that if the ball is presently with quarterback i_2, the mean number of throws before the ball is with players i_1, i_3, and i_4 is 1.4999, 7.0000, and 2.3639 throws, respectively. The mean number of throws before the ball returns to player i_2 is 6.8333, and so on.

We also use the fundamental matrix L to obtain the variance of the first-passage time for regular Markov chains.

Theorem 9.4.6

Let L be the fundamental matrix and F be the mean first-passage matrix for a regular Markov chain. The variance first-passage matrix V_2 is given by

$$V_2 = F\left(2L_{dg}Q - I\right) + 2\left[LF - H(LF)_{dg}\right],$$

where

1. L_{dg} is a diagonal matrix containing the main diagonal elements of the fundamental matrix L.
2. Q is a diagonal matrix with the diagonal elements $d_{ii} = 1/\gamma$.
3. $H = WQ$, W is the limiting matrix of the regular transition matrix P.
4. $(LF)_{dg}$ is a diagonal matrix containing the main diagonal of the product of the matrices L and F.

Example 9.4.8

The variance for the first-passage matrix of Example 9.4.7 is as follows:

$$LQ = \begin{pmatrix} 1.6828 & 0 & 0 & 0 \\ 0 & 6.0160 & 0 & 0 \\ 0 & 0 & 6.1828 & 0 \\ 0 & 0 & 0 & 2.9557 \end{pmatrix}$$

$$2LQ - I = \begin{pmatrix} 2.3656 & 0 & 0 & 0 \\ 0 & 11.0322 & 0 & 0 \\ 0 & 0 & 11.3656 & 0 \\ 0 & 0 & 0 & 4.9114 \end{pmatrix}$$

$$F(2LQ - I) = \begin{pmatrix} 5.3884 & 59.1778 & 62.5108 & 13.3944 \\ 3.5483 & 75.3863 & 79.5592 & 11.6101 \\ 3.1539 & 73.5480 & 77.6646 & 12.0108 \\ 2.3654 & 69.8680 & 73.8753 & 18.3063 \end{pmatrix}$$

$$LF = \begin{pmatrix} 1.6828 & 3.9630 & 4.0634 & 2.0149 \\ 1.3205 & 6.0160 & 6.1628 & 2.0812 \\ 1.2543 & 6.0319 & 6.1828 & 2.2127 \\ 0.7929 & 5.0221 & 5.1544 & 2.9557 \end{pmatrix}$$

$$H(LF)_{dg} = \begin{pmatrix} 1.6828 & 6.0160 & 6.1828 & 2.9557 \\ 1.6828 & 6.0160 & 6.1828 & 2.9557 \\ 1.6828 & 6.0160 & 6.1828 & 2.9557 \\ 1.6828 & 6.0160 & 6.1828 & 2.9557 \end{pmatrix};$$

$$2\left[LF - H(LF)_{dg}\right] = \begin{pmatrix} 0 & -4.1060 & -4.2388 & -1.8816 \\ -0.7246 & 0 & -0.0400 & -1.7490 \\ -0.8570 & 0.0318 & 0 & -1.4860 \\ -1.7798 & -1.9878 & -2.0568 & 0 \end{pmatrix}.$$

Thus, the variance matrix is

$$V_2 = F\left(2L_{dg}Q - I\right) + 2\left[LF - H(LF)_{gd}\right] = \begin{pmatrix} 5.3884 & 55.0718 & 58.2720 & 11.5128 \\ 2.8237 & 75.3863 & 79.5192 & 9.8611 \\ 2.2969 & 73.5798 & 77.6646 & 10.5248 \\ 0.5856 & 67.8802 & 71.8185 & 18.3063 \end{pmatrix}.$$

Therefore, the variance for the first-passage times is largest for the "middle states." The minimum variance for the mean first-passage time occurs when player i_4 is initiating the throwing to player i_1. In general, the variances are quite large for the given random process.

Applied Problems

9.1 Draw the state diagram for each of the following transition matrices:

a. $P = \begin{pmatrix} & i_1 & i_2 & i_3 & i_4 \\ i_1 & 1 & 0 & 0 & 0 \\ i_2 & \frac{1}{2} & \frac{1}{2} & 0 & 0 \\ i_3 & \frac{1}{4} & \frac{1}{2} & 0 & \frac{1}{4} \\ i_4 & \frac{1}{4} & \frac{1}{4} & \frac{1}{4} & \frac{1}{4} \end{pmatrix}$

b. $P = \begin{pmatrix} & i_1 & i_2 & i_3 & i_4 \\ i_1 & 0 & q & 0 & p \\ i_2 & 0 & q & 0 & p \\ i_3 & q & 0 & 0 & p \\ i_4 & 0 & 0 & 1 & 0 \end{pmatrix}$, $p + q = 1$

9.2 Write the transition matrix for each of the following state diagrams:

a. $P = \begin{array}{c} \\ i_1 \\ i_6 \\ i_2 \\ i_3 \\ i_4 \\ i_5 \end{array} \begin{array}{c} i_1 \ i_6 \quad i_2 \ i_3 \ i_4 \ i_5 \end{array} \left[\begin{array}{cc|cccc} 1 & 0 & 0 & 0 & 0 & 0 \\ 0 & 1 & 0 & 0 & 0 & 0 \\ \hline q & 0 & 0 & p & 0 & 0 \\ 0 & 0 & q & 0 & p & 0 \\ 0 & 0 & 0 & q & 0 & p \\ 0 & p & 0 & 0 & 0 & 0 \end{array} \right],$

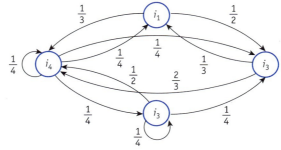

b.

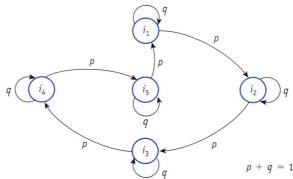

$p + q = 1$

9.3 A woman can get to her office by car, train, or bus. She will take either the train or the bus the day after she drives. If she takes the train to her office, then the next day she is just as likely to drive as to take the bus. But if she takes the bus, then the next day chances are three to one that she will take the train rather than drive.

   a. What is the state space of the process?
   b. Write the transition matrix of the Markov chain.
   c. Draw the state diagram of the transition matrix.

9.4 A quarterback usually throws the ball to only three players, E_1, E_2, and E_3. He never throws the ball to the same player on successive plays. If he throws to player E_1, then on the next play he will throw to E_2. However, if he throws to either E_2 or E_3, then on the next play he is twice as likely to throw to player E_1 as to the other remaining player. Does this process constitute a Markov chain?

9.5 A cage consists of seven compartments that open into one another, as shown in the accompanying diagram. A mouse is placed in one of the compartments, and it begins moving from compartment to compartment in search of food. If each exit of any given compartment has an equally likely chance of being selected by the mouse, find the transition probability matrix of the Markov chain.

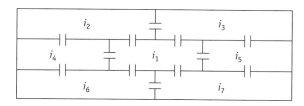

9.6 A player has $3 with which to i_3 play a game. On each play of the game, it is possible for her to win $2 with a probability of 0.40 or to lose $1 with a probability of 0.60. The game stops when she has either won or lost $3. Determine the transition matrix for this special case of the gambler's ruin problem.

9.7 (Random walk with reflecting barriers) Consider the random-walk problem, which was formulated in Section 9.1, under the following additional condition: When the woman reaches a boundary state, she is required to take a step backward to the state from which she came. Thus, in Example 9.1.1, when the woman enters state i_1, she is on the next step "reflected" back to state i_2. Similarly, if she enters state i_6, she goes back to i_5 on the next step.
   a. Write the transition matrix of the random walk.
   b. Assume that the process starts with state; draw a tree diagram showing the one-, two-, three-, and four-step transitions.
   c. Write the first four probability distribution vectors directly from the tree diagram.
   d. Interpret the meaning of π_3.

9.8 Use Equation 9.1.9 to calculate the first four probability distribution vectors of Problem 9.7, and compare your answers with Problem 9.7(c).

9.9 In Problem 9.6, calculate the following:
   a. The probability that the player has lost her money at the end of at most four steps.
   b. The probability that the game will last more than six plays, that is, that the process has not entered states i_1 or i_7.

9.10 Referring back to Problem 9.3, suppose that the process begins with the woman taking the train to her office. What is the probability that she will be taking the train again 5 days later?

9.11 Using mathematical induction, prove that Equation (9.2.4) is true for all m.

9.12 Matrices in general do not commute, but probability transition matrices do commute. Show that if P is an $h \times h$ transition matrix and m and n are positive integers, then

$$P^{(m+n)} = P^{(m)} \cdot P^{(n)} = P^{(n)} P^{(m)}.$$

9.13 a. Calculate the three-step transition matrix of Problem 9.5.
   b. Suppose that the mouse is initially placed in compartment i_3. What are the probabilities that, after three steps, the mouse will be in states $i_1, i_2, i_3, i_4, i_5, i_6$, and i_7, respectively?

9.14 Consider a sequence of Bernoulli trials that generates a chain, the links of which are either a success with the probability p or a failure with the probability $1-p$. The transition matrix is given by

$$P = \begin{pmatrix} p & 1-p \\ p & 1-p \end{pmatrix}.$$

Find the eigenvalues of matrix P.

9.15 Let the transition matrix of a simple random-walk problem be given by

$$P = \begin{pmatrix} \frac{1}{4} & \frac{1}{2} & \frac{1}{4} \\ 0 & \frac{3}{4} & \frac{1}{4} \\ 0 & 0 & 1 \end{pmatrix}.$$

Calculate the eigenvalues of matrix P.

9.16 Apply Theorem 9.3.1 to find P^r of the transition matrix given in Problem 9.15, and evaluate the resulting matrix for $r = 10$. Interpret the third row of P^{10}.

9.17 Use the *Jordan canonical form* to raise the transition matrix P, given by

$$P = \begin{pmatrix} 1 & 0 & 0 & 0 \\ \frac{1}{3} & \frac{2}{3} & 0 & 0 \\ \frac{1}{3} & \frac{1}{3} & \frac{1}{3} & 0 \\ \frac{1}{6} & \frac{1}{4} & \frac{1}{4} & \frac{1}{3} \end{pmatrix},$$

to the kth power.

9.18 Are the eigenvalues of the transition matrix P, given by

$$P = \begin{pmatrix} \frac{1}{3} & 0 & \frac{1}{4} & \frac{5}{12} \\ 0 & 0 & \frac{1}{6} & \frac{1}{2} & \frac{1}{3} \\ 0 & 0 & \frac{1}{3} & \frac{2}{3} \\ 0 & 0 & 0 & 1 \end{pmatrix},$$

all distinct? Use accordingly either Theorem 9.3.1 or the *Jordan canonical form* to obtain P^{15} and to interpret the first and last row of the resulting matrix. Does matrix P^{15} seem to be approaching a steady state?

9.19 Let the transition matrix P be given by

$$P = \begin{pmatrix} \frac{1}{2} & 0 & \frac{1}{2} \\ \frac{1}{2} & \frac{1}{2} & 0 \\ 0 & 0 & 1 \end{pmatrix}.$$

Raise matrix P to the nth power, and evaluate the resulting matrix for $n = 20$.

9.20 Construct two 3×3 transition matrices so that one will have distinct eigenvalues and the other will have two or more identical eigenvalues.

9.21 Find the necessary number of steps that converts the transition matrix P, given by

$$P = \begin{pmatrix} 0 & 1 & 0 & 0 & 0 \\ \frac{1}{3} & 0 & \frac{2}{3} & 0 & 0 \\ 0 & \frac{1}{3} & 0 & \frac{2}{3} & 0 \\ 0 & 0 & \frac{1}{3} & 0 & \frac{2}{3} \\ 0 & 0 & 0 & 0 & 1 \end{pmatrix},$$

into a positive transition matrix. What is the physical interpretation of the resulting matrix?

9.22 Consider a Markov chain, the transition probability matrix of which is given by

$$P = \begin{matrix} & \begin{matrix} i_1 & i_2 & i_3 & i_4 & i_5 & i_6 \end{matrix} \\ \begin{matrix} i_2 \\ i_2 \\ i_2 \\ i_2 \\ i_2 \\ i_2 \end{matrix} & \begin{pmatrix} 0 & \frac{1}{3} & 0 & \frac{1}{3} & 0 & \frac{1}{3} \\ \frac{1}{4} & \frac{1}{4} & 0 & 0 & 0 & \frac{1}{2} \\ 0 & \frac{1}{2} & 0 & 0 & \frac{1}{3} & \frac{1}{6} \\ 0 & 0 & 0 & \frac{1}{2} & \frac{1}{2} & 0 \\ 0 & 0 & 0 & \frac{1}{2} & \frac{1}{2} & 0 \\ 0 & 0 & 0 & 0 & 0 & 1 \end{pmatrix} \end{matrix}.$$

Classify which state or states are absorbing, periodic, persistent, and transient.

9.23 Rearrange the transition matrix of the Markov chain given in Problem 9.7 into the form of Equation (9.4.1), and calculate the normal matrix of the random walk. What is the physical interpretation of the fundamental matrix?

9.24 Referring to Problem 9.6, find the mean of the total number of times the gambler is in a given transient state.

9.25 In Problem 9.7, let $p=2/3$ and $q=1/3$; that is, let the transition matrix of the Markov chain be given by

$$P = \begin{pmatrix} & i_1 & i_2 & i_3 & i_4 & i_5 & i_6 \\ i_1 & 0 & 1 & 0 & 0 & 0 & 0 \\ i_2 & \frac{1}{3} & 0 & \frac{2}{3} & 0 & 0 & 0 \\ i_3 & 0 & \frac{1}{3} & 0 & \frac{2}{3} & 0 & 0 \\ i_4 & 0 & 0 & \frac{1}{3} & 0 & \frac{2}{3} & 0 \\ i_5 & 0 & 0 & 0 & \frac{1}{3} & 0 & \frac{2}{3} \\ i_6 & 0 & 0 & 0 & 0 & 1 & 0 \end{pmatrix}.$$

a. Find the mean of the total number of times that the random walk with reflecting barriers is in a given transient state, $i_n, n=1,3,4,6$.
b. What is the variance of the number of times that the process (random walk) is in $i_n, n=1,3,4,6$, given that it started in one of the transient states?
c. Given that the process started in one of the transient states, what is the expected number of transitions before absorption?
d. What is the variance of the number of transitions before absorption, given that the process started in one of the following states: i_1, i_3, i_4, or i_6?
e. What is the probability that the process will be in an absorbing state, given that it started in a transient state?

9.26 Answer Problem 9.25(a) through (e) for the transition of the Markov chain obtained in Problem 9.5.

9.27 Let T be an $n \times n$ matrix of transition probabilities between transient states. Show that

$$(I-T)^{-1} = I + T + T^2 + \cdots + T^n + \cdots$$

converges. This will assure the existence of the fundamental matrix of absorbing Markov chains.

9.28 Does the Markov chain given in Problem 9.3 constitute a regular Markov chain? What is the necessary number of transitions for the stochastic matrix to be regular?

9.29 Suppose that a mouse is placed in compartment i_1 in the process described in Problem 9.5.
a. Draw a state diagram of the Markov chain.
b. Which of the states form an ergodic set, and which are transient?

9.30 Obtain the limiting matrix W of a Markov chain, the transition matrix P of which is given by

$$P = \begin{pmatrix} & i_1 & i_2 & i_3 & i_4 \\ i_1 & 0 & \frac{1}{2} & \frac{1}{4} & \frac{1}{4} \\ i_2 & \frac{1}{2} & 0 & \frac{1}{4} & \frac{1}{4} \\ i_3 & \frac{1}{2} & \frac{1}{2} & 0 & 0 \\ i_4 & \frac{1}{2} & \frac{1}{4} & \frac{1}{4} & 0 \end{pmatrix}.$$

What is the physical interpretation of matrix W?

9.31 Using the methods discussed in Section 9.3, show that the probability transition matrix given in Problem 9.29 reaches a steady state after 10 transitions.

9.32 Prove Corollary 9.4.1.

9.33 Calculate the fundamental matrix for the regular transition matrix given in Problem 9.29.

9.34 Compute the mean first-passage matrix for the regular stochastic matrix given in Problem 9.29. Give a physical interpretation of the resulting matrix.

9.35 Compute and interpret the variance of the first-passage time matrix of Problem 9.34.

References

[1] Parzen, E. *Stochastic Processes*. San Francisco: Holden-Day, Inc., 1962.

[2] Kemeny, J. G. and J. L. Snell. *Finite Markov Chains*. New York: D. Van Nostrand Company, Inc., 1960.

[3] Bailey, N. T. J. *The Elements of Stochastic Process*. New York: John Wiley & Sons, Inc., 1964.

[4] Bartlett, M. S. *An Introduction to Stochastic Processes*. New York: Cambridge University Press, 1955.

[5] Bharucha-Reid, A. T. *Elements of the Theory of Markov Processes and Their Applications*. New York: McGraw-Hill Book Company, Inc., 1960.

[6] Breiman, L. *Probability and Stochastic Processes*. Boston: Houghton-Mifflin Company, 1969.

[7] Cox, D. R. and H. D. Miller. *The Theory of Stochastic Processes*. New York: John Wiley & Sons, Inc., 1965.

[8] Feller, W. *An Introduction to Probability Theory and Its Applications*, 2nd ed. New York: John Wiley & Sons, Inc., 1957.

List of Tables

Individual Terms of the Binomial Distribution:

$$f(x) = \binom{n}{x} p^x q^{n-x}, \ x = 0,1,2,\ldots,n, \ q = 1-p, \ p+q = 1$$

Cumulative Terms of the Binomial Distribution:

$$f(x) = \binom{n}{x} p^x q^{n-x}, \ x = 0,1,2,\ldots,n, \ q = 1-p, \ p+q = 1$$

$$Pr(X \geq x') = \sum_{x=x'}^{n} \binom{n}{x} f^x q^{n-z}, \ x = 0,1,2,\ldots,n, \ p+q = 1$$

Individual Terms of the Poisson Distribution:

$$f(x) = \binom{n}{x} p^x q^{n-x}, \ x = 0,1,2,\ldots,n, \ q = 1-p, \ p+q = 1$$

$$f(x) = \frac{e^{-m} m^x}{x!}, \ x = 0,1,2,\ldots, m > 0$$

Cumulative Terms of the Poisson Distribution:

$$f(x) = \binom{n}{x} p^x q^{n-x}, \ x = 0,1,2,\ldots,n, \ q = 1-p, \ p+q = 1$$

$$Pr(X \geq x') = \sum_{x=x'}^{n} \frac{e^{-m} m^x}{x!}, \ x = 0,1,2,\ldots, m > 0$$

Hypergeometric Distribution:

$$f(x) = \binom{n}{x} p^x q^{n-x}, \ x = 0,1,2,\ldots,n, \ q = 1-p, \ p+q = 1$$

$$f(x) = \frac{\binom{n}{x}\binom{N-n}{n-x}}{\binom{N}{n}}$$

Negative Binomial Distribution:

$$f(x) = \binom{n}{x} p^x q^{n-x}, \ x = 0, 1, 2, \ldots, n, \ q = 1-p, \ p+q = 1$$

$$f(x) = \binom{x+r-1}{r-1} p^r q^x, \ x = 0, 1, 2, \ldots; \ p+q = 1$$

Exponential Functions
Normal Distribution:

$$f(x) = \binom{n}{x} p^x q^{n-x}, \ x = 0, 1, 2, \ldots, n, \ q = 1-p, \ p+q = 1$$

$$Pr(X \leq x') = \frac{1}{\sqrt{2\pi\sigma^2}} \int_{-\infty}^{x'} e^{-\frac{1}{2}\left(\frac{x-\mu}{\sigma}\right)^2} dx$$

Table 1
Individual Terms of the Binomial Distribution

The $(x+1)^{st}$ term in the expansion of the binomial $[\theta + (1-\theta)]^n$ is given by

$$f(x; n, \theta) = \binom{n}{x} \theta^x (1-\theta)^{n-x}, \ x = 0, 1, 2, \ldots, n.$$

This is the probability of exactly x successes in n independent binomial trials with probability of successes on a single trial equal to θ.

For $\theta > 0.5$, the value of $\binom{n}{x} \theta^x (1-\theta)^{n-x}$ is found by using the table entry for $\binom{n}{n-x} (1-\theta)^{n-x} \theta^x$.

Source: Beyer, W. H. (ed). *Handbook of Tables for Probability and Statistics.* Cleveland: The Chemical Rubber Company, 1966. Reprinted by permission of the publisher.

		θ									
n	x	0.05	0.10	0.15	0.20	0.25	0.30	0.35	0.40	0.45	0.50
1	0	0.9500	0.9000	0.8500	0.8000	0.7500	0.7000	0.6500	0.6000	0.5500	0.5000
	1	0.0500	0.1000	0.1500	0.2000	0.2500	0.3000	0.3500	0.4000	0.4500	0.5000
2	0	0.9025	0.8100	0.7225	0.6400	0.5625	0.4900	0.4225	0.3600	0.3025	0.2500
	1	0.0950	0.1800	0.2550	0.3200	0.3750	0.4200	0.4550	0.4800	0.4950	0.5000
	2	0.0025	0.0100	0.0225	0.0400	0.0625	0.0900	0.1225	0.1600	0.2025	0.2500
3	0	0.8574	0.7290	0.6141	0.5120	0.4219	0.3430	0.2746	0.2160	0.1664	0.1250
	1	0.1354	0.2430	0.3251	0.3840	0.4219	0.4410	0.4436	0.4320	0.4084	0.3750
	2	0.0071	0.0270	0.0574	0.0960	0.1406	0.1890	0.2389	0.2880	0.3341	0.3750
	3	0.0001	0.0010	0.0034	0.0080	0.0156	0.0270	0.0429	0.0640	0.0911	0.1250
4	0	0.8145	0.6561	0.5220	0.4096	0.3164	0.2401	0.1785	0.1296	0.0915	0.0625
	1	0.1715	0.2916	0.3685	0.4096	0.4219	0.4116	0.3845	0.3456	0.2995	0.2500
	2	0.0135	0.0486	0.0975	0.1536	0.2109	0.2646	0.3105	0.3456	0.3675	0.3750
	3	0.0005	0.0036	0.0115	0.0256	0.0469	0.0756	0.1115	0.1536	0.2005	0.2500
	4	0.0000	0.0001	0.0005	0.0016	0.0039	0.0081	0.0150	0.0256	0.0410	0.0625
5	0	0.7738	0.5905	0.4437	0.3277	0.2373	0.1681	0.1160	0.0778	0.0503	0.0312
	1	0.2036	0.3280	0.3915	0.4096	0.3955	0.3602	0.3124	0.2592	0.2059	0.1562
	2	0.0214	0.0729	0.1382	0.2048	0.2637	0.3087	0.3364	0.3456	0.3369	0.3125
	3	0.0011	0.0081	0.0244	0.0512	0.0879	0.1323	0.1811	0.2304	0.2757	0.3125
	4	0.0000	0.0004	0.0022	0.0064	0.0146	0.0284	0.0488	0.0768	0.1128	0.1562
	5	0.0000	0.0000	0.0001	0.0003	0.0010	0.0024	0.0053	0.0102	0.0185	0.0312
6	0	0.7351	0.5314	0.3771	0.2621	0.1780	0.1176	0.0754	0.0467	0.0277	0.0156
	1	0.2321	0.3543	0.3993	0.3932	0.3560	0.3025	0.2437	0.1866	0.1359	0.0938
	2	0.0305	0.0984	0.1762	0.2458	0.2966	0.3241	0.3280	0.3110	0.2780	0.2344
	3	0.0021	0.0146	0.0415	0.0819	0.1318	0.1852	0.2355	0.2765	0.3032	0.3125
	4	0.0001	0.0012	0.0055	0.0154	0.0330	0.0595	0.0951	0.1382	0.1861	0.2344
	5	0.0000	0.0001	0.0004	0.0015	0.0044	0.0102	0.0205	0.0369	0.0609	0.0938
	6	0.0000	0.0000	0.0000	0.0001	0.0002	0.0007	0.0018	0.0041	0.0083	0.0156
7	0	0.6983	0.4783	0.3206	0.2097	0.1335	0.0824	0.0490	0.0280	0.0152	0.0078
	1	0.2573	0.3720	0.3960	0.3670	0.3115	0.2471	0.1848	0.1306	0.0872	0.0547
	2	0.0406	0.1240	0.2097	0.2753	0.3115	0.3177	0.2985	0.2613	0.2140	0.1641
	3	0.0036	0.0230	0.0617	0.1147	0.1730	0.2269	0.2679	0.2903	0.2918	0.2734
	4	0.0002	0.0026	0.0109	0.0287	0.0577	0.0972	0.1442	0.1935	0.2388	0.2734

Linear interpolation with respect to θ will in general be accurate at most to two decimal points.

continued

		θ									
n	x	0.05	0.10	0.15	0.20	0.25	0.30	0.35	0.40	0.45	0.50
7	5	0.0000	0.0002	0.0012	0.0043	0.0115	0.0250	0.0466	0.0774	0.1172	0.1641
	6	0.0000	0.0000	0.0001	0.0004	0.0013	0.0036	0.0084	0.0172	0.0320	0.0547
	7	0.0000	0.0000	0.0000	0.0000	0.0001	0.0002	0.0006	0.0016	0.0037	0.0078
8	0	0.6634	0.4305	0.2725	0.1678	0.1001	0.0576	0.0319	0.0168	0.0084	0.0039
	1	0.2793	0.3826	0.3847	0.3355	0.2670	0.1977	0.1373	0.0896	0.0548	0.0312
	2	0.0515	0.1488	0.2376	0.2936	0.3115	0.2965	0.2587	0.2090	0.1569	0.1094
	3	0.0054	0.0331	0.0839	0.1468	0.2076	0.2541	0.2786	0.2787	0.2568	0.2188
	4	0.0004	0.0046	0.0185	0.0459	0.0865	0.1361	0.1875	0.2322	0.2627	0.2734
	5	0.0000	0.0004	0.0026	0.0092	0.0231	0.0467	0.0808	0.1239	0.1719	0.2188
	6	0.0000	0.0000	0.0002	0.0011	0.0038	0.0100	0.0217	0.0413	0.0703	0.1094
	7	0.0000	0.0000	0.0000	0.0001	0.0004	0.0012	0.0033	0.0079	0.0164	0.0312
	8	0.0000	0.0000	0.0000	0.0000	0.0000	0.0001	0.0002	0.0007	0.0017	0.0039
9	0	0.6302	0.3874	0.2316	0.1342	0.0751	0.0404	0.0207	0.0101	0.0046	0.0020
	1	0.2985	0.3874	0.3679	0.3020	0.2253	0.1556	0.1004	0.0605	0.0339	0.0176
	2	0.0629	0.1722	0.2597	0.3020	0.3003	0.2668	0.2162	0.1612	0.1110	0.0703
	3	0.0077	0.0446	0.1069	0.1762	0.2336	0.2668	0.2716	0.2508	0.2119	0.1641
	4	0.0006	0.0074	0.0283	0.0661	0.1168	0.1715	0.2194	0.2508	0.2600	0.2461
	5	0.0000	0.0008	0.0050	0.0165	0.0389	0.0735	0.1181	0.1672	0.2128	0.2461
	6	0.0000	0.0001	0.0006	0.0028	0.0087	0.0210	0.0424	0.0743	0.1160	0.1641
	7	0.0000	0.0000	0.0000	0.0003	0.0012	0.0039	0.0098	0.0212	0.0407	0.0703
	8	0.0000	0.0000	0.0000	0.0000	0.0001	0.0004	0.0013	0.0035	0.0083	0.0176
	9	0.0000	0.0000	0.0000	0.0000	0.0000	0.0000	0.0001	0.0003	0.0008	0.0020
10	0	0.5987	0.2487	0.1969	0.1074	0.0563	0.0282	0.0135	0.0060	0.0025	0.0010
	1	0.3151	0.3874	0.3474	0.2684	0.1877	0.1211	0.0725	0.0403	0.0207	0.0098
	2	0.0746	0.1937	0.2759	0.3020	0.2816	0.2335	0.1757	0.1209	0.0763	0.0439
	3	0.0105	0.0574	0.1298	0.2013	0.2503	0.2668	0.2522	0.2150	0.1665	0.1172
	4	0.0010	0.0112	0.0401	0.0881	0.1460	0.2001	0.2377	0.2508	0.2384	0.2051
	5	0.0001	0.0015	0.0085	0.0264	0.0584	0.1029	0.1536	0.2007	0.2340	0.2461
	6	0.0000	0.0001	0.0012	0.0055	0.0162	0.0368	0.0689	0.1115	0.1596	0.2051
	7	0.0000	0.0000	0.0001	0.0008	0.0031	0.0090	0.0212	0.0425	0.0746	0.1172
	8	0.0000	0.0000	0.0000	0.0001	0.0004	0.0014	0.0043	0.0106	0.0229	0.0439
	9	0.0000	0.0000	0.0000	0.0000	0.0000	0.0001	0.0005	0.0016	0.0042	0.0098
	10	0.0000	0.0000	0.0000	0.0000	0.0000	0.0000	0.0000	0.0001	0.0003	0.0010
11	0	0.5688	0.3138	0.1673	0.0859	0.0422	0.0198	0.0088	0.0036	0.0014	0.0004
	1	0.3293	0.3835	0.3248	0.2362	0.1549	0.0932	0.0518	0.0266	0.0125	0.0055

		θ									
n	x	0.05	0.10	0.15	0.20	0.25	0.30	0.35	0.40	0.45	0.50
11	2	0.0867	0.2131	0.2866	0.2953	0.2581	0.1998	0.1395	0.0887	0.0513	0.0269
	3	0.0137	0.0710	0.1517	0.2215	0.2581	0.2568	0.2254	0.1774	0.1259	0.0806
	4	0.0014	0.0158	0.0536	0.1107	0.1721	0.2201	0.2428	0.2365	0.2060	0.1611
	5	0.0001	0.0025	0.0132	0.0388	0.0803	0.1321	0.1830	0.2207	0.2360	0.2256
	6	0.0000	0.0003	0.0023	0.0097	0.0268	0.0566	0.0985	0.1471	0.1931	0.2256
	7	0.0000	0.0000	0.0003	0.0017	0.0064	0.0173	0.0379	0.0701	0.1128	0.1611
	8	0.0000	0.0000	0.0000	0.0002	0.0011	0.0037	0.0102	0.0234	0.0462	0.0806
	9	0.0000	0.0000	0.0000	0.0000	0.0001	0.0005	0.0018	0.0052	0.0126	0.0269
	10	0.0000	0.0000	0.0000	0.0000	0.0000	0.0000	0.0002	0.0007	0.0021	0.0054
	11	0.0000	0.0000	0.0000	0.0000	0.0000	0.0000	0.0000	0.0000	0.0002	0.0005
12	0	0.5404	0.2824	0.1422	0.0687	0.0317	0.0138	0.0057	0.0022	0.0008	0.0002
	1	0.3413	0.3766	0.3012	0.2062	0.1267	0.0712	0.0368	0.0174	0.0075	0.0029
	2	0.0988	0.2301	0.2924	0.2835	0.2323	0.1678	0.1088	0.0639	0.0339	0.0161
	3	0.0173	0.0852	0.1720	0.2362	0.2581	0.2397	0.1954	0.1419	0.0923	0.0537
	4	0.0021	0.0213	0.0683	0.1329	0.1936	0.2311	0.2367	0.2128	0.1700	0.1208
	5	0.0002	0.0038	0.0193	0.0532	0.1032	0.1585	0.2039	0.2270	0.2225	0.1934
	6	0.0000	0.0005	0.0040	0.0155	0.0401	0.0792	0.1281	0.1766	0.2124	0.2256
	7	0.0000	0.0000	0.0006	0.0033	0.0115	0.0291	0.0591	0.1009	0.1489	0.1934
	8	0.0000	0.0000	0.0001	0.0005	0.0024	0.0078	0.0199	0.0420	0.0762	0.1208
	9	0.0000	0.0000	0.0000	0.0001	0.0004	0.0015	0.0048	0.0125	0.0277	0.0537
	10	0.0000	0.0000	0.0000	0.0000	0.0000	0.0002	0.0008	0.0025	0.0068	0.0161
	11	0.0000	0.0000	0.0000	0.0000	0.0000	0.0000	0.0001	0.0003	0.0010	0.0029
	12	0.0000	0.0000	0.0000	0.0000	0.0000	0.0000	0.0000	0.0000	0.0001	0.0002
13	0	0.5133	0.2542	0.1209	0.0550	0.0238	0.0097	0.0037	0.0013	0.0004	0.0001
	1	0.3512	0.3672	0.2774	0.1787	0.1029	0.0540	0.0259	0.0113	0.0045	0.0016
	2	0.1109	0.2448	0.2937	0.2680	0.2059	0.1388	0.0836	0.0453	0.0220	0.0095
	3	0.0214	0.0997	0.1900	0.2457	0.2517	0.2181	0.1651	0.1107	0.0660	0.0349
	4	0.0028	0.0277	0.0838	0.1535	0.2097	0.2337	0.2222	0.1845	0.1350	0.0873
	5	0.0003	0.0055	0.0266	0.0691	0.1258	0.1803	0.2154	0.2214	0.1989	0.1571
	6	0.0000	0.0008	0.0063	0.0230	0.0559	0.1030	0.1546	0.1968	0.2169	0.2095
	7	0.0000	0.0001	0.0011	0.0058	0.0186	0.0442	0.0833	0.1312	0.1775	0.2095
	8	0.0000	0.0000	0.0001	0.0011	0.0047	0.0142	0.0336	0.0656	0.1089	0.1571
	9	0.0000	0.0000	0.0000	0.0001	0.0009	0.0034	0.0101	0.0243	0.0495	0.0873

continued

		θ									
n	x	0.05	0.10	0.15	0.20	0.25	0.30	0.35	0.40	0.45	0.50
13	10	0.0000	0.0000	0.0000	0.0000	0.0001	0.0006	0.0022	0.0065	0.0162	0.0349
	11	0.0000	0.0000	0.0000	0.0000	0.0000	0.0001	0.0003	0.0012	0.0036	0.0095
	12	0.0000	0.0000	0.0000	0.0000	0.0000	0.0000	0.0000	0.0001	0.0005	0.0016
	13	0.0000	0.0000	0.0000	0.0000	0.0000	0.0000	0.0000	0.0000	0.0000	0.0001
14	0	0.4877	0.2288	0.1028	0.0440	0.0178	0.0068	0.0024	0.0008	0.0002	0.0001
	1	0.3593	0.3559	0.2539	0.1539	0.0832	0.0407	0.0181	0.0073	0.0027	0.0009
	2	0.1229	0.2570	0.2912	0.2501	0.1802	0.1134	0.0634	0.0317	0.0141	0.0056
	3	0.0259	0.1142	0.2056	0.2501	0.2402	0.1943	0.1366	0.0845	0.0462	0.0222
	4	0.0037	0.0349	0.0998	0.1720	0.2202	0.2290	0.2022	0.1549	0.1040	0.0611
	5	0.0004	0.0078	0.0352	0.0860	0.1468	0.1963	0.2178	0.2066	0.1701	0.1222
	6	0.0000	0.0013	0.0093	0.0322	0.0734	0.1262	0.1759	0.2066	0.2088	0.1833
	7	0.0000	0.0002	0.0019	0.0092	0.0280	0.0618	0.1082	0.1574	0.1952	0.2095
	8	0.0000	0.0000	0.0003	0.0020	0.0082	0.0232	0.0510	0.0918	0.1398	0.1833
	9	0.0000	0.0000	0.0000	0.0003	0.0018	0.0066	0.0183	0.0408	0.0762	0.1222
	10	0.0000	0.0000	0.0000	0.0000	0.0003	0.0014	0.0049	0.0136	0.0312	0.0611
	11	0.0000	0.0000	0.0000	0.0000	0.0000	0.0002	0.0010	0.0033	0.0093	0.0222
	12	0.0000	0.0000	0.0000	0.0000	0.0000	0.0000	0.0001	0.0005	0.0019	0.0056
	13	0.0000	0.0000	0.0000	0.0000	0.0000	0.0000	0.0000	0.0001	0.0002	0.0009
	14	0.0000	0.0000	0.0000	0.0000	0.0000	0.0000	0.0000	0.0000	0.0000	0.0001
15	0	0.4633	0.2059	0.0874	0.0352	0.0134	0.0047	0.0016	0.0005	0.0001	0.0000
	1	0.3658	0.3432	0.2312	0.1319	0.0668	0.0305	0.0126	0.0047	0.0016	0.0005
	2	0.1348	0.2669	0.2856	0.2309	0.1559	0.0916	0.0476	0.0219	0.0090	0.0032
	3	0.0307	0.1285	0.2184	0.2501	0.2252	0.1700	0.1110	0.0634	0.0318	0.0139
	4	0.0049	0.0428	0.1156	0.1876	0.2252	0.2186	0.1792	0.1268	0.0780	0.0417
	5	0.0006	0.0105	0.0449	0.1032	0.1651	0.2061	0.2123	0.1859	0.1404	0.0916
	6	0.0000	0.0019	0.0132	0.0430	0.0917	0.1472	0.1906	0.2066	0.1914	0.1527
	7	0.0000	0.0003	0.0030	0.0138	0.0393	0.0811	0.1319	0.1771	0.2013	0.1964
	8	0.0000	0.0000	0.0005	0.0035	0.0313	0.0348	0.0710	0.1181	0.1647	0.1964
	9	0.0000	0.0000	0.0001	0.0007	0.0034	0.0116	0.0298	0.0612	0.1048	0.1527
	10	0.0000	0.0000	0.0000	0.0001	0.0007	0.0030	0.0096	0.0245	0.0515	0.0916
	11	0.0000	0.0000	0.0000	0.0000	0.0001	0.0006	0.0024	0.0074	0.0191	0.0417
	12	0.0000	0.0000	0.0000	0.0000	0.0000	0.0001	0.0004	0.0016	0.0052	0.0139
	13	0.0000	0.0000	0.0000	0.0000	0.0000	0.0000	0.0001	0.0003	0.0010	0.0032
	14	0.0000	0.0000	0.0000	0.0000	0.0000	0.0000	0.0000	0.0000	0.0001	0.0005
	15	0.0000	0.0000	0.0000	0.0000	0.0000	0.0000	0.0000	0.0000	0.0000	0.0000

		θ									
n	x	0.05	0.10	0.15	0.20	0.25	0.30	0.35	0.40	0.45	0.50
16	0	0.4401	0.1853	0.0743	0.0281	0.0100	0.0033	0.0010	0.0003	0.0001	0.0000
	1	0.3706	0.3294	0.2097	0.1126	0.0535	0.0228	0.0087	0.0030	0.0009	0.0002
	2	0.1463	0.2745	0.2775	0.2111	0.1336	0.0732	0.0353	0.0150	0.0056	0.0018
	3	0.0359	0.1423	0.2285	0.2463	0.2079	0.1465	0.0888	0.0468	0.0215	0.0085
	4	0.0061	0.0514	0.1311	0.2001	0.2252	0.2040	0.1553	0.1014	0.0572	0.0278
	5	0.0008	0.0137	0.0555	0.1201	0.1802	0.2099	0.2008	0.1623	0.1123	0.0667
	6	0.0001	0.0028	0.0180	0.0550	0.1101	0.1649	0.1982	0.1983	0.1684	0.1222
	7	0.0000	0.0004	0.0045	0.0197	0.0524	0.1010	0.1524	0.1889	0.1969	0.1746
	8	0.0000	0.0001	0.0009	0.0055	0.0197	0.0487	0.0923	0.1417	0.1812	0.1964
	9	0.0000	0.0000	0.0001	0.0012	0.0058	0.0185	0.0442	0.0840	0.1318	0.1746
	10	0.0000	0.0000	0.0000	0.0002	0.0014	0.0056	0.0167	0.0392	0.0755	0.1222
	11	0.0000	0.0000	0.0000	0.0000	0.0002	0.0013	0.0049	0.0142	0.0337	0.0667
	12	0.0000	0.0000	0.0000	0.0000	0.0000	0.0002	0.0011	0.0040	0.0115	0.0278
	13	0.0000	0.0000	0.0000	0.0000	0.0000	0.0000	0.0002	0.0008	0.0029	0.0085
	14	0.0000	0.0000	0.0000	0.0000	0.0000	0.0000	0.0000	0.0001	0.0005	0.0018
	15	0.0000	0.0000	0.0000	0.0000	0.0000	0.0000	0.0000	0.0000	0.0001	0.0002
	16	0.0000	0.0000	0.0000	0.0000	0.0000	0.0000	0.0000	0.0000	0.0000	0.0000
17	0	0.4181	0.1668	0.0631	0.0225	0.0075	0.0023	0.0007	0.0002	0.0000	0.0000
	1	0.3741	0.3150	0.1893	0.0957	0.0426	0.0169	0.0060	0.0019	0.0005	0.0001
	2	0.1575	0.2800	0.2673	0.1914	0.1136	0.0581	0.026	0.0102	0.0035	0.0010
	3	0.0415	0.1556	0.2359	0.2393	0.1893	0.1245	0.0701	0.0341	0.0144	0.0052
	4	0.9076	0.0605	0.1457	0.2093	0.2209	0.1868	0.1320	0.0796	0.0411	0.0182
	5	0.0010	0.0175	0.0668	0.1361	0.1914	0.2081	0.1849	0.1379	0.0875	0.0472
	6	0.0001	0.0039	0.0236	0.0680	0.1276	0.1784	0.1991	0.1839	0.1432	0.0944
	7	0.0000	0.0007	0.0065	0.0267	0.0668	0.1201	0.1685	0.1927	0.1841	0.1484
	8	0.0000	0.0001	0.0014	0.0084	0.0279	0.0644	0.1134	0.1606	0.1883	0.1855
	9	0.0000	0.0000	0.0003	0.0021	0.0093	0.0276	0.0611	0.1070	0.1540	0.1855
	10	0.0000	0.0000	0.0000	0.0004	0.0025	0.0095	0.0263	0.0571	0.1008	0.1484
	11	0.0000	0.0000	0.0000	0.0001	0.0005	0.0026	0.0090	0.0242	0.0525	0.0944
	12	0.0000	0.0000	0.0000	0.0000	0.0001	0.0006	0.0024	0.0081	0.0215	0.0472
	13	0.0000	0.0000	0.0000	0.0000	0.0000	0.0001	0.0005	0.0021	0.0068	0.0182
	14	0.0000	0.0000	0.0000	0.0000	0.0000	0.0000	0.0001	0.0004	0.0016	0.0052
	15	0.0000	0.0000	0.0000	0.0000	0.0000	0.0000	0.0000	0.0001	0.0003	0.0010
	16	0.0000	0.0000	0.0000	0.0000	0.0000	0.0000	0.0000	0.0000	0.0000	0.0001
	17	0.0000	0.0000	0.0000	0.0000	0.0000	0.0000	0.0000	0.0000	0.0000	0.0000

continued

		θ									
n	x	0.05	0.10	0.15	0.20	0.25	0.30	0.35	0.40	0.45	0.50
18	0	0.3972	0.1501	0.0536	0.0180	0.0056	0.0016	0.0004	0.0001	0.0000	0.0000
	1	0.3763	0.3002	0.1704	0.0811	0.0338	0.0126	0.0042	0.0012	0.0003	0.0001
	2	0.1683	0.2835	0.2556	0.1723	0.0958	0.0458	0.0190	0.0069	0.0022	0.0006
	3	0.0473	0.1680	0.2406	0.2297	0.1704	0.1046	0.0547	0.0246	0.0095	0.0031
	4	0.0093	0.0700	0.1592	0.2153	0.2130	0.1681	0.1104	0.0614	0.0291	0.0177
	5	0.0014	0.0218	0.0787	0.1507	0.1988	0.2017	0.1664	0.1146	0.0666	0.0327
	6	0.0002	0.0052	0.0301	0.0816	0.1436	0.1873	0.1941	0.1655	0.1181	0.0708
	7	0.0000	0.0010	0.0091	0.0350	0.0820	0.1376	0.1792	0.1892	0.1657	0.1214
	8	0.0000	0.0002	0.0022	0.0120	0.0376	0.0811	0.1327	0.1734	0.1864	0.1669
	9	0.0000	0.0000	0.0004	0.0033	0.0139	0.0386	0.0794	0.1284	0.1694	0.1855
	10	0.0000	0.0000	0.0001	0.0008	0.0042	0.0149	0.0385	0.0771	0.1248	0.1669
	11	0.0000	0.0000	0.0000	0.0001	0.0010	0.0046	0.0151	0.0374	0.0742	0.1214
	12	0.0000	0.0000	0.0000	0.0000	0.0002	0.0012	0.0047	0.0145	0.0354	0.0708
	13	0.0000	0.0000	0.0000	0.0000	0.0000	0.0002	0.0012	0.0045	0.0134	0.0327
	14	0.0000	0.0000	0.0000	0.0000	0.0000	0.0000	0.0002	0.0011	0.0039	0.0117
	15	0.0000	0.0000	0.0000	0.0000	0.0000	0.0000	0.0000	0.0002	0.0009	0.0031
	16	0.0000	0.0000	0.0000	0.0000	0.0000	0.0000	0.0000	0.0000	0.0001	0.0006
	17	0.0000	0.0000	0.0000	0.0000	0.0000	0.0000	0.0000	0.0000	0.0000	0.0001
	18	0.0000	0.0000	0.0000	0.0000	0.0000	0.0000	0.0000	0.0000	0.0000	0.0000
19	0	0.3774	0.1351	0.0456	0.0144	0.0042	0.0011	0.0003	0.0001	0.0000	0.0000
	1	0.3774	0.2852	0.1529	0.0685	0.0268	0.0093	0.0029	0.0008	0.0002	0.0000
	2	0.1787	0.2852	0.2428	0.1540	0.0803	0.0358	0.0138	0.0046	0.0013	0.0003
	3	0.0533	0.1796	0.2428	0.2182	0.1517	0.0869	0.0422	0.0175	0.0062	0.0018
	4	0.0112	0.0798	0.1714	0.2182	0.2023	0.1491	0.0909	0.0467	0.0203	0.0074
	5	0.0018	0.0266	0.0907	0.1636	0.2023	0.1916	0.1468	0.0933	0.0497	0.0222
	6	0.0002	0.0069	0.0374	0.0955	0.1574	0.1916	0.1844	0.1451	0.0949	0.0518
	7	0.0000	0.0014	0.0122	0.0443	0.0974	0.1525	0.1844	0.1797	0.1443	0.0961
	8	0.0000	0.0002	0.0032	0.0166	0.0487	0.0981	0.1489	0.1797	0.1771	0.1442
	9	0.0000	0.0000	0.0007	0.0051	0.0198	0.0514	0.0980	0.1464	0.1771	0.1762
	10	0.0000	0.0000	0.0001	0.0013	0.0066	0.0220	0.0528	0.0976	0.1449	0.1762
	11	0.0000	0.0000	0.0000	0.0003	0.0018	0.0077	0.0233	0.0532	0.0970	0.1442
	12	0.0000	0.0000	0.0000	0.0000	0.0004	0.0022	0.0083	0.0237	0.0529	0.0961
	13	0.0000	0.0000	0.0000	0.0000	0.0001	0.0005	0.0024	0.0085	0.0233	0.0518
	14	0.0000	0.0000	0.0000	0.0000	0.0000	0.0001	0.0006	0.0024	0.0082	0.0222

						θ					
n	x	0.05	0.10	0.15	0.20	0.25	0.30	0.35	0.40	0.45	0.50
19	15	0.0000	0.0000	0.0000	0.0000	0.0000	0.0000	0.0001	0.0005	0.0022	0.0074
	16	0.0000	0.0000	0.0000	0.0000	0.0000	0.0000	0.0000	0.0001	0.0005	0.0018
	17	0.0000	0.0000	0.0000	0.0000	0.0000	0.0000	0.0000	0.0000	0.0001	0.0003
	18	0.0000	0.0000	0.0000	0.0000	0.0000	0.0000	0.0000	0.0000	0.0000	0.0000
	19	0.0000	0.0000	0.0000	0.0000	0.0000	0.0000	0.0000	0.0000	0.0000	0.0000
20	0	0.3585	0.1216	0.0388	0.0115	0.0032	0.0008	0.0002	0.0000	0.0000	0.0000
	1	0.3774	0.2702	0.1368	0.0576	0.0211	0.0068	0.0020	0.0005	0.0001	0.0000
	2	0.1887	0.2852	0.2293	0.1369	0.0669	0.0278	0.0100	0.0031	0.0008	0.0002
	3	0.0596	0.1901	0.2428	0.2054	0.1339	0.0716	0.0323	0.0123	0.0040	0.0011
	4	0.0133	0.0898	0.1821	0.2182	0.1897	0.1304	0.0738	0.0350	0.0139	0.0046
	5	0.0022	0.0319	0.1028	0.1746	0.2023	0.1789	0.1272	0.0746	0.0365	0.0148
	6	0.0003	0.0089	0.0454	0.1091	0.1686	0.1916	0.1712	0.1244	0.0746	0.0370
	7	0.0000	0.0020	0.0160	0.0545	0.1124	0.1643	0.1844	0.1659	0.1221	0.0739
	8	0.0000	0.0004	0.0046	0.0222	0.0609	0.1144	0.1614	0.1797	0.1623	0.1201
	9	0.0000	0.0001	0.0011	0.0074	0.0271	0.0654	0.1158	0.1597	0.1771	0.1602
	10	0.0000	0.0000	0.0002	0.0020	0.0099	0.0308	0.0686	0.1171	0.1593	0.1762
	11	0.0000	0.0000	0.0000	0.0005	0.0030	0.0120	0.0336	0.0710	0.1185	0.1602
	12	0.0000	0.0000	0.0000	0.0001	0.0008	0.0039	0.0136	0.0355	0.0727	0.1201
	13	0.0000	0.0000	0.0000	0.0000	0.0002	0.0010	0.0045	0.0146	0.0366	0.0739
	14	0.0000	0.0000	0.0000	0.0000	0.0000	0.0002	0.0012	0.0049	0.0150	0.0370
	15	0.0000	0.0000	0.0000	0.0000	0.0000	0.0000	0.0003	0.0013	0.0049	0.0148
	16	0.0000	0.0000	0.0000	0.0000	0.0000	0.0000	0.0000	0.0003	0.0013	0.0046
	17	0.0000	0.0000	0.0000	0.0000	0.0000	0.0000	0.0000	0.0000	0.0002	0.0011
	18	0.0000	0.0000	0.0000	0.0000	0.0000	0.0000	0.0000	0.0000	0.0000	0.0002
	19	0.0000	0.0000	0.0000	0.0000	0.0000	0.0000	0.0000	0.0000	0.0000	0.0000
	20	0.0000	0.0000	0.0000	0.0000	0.0000	0.0000	0.0000	0.0000	0.0000	0.0000

Table 2
Cumulative Terms of the Binomial Distribution

For the binomial probability function $f(x;n,\theta)$ the probability of observing x' or more successes is given by

$$\sum_{x=x'}^{n}\binom{n}{x}\theta^x(1-\theta)^{n-x},$$

This table contains the values of $\sum_{x=x'}^{n}\binom{n}{x}\theta^x(1-\theta)^{n-x}$ for specified values of n, x', and θ.

If $\theta > 0.5$, the values for $\sum_{x=x'}^{n}\binom{n}{x}\theta^x(1-\theta)^{n-x}$ are obtained using the corresponding results obtained from

$$1-\sum_{x=n-x'+1}^{n}\binom{n}{x}(1-\theta)^x\theta^{n-x}$$

The cumulative binomial distribution is related to the incomplete beta function as follows:

$$\sum_{x=x'}^{n}\binom{n}{x}\theta^x(1-\theta)^{n-x} = I_\theta(x',n-x'+1),$$

$$\sum_{x=0}^{x'-1}\binom{n}{x}\theta^x(1-\theta)^{n-x} = 1-I_\theta(x',n-x'+1)$$

$$=-1\int_0^\theta u^{x'-1}(1-u)^{n-x'}du \Big/ \int_0^1 u^{x'-1}(1-u)^{n-x'}du.$$

The cumulative binomial distribution is related to the cumulative negative binomial distribution as follows:

$$1-\sum_{x'=0}^{r-1}\binom{x+r}{x'}\theta^{x'}(1-\theta)^{x+r-x'} = \sum_{x'=0}^{x}\binom{x'+r-1}{r-1}\theta^r(1-\theta)^{x'}$$

or

$$\sum_{x'=r}^{x+r}\binom{x+r}{x'}\theta^{x'}(1-\theta)^{x+r-x'} = \sum_{x'=0}^{x}\binom{x'+r-1}{r-1}\theta^r(1-\theta)^{x'}.$$

Source: Beyer, W. H. (ed). *Handbook of Tables for Probability and Statistics.* Cleveland: The Chemical Rubber Company, 1966. Reprinted by permission of the publisher.

		θ									
n	x	0.05	0.10	0.15	0.20	0.25	0.30	0.35	0.40	0.45	0.50
2	1	0.0975	0.1900	0.2775	0.3600	0.4375	0.5100	0.5775	0.6400	0.6975	0.7500
	2	0.0025	0.0100	0.0225	0.0400	0.0625	0.0900	0.1225	0.1600	0.2025	0.2500
3	1	0.1426	0.2710	0.3859	0.4880	0.5781	0.6570	0.7254	0.7840	0.8336	0.8750
	2	0.0072	0.0280	0.0608	0.1040	0.1562	0.2160	0.2818	0.3520	0.4252	0.5000
	3	0.0001	0.0010	0.0034	0.0080	0.0156	0.0270	0.0429	0.0640	0.0911	0.1250
4	1	0.1855	0.3439	0.4780	0.5904	0.6836	0.7599	0.8215	0.8704	0.9085	0.9375
	2	0.0140	0.0523	0.1095	0.1808	0.2716	0.3483	0.4307	0.5248	0.6090	0.6875
	3	0.0005	0.0037	0.0120	0.0272	0.0508	0.0837	0.1265	0.1792	0.2415	0.3125
	4	0.0000	0.0001	0.0005	0.0016	0.0039	0.0081	0.0150	0.0256	0.0410	0.0625
5	1	0.2262	0.4095	0.5563	0.6723	0.7627	0.8319	0.8840	0.9222	0.9497	0.9688
	2	0.0226	0.0815	0.1648	0.2627	0.3672	0.4718	0.5716	0.6630	0.7438	0.8125
	3	0.0012	0.0086	0.0266	0.0579	0.1035	0.1631	0.2352	0.3174	0.4069	0.5000
	4	0.0000	0.0005	0.0022	0.0067	0.0156	0.0308	0.0540	0.0870	0.1312	0.1875
	5	0.0000	0.0000	0.0001	0.0003	0.0010	0.0024	0.0053	0.0102	0.0185	0.0312
6	1	0.2649	0.4686	0.6229	0.7379	0.8220	0.8824	0.9246	0.9533	0.9723	0.9844
	2	0.0328	0.1143	0.2235	0.3447	0.4166	0.5798	0.6809	0.7667	0.8364	0.8906
	3	0.0022	0.0158	0.0473	0.0989	0.1694	0.2557	0.3529	0.4557	0.5585	0.6562
	4	0.0001	0.0013	0.0059	0.0170	0.0376	0.0705	0.1174	0.1792	0.2553	0.3438
	5	0.0000	0.0001	0.0004	0.0016	0.0046	0.0109	0.0223	0.0410	0.0692	0.1094
	6	0.0000	0.0000	0.0000	0.0001	0.0002	0.0007	0.0018	0.0041	0.0083	0.0156
7	1	0.3017	0.5217	0.6794	0.7903	0.8665	0.9176	0.9510	0.9720	0.9848	0.9922
	2	0.0444	0.1497	0.2834	0.4233	0.5551	0.6707	0.7662	0.8414	0.8976	0.9375
	3	0.0038	0.0257	0.0738	0.1480	0.2436	0.3529	0.4677	0.5801	0.6836	0.7734
	4	0.0002	0.0027	0.0121	0.0333	0.0706	0.1260	0.1998	0.2898	0.3917	0.5000
	5	0.0000	0.0002	0.0012	0.0047	0.0129	0.0288	0.0556	0.0963	0.1529	0.2266
	6	0.0000	0.0000	0.0001	0.0004	0.0013	0.0038	0.0090	0.0188	0.0357	0.0625
	7	0.0000	0.0000	0.0000	0.0000	0.0001	0.0002	0.0006	0.0016	0.0037	0.0078
8	1	0.3366	0.5695	0.7275	0.8322	0.8999	0.9424	0.9681	0.9832	0.9916	0.9961
	2	0.0572	0.1869	0.3428	0.4967	0.6329	0.7447	0.8309	0.8936	0.9368	0.9648
	3	0.0058	0.0381	0.1052	0.2031	0.3215	0.4482	0.5722	0.6846	0.7799	0.8555
	4	0.0004	0.0050	0.0214	0.0563	0.1138	0.1941	0.2936	0.4059	0.5230	0.6367
	5	0.0000	0.0004	0.0029	0.0104	0.0273	0.0580	0.1061	0.1737	0.2604	0.3633

Linear interpolation will be accurate at most to two decimal points.

continued

		θ									
n	x	0.05	0.10	0.15	0.20	0.25	0.30	0.35	0.40	0.45	0.50
8	6	0.0000	0.0000	0.0002	0.0012	0.0042	0.0113	0.0253	0.0498	0.0885	0.1445
	7	0.0000	0.0000	0.0000	0.0001	0.0004	0.0013	0.0036	0.0085	0.0181	0.0352
	8	0.0000	0.0000	0.0000	0.0000	0.0000	0.0001	0.0002	0.0007	0.0017	0.0039
9	1	0.3698	0.6126	0.7684	0.8658	0.9249	0.9596	0.9793	0.9899	0.9954	0.9980
	2	0.0712	0.2252	0.4005	0.5638	0.6997	0.8040	0.8789	0.9295	0.9615	0.9805
	3	0.0084	0.0530	0.1409	0.2618	0.3993	0.5372	0.6627	0.7682	0.8505	0.9102
	4	0.0006	0.0083	0.0339	0.0856	0.1657	0.2703	0.3911	0.5174	0.6386	0.7461
	5	0.0000	0.0009	0.0056	0.0196	0.0489	0.0988	0.1717	0.2666	0.3786	0.5000
	6	0.0000	0.0001	0.0006	0.0031	0.0100	0.0253	0.0536	0.0994	0.1658	0.2539
	7	0.0000	0.0000	0.0000	0.0003	0.0013	0.0043	0.0112	0.0250	0.0498	0.0898
	8	0.0000	0.0000	0.0000	0.0000	0.0001	0.0004	0.0014	0.0038	0.0091	0.0195
	9	0.0000	0.0000	0.0000	0.0000	0.0000	0.0000	0.0001	0.0003	0.0008	0.0020
10	1	0.3151	0.3874	0.3474	0.2684	0.1877	0.1211	0.0725	0.0403	0.0207	0.0098
	2	0.0746	0.1937	0.2759	0.3020	0.2816	0.2335	0.1757	0.1209	0.0763	0.0439
	3	0.0105	0.0574	0.1298	0.2013	0.2503	0.2668	0.2522	0.2150	0.1665	0.1172
	4	0.0010	0.0112	0.0401	0.0881	0.1460	0.2001	0.2377	0.2508	0.2384	0.2051
	5	0.0001	0.0015	0.0085	0.0264	0.0584	0.1029	0.1536	0.2007	0.2340	0.2461
	6	0.0000	0.0001	0.0012	0.0055	0.0162	0.0368	0.0689	0.1115	0.1596	0.2051
	7	0.0000	0.0000	0.0001	0.0008	0.0031	0.0090	0.0212	0.0425	0.0746	0.1172
	8	0.0000	0.0000	0.0000	0.0001	0.0004	0.0014	0.0043	0.0106	0.0229	0.0439
	9	0.0000	0.0000	0.0000	0.0000	0.0000	0.0001	0.0005	0.0016	0.0042	0.0098
	10	0.0000	0.0000	0.0000	0.0000	0.0000	0.0000	0.0000	0.0001	0.0003	0.0010
11	1	0.3293	0.3835	0.3248	0.2362	0.1549	0.0932	0.0518	0.0266	0.0125	0.0055
	2	0.0867	0.2131	0.2866	0.2953	0.2581	0.1998	0.1395	0.0887	0.0513	0.0269
	3	0.0137	0.0710	0.1517	0.2215	0.2581	0.2568	0.2254	0.1774	0.1259	0.0806
	4	0.0014	0.0158	0.0536	0.1107	0.1721	0.2201	0.2428	0.2365	0.2060	0.1611
	5	0.0001	0.0025	0.0132	0.0388	0.0803	0.1321	0.1830	0.2207	0.2360	0.2256
	6	0.0000	0.0003	0.0023	0.0097	0.0268	0.0566	0.0985	0.1471	0.1931	0.2256
	7	0.0000	0.0000	0.0003	0.0017	0.0064	0.0173	0.0379	0.0701	0.1128	0.1611
	8	0.0000	0.0000	0.0000	0.0002	0.0011	0.0037	0.0102	0.0234	0.0462	0.0806
	9	0.0000	0.0000	0.0000	0.0000	0.0001	0.0005	0.0018	0.0052	0.0126	0.0269
	10	0.0000	0.0000	0.0000	0.0000	0.0000	0.0000	0.0002	0.0007	0.0021	0.0054
	11	0.0000	0.0000	0.0000	0.0000	0.0000	0.0000	0.0000	0.0000	0.0002	0.0005

		θ									
n	x	0.05	0.10	0.15	0.20	0.25	0.30	0.35	0.40	0.45	0.50
12	1	0.3413	0.3766	0.3012	0.2062	0.1267	0.0712	0.0368	0.0174	0.0075	0.0029
	2	0.0988	0.2301	0.2924	0.2835	0.2323	0.1678	0.1088	0.0639	0.0339	0.0161
	3	0.0173	0.0852	0.1720	0.2362	0.2581	0.2397	0.1954	0.1419	0.0923	0.0537
	4	0.0021	0.0213	0.0683	0.1329	0.1936	0.2311	0.2367	0.2128	0.1700	0.1208
	5	0.0002	0.0038	0.0193	0.0532	0.1032	0.1585	0.2039	0.2270	0.2225	0.1934
	6	0.0000	0.0005	0.0040	0.0155	0.0401	0.0792	0.1281	0.1766	0.2124	0.2256
	7	0.0000	0.0000	0.0006	0.0033	0.0115	0.0291	0.0591	0.1009	0.1489	0.1934
	8	0.0000	0.0000	0.0001	0.0005	0.0024	0.0078	0.0199	0.0420	0.0762	0.1208
	9	0.0000	0.0000	0.0000	0.0001	0.0004	0.0015	0.0048	0.0125	0.0277	0.0537
	10	0.0000	0.0000	0.0000	0.0000	0.0000	0.0002	0.0008	0.0025	0.0068	0.0161
	11	0.0000	0.0000	0.0000	0.0000	0.0000	0.0000	0.0001	0.0003	0.0010	0.0029
	12	0.0000	0.0000	0.0000	0.0000	0.0000	0.0000	0.0000	0.0000	0.0001	0.0002
13	1	0.4867	0.7458	0.8791	0.9450	0.9762	0.9903	0.9963	0.9987	0.9996	0.9999
	2	0.1354	0.3787	0.6017	0.7664	0.8733	0.9363	0.9704	0.9874	0.9951	0.9983
	3	0.0245	0.1339	0.2704	0.4983	0.6674	0.7975	0.8868	0.9421	0.9731	0.9888
	4	0.0031	0.0342	0.0967	0.2527	0.4157	0.5794	0.7217	0.8314	0.9071	0.9539
	5	0.0003	0.0065	0.0260	0.0991	0.2060	0.3457	0.4995	0.6470	0.7721	0.8666
	6	0.0000	0.0009	0.0053	0.0300	0.0802	0.1654	0.2841	0.4256	0.5732	0.7095
	7	0.0000	0.0001	0.0013	0.0070	0.0243	0.0624	0.1295	0.2288	0.3563	0.5000
	8	0.0000	0.0000	0.0002	0.0012	0.0056	0.0182	0.0462	0.0977	0.1788	0.2905
	9	0.0000	0.0000	0.0000	0.0002	0.0010	0.0040	0.0126	0.0321	0.0698	0.1334
	10	0.0000	0.0000	0.0000	0.0000	0.0001	0.0007	0.0025	0.0078	0.0203	0.0461
	11	0.0000	0.0000	0.0000	0.0000	0.0000	0.0001	0.0003	0.0013	0.0041	0.0112
	12	0.0000	0.0000	0.0000	0.0000	0.0000	0.0000	0.0000	0.0001	0.0005	0.0017
	13	0.0000	0.0000	0.0000	0.0000	0.0000	0.0000	0.0000	0.0000	0.0000	0.0001
14	1	0.5123	0.7712	0.8972	0.9560	0.9822	0.9932	0.9976	0.9992	0.9998	0.9999
	2	0.1530	0.4154	0.6433	0.8021	0.8990	0.9525	0.9795	0.9919	0.9971	0.9991
	3	0.0301	0.1584	0.3521	0.5519	0.7189	0.8392	0.9161	0.9602	0.9830	0.9935
	4	0.0042	0.0441	0.1465	0.3018	0.4787	0.6448	0.7795	0.8757	0.9368	0.9713
	5	0.0004	0.0092	0.0467	0.1298	0.2585	0.4158	0.5773	0.7207	0.8328	0.9102
	6	0.0000	0.0015	0.0115	0.0439	0.1117	0.2195	0.3595	0.5141	0.6627	0.7880
	7	0.0000	0.0002	0.0022	0.0116	0.0383	0.0933	0.1836	0.3075	0.4539	0.6047
	8	0.0000	0.0000	0.0003	0.0024	0.0103	0.0315	0.0753	0.1501	0.2586	0.3953
	9	0.0000	0.0000	0.0000	0.0004	0.0022	0.0083	0.0243	0.0583	0.1189	0.2120
	10	0.0000	0.0000	0.0000	0.0000	0.0003	0.0017	0.0060	0.0175	0.0426	0.0898

continued

		θ									
n	x	0.05	0.10	0.15	0.20	0.25	0.30	0.35	0.40	0.45	0.50
14	11	0.0000	0.0000	0.0000	0.0000	0.0000	0.0002	0.0011	0.0039	0.0114	0.0287
	12	0.0000	0.0000	0.0000	0.0000	0.0000	0.0000	0.0001	0.0006	0.0022	0.0065
	13	0.0000	0.0000	0.0000	0.0000	0.0000	0.0000	0.0000	0.0001	0.0003	0.0009
	14	0.0000	0.0000	0.0000	0.0000	0.0000	0.0000	0.0000	0.0000	0.0000	0.0001
15	1	0.5367	0.7941	0.9126	0.9648	0.9866	0.9953	0.9984	0.9995	0.9999	1.0000
	2	0.1710	0.4510	0.6814	0.8329	0.9198	0.9647	0.9858	0.9948	0.9983	0.9995
	3	0.0362	0.1841	0.3958	0.6020	0.7639	0.8732	0.9383	0.9729	0.9893	0.9963
	4	0.0055	0.0556	0.1773	0.3518	0.5387	0.7031	0.8273	0.9095	0.9576	0.9824
	5	0.0006	0.0127	0.0617	0.1642	0.3135	0.4845	0.6481	0.7827	0.8796	0.9408
	6	0.0001	0.0022	0.0168	0.0611	0.1484	0.2784	0.4357	0.5968	0.7392	0.8491
	7	0.0000	0.0003	0.0036	0.0181	0.0566	0.1311	0.2452	0.3902	0.5478	0.6964
	8	0.0000	0.0000	0.0006	0.0042	0.0173	0.0500	0.1132	0.2131	0.3465	0.5000
	9	0.0000	0.0000	0.0001	0.0008	0.0042	0.0152	0.0422	0.0950	0.1818	0.3036
	10	0.0000	0.0000	0.0000	0.0001	0.0008	0.0037	0.0124	0.0338	0.0769	0.1509
	11	0.0000	0.0000	0.0000	0.0000	0.0001	0.0007	0.0028	0.0093	0.0255	0.0592
	12	0.0000	0.0000	0.0000	0.0000	0.0000	0.0001	0.0005	0.0019	0.0063	0.0176
	13	0.0000	0.0000	0.0000	0.0000	0.0000	0.0000	0.0001	0.0003	0.0011	0.0037
	14	0.0000	0.0000	0.0000	0.0000	0.0000	0.0000	0.0000	0.0000	0.0001	0.0005
	15	0.0000	0.0000	0.0000	0.0000	0.0000	0.0000	0.0000	0.0000	0.0000	0.0000
16	1	0.5599	0.8147	0.9257	0.9719	0.9900	0.9967	0.9990	0.9997	0.9999	1.0000
	2	0.1892	0.4853	0.7161	0.8593	0.9365	0.9739	0.9902	0.9967	0.9990	0.9997
	3	0.0429	0.2108	0.4386	0.6482	0.8029	0.9006	0.9549	0.9817	0.9934	0.9979
	4	0.0070	0.0684	0.2101	0.4019	0.5950	0.7541	0.8661	0.9349	0.9719	0.9894
	5	0.0009	0.0170	0.0791	0.2018	0.3698	0.5501	0.7108	0.8334	0.9147	0.9616
	6	0.0001	0.0033	0.0235	0.0817	0.1897	0.3402	0.5100	0.6712	0.8024	0.8949
	7	0.0000	0.0005	0.0056	0.0267	0.0796	0.1753	0.3119	0.4728	0.6340	0.7228
	8	0.0000	0.0001	0.0011	0.0070	0.0271	0.0744	0.1594	0.2839	0.4371	0.5982
	9	0.0000	0.0000	0.0002	0.0015	0.0075	0.0257	0.0671	0.1423	0.2559	0.4018
	10	0.0000	0.0000	0.0000	0.0002	0.0016	0.0071	0.0229	0.0583	0.1241	0.2272
	11	0.0000	0.0000	0.0000	0.0000	0.0003	0.0016	0.0062	0.0191	0.0486	0.1051
	12	0.0000	0.0000	0.0000	0.0000	0.0000	0.0003	0.0013	0.0049	0.0149	0.0384
	13	0.0000	0.0000	0.0000	0.0000	0.0000	0.0000	0.0002	0.0009	0.0035	0.0106
	14	0.0000	0.0000	0.0000	0.0000	0.0000	0.0000	0.0000	0.0001	0.0003	0.0021
	15	0.0000	0.0000	0.0000	0.0000	0.0000	0.0000	0.0000	0.0000	0.0001	0.0003
	16	0.0000	0.0000	0.0000	0.0000	0.0000	0.0000	0.0000	0.0000	0.0000	0.0000

		θ									
n	x	0.05	0.10	0.15	0.20	0.25	0.30	0.35	0.40	0.45	0.50
17	1	0.5819	0.8332	0.9369	0.9775	0.9925	0.9977	0.9993	0.9998	1.0000	1.0000
	2	0.2078	0.5182	0.7475	0.8818	0.9499	0.9807	0.9933	0.9979	0.9994	0.9999
	3	0.0503	0.2382	0.4802	0.6904	0.8363	0.9226	0.9673	0.9877	0.9959	0.9988
	4	0.0088	0.0826	0.2444	0.4511	0.6470	0.7981	0.8972	0.9536	0.9816	0.9936
	5	0.0012	0.0221	0.0987	0.2418	0.4261	0.6113	0.7652	0.8740	0.9404	0.9755
	6	0.0001	0.0047	0.0319	0.1057	0.2347	0.4032	0.5803	0.7361	0.8529	0.9283
	7	0.0000	0.0008	0.0083	0.0377	0.1071	0.2248	0.3812	0.5522	0.7098	0.8338
	8	0.0000	0.0001	0.0017	0.0109	0.0402	0.1046	0.2128	0.3595	0.5257	0.6855
	9	0.0000	0.0000	0.0003	0.0026	0.0124	0.0403	0.0994	0.1989	0.3374	0.5000
	10	0.0000	0.0000	0.0000	0.0005	0.0031	0.0127	0.0383	0.0919	0.1834	0.3145
	11	0.0000	0.0000	0.0000	0.0001	0.0006	0.0032	0.0120	0.0348	0.0826	0.1662
	12	0.0000	0.0000	0.0000	0.0000	0.0001	0.0007	0.0030	0.0106	0.0301	0.0717
	13	0.0000	0.0000	0.0000	0.0000	0.0000	0.0001	0.0006	0.0025	0.0086	0.0245
	14	0.0000	0.0000	0.0000	0.0000	0.0000	0.0000	0.0000	0.0005	0.0019	0.0064
	15	0.0000	0.0000	0.0000	0.0000	0.0000	0.0000	0.0000	0.0001	0.0003	0.0012
	16	0.0000	0.0000	0.0000	0.0000	0.0000	0.0000	0.0000	0.0000	0.0000	0.0001
	17	0.0000	0.0000	0.0000	0.0000	0.0000	0.0000	0.0000	0.0000	0.0000	0.0000
18	1	0.6028	0.8499	0.9464	0.9820	0.9944	0.9984	0.9996	0.9999	1.0000	1.0000
	2	0.2265	0.5497	0.7759	0.9009	0.9605	0.9858	0.9954	0.9987	0.9997	0.9999
	3	0.0581	0.2662	0.5203	0.7287	0.8647	0.9400	0.9764	0.9918	0.9975	0.9993
	4	0.0109	0.0982	0.2798	0.4990	0.6943	0.8354	0.9217	0.9672	0.9880	0.9962
	5	0.0015	0.0282	0.1206	0.2836	0.4813	0.6673	0.8114	0.9058	0.9589	0.9846
	6	0.0002	0.0064	0.0419	0.1329	0.2825	0.4656	0.6450	0.7912	0.8923	0.9519
	7	0.0000	0.0012	0.0118	0.0513	0.1390	0.2783	0.4509	0.6257	0.7742	0.8811
	8	0.0000	0.0002	0.0027	0.0163	0.0569	0.1407	0.2717	0.4366	0.6085	0.7597
	9	0.0000	0.0000	0.0005	0.0043	0.0193	0.0596	0.1391	0.2632	0.4222	0.5927
	10	0.0000	0.0000	0.0001	0.0009	0.0054	0.0210	0.0597	0.1347	0.2527	0.4073
	11	0.0000	0.0000	0.0000	0.0002	0.0012	0.0061	0.0212	0.0576	0.1280	0.2403
	12	0.0000	0.0000	0.0000	0.0000	0.0002	0.0014	0.0062	0.0203	0.0537	0.1189
	13	0.0000	0.0000	0.0000	0.0000	0.0000	0.0003	0.0014	0.0058	0.0183	0.0481
	14	0.0000	0.0000	0.0000	0.0000	0.0000	0.0000	0.0003	0.0013	0.0049	0.0154
	15	0.0000	0.0000	0.0000	0.0000	0.0000	0.0000	0.0000	0.0002	0.0010	0.0038
	16	0.0000	0.0000	0.0000	0.0000	0.0000	0.0000	0.0000	0.0000	0.0001	0.0007
	17	0.0000	0.0000	0.0000	0.0000	0.0000	0.0000	0.0000	0.0000	0.0000	0.0001
	18	0.0000	0.0000	0.0000	0.0000	0.0000	0.0000	0.0000	0.0000	0.0000	0.0000

continued

		θ									
n	x	0.05	0.10	0.15	0.20	0.25	0.30	0.35	0.40	0.45	0.50
19	1	0.6226	0.8649	0.9544	0.9856	0.9958	0.9989	0.9997	0.9999	1.0000	1.0000
	2	0.2453	0.5797	0.8015	0.9171	0.9690	0.9896	0.9969	0.9992	0.9998	1.0000
	3	0.0665	0.2946	0.5587	0.7631	0.8887	0.9538	0.9830	0.9945	0.9985	0.9996
	4	0.0132	0.1150	0.3159	0.5449	0.7369	0.8668	0.9409	0.9770	0.9923	0.9978
	5	0.0020	0.0352	0.1444	0.3267	0.5346	0.7178	0.8500	0.9304	0.9720	0.9904
	6	0.0002	0.0086	0.0537	0.1631	0.3322	0.5261	0.7032	0.8371	0.9223	0.9682
	7	0.0000	0.0017	0.0163	0.0676	0.1749	0.3345	0.5188	0.6919	0.8273	0.9165
	8	0.0000	0.0003	0.0041	0.0233	0.0775	0.1820	0.3344	0.5122	0.6831	0.8204
	9	0.0000	0.0000	0.0008	0.0067	0.0287	0.0839	0.1855	0.3325	0.5060	0.6762
	10	0.0000	0.0000	0.0001	0.0016	0.0089	0.0326	0.0875	0.1861	0.3290	0.5000
	11	0.0000	0.0000	0.0000	0.0003	0.0023	0.0105	0.0347	0.0885	0.1841	0.3238
	12	0.0000	0.0000	0.0000	0.0000	0.0005	0.0028	0.0114	0.0352	0.0871	0.1796
	13	0.0000	0.0000	0.0000	0.0000	0.0001	0.0006	0.0031	0.0116	0.0342	0.0835
	14	0.0000	0.0000	0.0000	0.0000	0.0000	0.0001	0.0007	0.0031	0.0109	0.0318
	15	0.0000	0.0000	0.0000	0.0000	0.0000	0.0000	0.0001	0.0006	0.0028	0.0096
	16	0.0000	0.0000	0.0000	0.0000	0.0000	0.0000	0.0000	0.0001	0.0005	0.0022
	17	0.0000	0.0000	0.0000	0.0000	0.0000	0.0000	0.0000	0.0000	0.0001	0.0004
	18	0.0000	0.0000	0.0000	0.0000	0.0000	0.0000	0.0000	0.0000	0.0000	0.0000
	19	0.0000	0.0000	0.0000	0.0000	0.0000	0.0000	0.0000	0.0000	0.0000	0.0000
20	1	0.6415	0.8784	0.9612	0.9885	0.9968	0.9992	0.9998	1.0000	1.0000	1.0000
	2	0.2642	0.6083	0.8244	0.9308	0.9757	0.9924	0.9979	0.9995	0.9999	1.0000
	3	0.0755	0.3231	0.5951	0.7939	0.9087	0.9645	0.9879	0.9964	0.9991	0.9998
	4	0.0159	0.1330	0.3523	0.5886	0.7748	0.8929	0.9556	0.9840	0.9951	0.9987
	5	0.0026	0.0432	0.1702	0.3704	0.5852	0.7625	0.8818	0.9490	0.9811	0.9941
	6	0.0003	0.0113	0.0673	0.1958	0.3828	0.5836	0.7546	0.8744	0.9447	0.9793
	7	0.0000	0.0024	0.0219	0.0867	0.2142	0.3920	0.5834	0.7500	0.8701	0.9423
	8	0.0000	0.0004	0.0059	0.0321	0.1018	0.2277	0.3990	0.5841	0.7480	0.8684
	9	0.0000	0.0001	0.0013	0.0100	0.0409	0.1123	0.2376	0.4044	0.5857	0.7483
	10	0.0000	0.0000	0.0002	0.0026	0.0139	0.0480	0.1218	0.2447	0.4086	0.5881
	11	0.0000	0.0000	0.0000	0.0006	0.0039	0.0171	0.0532	0.1275	0.2493	0.4119
	12	0.0000	0.0000	0.0000	0.0001	0.0009	0.0051	0.0196	0.0565	0.1308	0.2517
	13	0.0000	0.0000	0.0000	0.0000	0.0002	0.0013	0.0060	0.0210	0.0580	0.1316
	14	0.0000	0.0000	0.0000	0.0000	0.0000	0.0003	0.0015	0.0065	0.0214	0.0577
	15	0.0000	0.0000	0.0000	0.0000	0.0000	0.0000	0.0003	0.0016	0.0064	0.0207

		θ									
n	x	0.05	0.10	0.15	0.20	0.25	0.30	0.35	0.40	0.45	0.50
20	16	0.0000	0.0000	0.0000	0.0000	0.0000	0.0000	0.0000	0.0003	0.0015	0.0059
	17	0.0000	0.0000	0.0000	0.0000	0.0000	0.0000	0.0000	0.0000	0.0003	0.0013
	18	0.0000	0.0000	0.0000	0.0000	0.0000	0.0000	0.0000	0.0000	0.0000	0.0002
	19	0.0000	0.0000	0.0000	0.0000	0.0000	0.0000	0.0000	0.0000	0.0000	0.0000
	20	0.0000	0.0000	0.0000	0.0000	0.0000	0.0000	0.0000	0.0000	0.0000	0.0000
21	1	0.6594	0.8906	0.9671	0.9908	0.9976	0.9994	0.9999	1.0000	1.0000	1.0000
	2	0.2830	0.6353	0.8450	0.9424	0.9810	0.9944	0.9996	0.9997	0.9999	1.0000
	3	0.0849	0.3516	0.6295	0.8213	0.9255	0.9729	0.9914	0.9976	0.9994	0.9999
	4	0.0189	0.1520	0.3887	0.6296	0.8083	0.9144	0.9669	0.9890	0.9969	0.9993
	5	0.0032	0.0522	0.1975	0.4140	0.6326	0.8016	0.9076	0.9630	0.9874	0.9967
	6	0.0004	0.0144	0.0827	0.2307	0.4334	0.6373	0.7991	0.9043	0.9611	0.9867
	7	0.0000	0.0033	0.0287	0.1085	0.2564	0.4495	0.6433	0.7998	0.9036	0.9608
	8	0.0000	0.0006	0.0083	0.0431	0.1299	0.2770	0.4635	0.6505	0.8029	0.9054
	9	0.0000	0.0001	0.0020	0.0144	0.0561	0.1477	0.2941	0.4763	0.6587	0.8083
	10	0.0000	0.0000	0.0004	0.0041	0.0206	0.0676	0.1632	0.3086	0.4883	0.6682
	11	0.0000	0.0000	0.0001	0.0010	0.0064	0.0264	0.0772	0.1744	0.3210	0.5000
	12	0.0000	0.0000	0.0000	0.0002	0.0017	0.0087	0.0313	0.0849	0.1841	0.3318
	13	0.0000	0.0000	0.0000	0.0000	0.0004	0.0024	0.0108	0.0652	0.0908	0.1917
	14	0.0000	0.0000	0.0000	0.0000	0.0001	0.0006	0.0031	0.0123	0.0379	0.0946
	15	0.0000	0.0000	0.0000	0.0000	0.0000	0.0001	0.0007	0.0036	0.0132	0.0392
	16	0.0000	0.0000	0.0000	0.0000	0.0000	0.0000	0.0001	0.0008	0.0037	0.0133
	17	0.0000	0.0000	0.0000	0.0000	0.0000	0.0000	0.0000	0.0002	0.0008	0.0036
	18	0.0000	0.0000	0.0000	0.0000	0.0000	0.0000	0.0000	0.0000	0.0001	0.0007
	19	0.0000	0.0000	0.0000	0.0000	0.0000	0.0000	0.0000	0.0000	0.0000	0.0001

Table 3
Individual Terms of the Poisson Distribution

The Poisson probability function is given by

$$f(x;\lambda) = \frac{\lambda^x e^{-\lambda}}{x!}, \quad \lambda > 0, \; x = 0, 1, 2, \ldots.$$

This table contains the individual terms of $f(x;\lambda)$ for specified values of x and λ.

Source: Beyer, W. H. (ed). *Handbook of Tables for Probability and Statistics.* Cleveland: The Chemical Rubber Company, 1966. Reprinted by permission of the publisher.

					λ					
x	0.1	0.2	0.3	0.4	0.5	0.6	0.7	0.8	0.9	1.0
0	0.9048	0.8187	0.7408	0.6703	0.6065	0.5488	0.4966	0.4493	0.4066	0.3679
1	0.0905	0.1637	0.2222	0.2681	0.3033	0.3293	0.3476	0.3595	0.3659	0.3679
2	0.0045	0.0164	0.0333	0.0536	0.0758	0.0988	0.1217	0.1438	0.1647	0.1839
3	0.0002	0.0011	0.0033	0.0072	0.0126	0.0198	0.0284	0.0383	0.0494	0.0613
4	0.0000	0.0001	0.0003	0.0007	0.0016	0.0030	0.0050	0.0077	0.0111	0.0153
5	0.0000	0.0000	0.0000	0.0001	0.0002	0.0004	0.0007	0.0012	0.0020	0.0031
6	0.0000	0.0000	0.0000	0.0000	0.0000	0.0000	0.0001	0.0002	0.0003	0.0005
7	0.0000	0.0000	0.0000	0.0000	0.0000	0.0000	0.0000	0.0000	0.0000	0.0001

					λ					
x	1.1	1.2	1.3	1.4	1.5	1.6	1.7	1.8	1.9	2.0
0	0.3329	0.3012	0.2725	0.2466	0.2231	0.2019	0.1827	0.1653	0.1496	0.1353
1	0.3662	0.3614	0.3543	0.3452	0.3347	0.3230	0.3106	0.2975	0.2842	0.2707
2	0.2014	0.2169	0.2303	0.2417	0.2510	0.2584	0.2640	0.2678	0.2700	0.2707
3	0.0738	0.0867	0.0998	0.1128	0.1255	0.1378	0.1496	0.1607	0.1710	0.1804
4	0.0203	0.0260	0.0324	0.0395	0.0471	0.0551	0.0636	0.0723	0.0812	0.0902
5	0.0045	0.0062	0.0084	0.0111	0.0141	0.0176	0.0216	0.0260	0.0309	0.0361
6	0.0008	0.0012	0.0018	0.0026	0.0035	0.0047	0.0061	0.0078	0.0098	0.0120
7	0.0001	0.0002	0.0003	0.0005	0.0008	0.0011	0.0015	0.0020	0.0027	0.0034
8	0.0000	0.0000	0.0001	0.0001	0.0001	0.0002	0.0003	0.0005	0.0006	0.0009
9	0.0000	0.0000	0.0000	0.0000	0.0000	0.0000	0.0001	0.0001	0.0001	0.0002

	λ									
x	2.1	2.2	2.3	2.4	2.5	2.6	2.7	2.8	2.9	3.0
0	0.1225	0.1108	0.1003	0.0907	0.0821	0.0743	0.0672	0.0608	0.0550	0.0498
1	0.2572	0.2438	0.2306	0.2177	0.2052	0.1931	0.1815	0.1703	0.1596	0.1494
2	0.2700	0.2681	0.2652	0.2613	0.2565	0.2510	0.2450	0.2384	0.2314	0.2240
3	0.1890	0.1966	0.2033	0.2090	0.2138	0.2176	0.2205	0.2225	0.2237	0.2240
4	0.0992	0.1082	0.1169	0.1254	0.1336	0.1414	0.1488	0.1557	0.1622	0.1680
5	0.0417	0.0476	0.0538	0.0602	0.0668	0.0735	0.0804	0.0872	0.0940	0.1008
6	0.0146	0.0174	0.0206	0.0241	0.0278	0.0319	0.0362	0.0407	0.0455	0.0504
7	0.0044	0.0055	0.0068	0.0083	0.0099	0.0118	0.0139	0.0163	0.0188	0.0216
8	0.0011	0.0015	0.0019	0.0025	0.0031	0.0038	0.0047	0.0057	0.0068	0.0081
9	0.0003	0.0004	0.0005	0.0007	0.0009	0.0011	0.0014	0.0018	0.0022	0.0027
10	0.0001	0.0001	0.0001	0.0002	0.0002	0.0003	0.0004	0.0005	0.0006	0.0008
11	0.0000	0.0000	0.0000	0.0000	0.0000	0.0001	0.0001	0.0001	0.0002	0.0002
12	0.0000	0.0000	0.0000	0.0000	0.0000	0.0000	0.0000	0.0000	0.0000	0.0001

	λ									
x	3.1	3.2	3.3	3.4	3.5	3.6	3.7	3.0	3.9	4.0
0	0.0450	0.0408	0.0369	0.0334	0.0302	0.0273	0.0247	0.0224	0.0202	0.0183
1	0.1397	0.1304	0.1217	0.1135	0.1057	0.0984	0.0915	0.0850	0.0789	0.0733
2	0.2165	0.2087	0.2008	0.1929	0.1850	0.1771	0.1692	0.1615	0.1539	0.1465
3	0.2237	0.2226	0.2209	0.2186	0.2158	0.2125	0.2087	0.2046	0.2001	0.1954
4	0.1734	0.1781	0.1823	0.1858	0.1888	0.1912	0.1931	0.1944	0.1951	0.1954
5	0.1075	0.1140	0.1203	0.1264	0.1322	0.1377	0.1429	0.1477	0.1522	0.1563
6	0.0555	0.0608	0.0662	0.0716	0.0771	0.0826	0.0881	0.0936	0.0989	0.1042
7	0.0246	0.0278	0.0312	0.0348	0.0385	0.0425	0.0466	0.0508	0.0551	0.0595
8	0.0095	0.0111	0.0129	0.0148	0.0168	0.0191	0.0215	0.0241	0.0269	0.0298
9	0.0033	0.0040	0.0047	0.0056	0.0066	0.0076	0.0089	0.0102	0.1116	0.0132
10	0.0010	0.0013	0.0016	0.0019	0.0023	0.0028	0.0033	0.0039	0.0045	0.0053
11	0.0003	0.0004	0.0005	0.0006	0.0007	0.0009	0.0011	0.0013	0.0016	0.0019
12	0.0001	0.0001	0.0001	0.0002	0.0002	0.0003	0.0003	0.0004	0.0005	0.0006
13	0.0000	0.0000	0.0000	0.0000	0.0001	0.0001	0.0001	0.0001	0.0002	0.0002
14	0.0000	0.0000	0.0000	0.0000	0.0000	0.0000	0.0000	0.0000	0.0000	0.0001

continued

	λ									
x	4.1	4.2	4.3	4.4	4.5	4.6	4.7	4.8	4.9	5.0
0	0.0166	0.0150	0.0136	0.0123	0.0111	0.0101	0.0091	0.0082	0.0074	0.0067
1	0.0679	0.0630	0.0583	0.0540	0.0500	0.0462	0.0427	0.0395	0.0365	0.0337
2	0.1393	0.1323	0.1254	0.1188	0.1125	0.1063	0.1005	0.0948	0.0894	0.0842
3	0.1904	0.1852	0.1798	0.1743	0.1687	0.1631	0.1574	0.1517	0.1460	0.1404
4	0.1951	0.1944	0.1933	0.1917	0.1898	0.1875	0.1849	0.1820	0.1789	0.1755
5	0.1600	0.1633	0.1662	0.1687	0.1708	0.1725	0.1738	0.1747	0.1753	0.1755
6	0.1093	0.1143	0.1191	0.1237	0.1281	0.1323	0.1362	0.1398	0.1432	0.1462
7	0.0640	0.0686	0.0732	0.0778	0.0824	0.0869	0.0914	0.0959	0.1002	0.1044
8	0.0328	0.0360	0.0393	0.0428	0.0463	0.0500	0.0537	0.0575	0.0614	0.0653
9	0.0150	0.0168	0.0188	0.0209	0.0232	0.0255	0.0280	0.0307	0.0334	0.0363
10	0.0061	0.0071	0.0081	0.0092	0.0104	0.0118	0.0132	0.0147	0.0164	0.0181
11	0.0023	0.0027	0.0032	0.0037	0.0043	0.0049	0.0056	0.0064	0.0073	0.0082
12	0.0008	0.0009	0.0011	0.0014	0.0016	0.0019	0.0022	0.0026	0.0030	0.0034
13	0.0002	0.0003	0.0004	0.0005	0.0006	0.0007	0.0008	0.0009	0.0011	0.0013
14	0.0001	0.0001	0.0001	0.0001	0.0002	0.0002	0.0003	0.0003	0.0004	0.0005
15	0.0000	0.0000	0.0000	0.0000	0.0001	0.0001	0.0001	0.0001	0.0001	0.0002

	λ									
x	5.1	5.2	5.3	5.4	5.5	5.6	5.7	5.8	5.9	6.0
0	0.0061	0.0055	0.0050	0.0045	0.0041	0.0037	0.0033	0.0030	0.0027	0.0025
1	0.0311	0.0287	0.0265	0.0244	0.0225	0.0207	0.0191	0.0176	0.0162	0.0149
2	0.0793	0.0746	0.0701	0.0659	0.0618	0.0580	0.0544	0.0509	0.0477	0.0446
3	0.1348	0.1293	0.1239	0.1185	0.1133	0.1082	0.1033	0.0985	0.0938	0.0892
4	0.1719	0.1681	0.1641	0.1600	0.1558	0.1515	0.1472	0.1428	0.1383	0.1339
5	0.1753	0.1748	0.1740	0.1728	0.1714	0.1697	0.1678	0.1656	0.1632	0.1606
6	0.1490	0.1515	0.1537	0.1555	0.1571	0.1584	0.1594	0.1601	0.1605	0.1606
7	0.1086	0.1125	0.1163	0.1200	0.1234	0.1267	0.1298	0.1326	0.1353	0.1377
8	0.0692	0.0731	0.0771	0.0810	0.0849	0.0887	0.0925	0.0962	0.0098	0.1033
9	0.0392	0.0423	0.0454	0.0486	0.0519	0.0552	0.0586	0.0620	0.0654	0.0688
10	0.0200	0.0220	0.0241	0.0262	0.0285	0.0309	0.0334	0.0359	0.0386	0.0413
11	0.0093	0.0104	0.0116	0.0129	0.0143	0.0157	0.0173	0.0190	0.0207	0.0225
12	0.0039	0.0045	0.0051	0.0058	0.0065	0.0073	0.0082	0.0092	0.0102	0.0113
13	0.0015	0.0018	0.0021	0.00024	0.0028	0.0032	0.0036	0.0041	0.0046	0.0052

	λ									
x	5.1	5.2	5.3	5.4	5.5	5.6	5.7	5.8	5.9	6.0
14	0.0006	0.0007	0.0008	0.0009	0.0011	0.0013	0.0015	0.0017	0.0019	0.0022
15	0.0002	0.0002	0.0003	0.0003	0.0004	0.0005	0.0006	0.0007	0.0008	0.0009
16	0.0001	0.0001	0.0001	0.0001	0.0001	0.0002	0.0002	0.0002	0.0003	0.0003
17	0.0000	0.0000	0.0000	0.0000	0.0000	0.0000	0.0001	0.0001	0.0001	0.0001

	λ									
x	6.1	6.2	6.3	6.4	6.5	6.6	6.7	6.8	6.9	7.0
0	0.0022	0.0020	0.0018	0.0017	0.0015	0.0014	0.0012	0.0011	0.0010	0.0009
1	0.0137	0.0126	0.0116	0.0106	0.0098	0.0090	0.0082	0.0076	0.0070	0.0064
2	0.0417	0.0390	0.0364	0.0340	0.0318	0.0296	0.0276	0.0258	0.0240	0.0223
3	0.0848	0.0806	0.0765	0.0726	0.0688	0.0562	0.0617	0.0584	0.0552	0.0521
4	0.1294	0.1249	0.1205	0.1162	0.1118	0.1076	0.1034	0.0992	0.0952	0.0912
5	0.1579	0.1549	0.1519	0.1487	0.1454	0.1420	0.1385	0.1349	0.1314	0.1277
6	0.1605	0.1601	0.1595	0.1586	0.0575	0.1562	0.1546	0.1529	0.1511	0.1490
7	0.1399	0.1418	0.1435	0.1450	0.1462	0.1472	0.1480	0.1486	0.1489	0.1490
8	0.1066	0.1099	0.1130	0.1160	0.1188	0.1215	0.1240	0.1263	0.1284	0.1304
9	0.0723	0.0575	0.0791	0.0825	0.0858	0.0891	0.0923	0.0954	0.0985	0.1014
10	0.0441	0.0469	0.0498	0.0528	0.0558	0.0588	0.0618	0.0649	0.0679	0.0710
11	0.0245	0.0265	0.0285	0.0307	0.0330	0.0353	0.0377	0.0401	0.0426	0.0452
12	0.0124	0.0137	0.0105	0.0164	0.0179	0.0194	0.0210	0.0227	0.0245	0.0264
13	0.0058	0.0065	0.0073	0.0081	0.0089	0.0098	0.0108	0.0119	0.0130	0.0142
14	0.0025	0.0029	0.0033	0.0037	0.0041	0.0046	0.0052	0.0058	0.0064	0.0071
15	0.0010	0.0012	0.0014	0.0016	0.0018	0.0020	0.0023	0.0026	0.0029	0.0033
16	0.0004	0.0005	0.0005	0.0006	0.0007	0.0008	0.0010	0.0011	0.0013	0.0014
17	0.0001	0.0002	0.0002	0.0002	0.0003	0.0003	0.0004	0.0004	0.0005	0.0006
18	0.0000	0.0001	0.0001	0.0001	0.0001	0.0001	0.0001	0.0002	0.0002	0.0002
19	0.0000	0.0000	0.0000	0.0000	0.0000	0.0000	0.0000	0.0001	0.0001	0.0001

continued

	λ									
x	7.1	7.2	7.3	7.4	7.5	7.6	7.7	7.8	7.9	8.0
0	0.0008	0.0007	0.0007	0.0006	0.0006	0.0005	0.0005	0.0004	0.0004	0.0003
1	0.0059	0.0054	0.0049	0.0045	0.0041	0.0038	0.0035	0.0032	0.0029	0.0027
2	0.0208	0.0194	0.0180	0.0167	0.0156	0.0145	0.0134	0.0125	0.0116	0.0107
3	0.0492	0.0464	0.0438	0.0413	0.0389	0.0366	0.0345	0.0324	0.0305	0.0286
4	0.0874	0.0836	0.0799	0.0764	0.0729	0.0696	0.0663	0.0632	0.0602	0.0573
5	0.1241	0.1204	0.1167	0.1130	0.1094	0.1057	0.1021	0.0986	0.0951	0.0916
6	0.1468	0.1445	0.1420	0.1394	0.1367	0.1339	0.1311	0.1282	0.1252	0.1221
7	0.1489	0.1486	0.1481	0.1474	0.1465	0.1454	0.1442	0.1428	0.1413	0.1396
8	0.1321	0.1337	0.1351	0.1363	0.1373	0.1382	0.1388	0.1392	0.1395	0.1396
9	0.1042	0.1070	0.1096	0.1121	0.1144	0.1167	0.1187	0.1207	0.1224	0.1241
10	0.0740	0.0770	0.0800	0.0829	0.0858	0.0887	0.0914	0.0941	0.0967	0.0993
11	0.0478	0.0504	0.0531	0.0558	0.0585	0.0613	0.0640	0.0667	0.0695	0.0722
12	0283	0.0303	0.0323	0.0344	0.0366	0.0388	0.0411	0.0434	0.0457	0.0481
13	0.0154	0.0168	0.0181	0.0196	0.0211	0.0227	0.0243	0.0260	0.0278	0.0296
14	0.0078	0.0086	0.0095	0.0104	0.0113	0.0123	0.0134	0.0145	0.0157	0.0169
15	0.0037	0.0041	0.0046	0.0051	0.0057	0.0062	0.0069	0.0075	0.0083	0.0090
16	0.0016	0.0019	0.0021	0.0024	0.0026	0.0030	0.0033	0.0037	0.0041	0.0045
17	0.0007	0.0008	0.0009	0.0010	0.0012	0.0013	0.0015	0.0017	0.0019	0.0021
18	0.0003	0.0003	0.0004	0.0004	0.0005	0.0006	0.0006	0.0007	0.0008	0.0009
19	0.0001	0.0001	0.0001	0.0002	0.0002	0.0002	0.0003	0.0003	0.0003	0.0004
20	0.0000	0.0000	0.0001	0.0001	0.0001	0.0001	0.0001	0.0001	0.0001	0.0002
21	0.0000	0.0000	0.0000	0.0000	0.0000	0.0000	0.0000	0.0000	0.0001	0.0001

	λ									
x	8.1	8.2	8.3	8.4	8.5	8.6	8.7	8.8	8.9	9.0
0	0.0003	0.0003	0.0002	0.0002	0.0002	0.0002	0.0002	0.0002	0.0001	0.0001
1	0.0025	0.0023	0.0021	0.0019	0.0017	0.0016	0.0014	0.0013	0.0012	0.0011
2	0.0100	0.0092	0.0086	0.0079	0.0074	0.0068	0.0063	0.0058	0.0054	0.0050
3	0.0269	0.0252	0.0237	0.0222	0.0208	0.0195	0.0183	0.0171	0.0160	0.0150
4	0.0544	0.0517	0.0491	0.0466	0.0443	0.0420	0.0398	0.0377	0.0357	0.0337
5	0.0882	0.0849	0.0816	0.0784	0.0752	0.0722	0.0692	0.0663	0.0635	0.0607
6	0.1191	0.1160	0.1128	0.1097	0.1066	0.1034	0.1003	0.0972	0.0941	0.0911
7	0.1378	0.1358	0.1338	0.1317	0.1294	0.1271	0.1247	0.1222	0.1197	0.1171

	λ									
x	8.1	8.2	8.3	8.4	8.5	8.6	8.7	8.8	8.9	9.0
8	0.1395	0.1392	0.1388	0.1382	0.1375	0.1366	0.1356	0.1344	0.1332	0.1318
9	0.1256	0.1269	0.1280	0.1290	0.1299	0.1306	0.1311	0.1315	0.1317	0.1318
10	0.1017	0.1040	0.1063	0.1084	0.1104	0.1123	0.1140	0.1157	0.1172	0.1186
11	0.0749	0.0776	0.0802	0.0828	0.0853	0.0878	0.0902	0.0925	0.0948	0.0970
12	0.0505	0.0530	0.0555	0.0579	0.0604	0.0629	0.0654	0.0679	0.0703	0.0728
13	0.0315	0.0334	0.0354	0.0374	0.0395	0.0416	0.0438	0.0459	0.0481	0.0504
14	0.0182	0.0196	0.0210	0.0225	0.0240	0.0256	0.0272	0.0289	0.0306	0.0324
15	0.0098	0.0107	0.0116	0.0126	0.0136	0.0147	0.0158	0.0169	0.0182	0.0194
16	0.0050	0.0055	0.0060	0.0066	0.0072	0.0079	0.0086	0.0093	0.0101	0.0109
17	0.0024	0.0026	0.0029	0.0033	0.0036	0.0040	0.0044	0.0048	0.0053	0.0058
18	0.0011	0.0012	0.0014	0.0015	0.0017	0.0019	0.0021	0.0024	0.0026	0.0029
19	0.0005	0.0005	0.0006	0.0007	0.0008	0.0009	0.0010	0.0011	0.0012	0.0014
20	0.0002	0.0002	0.0002	0.0003	0.0003	0.0004	0.0004	0.0005	0.0005	0.0006
21	0.0001	0.0001	0.0001	0.0001	0.0001	0.0002	0.0002	0.0002	0.0002	0.0003
22	0.0000	0.0000	0.0000	0.0000	0.0001	0.0001	0.0001	0.0001	0.0001	0.0001

	λ									
x	9.1	9.2	9.3	9.4	9.5	9.6	9.7	9.8	9.9	1.0
0	0.0001	0.0001	0.0001	0.0001	0.0001	0.0001	0.0001	0.0001	0.0001	0.0000
1	0.0010	0.0009	0.0009	0.0008	0.0007	0.0007	0.0006	0.0005	0.0005	0.0005
2	0.0046	0.0043	0.0040	0.0037	0.0034	0.0031	0.0029	0.0027	0.0025	0.0023
3	0.0140	0.0131	0.0123	0.0115	0.0107	0.0100	0.0093	0.0087	0.0081	0.0076
4	0.0319	0.0302	0.0285	0.0269	0.0254	0.0240	0.0226	0.0213	0.0201	0.0189
5	0.0581	0.0555	0.0530	0.0506	0.0483	0.0460	0.0439	0.0418	0.0398	0.0378
6	0.0881	0.0851	0.0822	0.0793	0.0764	0.0736	0.0709	0.0682	0.0656	0.0631
7	0.1145	0.1118	0.1091	0.1064	0.1037	0.1010	0.0982	0.0955	0.0928	0.0901
8	0.1302	0.1286	0.1269	0.1251	0.1232	0.1212	0.1191	0.1170	0.1148	0.1126
9	0.1317	0.1315	0.1311	0.1306	0.1300	0.1293	0.1284	0.1274	0.1263	0.1251
10	0.1198	0.1210	0.1219	0.1228	0.1235	0.1241	0.1245	0.1249	0.1250	0.1251
11	0.0991	0.1012	0.1031	0.1049	0.1067	0.1083	0.1098	0.1112	0.1125	0.1137
12	0.0752	0.0776	0.0799	0.0822	0.0844	0.0866	0.0888	0.0908	0.0928	0.0948
13	0.0526	0.0549	0.0572	0.0594	0.0617	0.0640	0.0662	0.0685	0.0707	0.0729
14	0.0342	0.0361	0.0380	0.0399	0.0419	0.0439	0.0459	0.0479	0.0500	0.0521

continued

	λ									
x	9.1	9.2	9.3	9.4	9.5	9.6	9.7	9.8	9.9	1.0
15	0.0208	0.0221	0.0235	0.0250	0.0265	0.0281	0.0297	0.0313	0.0330	0.0347
16	0.0118	0.0127	0.0137	0.0147	0.0157	0.0168	0.0180	0.0192	30204	0.0217
17	0.0063	0.0069	0.0075	0.0081	0.0088	0.0095	0.0103	0.0111	0.0119	0.0128
18	0.0032	0.0035	0.0039	0.0042	0.0046	0.0051	0.0055	0.0060	0.0065	0.0071
19	0.0015	0.0017	0.0019	0.0021	0.0023	0.0026	0.0028	0.0031	0.0034	0.0037
20	0.0007	0.0008	0.0009	0.0010	0.0011	0.0012	0.0014	0.0015	0.0017	0.0019
21	0.0003	0.0003	0.0004	0.0004	0.0005	0.0006	0.0006	0.0007	0.0008	0.0009
22	0.0001	0.0001	0.0002	0.0002	0.0002	0.0002	0.0003	0.0003	0.0004	0.0004
23	0.0000	0.0001	0.0001	0.0001	0.0001	0.0001	0.0001	0.0001	0.0002	0.0002
24	0.0000	0.0000	0.0000	0.0000	0.0000	0.0000	0.0000	0.0001	0.0001	0.0001

	λ									
x	11	12	13	14	15	16	17	18	19	20
0	0.0000	0.0000	0.0000	0.0000	0.0000	0.0000	0.0000	0.0000	0.0000	0.0000
1	0.0002	0.0001	0.0000	0.0000	0.0000	0.0000	0.0000	0.0000	0.0000	0.0000
2	0.0010	0.0004	0.0002	0.0001	0.0000	0.0000	0.0000	0.0000	0.0000	0.0000
3	0.0037	0.0018	0.0008	0.0004	0.0002	0.0001	0.0000	0.0000	0.0000	0.0000
4	0.0102	0.0053	0.0027	0.0013	0.0006	0.0003	0.0001	0.0001	0.0000	0.0000
5	0.0224	0.0127	0.0070	0.0037	0.0019	0.0010	0.0005	0.0002	0.0001	0.0001
6	0.0411	0.0255	0.0152	0.0087	0.0048	0.0026	0.0014	0.0007	0.0004	0.0002
7	0.0646	0.0437	0.0281	0.0174	0.0104	0.0060	0.0034	0.0018	0.0010	0.0005
8	0.0888	0.0655	0.0457	0.0304	0.0194	0.0120	0.0072	0.0042	0.0024	0.0013
9	0.1085	0.0874	0.0661	0.0473	0.0324	0.0213	0.0135	0.0083	0.0050	0.0029
10	0.1194	0.1048	0.0859	0.0663	0.0486	0.0341	0.0230	0.0150	0.0095	0.0058
11	0.1194	0.1144	0.1015	0.0844	0.0663	0.0496	0.0355	0.0245	0.0164	0.0106
12	0.1094	0.1144	0.1099	0.0984	0.0829	0.0661	0.0504	0.0368	0.0259	0.0176
13	0.0926	0.1056	0.1099	0.1060	0.0956	0.0814	0.0658	0.0509	0.0378	0.0271
14	0.0728	0.0905	0.1021	0.1060	0.1024	0.0930	0.0800	0.0655	0.0514	0.0387
15	0.0534	0.0724	0.0885	0.0989	0.1024	0.0992	0.0906	0.0786	0.0650	0.0516
16	0.0367	0.0543	0.0719	0.0866	0.0960	0.0992	0.0963	0.0884	0.0772	0.0646
17	0.0237	0.0383	0.0550	0.0713	0.0847	0.0934	0.0963	0.0936	0.0863	0.0760
18	0.0145	0.0256	0.0397	0.0554	0.0706	0.0830	0.0909	0.0936	0.0911	0.0844
19	0.0084	0.0161	0.0272	0.0409	0.0557	0.0699	0.0814	0.0887	0.0911	0.0888

	λ									
x	11	12	13	14	15	16	17	18	19	20
20	0.0046	0.0097	0.0177	0.0286	0.0418	0.0559	0.0692	0.0798	0.0866	0.0888
21	0.0024	0.0055	0.0109	0.0191	0.0299	0.0426	0.0560	0.0684	0.0783	0.0846
22	0.0012	0.0030	0.0065	0.0121	0.0204	0.0310	0.0433	0.0560	0.0676	0.0769
23	0.0006	0.0016	0.0037	0.0074	0.0133	0.0216	0.0320	0.0438	0.0559	0.0669
24	0.0003	0.0008	0.0020	0.0043	0.0083	0.0144	0.0226	0.0328	0.0442	0.0557
25	0.0001	0.0004	0.0010	0.0024	0.0050	0.0092	0.0154	0.0237	0.0336	0.0446
26	0.0000	0.0002	0.0005	0.0013	0.0029	0.0057	0.0101	0.0164	0.0246	0.0343
27	0.0000	0.0001	0.0002	0.0007	0.0016	0.0034	0.0063	0.0109	0.0173	0.0254
28	0.0000	0.0000	0.0001	0.0003	0.0009	0.0019	0.0038	0.0070	0.0117	0.0181
29	0.0000	0.0000	0.0001	0.0002	0.0004	0.0011	0.0023	0.0044	0.0077	0.0125
30	0.0000	0.0000	0.0000	0.0001	0.0002	0.0006	0.0013	0.0026	0.0049	0.0083
31	0.0000	0.0000	0.0000	0.0000	0.0001	0.0003	0.0007	0.0015	0.0030	0.0054
32	0.0000	0.0000	0.0000	0.0000	0.0001	0.0001	0.0004	0.0009	0.0018	0.0034
33	0.0000	0.0000	0.0000	0.0000	0.0000	0.0001	0.0002	0.0005	0.0010	0.0020
34	0.0000	0.0000	0.0000	0.0000	0.0000	0.0000	0.0001	0.0002	0.0006	0.0012
35	0.0000	0.0000	0.0000	0.0000	0.0000	0.0000	0.0000	0.0001	0.0003	0.0007
36	0.0000	0.0000	0.0000	0.0000	0.0000	0.0000	0.0000	0.0001	0.0002	0.0004
37	0.0000	0.0000	0.0000	0.0000	0.0000	0.0000	0.0000	0.0000	0.0001	0.0002
38	0.0000	0.0000	0.0000	0.0000	0.0000	0.0000	0.0000	0.0000	0.0000	0.0001
39	0.0000	0.0000	0.0000	0.0000	0.0000	0.0000	0.0000	0.0000	0.0000	0.0001

Table 4
Cumulative Terms of the Poisson Distribution

This table contains the values of

$$\sum_{x=x'}^{\infty} \frac{e^{-\lambda}\lambda^x}{x!}.$$

For specified values of x' and λ. The cumulative Poisson distribution and the cumulative chi-square (χ^2) distribution are related as follows.

$$\sum_{x=0}^{x'-1} \frac{e^{-\lambda}\lambda^x}{x!} = 1 - F(\chi^2)$$

$$= \frac{1}{2^{\frac{n}{2}}\Gamma\left(\frac{n}{2}\right)} \int_{\chi^2}^{\infty} x^{\frac{n}{2}-1} e^{-\frac{x}{2}} dx$$

where $\lambda = \frac{1}{2}\chi^2$ and $x' = \frac{1}{2}n$.

Source: Beyer, W. H. (ed). *Handbook of Tables for Probability and Statistics*. Cleveland: The Chemical Rubber Company, 1966. Reprinted by permission of the publisher.

					λ					
x'	0.1	0.2	0.3	0.4	0.5	0.6	0.7	0.8	0.9	1.0
0	1.0000	1.0000	1.0000	1.0000	1.0000	1.0000	1.0000	1.0000	1.0000	1.0000
1	0.0952	0.1813	0.2592	0.3297	0.3935	0.4512	0.5034	0.5507	0.5934	0.6321
2	0.0047	0.0175	0.0369	0.0616	0.09020	0.1219	0.1558	0.1912	0.2275	0.2642
3	0.0002	0.0011	0.0036	0.0079	0.0144	0.0231	0.0341	0.0474	0.0629	0.0803
4	0.0000	0.0001	0.0003	0.0008	0.0018	0.0034	0.0058	0.0091	0.0135	0.0190
5	0.0000	0.0000	0.0000	0.0001	0.0002	0.0004	0.0008	0.0014	0.0023	0.0037
6	0.0000	0.0000	0.0000	0.0000	0.0000	0.0000	0.0001	0.0002	0.0003	0.0006
7	0.0000	0.0000	0.0000	0.0000	0.0000	0.0000	0.0000	0.0000	0.0000	0.0001

					λ					
x'	1.1	1.2	1.3	1.4	1.5	1.6	1.7	1.8	1.9	2.0
0	1.0000	1.0000	1.0000	1.0000	1.0000	1.0000	1.0000	1.0000	1.0000	1.0000
1	0.6671	0.6988	0.7275	0.7534	0.7769	0.7981	0.8173	0.8347	0.8504	0.8647
2	0.3010	0.3374	0.3732	0.4082	0.4422	0.4751	0.5068	0.5372	0.5663	0.5940

	λ									
x'	1.1	1.2	1.3	1.4	1.5	1.6	1.7	1.8	1.9	2.0
3	0.0996	0.1205	0.1429	0.1665	0.1912	0.2166	0.2428	0.2694	0.2963	0.3233
4	0.0257	0.0338	0.0431	0.0537	0.0656	0.0788	0.0932	0.1087	0.1253	0.1429
5	0.0054	0.0077	0.0107	0.0143	0.0186	0.0237	0.0296	0.0364	0.0441	0.0527
6	0.0010	0.0015	0.0022	0.0032	0.0045	0.0060	0.0080	0.0104	0.0132	0.0166
7	0.0001	0.0003	0.0004	0.0006	0.0009	0.0013	0.0019	0.0026	0.0034	0.0045
8	0.0000	0.0000	0.0001	0.0001	0.0002	0.0003	0.0004	0.0006	0.0008	0.0011
9	0.0000	0.0000	0.0000	0.0000	0.0000	0.0000	0.0001	0.0001	0.0002	0.0002

	λ									
x'	2.1	2.2	2.3	2.4	2.5	2.6	2.7	2.8	2.9	3.0
0	1.0000	1.0000	1.0000	1.0000	1.0000	1.0000	1.0000	1.0000	1.0000	1.0000
1	0.8775	0.8892	0.8997	0.9093	0.9179	0.9257	0.9328	0.9392	0.9450	0.9502
2	0.6204	0.6454	0.6691	0.6916	0.7127	0.7326	0.7513	0.7689	0.7854	0.8009
3	0.3504	0.3773	0.4040	0.4303	0.4562	0.4816	0.5064	0.5305	0.5540	0.5768
4	0.1614	0.1806	0.2007	0.2213	0.2424	0.2640	0.2859	0.3081	0.3304	0.3528
5	0.0621	0.0725	0.0838	0.0959	0.1088	0.1226	0.1371	0.1523	0.1682	0.1847
6	0.0204	0.0249	0.0300	0.0357	0.0420	0.0490	0.0567	0.0651	0.0742	0.0839
7	0.0059	0.0075	0.0094	0.0116	0.0142	0.0172	0.0206	0.0244	0.0287	0.0335
8	0.0015	0.0020	0.0026	0.0033	0.0042	0.0053	0.0066	0.0081	0.0099	0.0119
9	0.0003	0.0005	0.0006	0.0009	0.0011	0.0015	0.0019	0.0024	0.0031	0.0038
10	0.0001	0.0001	0.0001	0.0002	0.0003	0.0004	0.0005	0.0007	0.0009	0.0011
11	0.0000	0.0000	0.0000	0.0000	0.0001	0.0001	0.0001	0.0002	0.0002	0.0003
12	0.0000	0.0000	0.0000	0.0000	0.0000	0.0000	0.0000	0.0000	0.0001	0.0001

	λ									
x'	3.1	3.2	3.3	3.4	3.5	3.6	3.7	3.8	3.9	4.0
0	1.0000	1.0000	1.0000	1.0000	1.0000	1.0000	1.0000	1.0000	1.0000	1.0000
1	0.9550	0.9592	0.9631	0.9666	0.9698	0.9727	0.9753	0.9776	0.9798	0.9817
2	0.8153	0.8288	0.8414	0.8532	0.8641	0.8743	0.8838	0.8926	0.9008	0.9084
3	0.5988	0.6201	0.6406	0.6603	0.6792	0.6973	0.7146	0.7311	0.7469	0.7619
4	0.3752	0.3975	0.4197	0.4416	0.4634	0.4848	0.5058	0.5265	0.5468	0.5665

continued

	λ									
x'	3.1	3.2	3.3	3.4	3.5	3.6	3.7	3.8	3.9	4.0
5	0.2018	0.2194	0.2374	0.2558	0.2746	0.2936	0.3128	0.3322	0.3516	0.3712
6	0.0943	0.1054	0.1171	0.1295	0.1424	0.1559	0.1699	0.1844	0.1994	0.2149
7	0.0388	0.0446	0.0510	0.0579	0.0653	0.0733	0.0818	0.0909	0.1005	0.1107
8	0.0142	0.0168	0.0198	0.0231	0.0267	0.0308	0.0352	0.0401	0.0454	0.0511
9	0.0047	0.0057	0.0069	0.0083	0.0099	0.0117	0.0137	0.0160	0.0185	0.0214
10	0.0014	0.0018	0.0022	0.0027	0.0033	0.0040	0.0048	0.0058	0.0069	0.0081
11	0.0004	0.0005	0.0006	0.0008	0.0010	0.0013	0.0016	0.0019	0.0023	0.0028
12	0.0001	0.0001	0.0002	0.0002	0.0003	0.0004	0.0005	0.0006	0.0007	0.0009
13	0.0000	0.0000	0.0000	0.0001	0.0001	0.0001	0.0001	0.0002	0.0002	0.0003
14	0.0000	0.0000	0.0000	0.0000	0.0000	0.0000	0.0000	0.0000	0.0001	0.0001

	λ									
x'	4.1	4.2	4.3	4.4	4.5	4.6	4.7	4.8	4.9	5.0
0	1.0000	1.0000	1.0000	1.0000	1.0000	1.0000	1.0000	1.0000	1.0000	1.0000
1	0.9834	0.9850	0.9864	0.9877	0.9889	0.9899	0.9909	0.9918	0.9926	0.9933
2	0.9155	0.9220	0.9281	0.9337	0.9389	0.9437	0.9482	0.9523	0.9561	0.9596
3	0.7762	0.7898	0.8026	0.8149	0.8264	0.8374	0.8477	0.8575	0.8667	0.8753
4	0.5858	0.6046	0.6228	0.6406	0.6577	0.6743	0.6903	0.7058	0.7207	0.7350
5	0.3907	0.4102	0.4296	0.4488	0.4679	0.4868	0.5054	0.5237	0.5418	0.5595
6	0.2307	0.2469	0.2633	0.2801	0.2971	0.3142	0.3316	0.3490	0.3665	0.3740
7	0.1214	0.1325	0.1442	0.1564	0.1689	0.1820	0.1954	0.2092	0.2233	0.2378
8	0.0573	0.0639	0.0710	0.0786	0.0866	0.0951	0.1040	0.1133	0.1231	0.1334
9	0.0245	0.0279	0.0317	0.0358	0.0403	0.0451	0.0503	0.0558	0.0618	0.0681
10	0.0095	0.0111	0.0129	0.0149	0.0171	0.0195	0.0222	0.0251	0.0283	0.0318
11	0.0034	0.0041	0.0048	0.0057	0.0067	0.0078	0.0090	0.0104	0.0120	0.0137
12	0.0011	0.0014	0.0017	0.0020	0.0024	0.0029	0.0034	0.0040	0.0047	0.0055
13	0.0003	0.0004	0.0005	0.0007	0.0008	0.0010	0.0012	0.0014	0.0017	0.0020
14	0.0001	0.0001	0.0002	0.0002	0.0003	0.0003	0.0004	0.0005	0.0006	0.0007
15	0.0000	0.0000	0.0000	0.0000	0.0001	0.0001	0.0001	0.0001	0.0002	0.0002
16	0.0000	0.0000	0.0000	0.0000	0.0000	0.0000	0.0000	0.0000	0.0001	0.0001

	λ									
x'	5.1	5.2	5.3	5.4	5.5	5.6	5.7	5.8	5.9	6.0
0	1.0000	1.0000	1.0000	1.0000	1.0000	1.0000	1.0000	1.0000	1.0000	1.0000
1	0.9939	0.9945	0.9950	0.9955	0.9959	0.9963	0.9967	0.9970	0.9973	0.9975
2	0.9628	0.9658	0.9686	0.9711	0.9734	0.9756	0.9776	0.9794	0.9811	0.9826
3	0.8835	0.8912	0.8984	0.8052	0.9116	0.9176	0.9232	0.9285	0.9334	0.9380
4	0.7487	0.7619	0.7746	0.7867	0.7983	0.8094	0.8200	0.8300	0.8396	0.8488
5	0.5769	0.5939	0.6105	0.6267	0.6425	0.6579	0.6728	0.6873	0.7013	0.7149
6	0.4016	0.4191	0.4365	0.4539	0.4711	0.4881	0.5050	0.5217	0.5381	0.5543
7	0.2526	0.2676	0.2829	0.2983	0.3140	0.3297	0.3456	0.3616	0.3776	0.3937
8	0.1440	0.1551	0.1665	0.1783	0.1905	0.2030	0.2159	0.2290	0.2424	0.2560
9	0.0748	0.0819	0.0894	0.0974	0.1056	0.1143	0.1234	0.1328	0.1426	0.1528
10	0.0356	0.0397	0.0441	0.0488	0.0538	0.0591	0.0648	0.0708	0.0772	0.0839
11	0.0156	0.0177	0.0200	0.0225	0.0253	0.0282	0.0314	0.0349	0.0386	0.0426
12	0.0063	0.0073	0.0084	0.0096	0.0110	0.0125	0.0141	0.0160	0.0179	0.0201
13	0.0024	0.0028	0.0033	0.0038	0.0045	0.0051	0.0059	0.0068	0.0078	0.0088
14	0.0008	0.0010	0.0012	0.0014	0.0017	0.0020	0.0023	0.0027	0.0031	0.0036
15	0.0003	0.0003	0.0004	0.0005	0.0006	0.0007	0.0009	0.0010	0.0012	0.0014
16	0.0001	0.0001	0.0001	0.0002	0.0002	0.0002	0.0003	0.0004	0.0004	0.0005
17	0.0000	0.0000	0.0000	0.0001	0.0001	0.0001	0.0001	0.0001	0.0001	0.0002
18	0.0000	0.0000	0.0000	0.0000	0.0000	0.0000	0.0000	0.0000	0.0000	0.0001

	λ									
x'	6.1	6.2	6.3	6.4	6.5	6.6	6.7	6.8	6.9	7.0
0	1.0000	1.0000	1.0000	1.0000	1.0000	1.0000	1.0000	1.0000	1.0000	1.0000
1	0.9978	0.9980	0.9982	0.9983	0.9985	0.9986	0.9988	0.9989	0.9990	0.9991
2	0.9841	0.9854	0.9866	0.9877	0.9887	0.9897	0.9905	0.9913	0.9920	0.9927
3	0.9423	0.9464	0.9502	0.9537	0.9570	0.9600	0.9629	0.9656	0.9680	0.9704
4	0.8575	0.8658	0.8736	0.8811	0.8882	0.8948	0.9012	0.9072	0.9129	0.9182
5	0.7281	0.7408	0.7531	0.7649	0.7763	0.7843	0.7978	0.8080	0.8177	0.8270
6	0.5702	0.5859	0.6012	0.6163	0.6310	0.6453	0.6594	0.6730	0.6863	0.6993
7	0.4098	0.4258	0.4418	0.4577	0.4735	0.4892	0.5047	0.5201	0.5353	0.5503
8	0.2699	0.2840	0.2983	0.3127	0.3272	0.3419	0.3567	0.3715	0.3864	0.4013
9	0.1633	0.0741	0.1852	0.1967	0.2084	0.2204	0.2327	0.2452	0.2580	0.2709

continued

	λ									
x'	6.1	6.2	6.3	6.4	6.5	6.6	6.7	6.8	6.9	7.0
10	0.0910	0.0984	0.0161	0.1142	0.1226	0.1314	0.1404	0.1498	0.1505	0.1695
11	0.0469	0.0514	0.0563	0.0614	0.0668	0.0726	0.0786	0.0849	0.0916	0.0985
12	0.0224	0.0250	0.0277	0.0307	0.0339	0.0373	0.0409	0.0448	0.0490	0.0534
13	0.0100	0.0113	0.0127	0.0143	0.0160	0.0179	0.0199	0.0221	0.0245	0.0270
14	0.0042	0.0048	0.0055	0.0063	0.0071	0.0080	0.0091	0.0102	0.0115	0.0128
15	0.0016	0.0019	0.0022	0.0026	0.0030	0.0034	0.0039	0.0044	0.0050	0.0057
16	0.0006	0.0007	0.0008	0.0010	0.0012	0.0014	0.0016	0.0018	0.0021	0.0024
17	0.0002	0.0003	0.0003	0.0004	0.0004	0.0005	0.0006	0.0007	0.0008	0.0010
18	0.0001	0.0001	0.0001	0.0001	0.0002	0.0002	0.0002	0.0003	0.0003	0.0004
19	0.0000	0.0000	0.0000	0.0000	0.0001	0.0001	0.0001	0.0001	0.0001	0.0001

	λ									
x'	7.1	7.2	7.3	7.4	7.5	7.6	7.7	7.8	7.9	8.0
0	1.0000	1.0000	1.0000	1.0000	1.0000	1.0000	1.0000	1.0000	1.0000	1.0000
1	0.9992	0.9993	0.9993	0.9994	0.9994	0.9995	0.9995	0.9996	0.9996	0.9997
2	0.9933	0.9939	0.9944	0.9949	0.9953	0.9957	0.9961	0.9964	0.9967	0.9970
3	0.9725	0.9745	0.9764	0.0781	0.9797	0.9812	0.9826	0.9839	0.9851	0.9862
4	0.9233	0.9281	0.9326	0.9368	0.9409	0.9446	0.9482	0.9515	0.9547	0.9576
5	0.8359	0.8445	0.8527	0.8605	0.8679	0.8751	0.8819	0.8883	0.8945	0.9004
6	0.7119	0.7241	0.7360	0.7474	0.7586	0.7693	0.7797	0.7897	0.7994	0.8088
7	0.5651	0.5796	0.5940	0.6080	0.6218	0.6354	0.6486	0.6616	0.6743	0.6866
8	0.4162	0.4311	0.4459	0.4607	0.4754	0.4900	0.5044	0.5188	0.5330	0.5470
9	0.2840	0.2973	0.3108	0.3243	0.3380	0.3518	0.3657	0.3796	0.3935	0.4075
10	0.1798	0.1904	32012	0.2123	0.2236	0.2351	0.2469	0.2589	0.2710	0.2834
11	0.1058	0.1133	0.1212	0.1293	0.1378	0.1465	0.1555	0.1648	0.1743	0.1841
12	0.0580	0.0629	0.0681	0.0735	0.0792	0.0852	0.0915	0.0980	0.1048	0.1119
13	0.0297	0.0327	0.0358	0.0391	0.0427	0.0464	0.0504	0.0546	0.0591	0.0638
14	0.0143	0.0159	0.0176	0.0195	0.0216	0.0238	0.0261	0.0286	0.0313	0.0342
15	0.0065	0.0073	0.0082	0.0092	0.0103	0.0114	0.0127	0.0141	0.0156	0.0173
16	0.0028	0.0031	0.0036	0.0041	0.0046	0.0052	0.0059	0.0066	0.0074	0.0082
17	0.0011	0.0013	0.0015	0.0017	0.0020	0.0022	0.0026	0.0029	0.0033	0.0037
18	0.0004	0.0005	0.0006	0.0007	0.0008	0.0009	0.0011	0.0012	0.0014	0.0016
19	0.0002	0.0002	0.0002	0.0003	0.0003	0.0004	0.0004	0.0005	0.0006	0.0006

	λ									
x'	7.1	7.2	7.3	7.4	7.5	7.6	7.7	7.8	7.9	8.0
20	0.0001	0.0001	0.0001	0.0001	0.0001	0.0001	0.0002	0.0002	0.0002	0.0003
21	0.0000	0.0000	0.0000	0.0000	0.0000	0.0000	0.0001	0.0001	0.0001	0.0001

	λ									
x'	8.1	8.2	8.3	8.4	8.5	8.6	8.7	8.8	8.9	9.0
0	1.0000	1.0000	1.0000	1.0000	1.0000	1.0000	1.0000	1.0000	1.0000	1.0000
1	0.9997	0.9997	0.9998	0.9998	0.9998	0.9998	0.9998	0.9998	0.9999	0.9999
2	0.9972	0.9975	0.9977	0.9979	0.9981	0.9982	0.9984	0.9985	0.9987	0.9988
3	0.9873	0.9882	0.9891	0.9900	0.9907	0.9914	0.9921	0.9927	0.9932	0.9938
4	0.9604	0.9630	0.9654	0.9677	0.9699	0.9719	0.9738	0.9756	0.9772	0.9788
5	0.9060	0.9113	0.9163	0.9211	0.9256	0.9299	0.9340	0.9379	0.9416	0.9450
6	0.8178	0.8264	0.8347	0.8427	0.8504	0.8578	0.8648	0.8716	0.8781	0.8843
7	0.6987	0.7104	0.7219	0.7330	0.7438	0.7543	0.7645	0.7744	0.7840	0.7932
8	0.5609	0.5746	0.5881	0.6013	0.6144	0.6272	0.6398	0.6522	0.6643	0.6761
9	0.4214	0.4353	0.4493	0.4631	0.4769	0.4906	0.5042	0.5177	0.5311	0.5443
10	0.5959	0.3085	0.3212	0.3341	0.3470	0.3600	0.3731	0.3863	0.3994	0.4126
11	0.1942	0.2045	0.2150	0.2257	0.2366	0.2478	0.2591	0.2706	0.2822	0.2940
12	0.1193	0.1269	0.1348	0.1429	0.1513	0.1600	0.1689	0.1780	0.1874	0.1970
13	0.0687	0.0739	0.0793	0.0850	0.0909	0.0971	0.1035	0.1102	0.1171	0.1242
14	0.0372	0.0405	0.0439	0.0476	0.0514	0.0555	0.0597	0.0642	0.0689	0.0739
15	0.0190	0.0209	0.0229	0.0251	0.0274	0.0299	0.0325	0.03530	0.0383	0.0415
16	0.0092	0.0102	0.0113	0.0125	0.0138	0.0152	0.0168	0.0184	0.0202	0.0220
17	0.0042	0.0047	0.0053	0.0059	0.0066	0.0074	0.0082	0.0091	0.0101	0.0111
18	0.0018	0.0021	0.0023	0.0027	0.0030	0.0034	0.0038	0.0043	0.0048	0.0053
19	0.0008	0.0009	0.0010	0.0011	0.0013	0.0015	0.0017	0.0019	0.0022	0.0024
20	0.0003	0.0003	0.0004	0.0005	0.0005	0.0006	0.0007	0.0008	0.0009	0.0011
21	0.0001	0.0001	0.0002	0.0002	0.0002	0.0002	0.0003	0.0003	0.0004	0.0004
22	0.0000	0.0000	0.0001	0.0001	0.0001	0.0001	0.0001	0.0001	0.0002	0.0002
23	0.0000	0.0000	0.0000	0.0000	0.0000	0.0000	0.0000	0.0000	0.0001	0.0001

continued

	λ									
x'	9.1	9.2	9.3	9.4	9.5	9.6	9.7	9.8	9.9	10
0	1.0000	1.0000	1.0000	1.0000	1.0000	1.0000	1.0000	1.0000	1.0000	1.0000
1	0.9999	0.9999	0.9999	0.9999	0.9999	0.9999	0.9999	0.9999	1.0000	1.0000
2	0.9989	0.9990	0.9991	0.9991	0.9992	0.9993	0.9993	0.9994	0.9995	0.9995
3	0.9942	0.9947	0.9951	0.9955	0.9958	0.9962	0.9965	0.9967	0.9970	0.9972
4	0.9802	0.9816	0.9828	0.9840	0.9851	0.9862	0.9871	0.9880	0.9889	0.9897
5	0.8483	0.9514	0.9544	0.9571	0.9597	0.9622	0.9645	0.9667	0.9688	0.9707
6	0.8902	0.8959	0.9014	0.9065	0.9115	0.9162	0.9207	0.9250	0.9290	0.9329
7	0.8022	0.8108	0.8192	0.7283	0.8351	0.8426	0.8498	0.8567	0.8634	0.8699
8	0.6877	0.6990	0.7101	0.7208	0.7313	0.7416	0.7515	0.7612	0.7706	0.7798
9	0.5574	0.5704	0.5832	0.5958	0.6082	0.6204	0.6324	0.6442	0.6558	0.6672
10	0.4258	0.4389	0.4521	0.4651	0.4782	0.4911	0.5040	0.5168	0.5295	0.5421
11	0.0359	0.3180	0.3301	0.3424	0.3547	0.3671	0.3795	0.3920	0.4045	0.4170
12	0.0268	0.2168	0.2270	0.2374	0.2480	0.2588	0.2697	0.2807	0.2919	0.3032
13	0.1316	0.1393	0.1471	0.1552	0.1636	0.1721	0.1809	0.1899	0.1991	0.2084
14	0.0790	0.0844	0.0900	0.0958	0.1019	0.1081	0.1147	0.1214	0.1284	0.1355
15	0.0448	0.0483	0.0520	0.0559	0.0600	0.0643	0.0688	0.0735	0.0784	0.0835
16	0.0240	0.0262	0.0285	0.0309	0.0335	0.0362	0.0391	0.0421	0.0454	0.0487
17	0.0122	0.0135	0.0148	0.0162	0.0177	0.0194	0.0211	0.0230	0.0249	0.0270
18	0.0059	0.0066	0.0073	0.0081	0.0089	0.0098	0.0108	0.0119	0.0130	0.0143
19	0.0027	0.0031	0.0034	0.0038	0.0043	0.0048	0.0053	0.0059	0.0065	0.0072
20	0.0012	0.0014	0.0015	0.0017	0.0020	0.0022	0.0025	0.0028	0.0031	0.0035
21	0.0005	0.0006	0.0007	0.0008	0.0009	0.0010	0.0011	0.0013	0.0014	0.0016
22	0.0002	0.0002	0.0003	0.0003	0.0004	0.0004	0.0005	0.0005	0.0006	0.0007
23	0.0001	0.0001	0.0001	0.0001	0.0001	0.0002	0.0002	0.0002	0.0003	0.0003
24	0.0000	0.0000	0.0000	0.0000	0.0001	0.0001	0.0001	0.0001	0.0001	0.0001

	λ									
x'	11	12	13	14	15	16	17	18	19	20
0	1.0000	1.0000	1.0000	1.0000	1.0000	1.0000	1.0000	1.0000	1.0000	1.0000
1	1.0000	1.0000	1.0000	1.0000	1.0000	1.0000	1.0000	1.0000	1.0000	1.0000
2	0.9998	0.9999	1.0000	1.0000	1.0000	1.0000	1.0000	1.0000	1.0000	1.0000
3	0.9988	0.9995	0.9998	0.9999	1.0000	1.0000	1.0000	1.0000	1.0000	1.0000
4	0.9951	0.9977	0.9990	0.9995	0.9998	0.9999	1.0000	1.0000	1.0000	1.0000

	λ									
x′	11	12	13	14	15	16	17	18	19	20
5	0.9849	0.9924	0.9963	0.9982	0.9991	0.9996	0.9998	0.9999	1.0000	1.0000
6	0.9625	0.9797	0.9893	0.9945	0.9972	0.9986	0.9993	0.9997	0.9998	0.9999
7	0.9214	0.9542	0.9741	0.9858	0.9924	0.9960	0.9979	0.9990	0.9995	0.9997
8	0.8568	0.9105	0.9460	0.9684	0.9820	0.9900	0.9946	0.9971	0.9985	0.9992
9	0.7680	0.8450	0.9002	0.9379	0.9626	0.9780	0.9874	0.9929	0.9961	0.9979
10	0.6595	0.7576	0.8342	0.8906	0.9301	0.9567	0.9739	0.9846	0.9911	0.9950
11	0.5401	0.6528	0.7483	0.8243	0.8815	0.9226	0.9509	0.9696	0.9817	0.9892
12	0.4207	0.5384	0.6468	0.7400	0.8152	0.8730	0.9153	0.9451	0.9653	0.9786
13	0.3113	0.4240	0.5369	0.6415	0.7324	0.8069	0.8650	0.9083	0.9394	0.9610
14	0.2187	0.3185	0.4270	0.5356	0.6368	0.7255	0.7991	0.8574	0.9016	0.9339
15	0.1460	0.2280	0.3249	0.4296	0.5343	0.6325	0.7192	0.7919	0.8503	0.8951
16	0.0926	0.1556	0.2364	0.3306	0.4319	0.5333	0.6285	0.7133	0.7852	0.8435
17	0.0559	0.1013	0.1645	0.2441	0.3359	0.4340	0.5323	0.6250	0.7080	0.7789
18	0.0322	0.0630	0.1095	0.1728	0.2511	0.3407	0.4360	0.5314	0.6216	0.7030
19	0.0177	0.0374	0.0698	0.1174	0.1805	0.2577	0.3450	0.4378	0.5305	0.6186
20	0.0093	0.0213	0.0427	0.0765	0.1248	0.1878	0.2637	0.3491	0.4394	0.5297
21	0.0047	0.0116	0.0250	0.0479	0.0830	0.1318	0.1945	0.2693	0.3528	0.4409
22	0.0023	0.0061	0.0141	0.0288	0.0531	0.0892	0.1385	0.2009	0.2745	0.3563
23	0.0010	0.0030	0.0076	0.0167	0.0327	0.0582	0.0953	0.1449	0.2069	0.2794
24	0.0005	0.0015	0.0040	0.0093	0.0195	0.0367	0.0633	0.1011	0.1510	0.2125
25	0.0002	0.0007	0.0020	0.0050	0.0112	0.0223	0.0406	0.0683	0.1067	0.1568
26	0.0001	0.0003	0.0010	0.0026	0.0062	0.0131	0.0252	0.0446	0.0731	0.1122
27	0.0000	0.0001	0.0005	0.0013	0.0033	0.0075	0.0152	0.0282	0.0486	0.0779
28	0.0000	0.0001	0.0002	0.0006	0.0017	0.0041	0.0088	0.0173	0.0313	0.0525
29	0.0000	0.0000	0.0001	0.0003	0.0009	0.0022	0.0050	0.0103	0.0195	0.0343
30	0.0000	0.0000	0.0000	0.0001	0.0004	0.0011	0.0027	0.0059	0.0118	0.0218
31	0.0000	0.0000	0.0000	0.0001	0.0002	0.0006	0.0014	0.0033	0.0070	0.0135
32	0.0000	0.0000	0.0000	0.0000	0.0001	0.0003	0.0007	0.0018	0.0040	0.0081
33	0.0000	0.0000	0.0000	0.0000	0.0000	0.0001	0.0004	0.0010	0.0022	0.0047
34	0.0000	0.0000	0.0000	0.0000	0.0000	0.0001	0.0002	0.0005	0.0012	0.0027
35	0.0000	0.0000	0.0000	0.0000	0.0000	0.0000	0.0001	0.0002	0.0006	0.0015
36	0.0000	0.0000	0.0000	0.0000	0.0000	0.0000	0.0000	0.0001	0.0003	0.0008
37	0.0000	0.0000	0.0000	0.0000	0.0000	0.0000	0.0000	0.0001	0.0002	0.0004
38	0.0000	0.0000	0.0000	0.0000	0.0000	0.0000	0.0000	0.0000	0.0001	0.0002
39	0.0000	0.0000	0.0000	0.0000	0.0000	0.0000	0.0000	0.0000	0.0000	0.0001
40	0.0000	0.0000	0.0000	0.0000	0.0000	0.0000	0.0000	0.0000	0.0000	0.0001

Table 5
Hypergeometric Distribution

The hypergeometric probability function is given by

$$f(x; N,n,k) = \frac{\binom{k}{x}\binom{N-k}{n-x}}{\binom{N}{n}} = \frac{\dfrac{k!}{x!(k-x)!} \cdot \dfrac{(N-k)!}{(n-x)!(N-k-n+z)!}}{\dfrac{N!}{n!(N-n)!}}$$

$$= \frac{k\ln!}{x!(k-x)!(n-x)!} \cdot \frac{(N-k)!(N-n)!}{N!(N-k-n+z)!},$$

where N = number of items in a finite population consisting of A successes and B failures $(A+B=N)$.

n = number of items drawn in sample without replacement, from the N items,
k = number of failures in finite population = B,
x = number of failures in sample.
$f(x; N,n,k)$ gives the probability of exactly x failures and $n-x$ successes in the sample of n items.

$$F(x; N,n,k) = \sum_{r=0}^{x} \frac{\binom{k}{r}\binom{N-k}{n-r}}{\binom{N}{n}}.$$

$F(x; N,n,k)$ gives the probability of x or fewer failures in the sample of n items.

Source: Lieberman, G. J., Owen, D. B. *Hypergeometric Probability Distribution.* Stanford, California: Stanford University Press, 1961. Reprinted by permission of the publishers.

$$f(x; N,n,k) = \frac{\binom{k}{x}\binom{N-k}{n-x}}{\binom{N}{n}} \qquad F(x; N,n,k) = \sum_{r=0}^{x} \frac{\binom{k}{r}\binom{N-k}{n-r}}{\binom{N}{n}}$$

N	n	k	x	F(x)	f(x)	N	n	k	x	F(x)	f(x)
2	1	1	0	0.500000	0.500000	6	2	2	2	1.000000	0.666667
2	1	1	1	1.000000	0.500000	6	3	1	0	0.500000	0.500000
3	1	1	0	0.666667	0.666667	6	3	1	1	1.000000	0.500000
3	1	1	1	1.000000	0.333333	6	3	2	0	0.200000	0.200000
3	2	1	0	0.333333	0.333333	6	3	2	1	0.800000	0.600000
3	2	1	1	1.000000	0.666667	6	3	2	2	1.000000	0.200000
3	2	2	0	0.666667	0.666667	6	3	3	0	0.050000	0.050000

N	n	k	x	F(x)	f(x)	N	n	k	x	F(x)	f(x)
3	2	2	1	1.000000	0.333333	6	3	3	1	0.500000	0.450000
4	1	1	0	0.750000	0.750000	6	3	3	2	0.950000	0.450000
4	1	1	1	1.000000	0.250000	6	3	3	3	1.000000	0.050000
4	2	1	0	0.500000	0.500000	6	4	1	0	0.333333	0.333333
4	2	1	1	1.000000	0.500000	6	4	1	1	1.000000	0.666667
4	2	2	0	0.166667	0.166667	6	4	2	0	0.066667	0.066667
4	2	2	1	0.833333	0.666667	6	4	2	1	0.600000	0.533333
4	2	2	2	1.000000	0.166667	6	4	2	2	1.000000	0.400000
4	3	1	0	0.250000	0.250000	6	4	3	1	0.200000	0.200000
4	3	1	1	1.000000	0.750000	6	4	3	2	0.800000	0.600000
4	3	2	1	0.500000	0.500000	6	4	3	3	1.000000	0.200000
4	3	2	2	1.000000	0.500000	6	4	4	2	0.400000	0.400000
4	3	3	2	0.750000	0.750000	6	4	4	3	0.933333	0.533333
4	3	3	3	1.000000	0.250000	6	4	4	4	1.000000	0.066667
5	1	1	0	0.800000	0.800000	6	5	1	0	0.166667	0.166667
5	1	1	1	1.000000	0.200000	6	5	1	1	1.000000	0.833333
5	2	1	0	0.600000	0.600000	6	5	2	1	0.333333	0.333333
5	2	1	1	1.000000	0.400000	6	5	2	2	1.000000	0.666667
5	2	2	0	0.300000	0.300000	6	5	3	2	0.500000	0.500000
5	2	2	1	0.900000	0.600000	6	5	3	3	1.000000	0.500000
5	2	2	2	1.000000	0.100000	6	5	4	3	0.666667	0.666667
5	3	1	0	0.400000	0.400000	6	5	4	4	1.000000	0.333333
5	3	1	1	1.000000	0.600000	6	5	5	4	0.833333	0.833333
5	3	2	0	0.100000	0.100000	6	5	5	5	1.000000	0.166667
5	3	2	1	0.700000	0.600000	7	1	1	0	0.857143	0.857143
5	3	2	2	1.000000	0.300000	7	1	1	1	1.000000	0.142857
5	3	3	1	0.300000	0.300000	7	2	1	0	0.714286	0.714286
5	3	3	2	0.900000	0.600000	7	2	1	1	1.000000	0.285714
5	3	3	3	1.000000	0.100000	7	2	2	0	0.476190	0.476190
5	4	1	0	0.200000	0.200000	7	2	2	1	0.952381	0.476190
5	4	1	1	1.000000	0.800000	7	2	2	2	1.000000	0.047619
5	4	2	1	0.400000	0.400000	7	3	1	0	0.571429	0.571429
5	4	2	2	0.000000	0.600000	7	3	1	1	1.000000	0.428571
5	4	3	2	0.600000	0.600000	7	3	2	0	0.285714	0.285714
5	4	3	3	1.000000	0.400000	7	3	2	1	0.857143	0.571429
5	4	4	3	0.800000	0.800000	7	3	2	2	1.000000	0.142857
5	4	4	4	1.000000	0.200000	7	3	3	0	0.114286	0.114286
6	1	1	0	0.833333	0.833333	7	3	3	1	0.628571	0.514286
6	1	1	1	1.000000	0.166667	7	3	3	2	0.971428	0.342857
6	2	1	0	0.666667	0.666667	7	3	3	3	1.000000	0.028571

continued

N	n	k	x	F(x)	f(x)	N	n	k	x	F(x)	f(x)
6	2	1	1	1.000000	0.333333	7	4	1	0	0.428571	0.428571
6	2	2	0	0.400000	0.400000	7	4	1	1	1.000000	0.571429
6	2	2	1	0.933333	0.533333	7	4	2	0	0.142857	0.142857
7	4	2	1	0.714286	0.571429	8	3	3	2	0.985143	0.267857
7	4	2	2	1.000000	0.285714	8	3	3	3	1.000000	0.017857
7	4	3	0	0.028571	0.028571	8	4	1	0	0.500000	0.500000
7	4	3	1	0.371429	0.342857	8	4	1	1	1.000000	0.500000
7	4	3	2	0.885714	0.514286	8	4	2	0	0.214286	0.214286
7	4	3	3	1.000000	0.114286	8	4	2	1	0.785714	0.571429
7	4	4	1	0.114286	0.114286	8	4	2	2	1.000000	0.214286
7	4	4	2	0.628571	0.514286	8	4	3	0	0.071429	0.071429
7	4	4	3	0.971428	0.342857	8	4	3	1	0.500000	0.428571
7	4	4	4	1.000000	0.028571	8	4	3	2	0.928571	0.428571
7	5	1	0	0.285714	0.285714	8	4	3	3	1.000000	0.071429
7	5	1	1	1.000000	0.714286	8	4	4	0	0.014286	0.014286
7	5	2	0	0.047619	0.047619	8	4	4	1	0.242857	0.228571
7	5	2	1	0.523809	0.476190	8	4	4	2	0.757143	0.514286
7	5	2	2	1.000000	0.476190	8	4	4	3	0.985714	0.228571
7	5	3	1	0.142857	0.142857	8	4	4	4	1.000000	0.014286
7	5	3	2	0.714286	0.571429	8	5	1	0	0.375000	0.375000
7	5	3	3	1.000000	0.285714	8	5	1	1	1.000000	0.625000
7	5	4	2	0.285714	0.285714	8	5	2	0	0.107143	0.107143
7	5	4	3	0.857143	0.571429	8	5	2	1	0.642857	0.535714
7	5	4	4	1.000000	0.142857	8	5	2	2	1.000000	0.357143
7	5	5	3	0.476190	0.476190	8	5	3	0	0.017857	0.017857
7	5	5	4	0.952381	0.476190	8	5	3	1	0.285714	0.267857
7	5	5	5	1.000000	0.047619	8	5	3	2	0.821429	0.535714
7	6	1	0	0.142857	0.142857	8	5	3	3	1.000000	0.178571
7	6	1	1	1.000000	0.857143	8	5	4	2	0.071429	0.071429
7	6	2	1	0.285714	0.285714	8	5	4	1	0.500000	0.428571
7	6	2	2	1.000000	0.714286	8	5	4	2	0.928571	0.428571
7	6	3	2	0.428571	0.428571	8	5	4	3	1.000000	0.071429
7	6	3	3	1.000000	0.571429	8	5	5	4	0.178571	0.178571
7	6	4	3	0.571429	0.571429	8	5	5	3	0.714286	0.535714
7	6	4	4	1.000000	0.428571	8	5	5	4	0.982143	0.267857
7	6	5	4	0.714286	0.714286	8	5	5	5	1.000000	0.017857
7	6	5	5	1.000000	0.285714	8	6	1	0	0.250000	0.250000
7	6	6	5	0.857143	0.857143	8	6	1	1	1.000000	0.750000
7	6	6	6	1.000000	0.142857	8	6	2	0	0.035714	0.035714
8	1	1	0	0.875000	0.875000	8	6	2	1	0.464286	0.428571

N	n	k	x	F(x)	f(x)	N	n	k	x	F(x)	f(x)
8	1	1	1	1.000000	0.125000	8	6	2	2	1.000000	0.535714
8	2	1	0	0.750000	0.750000	8	6	3	1	0.107143	0.107143
8	2	1	1	1.000000	0.250000	8	6	3	2	0.642857	0.535714
8	2	2	0	0.535714	0.535714	8	6	3	3	1.000000	0.357143
8	2	2	1	0.964286	0.428571	8	6	4	2	0.214286	0.214286
8	2	2	2	1.000000	0.035714	8	6	4	3	0.785714	0.571429
8	3	1	0	0.625000	0.625000	8	6	4	4	1.000000	0.214286
8	3	1	1	1.000000	0.375000	8	6	5	3	0.357143	0.357143
8	3	2	0	0.357143	0.357143	8	6	5	4	0.892857	0.535714
8	3	2	1	0.892857	0.535714	8	6	5	5	1.000000	0.107143
8	3	2	2	1.000000	0.107143	8	6	6	4	0.535714	0.535714
8	3	3	0	0.178571	0.178571	8	6	6	5	0.964286	0.428571
8	3	3	1	0.714286	0.535714	8	6	6	6	1.000000	0.035714
8	7	1	0	0.125000	0.125000	9	5	3	1	0.404762	0.357143
8	7	1	1	1.000000	0.875000	9	5	3	2	0.880952	0.476190
8	7	2	1	0.250000	0.250000	9	5	3	3	1.000000	0.119048
8	7	2	2	1.000000	0.750000	9	5	4	0	0.007936	0.007936
8	7	3	2	0.375000	0.375000	9	5	4	1	0.166667	0.158730
8	7	3	3	1.000000	0.625000	9	5	4	2	0.642857	0.476190
8	7	4	3	0.500000	0.500000	9	5	4	3	0.960317	0.317460
8	7	4	4	1.000000	0.500000	9	5	4	4	1.000000	0.039683
8	7	5	4	0.625000	0.625000	9	5	5	1	0.039683	0.039683
8	7	5	5	1.000000	0.375000	9	5	5	2	0.357143	0.317460
8	7	6	5	0.750000	0.750000	9	5	5	3	0.833333	0.476190
8	7	6	6	1.000000	0.250000	9	5	5	4	0.992063	0.158730
8	7	7	6	0.875000	0.875000	9	5	5	5	1.000000	0.007936
8	7	7	7	1.000000	0.125000	9	6	1	0	0.333333	0.333333
9	1	1	0	0.888889	0.888889	9	6	1	1	1.000000	0.666667
9	1	1	1	1.000000	0.111111	9	6	2	0	0.083333	0.083333
9	2	1	0	0.777778	0.777778	9	6	2	1	0.583333	0.500000
9	2	1	1	1.000000	0.222222	9	6	2	2	1.000000	0.416667
9	2	2	0	0.583333	0.583333	9	6	3	0	0.011905	0.011905
9	2	2	1	0.972222	0.388889	9	6	3	1	0.226190	0.214286
9	2	2	2	1.000000	0.277778	9	6	3	2	0.761905	0.535714
9	3	1	0	0.666667	0.666667	9	6	3	3	1.000000	0.238095
9	3	1	1	1.000000	0.333333	9	6	4	1	0.047619	0.047619
9	3	2	0	0.416667	0.416667	9	6	4	2	0.404762	0.357143
9	3	2	1	0.916667	0.500000	9	6	4	3	0.880952	0.476190
9	3	2	2	1.000000	0.083333	9	6	4	4	1.000000	0.119048
9	3	3	0	0.238095	0.238095	9	6	5	2	0.119048	0.119048

continued

N	n	k	x	F(x)	f(x)	N	n	k	x	F(x)	f(x)
9	3	3	1	0.773809	0.535714	9	6	5	3	0.595238	0.476190
9	3	3	2	0.988095	0.214286	9	6	5	4	0.952381	0.357143
9	3	3	3	1.000000	0.011905	9	6	5	5	1.000000	0.047619
9	4	1	0	0.555556	0.555556	9	6	6	3	0.238095	0.238095
9	4	1	1	1.000000	0.444444	9	6	6	4	0.773809	0.535714
9	4	2	0	0.277778	0.277778	9	6	6	5	0.988095	0.214286
9	4	2	1	0.833333	0.555556	9	6	6	6	1.000000	0.011905
9	4	2	2	1.000000	0.166667	9	7	1	0	0.222222	0.222222
9	4	3	0	0.119048	0.119048	9	7	1	1	1.000000	0.777778
9	4	3	1	0.595238	0.476190	9	7	2	0	0.027778	0.027778
9	4	3	2	0.952381	0.357143	9	7	2	1	0.416667	0.388889
9	4	3	3	1.000000	0.047619	9	7	3	2	1.000000	0.583333
9	4	4	0	0.039683	0.039683	9	7	3	1	0.083333	0.083333
9	4	4	1	0.357143	0.317460	9	7	3	2	0.583333	0.500000
9	4	4	2	0.833333	0.476190	9	7	3	3	1.000000	0.416667
9	4	4	3	0.992063	0.158730	9	7	4	2	0.166667	0.166667
9	4	4	4	1.000000	0.007936	9	7	4	3	0.722222	0.555556
9	5	1	0	0.444444	0.444444	9	7	4	4	1.000000	0.277778
9	5	1	1	1.000000	0.555556	9	7	5	3	0.277778	0.277778
9	5	2	0	0.166667	0.166667	9	7	5	4	0.833333	0.555556
9	5	2	1	0.722222	0.555556	9	7	5	5	1.000000	0.166667
9	5	2	2	1.000000	0.277778	9	7	6	4	0.416667	0.416667
9	5	3	0	0.047619	0.047619	9	7	6	5	0.916667	0.500000
9	7	6	6	1.000000	0.083333	10	5	1	0	0.500000	0.500000
9	7	7	5	0.583333	0.583333	10	5	1	1	1.000000	0.500000
9	7	7	6	0.972222	0.388889	10	5	2	0	0.222222	0.222222
9	7	7	7	1.000000	0.027778	10	5	2	1	0.777778	0.555556
9	8	1	0	0.111111	0.111111	10	5	2	2	1.000000	0.222222
9	8	1	1	1.000000	0.888889	10	5	3	0	0.083333	0.083333
9	8	2	1	0.222222	0.222222	10	5	3	1	0.500000	0.416667
9	8	2	2	1.000000	0.777778	10	5	3	2	0.916667	0.416667
9	8	3	2	0.333333	0.333333	10	5	3	3	1.000000	0.083333
9	8	3	3	1.000000	0.666667	10	5	4	0	0.023810	0.023810
9	8	4	3	0.444444	0.444444	10	5	4	1	0.261905	0.238095
9	8	4	4	1.000000	0.555556	10	5	4	2	0.738095	0.476190
9	8	5	4	0.555556	0.555556	10	5	4	3	0.976190	0.238095
9	8	5	5	1.000000	0.444444	10	5	4	4	1.000000	0.023810
9	8	6	5	0.666667	0.666667	10	5	5	0	0.003968	0.003968
9	8	6	6	1.000000	0.333333	10	5	5	1	0.103175	0.099206
9	8	7	6	0.777778	0.777778	10	5	5	2	0.500000	0.396825

N	n	k	x	F(x)	f(x)	N	n	k	x	F(x)	f(x)
9	8	7	7	1.000000	0.222222	10	5	5	3	0.896825	0.396825
9	8	8	7	0.888889	0.888889	10	5	5	4	0.996032	0.099206
9	8	8	8	1.000000	0.111111	10	5	5	5	1.000000	0.003968
10	1	1	0	0.900000	0.900000	10	6	1	0	0.400000	0.400000
10	1	1	1	1.000000	0.100000	10	6	1	1	1.000000	0.600000
10	2	1	0	0.800000	0.800000	10	6	2	0	0.133333	0.133333
10	2	1	1	1.000000	0.200000	10	6	2	1	0.666667	0.533333
10	2	2	0	0.622222	0.622222	10	6	2	2	1.000000	0.333333
10	2	2	1	0.977778	0.355556	10	6	3	0	0.033333	0.033333
10	2	2	2	1.000000	0.022222	10	6	3	1	0.33333	0.300000
10	3	1	0	0.700000	0.700000	10	6	3	2	0.833333	0.500000
10	3	1	1	1.000000	0.300000	10	6	3	3	1.000000	0.166667
10	3	2	0	0.466667	0.466667	10	6	4	0	0.004762	0.004762
10	3	2	1	0.933333	0.466667	10	6	4	1	0.119048	0.114286
10	3	2	2	1.000000	0.066667	10	6	4	2	0.547619	0.428571
10	3	3	0	0.291667	0.291667	10	6	4	3	0.928571	0.380952
10	3	3	1	0.816667	0.525000	10	6	4	4	1.000000	0.071429
10	3	3	2	0.991667	0.175000	10	6	5	1	0.023810	0.023810
10	3	3	3	1.000000	0.008333	10	6	5	2	0.261905	0.238095
10	4	1	0	0.600000	0.600000	10	6	5	3	0.738095	0.476190
10	4	1	1	1.000000	0.400000	10	6	5	4	0.976190	0.238095
10	4	2	0	0.333333	0.333333	10	6	5	5	1.000000	0.023810
10	4	2	1	0.866667	0.533333	10	6	6	2	0.071429	0.071429
10	4	2	2	1.000000	0.133333	10	6	6	3	0.452381	0.380952
10	4	3	0	0.166667	0.166667	10	6	6	4	0.880952	0.428571
10	4	3	1	0.666667	0.500000	10	6	6	5	0.995238	0.114286
10	4	3	2	0.966667	0.300000	10	6	6	6	1.000000	0.004762
10	4	3	3	1.000000	0.033333	10	7	1	0	0.300000	0.300000
10	4	4	0	0.071429	0.071429	10	7	1	1	1.000000	0.700000
10	4	4	1	0.452381	0.380952	10	7	2	0	0.066667	0.066667
10	4	4	2	0.880952	0.428571	10	7	2	1	0.533333	0.466667
10	4	4	3	0.995238	0.114286	10	7	2	2	1.000000	0.466667
10	4	4	4	1.000000	0.004762	10	7	3	0	0.008333	0.008333

Table 6
Negative Binomial Distribution

The negative binomial probability function is given by

$$f(x;r,\theta) = \binom{x+r-1}{r-1} \theta^r (1-\theta)^x, \qquad x = 0, 1, 2, \ldots ;.$$

where θ is the probability of success and $1-\theta$ the probability of failure of a given event. $f(x)$ is the probability that exactly $x+r$ trials will be required to produce r successes. The cumulative distribution is given by

$$F(x;r,\theta) = \sum_{x'=0}^{x} \binom{x'+r-1}{r-1} \theta^r (1-\theta)^{x'}.$$

The cumulative negative binomial distribution is related to the cumulative binomial distribution as follows:

$$\sum_{x'=0}^{x} \binom{x'+r-1}{r-1} \theta^r (1-\theta)^{x'} = \sum_{x'=r}^{x+r} \binom{x+r}{x'} \theta^{x'} (1-\theta)^{x+r-x'}.$$

Negative Binomial Probability and Distribution Functions

$$f(x;r,\theta) = \binom{x+r-1}{r-1} \theta^r (1-\theta)^x \qquad F(x;r,\theta) = \sum_{x'=0}^{x} \binom{x'+r-1}{r-1} \theta^r (1-\theta)^{x'}$$

$\theta = 0.9000, r = 1$

$x+r$	$f(x)$	$F(x)$
1	0.90000	0.9000
2	0.09000	0.9900

$\theta = 0.900, r = 4$

$x+r$	$f(x)$	$F(x)$
4	0.65610	0.6561
5	0.26244	0.9185
6	0.06561	0.9841
7	0.01312	0.9973

$\theta = 0.9000, r = 2$

$x+r$	$f(x)$	$F(x)$
2	0.81000	0.8100
3	0.16200	0.9720
4	0.02430	0.9963

$\theta = 0.900, r = 5$

$x+r$	$f(x)$	$F(x)$
5	0.59049	0.5905
6	0.29524	0.8857
7	0.08857	0.9743
8	0.02067	0.9950

$\theta = 0.9000, r = 3$		
$x + r$	$f(x)$	$F(x)$
3	0.72900	0.7290
4	0.21870	0.9477
5	0.04374	0.9914

$\theta = 0.9000, r = 6$		
$x + r$	$f(x)$	$F(x)$
6	0.53144	0.5314
7	0.31886	0.8503
8	0.11160	0.9619
9	0.02976	0.9917

Source: Beyer, W. H. (ed). *Handbook of Tables for Probability and Statistics.* Cleveland: The Chemical Rubber Company, 1966. Reprinted by permission of the publisher.

Table 7
Exponential Functions

Values of e^x, $\log_e x$ and e^{-x} where e is the base of the natural system of logarithms 2.71828... and x has values from 0 to 10. Facilitating the solution of exponential equations, these tables also serve as a table of natural or Naperian antilogarithms. For instance, if the logarithm or exponent $x = 3.26$, the corresponding number or value of e^x is 26.050. Its reciprocal e^{-x} is .038388.

Source: Beyer, W. H. (ed). *Handbook of Tables for Probability and Statistics.* Cleveland: The Chemical Rubber Company, 1966. Reprinted by permission of the publisher.

x	e^x	$\log_{10} e^x$	e^{-x}	x	e^x	$\log_{10} e^x$	e^{-x}
0.00	1.0000	0.00000	1.000000	0.50	1.6487	0.21715	0.606531
0.01	1.0101	0.00434	0.990050	0.51	1.6653	0.22149	0.600496
0.02	1.0202	0.00869	0.980199	0.52	1.6820	0.22583	0.594521
0.03	1.0305	0.01303	0.970446	0.53	1.6989	0.23018	0.588605
0.04	1.0408	0.01737	0.960789	0.54	1.7160	0.23452	0.582748
0.05	1.0513	0.02171	0.951229	0.55	1.7333	0.23886	0.576950
0.06	1.0618	0.02606	0.941765	0.56	1.7507	0.24320	0.571209
0.07	1.0725	0.03040	0.932394	0.57	1.7683	0.24755	0.565525
0.08	1.0833	0.03474	0.923116	0.58	1.7860	0.25189	0.569898
0.09	1.0942	0.03909	0.913931	0.59	1.8040	0.25623	0.554327
0.10	1.1052	0.14343	0.904837	0.60	1.8221	0.26058	0.548812
0.11	1.1163	0.04777	0.895834	0.61	1.8404	0.26492	0.543351
0.12	1.1275	0.05212	0.886920	0.62	1.8589	0.26926	0.537944
0.13	1.1388	0.05646	0.878095	0.63	1.8776	0.27361	0.532592
0.14	1.1503	0.06080	0.869358	0.64	1.8965	0.27795	0.527292
0.15	1.1618	0.06514	0.860708	0.65	1.9155	0.28229	0.522046
0.16	1.1735	0.06949	0.852144	0.66	1.9348	0.28663	0.516851
0.17	1.1853	0.07383	0.843665	0.67	1.9542	0.29098	0.511709
0.18	1.1972	0.07817	0.835270	0.68	1.9739	0.29532	0.506617
0.19	1.2092	0.08252	0.826959	0.69	1.9937	0.29966	0.501576
0.20	1.2214	0.08686	0.818731	0.70	2.0138	0.30401	0.496585
0.21	1.2337	0.09120	0.810584	0.71	2.0340	0.30835	0.491644
0.22	1.2461	0.09554	0.802519	0.72	2.0544	0.31269	0.486752
0.23	1.2586	0.09989	0.794534	0.73	2.0751	0.31703	0.481909
0.24	1.2712	0.10423	0.786628	0.74	2.0959	0.32138	0.477114
0.25	1.2840	0.10857	0.778801	0.75	2.1170	0.32572	0.472367
0.26	1.2969	0.11292	0.771052	0.76	2.1383	0.33006	0.467666

x	e^x	$\log_{10}e^x$	e^{-x}	x	e^x	$\log_{10}e^x$	e^{-x}
0.27	1.3100	0.11726	0.763379	0.77	2.1598	0.33441	0.463013
0.28	1.3231	0.12160	0.755784	0.78	2.1815	0.33875	0.458406
0.29	1.3364	0.12595	0.748264	0.79	2.2034	0.34309	0.452845
0.30	1.3499	0.13029	0.740818	0.80	2.2255	0.34744	0.449329
0.31	1.3634	0.13463	0.733447	0.81	2.2479	0.35178	0.444858
0.32	1.3771	0.13897	0.726149	0.82	2.2705	0.35612	0.440432
0.33	1.3910	0.14332	0.718924	0.83	2.2933	0.36046	0.436049
0.34	1.4049	0.14766	0.711770	0.84	2.3164	0.36481	0.431711
0.35	1.4191	0.15200	0.704688	0.85	2.3396	0.36915	0.427415
0.36	1.4333	0.15635	0.697676	0.86	2.3632	0.37349	0.423162
0.37	1.4477	0.16069	0.690734	0.87	2.3869	0.37784	0.418952
0.37	1.4623	0.16503	0.683861	0.88	2.4109	0.38218	0.414783
0.39	1.4770	0.16937	0.677057	0.89	2.4351	0.38652	0.410656
0.40	1.4918	0.17372	0.670320	0.90	2.4596	0.39087	0.40657
0.41	1.5068	0.17806	0.663650	0.91	2.4843	0.39521	0.402524
0.42	1.5220	0.18240	0.657047	0.92	2.5093	0.39955	0.398519
0.43	1.5373	0.18675	0.650509	0.93	2.5345	0.40389	0.394554
0.44	1.5527	0.19109	0.644036	0.94	2.5600	0.40824	0.390628
0.45	1.5683	0.19543	0.637628	0.95	2.5857	0.41258	0.386741
0.46	1.5841	0.19978	0.631284	0.96	2.6117	0.41692	0.382893
0.47	1.600	0.20412	0.625002	0.97	2.6379	0.42127	0.379083
0.48	1.6161	0.20846	0.618783	0.98	2.6645	0.42561	0.375311
0.49	1.6323	0.21280	0.612626	0.99	2.6912	0.42995	0.371577
0.50	1.6487	0.21715	0.606531	1.00	2.7183	0.43429	0.367879
1.00	2.7183	0.43429	0.367879	1.50	4.4817	0.65144	0.223130
1.01	2.7456	0.43864	0.364219	1.51	1.5267	0.65578	0.220910
1.02	2.7732	0.44298	0.360595	1.52	4.5722	0.66013	0.218712
1.03	2.8011	0.44732	0.357007	1.53	4.6182	0.66447	0.216536
1.04	2.8292	0.45167	0.353455	1.54	4.6646	0.66881	0.214381
1.05	2.8577	0.45601	0.349938	1.55	4.7115	0.67316	0.212248
1.06	2.8864	0.46035	0.346456	1.56	4.7588	0.67750	0.210136
1.07	2.9154	0.46470	0.343009	1.57	4.8066	0.68184	0.208045
1.08	2.9447	0.46904	0.339596	1.58	4.8550	0.68619	0.205975
1.09	2.9743	0.47338	0.336216	1.59	4.9037	0.69053	0.203926
1.10	3.0042	0.47772	0.332871	1.60	4.9530	0.69487	0.201897
1.11	3.0344	0.48207	0.329559	1.61	5.0028	0.69921	0.199888

continued

x	e^x	$\log_{10}e^x$	e^{-x}	x	e^x	$\log_{10}e^x$	e^{-x}
1.12	3.0649	0.48641	0.326280	1.62	5.0531	0.70356	0.197899
1.13	3.0957	0.49075	0.323033	1.63	5.1039	0.70790	0.195930
1.14	3.1268	0.49510	0.319819	1.64	5.1552	0.71224	0.193980
1.15	3.1582	0.49944	0.316637	1.65	5.2070	0.71659	0.192050
1.16	3.1899	0.50378	0.313486	1.66	5.2593	0.72093	0.190139
1.17	3.2220	0.50812	0.310367	1.67	5.3122	0.72527	0.188247
1.18	3.2544	0.51247	0.307279	1.68	5.3656	0.72961	0.186374
1.19	3.2871	0.51681	0.304221	1.69	5.4195	0.73396	0.184520
1.20	3.3201	0.52115	0.301194	1.70	5.4739	0.73830	0.182684
1.21	3.3535	0.52550	0.298197	1.71	5.5290	0.74264	0.180866
1.22	3.3872	0.52984	0.295230	1.72	5.5845	0.74699	0.179066
1.23	3.4212	0.53418	0.292293	1.73	5.6407	0.75133	0.177284
1.24	3.4556	0.53853	0.289384	1.74	5.6973	0.75567	0.175520
1.25	3.4903	0.54287	0.286505	1.75	5.7546	0.76002	0.173774
1.26	3.5254	0.54721	0.283654	1.76	5.8124	0.76436	0.172045
1.27	3.5609	0.55155	0.280832	1.77	5.8709	0.76870	0.170333
1.28	3.5966	0.55590	0.278037	1.78	5.9299	0.77304	0.168638
1.29	3.6328	0.56024	0.275271	1.79	5.9895	0.77739	0.166960
1.30	3.6693	0.56458	0.272532	1.80	6.0496	0.78173	0.165299
1.31	3.7062	0.56893	0.269820	1.81	6.1104	0.78607	0.163654
1.32	3.7434	0.57327	0.267135	1.82	6.1719	0.79042	0.162026
1.33	3.7810	0.57761	0.264447	1.83	6.2339	0.79476	0.160414
1.34	3.8190	0.58195	0.261846	1.84	6.2965	0.79910	0.158817
1.35	3.8574	0.58630	0.259240	1.85	6.3598	0.80344	0.157237
1.36	3.8962	0.59064	0.256661	1.86	6.4237	0.80779	0.155673
1.37	3.9354	0.59498	0.254107	1.87	6.4883	0.81213	0.154124
1.37	3.9749	0.59933	0.251579	1.88	6.5535	0.81647	0.152590
1.39	4.0149	0.60367	0.249075	1.89	6.6194	0.82082	0.151072
1.40	4.0552	0.60801	0.246597	1.90	6.6859	0.82516	0.149569
1.41	4.0960	0.61236	0.244143	1.91	6.7531	0.82950	0.148080
1.42	4.1371	0.61670	0.241714	1.92	6.8210	0.83385	0.146607
1.43	4.1787	0.62104	0.239309	1.93	6.8895	0.83819	0.145148
1.44	4.2207	0.62538	0.236928	1.94	6.9588	0.84253	0.143704
1.45	4.2631	0.62973	0.234570	1.95	7.0287	0.84687	0.142274
1.46	4.3060	0.63407	0.232236	1.96	7.0993	0.85122	0.140858
1.47	4.3492	0.63841	0.229925	1.97	7.1707	0.85556	0.139457
1.48	4.3929	0.64276	0.227638	1.98	7.2427	0.85990	0.138069

x	e^x	$\log_{10}e^x$	e^{-x}	x	e^x	$\log_{10}e^x$	e^{-x}
1.49	4.4371	0.64710	0.225373	1.99	7.3155	0.86425	0.136695
1.50	4.4817	0.65144	0.223130	2.00	7.3891	0.86859	0.135335
2.00	7.3891	0.86859	0.135335	2.50	12.182	1.08574	0.082085
2.01	7.4633	0.87293	0.133989	2.51	12.305	1.09008	0.081268
2.02	7.5383	0.87727	0.132655	2.52	12.429	1.09442	0.080460
2.03	7.6141	0.88162	0.131336	2.53	12.554	1.09877	0.079659
2.04	7.6906	0.88596	0.130029	2.54	12.680	1.10311	0.078866
2.05	7.7679	0.89030	0.128735	2.55	12.807	1.10745	0.078082
2.06	7.8460	0.89465	0.127454	2.56	12.936	1.11179	0.077305
2.07	7.9248	0.89899	0.126186	2.57	13.066	1.11614	0.076536
2.08	8.0045	0.90333	0.124930	2.58	13.197	1.12048	0.075774
2.09	8.0849	0.90768	0.123687	2.59	13.330	1.12482	0.075020
2.10	8.1662	0.91202	0.122456	2.60	13.464	1.12917	0.074274
2.11	8.2482	0.91636	0.121238	2.61	13.599	1.13351	0.073535
2.12	8.3311	0.92070	0.120032	2.62	13.736	1.13785	0.072803
2.13	8.4149	0.92505	0.118837	2.63	13.874	1.14219	0.072078
2.14	8.4994	0.92939	0.117655	2.64	14.013	1.14654	0.071361
2.15	8.5849	0.93373	0.116484	2.65	14.154	1.15088	0.070651
2.16	8.6711	0.93808	0.115325	2.66	14.296	1.15522	0.069948
2.17	8.7583	0.94242	0.114178	2.67	14.440	1.15957	0.069252
2.18	8.8463	0.94676	0.113042	2.68	14.585	1.16391	0.068563
2.19	8.9352	0.95110	0.111917	2.69	14.732	1.16825	0.067881
2.20	9.0250	0.95545	0.110803	2.70	14.880	1.17260	0.067206
2.21	9.1157	0.95979	0.109701	2.71	15.029	1.17694	0.066537
2.22	9.2073	0.96413	0.108609	2.72	15.180	1.18125	0.065875
2.23	9.2999	0.96848	0.107528	2.73	15.333	1.18562	0.065219
2.24	9.3933	0.97282	0.106459	2.74	15.487	1.18997	0.064570
2.25	9.4877	0.97716	0.105399	2.75	15.643	1.19431	0.063928
2.26	9.5831	0.98151	0.104350	2.76	15.800	1.19865	0.063292
2.27	9.6794	0.98585	0.103312	2.77	15.959	1.20300	0.062662
2.28	9.7767	0.99019	0.102284	2.78	16.119	1.20734	0.062039
2.29	9.8749	0.99453	0.101266	2.79	16.281	1.21168	0.061421
2.30	9.9742	0.99888	0.100259	2.80	16.445	1.21603	0.060810
2.31	10.074	1.00322	0.099261	2.81	16.610	1.22037	0.060205
2.32	10.176	1.00756	0.098274	2.82	16.777	1.22471	0.059606
2.33	10.278	1.01191	0.097296	2.83	16.945	1.22905	0.059013

continued

x	e^x	$\log_{10}e^x$	e^{-x}	x	e^x	$\log_{10}e^x$	e^{-x}
2.34	10.381	1.01625	0.096328	2.84	17.116	1.23340	0.058426
2.35	10.486	1.02059	0.095369	2.85	17.288	1.23774	0.057844
2.36	10.591	1.02493	0.094420	2.86	17.462	1.24208	0.057269
2.37	10.697	1.02928	0.093481	2.87	17.637	1.24643	0.056699
2.37	10.805	1.03362	0.092551	2.88	17.814	1.25077	0.056135
2.39	10.913	1.03796	0.091630	2.89	17.993	1.25511	0.055576
2.40	11.023	1.04231	0.090718	2.90	18.174	1.25945	0.055023
2.41	11.134	1.04665	0.089815	2.91	18.357	1.26380	0.054476
2.42	11.246	1.05099	0.088922	2.92	18.541	1.26814	0.053934
2.43	11.359	1.05534	0.088037	2.93	18.728	1.27248	0.053397
2.44	11.473	1.05968	0.087161	2.94	18.916	1.27683	0.052866
2.45	11.588	1.06402	0.086294	2.95	19.106	1.28117	0.052340
2.46	11.705	1.06836	0.085435	2.96	19.298	1.28551	0.051819
2.47	11.822	1.07271	0.084585	2.97	19.492	1.28985	0.051303
2.48	11.941	1.07705	0.083743	2.98	19.688	1.29420	0.050793
2.49	12.061	1.08139	0.082910	2.99	19.886	1.29854	0.050287
2.50	12.182	1.08574	0.082085	3.00	20.086	1.30288	0.049787
3.00	20.086	1.30288	0.049787	3.50	33.115	1.52003	0.030197
3.01	20.287	1.30723	0.049292	3.51	33.448	1.52437	0.029897
3.02	20.491	1.31157	0.048801	3.52	33.784	1.52872	0.029599
3.03	20.697	1.31591	0.048316	3.54	34.124	1.53306	0.029305
3.04	20.905	1.32026	0.047835	3.55	34.467	1.53740	0.029013
3.05	21.115	1.32460	0.047359	3.55	34.813	1.54175	0.028725
3.06	21.328	1.32894	0.046888	3.56	35.163	1.54609	0.028439
3.07	21.542	1.33328	0.046421	3.57	35.517	1.55043	0.028156
3.08	21.758	1.33763	0.045959	3.58	35.874	1.55477	0.027876
3.09	21.977	1.34197	0.045502	3.59	36.234	1.55912	0.027598
3.10	22.198	1.34631	0.045049	3.60	36.598	1.56346	0.027324
3.11	22.421	1.35066	0.044601	3.61	36.966	1.56780	0.027052
3.12	22.646	1.35500	0.044157	3.62	37.338	1.57215	0.026783
3.13	22.874	1.35934	0.043718	3.63	37.713	1.57649	0.026516
3.14	23.104	1.36368	0.043283	3.64	38.092	1.58083	0.026252
3.15	23.336	1.36803	0.042852	3.65	38.475	1.58517	0.025991
3.16	23.571	1.37237	0.042426	3.66	38.861	1.58952	0.025733
3.17	23.807	1.37671	0.042004	3.67	39.252	1.59386	0.025476

x	e^x	$\log_{10}e^x$	e^{-x}	x	e^x	$\log_{10}e^x$	e^{-x}
3.18	24.047	1.38106	0.041586	3.68	39.646	1.59820	0.025223
3.19	24.288	1.38540	0.041172	3.69	40.045	1.60255	0.024972
3.20	24.533	1.38974	0.040762	3.70	40.447	1.60689	0.024724
3.21	24.779	1.39409	0.040357	3.71	40.854	1.61123	0.024478
3.22	25.028	1.39843	0.039955	3.72	41.264	1.61558	0.024234
3.23	25.280	1.40277	0.039557	3.73	41.679	1.61992	0.023993
3.24	25.534	1.40711	0.039164	3.74	42.098	1.62426	0.023754
3.25	25.790	1.41146	0.038774	3.75	42.521	1.62860	0.023518
3.26	26.050	1.41580	0.038388	3.76	42.948	1.63295	0.023284
3.27	26.311	1.42014	0.038006	3.77	43.380	1.63729	0.023052
3.28	26.576	1.42449	0.037628	3.78	43.816	1.64163	0.022823
3.29	26.843	1.42883	0.037254	3.79	44.256	1.64598	0.022596
3.30	27.113	1.43317	0.036883	3.80	44.701	1.65032	0.022371
3.31	27.385	1.43751	0.036516	3.81	45.150	1.65466	0.022148
3.32	27.660	1.44186	0.036153	3.82	45.604	1.65900	0.021928
3.33	27.938	1.44620	0.035793	3.83	46.063	1.66335	0.021710
3.34	28.219	1.45054	0.035437	3.84	46.525	1.66769	0.021494
3.35	28.503	1.45489	0.035084	3.85	46.993	1.67203	0.021280
3.36	28.789	1.45923	0.034735	3.86	47.465	1.67638	0.021068
3.37	29.079	1.46357	0.034390	3.87	47.942	1.68072	0.020858
3.38	29.371	1.46792	0.034047	3.88	48.424	1.68506	0.020651
3.39	29.666	1.47226	0.033709	3.89	48.911	1.68941	0.020445
3.40	29.964	1.47660	0.033373	3.90	49.402	1.69375	0.020242
3.41	30.265	1.48094	0.033041	3.91	49.899	1.69809	0.020041
3.42	30.569	1.48529	0.032714	3.92	50.400	1.70243	0.019841
3.43	30.877	1.48063	0.032387	3.93	50.907	1.70678	0.019644
3.44	31.187	1.49397	0.032065	3.94	51.419	1.71112	0.019448
3.45	31.500	1.49832	0.031746	3.95	51.935	1.71546	0.019255
3.46	31.817	1.50266	0.031430	3.96	52.457	1.71981	0.019063
3.47	32.137	1.50700	0.031117	3.97	52.985	1.72415	0.018873
3.48	32.460	1.51134	0.030807	3.98	53.517	1.72849	0.018686
3.49	32.786	1.51569	0.030501	3.99	54.055	1.73283	0.018500
3.50	33.115	1.52003	0.030197	4.00	54.598	1.73718	0.018316
4.00	54.598	1.73718	0.018316	4.50	90.017	1.95433	0.011109
4.01	55.147	1.74152	0.018133	4.51	90.922	1.95867	0.010998

continued

x	e^x	$\log_{10}e^x$	e^{-x}	x	e^x	$\log_{10}e^x$	e^{-x}
4.02	55.701	1.74586	0.017953	4.52	91.836	1.96301	0.010889
4.03	56.261	1.75021	0.017774	4.53	92.759	1.96735	0.010781
4.04	56.826	1.75455	0.017597	4.54	93.691	1.97170	0.010673
4.05	57.397	1.75889	0.017422	4.55	94.632	1.97604	0.010567
4.06	57.974	1.76324	0.017249	4.56	95.583	1.98038	0.010462
4.07	58.557	1.76758	0.017077	4.57	96.544	1.98473	0.010358
4.08	59.145	1.77192	0.016907	4.58	97.514	1.98907	0.010255
4.09	59.740	1.77626	0.016739	4.59	98.494	1.99341	0.010153
4.10	60.340	1.78061	0.016573	4.60	99.484	1.99775	0.010052
4.11	60.947	1.78495	0.016408	4.61	100.48	2.00210	0.009952
4.12	61.559	1.78929	0.016245	4.62	101.49	2.00644	0.009853
4.13	62.178	1.79364	0.016083	4.63	102.51	2.01078	0.009755
4.14	62.803	1.79798	0.015923	4.64	103.54	2.01513	0.009658
4.15	63.434	1.80232	0.015764	4.65	104.58	2.01947	0.009562
4.16	64.072	1.80667	0.015608	4.66	105.64	2.02381	0.009466
4.17	64.715	1.81101	0.015452	4.67	106.70	2.02816	0.009372
4.18	65.366	1.81535	0.015299	4.68	107.77	2.03250	0.009279
4.19	66.023	1.81969	0.015146	4.69	108.85	2.03684	0.009187
4.20	66.686	1.82404	0.014996	4.70	109.95	2.04118	0.009095
4.21	67.357	1.82838	0.014846	4.71	111.05	2.04553	0.009005
4.22	68.033	1.83272	0.014699	4.72	112.17	2.04987	0.008915
4.23	68.717	1.83707	0.014552	4.73	113.30	2.05421	0.008826
4.24	69.408	1.84141	0.014408	4.74	114.43	2.05856	0.008739
4.25	70.105	1.84575	0.014264	4.75	115.58	2.06290	0.008652
4.26	70.810	1.85009	0.014122	4.76	116.75	2.06724	0.008566
4.27	71.522	1.85444	0.013982	4.77	117.92	2.07158	0.008480
4.28	72.240	1.85878	0.013843	4.78	119.10	2.07593	0.008396
4.29	72.966	1.86312	0.013705	4.79	120.30	2.08027	0.008312
4.30	73.700	1.86747	0.013569	4.80	121.51	2.08461	0.008230
4.31	74.440	1.87181	0.013434	4.81	122.73	2.08896	0.008148
4.32	75.189	1.87615	0.013300	4.82	123.97	2.09330	0.008067
4.33	75.944	1.88050	0.013168	4.38	125.21	2.09764	0.007987
4.34	76.708	1.88484	0.013037	4.84	126.47	2.10199	0.007907
4.35	77.478	1.88918	0.012907	4.85	127.74	2.10633	0.007828
4.36	78.257	1.89352	0.012778	4.86	129.02	2.11067	0.007750
4.37	79.044	1.89787	0.012651	4.87	130.32	2.11501	0.007673
4.38	79.838	1.90221	0.012525	4.88	131.63	2.11936	0.007597

x	e^x	$\log_{10}e^x$	e^{-x}	x	e^x	$\log_{10}e^x$	e^{-x}
4.39	80.640	1.90655	0.012401	4.89	132.95	2.12370	0.007521
4.40	81.451	1.91090	0.012277	4.90	134.29	2.12804	0.007447
4.41	82.269	1.91524	0.012155	4.91	135.64	2.13239	0.007372
4.42	83.096	1.91958	0.012034	4.92	137.00	2.13673	0.007299
4.43	83.931	1.92392	0.011914	4.93	138.38	2.14107	0.007227
4.44	84.775	1.92827	0.011796	4.94	139.77	2.14541	0.007155
4.45	85.267	1.93261	0.011679	4.95	141.17	2.14976	0.007083
4.46	86.488	1.93695	0.011562	4.96	142.59	2.15410	0.007013
4.47	87.357	1.94130	0.011447	4.97	144.03	2.15844	0.006943
4.48	88.235	1.94564	0.011333	4.98	145.47	2.16279	0.006874
4.49	89.121	1.94998	0.011221	4.99	146.94	2.16713	0.006806
4.50	90.017	1.95433	0.011109	5.00	148.41	2.17147	0.006738
5.00	148.41	2.17147	0.006738	5.50	244.69	2.38862	0.0040868
5.01	149.90	2.17582	0.006671	5.55	257.24	2.41033	0.0038875
5.02	151.41	2.18016	0.006605	5.60	270.43	2.43205	0.0036979
5.02	152.93	2.18450	0.006539	5.65	284.29	2.45376	0.0035175
5.04	154.47	2.18884	0.006474	5.70	298.57	2.47548	.00334600
5.05	156.02	2.19319	0.006409	5.75	314.19	2.49719	0.0031828
5.06	157.59	2.19753	0.006346	5.80	330.30	2.51891	0.0030276
5.07	159.17	2.20187	0.006282	5.85	347.23	2.54062	0.0028799
5.08	160.77	2.20622	0.006220	5.90	365.04	2.56234	0.0027394
5.09	162.39	2.21056	0.006158	5.95	383.75	2.58405	0.0026058
5.10	164.02	2.21490	0.006097	6.00	403.43	2.60577	0.0024788
5.11	165.67	2.21924	0.006036	6.05	424.11	2.62748	0.0023579
5.12	167.34	2.22359	0.005976	6.10	445.86	2.64920	0.0022429
5.13	169.02	2.22793	0.005917	6.15	468.72	2.67091	0.0021335
5.14	170.72	2.23227	0.005858	6.20	492.75	2.69263	0.0020294
5.15	172.43	2.23662	0.005799	6.25	518.01	2.71434	0.0019305
5.16	174.16	2.24096	0.005742	6.30	544.57	2.73606	0.0018363
5.17	175.91	2.24530	0.005685	6.35	572.49	2.75777	0.0017467
5.18	177.68	2.24965	0.005628	6.40	601.85	2.77948	0.0016616
5.19	179.47	2.25399	0.005572	6.45	632.70	2.80120	0.0015805
5.20	181.27	2.25833	0.005517	6.50	665.14	2.82291	0.0015034
5.21	183.09	2.26267	0.005462	6.55	699.24	2.84463	0.0014301
5.22	184.93	2.26702	0.005407	6.60	735.10	2.86634	0.0013604
5.23	186.79	2.27136	0.005354	6.65	772.78	2.88806	0.0012940

continued

x	e^x	$\log_{10}e^x$	e^{-x}	x	e^x	$\log_{10}e^x$	e^{-x}
5.24	188.67	2.27570	0.005300	6.70	812.41	2.90977	0.0012309
5.25	190.57	2.28005	0.005248	6.75	854.06	2.93149	0.0011709
5.26	192.48	2.28439	0.005195	6.80	897.85	2.95320	0.0011138
5.27	194.42	2.28873	0.005144	6.85	943.88	2.97492	0.0010595
5.28	196.37	2.29307	0.005092	6.90	992.27	2.99663	0.0010078
5.29	198.34	2.29742	0.005042	6.95	1043.1	3.01835	0.00095860
5.30	200.34	2.30176	0.004992	7.00	1096.6	3.04006	0.0009119
5.31	202.35	2.30610	0.004942	7.05	1152.9	3.06178	0.0008674
5.32	204.38	2.31045	0.004893	7.10	1212.0	3.08349	0.0008251
5.33	206.44	2.31479	0.004844	7.15	1274.1	3.10521	0.0007849
5.34	208.51	2.31913	0.004796	7.20	1339.4	3.12692	0.0007466
5.35	210.61	2.32348	0.004748	7.25	1408.1	3.14863	0.007102
5.36	212.72	2.32782	0.004701	7.30	1480.3	3.17035	0.0006755
5.37	214.86	2.33216	0.004654	7.35	1556.2	3.19206	0.0006426
5.38	217.02	2.33650	0.004608	7.40	1636.0	3.21378	0.0006113
5.39	219.20	2.34085	0.004562	7.45	1719.9	3.23549	0.0005814
5.40	221.41	2.34519	0.004517	7.50	1808.0	3.25721	0.0005531
5.41	223.63	2.34953	0.004472	7.55	1900.7	3.27892	0.0005261
5.42	225.88	2.35388	0.004427	7.60	1998.2	3.30064	0.0005005
5.43	228.15	2.35822	0.004383	7.65	2100.6	3.32235	0.0004760
5.44	230.44	2.36256	0.004339	7.70	2208.3	3.34407	0.0004528
5.45	232.76	2.36690	0.004296	7.75	2321.6	3.36578	0.0004307
5.46	235.10	2.37125	0.004254	7.80	2440.6	3.38750	0.0004097
5.47	237.46	2.37559	0.004211	7.85	2565.7	3.40921	0.0003898
5.48	239.85	2.37993	0.004169	7.90	2697.3	3.43093	0.0003707
5.49	242.26	2.38428	0.004128	7.95	2835.6	3.45264	0.003527
5.50	244.69	2.38862	0.004087	8.00	2981.0	3.47436	0.0003355
8.00	2981.0	3.47436	0.0003355	9.00	8103.1	3.90865	0.0001234
8.05	3133.8	3.49607	0.0003191	9.05	8518.5	3.93037	0.0001174
8.10	3294.5	3.51779	0.0003035	9.10	8955.3	3.95208	0.0001117
8.15	3463.4	3.53950	0.0002887	9.15	9414.4	3.97379	0.0001062
8.20	3641.0	3.56121	0.0002747	9.20	9897.1	3.99551	0.0001010
8.25	3827.6	3.58293	0.0002613	9.25	10405	4.01722	0.0000961
8.30	4023.9	3.60464	0.0002485	9.30	10938	4.03894	0.0000914
8.35	4230.2	3.62636	0.0002364	9.35	11499	4.06065	0.0000870
8.40	4447.1	3.64807	0.0002249	9.40	12088	4.08237	0.0000827

x	e^x	$\log_{10}e^x$	e^{-x}	x	e^x	$\log_{10}e^x$	e^{-x}
8.45	4675.1	3.66979	0.0002139	9.45	12708	4.10408	0.0000787
8.50	4914.8	3.69150	0.0002035	9.50	13360	4.12580	0.0000749
8.55	5166.8	3.71322	0.0001935	9.55	14045	4.14751	0.0000712
8.60	5431.7	3.73493	0.0001841	9.60	14765	4.16923	0.0000677
8.65	5710.1	3.75665	0.0001751	9.65	15522	4.19094	0.0000644
8.70	6002.9	3.77836	0.0001666	9.70	16318	4.21266	0.0000613
8.75	6310.7	3.80008	0.0001585	9.75	17154	4.23437	0.0000583
8.80	6634.2	3.82179	0.0001507	9.80	18034	4.25609	0.0000555
8.85	6974.4	3.84351	0.0001434	9.85	18958	4.27780	0.0000527
8.90	7332.0	3.86522	0.0001364	9.90	19930	4.29952	0.0000502
8.95	7707.9	3.88694	0.0001297	9.95	20952	4.32123	0.0000477
9.00	8103.1	3.90865	0.0001234	10.00	22026	4.34294	0.0000454

Table 8
Normal Distribution

This table gives values of :

(a) $f(x)$ = the probability density of a standardized random variable

$$= \frac{1}{\sqrt{2\pi}} e^{-\frac{1}{2}x^2}.$$

For negative values of x, one uses the fact that $f(-x) = f(x)$.

(b) $F(x)$ = the cumulative distribution function of a standardized normal random variable

$$= \int_{-\infty}^{x} \frac{1}{\sqrt{2\pi}} e^{-\frac{1}{2}t^2} dt.$$

For negative values of x, one uses the relationship $F(-x) = 1 - F(x)$. Values of x corresponding to a few special values of $F(x)$ are given in a separate table following the main table.

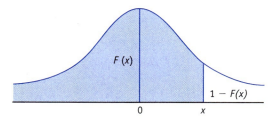

(c) $f'(x)$ = the first derivative of $f(x)$ with respect to x

$$= \frac{x}{\sqrt{2\pi}} e^{-\frac{1}{2}x^2} = -x f(x)$$

(d) $f''(x)$ = the second derivative of $f(x)$ with respect to x

$$= \frac{(x^2-1)}{\sqrt{2\pi}} e^{-\frac{1}{2}x^2} = (x^2-1) f(x)$$

(e) $f'''(x)$ = the third derivative of $f(x)$ with respect to x

$$= \frac{3x - x^3}{\sqrt{2\pi}} e^{-\frac{1}{2}x^2} = (3x - x^3) f(x)$$

(f) $f^{iv}(x)$ = the fourth derivative of $f(x)$ with respect to x

$$= \frac{x^4 - 6x^2 + 3}{\sqrt{2\pi}} e^{-\frac{1}{2}x^2} = (x^4 - 6x^2 + 3) f(x).$$

It should be noted that other probability integrals can be evaluated by the use of these tables. For example

$$\int_0^x f(t)\,dt = \frac{1}{2}\mathrm{erf}\left(\frac{x}{\sqrt{2}}\right),$$

where $\mathrm{erf}\left(\dfrac{x}{\sqrt{2}}\right)$ represents the error function associated with the normal curve.

To evaluate $\mathrm{erf}(2.3)$ one proceeds as follows: Since $\dfrac{x}{\sqrt{2}} = 2.3$, one finds $x = (2.3)(\sqrt{2}) = 3.25$. In the entry opposite $x = 3.25$, the value 0.9994 is given. Subtracting 0.5000 from the tabular value, one finds the value 0.4994. Thus $\mathrm{erf}(2.3) = 2(0.4994) = 0.9988$.

Source: Beyer, W. H. (ed). *Handbook of Tables for Probability and Statistics.* Cleveland: The Chemical Rubber Company, 1966. Reprinted by permission of the publisher.

x	$F(x)$	$1-F(x)$	$f(x)$	$f'(x)$	$f''(x)$	$f'''(x)$	$f^{iv}(x)$
0.00	0.5000	0.5000	0.3989	−0.0000	−0.3989	0.0000	1.1968
0.01	0.5040	0.4960	0.3989	−0.0040	−0.3989	0.0120	1.1965
0.02	0.5080	0.4920	0.3989	−0.0080	−0.3987	0.0239	1.1956
0.03	0.5120	0.4880	0.3988	−0.0120	−0.3984	0.0359	1.1941
0.04	0.5160	0.4840	0.3986	−0.0159	−0.3980	0.0478	1.1920
0.05	0.5199	0.4801	0.3984	−0.0199	−0.3975	0.0597	1.1894
0.06	0.5239	0.4761	0.3982	−0.0239	−0.3968	0.0716	1.1861
0.07	0.5279	0.4721	0.3980	−0.0279	−0.3960	0.0834	1.1822
0.08	0.5319	0.4681	0.3977	−0.0318	−0.3951	0.0952	1.1778
0.09	0.5359	0.4641	0.3973	−0.0358	−0.3941	0.1070	1.1727
0.10	0.5398	0.4602	0.3970	−0.0397	−0.3930	0.1187	1.1671
0.11	0.5438	0.4562	0.3965	−0.0436	−0.3917	0.1303	1.1609
0.12	0.5478	0.4522	0.3961	−0.0475	−0.3904	0.1419	1.1541
0.13	0.5517	0.4483	0.3956	−0.0514	−0.3889	0.1534	1.1468
0.14	0.5557	0.4443	0.3951	−0.0553	−0.3873	0.1648	1.1389
0.15	0.5596	0.4404	0.3945	−0.0592	−0.3856	0.1762	1.1304
0.16	0.5636	0.4364	0.3939	−0.0630	−0.3838	0.1874	1.1214
0.17	0.5675	0.4325	0.3932	−0.6688	−0.3819	0.1986	1.1118
0.18	0.5714	0.4286	0.3925	−0.0707	−0.3798	0.2097	1.1017
0.19	0.5753	0.4247	0.3918	−0.0744	−0.3777	0.2206	1.0911
0.20	0.5793	0.4207	0.3910	−0.0782	−0.3754	0.2315	1.0799
0.21	0.5832	0.4168	0.3902	−0.0820	−0.3730	0.2422	1.0682

continued

x	$F(x)$	$1-F(x)$	$f(x)$	$f'(x)$	$f''(x)$	$f'''(x)$	$f^{iv}(x)$
0.22	0.5871	0.4129	0.3894	−0.0857	−0.3706	0.2529	1.0560
0.23	0.5910	0.4090	0.3885	−0.0894	−0.3680	0.2634	1.0434
0.24	0.5948	0.4052	0.3876	−0.0930	−0.3653	0.2737	1.0302
0.25	0.5987	0.4013	0.3867	−0.0967	−0.3625	0.2840	1.0165
0.26	0.6026	0.3974	0.3857	−0.1003	−0.3596	0.2941	1.0024
0.27	0.6064	0.3936	0.3847	−0.1039	−0.3566	0.3040	0.9878
0.28	0.6103	0.3897	0.3836	−0.1074	−0.3535	0.3138	0.9727
0.29	0.6141	0.3859	0.3825	−0.1109	−0.3504	0.3235	0.9275
0.30	0.6179	0.3821	0.3814	−0.1144	−0.3471	0.3330	0.9413
0.31	0.6217	0.3783	0.3802	−0.1179	−0.3437	0.3423	0.9250
0.32	0.6255	0.3745	0.3790	−0.1213	−0.3402	0.3515	0.9082
0.33	0.6293	0.3707	0.3778	−0.1247	−0.3367	0.3605	0.8910
0.34	0.6331	0.3669	0.3765	−0.1280	−0.3330	0.3693	0.8735
0.35	0.6368	0.3632	0.3752	−0.1313	−0.3293	0.3779	0.8556
0.36	0.6406	0.3594	0.3739	−0.1346	−0.3255	0.3864	0.8373
0.37	0.6443	0.3557	0.3725	−0.1378	−0.3216	0.3947	0.8186
0.38	0.6480	0.3520	0.3712	−0.1410	−0.3176	0.4028	0.7996
0.39	0.6517	0.3483	0.3697	−0.1442	−0.3135	0.4107	0.7803
0.40	0.6554	0.3446	0.3683	−0.1473	−0.3094	0.4184	0.7607
0.41	0.6591	0.3409	0.3668	−0.1504	−0.3051	0.4259	0.7408
0.42	0.6628	0.3372	0.3653	−0.1534	−0.3008	0.4332	0.7206
0.43	0.6664	0.3336	0.3637	−0.1564	−0.2965	0.4403	0.7001
0.44	0.6700	0.3300	0.3621	−0.1593	−0.2920	0.4472	0.6793
0.45	0.6736	0.3264	0.3605	−0.1622	−0.2875	0.4539	0.6583
0.46	0.6772	0.3228	0.3589	−0.1651	−0.2830	0.4603	0.6371
0.47	0.6808	0.3192	0.3572	−0.1679	−0.2783	0.4666	0.6156
0.48	0.6844	0.3156	0.3555	−0.1707	−0.2736	0.4727	0.5940
0.49	0.6879	0.3121	0.3538	−0.1734	−0.2689	0.4785	0.5721
0.50	0.6915	0.3085	0.3521	−0.1760	−0.2641	0.4841	0.5501
0.51	0.6950	0.3050	0.353	−0.1787	−0.2592	0.4895	0.5279
0.52	0.6985	0.3015	0.3485	−0.1812	−0.2543	0.4947	0.5056
0.53	0.7019	0.2981	0.3467	−0.1837	−0.2493	0.4996	0.4831
0.54	0.7054	0.2946	0.3448	−0.1862	−0.2443	0.5043	0.4605
0.55	0.7088	0.2912	0.3429	−0.1886	−0.2392	0.5088	0.4378
0.56	0.7123	0.2877	0.3410	−0.1920	−0.2341	0.5131	0.4150
0.57	0.7157	0.2843	0.3391	−0.1933	−0.2289	0.5171	0.3921

x	$F(x)$	$1-F(x)$	$f(x)$	$f'(x)$	$f''(x)$	$f'''(x)$	$f^{iv}(x)$
0.58	0.7190	0.2810	0.3372	−0.1956	−0.2238	0.5209	0.3691
0.59	0.7224	0.2776	0.3352	−0.1978	−0.2185	0.5245	0.3461
0.60	0.7257	0.2743	0.3332	−0.1999	−0.2133	0.5278	0.3231
0.61	0.7291	0.2709	0.3312	−0.2020	−0.2080	0.5309	0.3000
0.62	0.7324	0.2676	0.3292	−0.2041	−0.2027	0.5338	0.2770
0.63	0.7357	0.2643	0.3271	−0.2061	−0.1973	0.5364	0.2539
0.64	0.7389	0.2611	0.3251	−0.2080	−0.1919	0.5389	0.2309
0.65	0.7422	0.2578	0.3230	−0.2099	−0.1865	0.5411	0.2078
0.66	0.7454	0.2546	0.3209	−0.2118	−0.1811	0.5431	0.1849
0.67	0.7486	0.2514	0.3187	−0.2136	−0.1757	0.5448	0.1620
0.68	0.7517	0.2483	0.3166	−0.2153	−0.1702	0.5463	0.1391
0.69	0.7549	0.2451	0.3144	−0.2170	−0.1647	0.5476	0.1164
0.70	0.7580	0.2420	0.3123	−0.2186	−0.1593	0.5486	0.0937
0.71	0.7611	0.2389	0.3101	−0.2201	−0.1538	0.5495	0.0712
0.72	0.7642	0.2358	0.3079	−0.2217	−0.1483	0.5501	0.0487
0.73	0.7673	0.2327	0.3056	−0.2231	−0.1428	0.5504	0.0265
0.74	0.7704	0.2296	0.3034	−0.2245	−0.1373	0.5506	0.0043
0.75	0.7734	0.2266	0.3011	−0.2259	−0.1318	0.5505	−0.0176
0.76	0.7764	0.2236	0.2989	−0.2271	−0.1262	0.5502	−0.0394
0.77	0.7794	0.2206	0.2966	−0.2284	−0.1207	0.5497	−0.0611
0.78	0.7823	0.2177	0.2943	−0.2296	−0.1153	0.5490	−0.0825
0.79	0.7852	0.2148	0.2920	−0.2307	−0.1098	0.5481	−0.1037
0.80	0.7881	0.2119	0.2897	−0.2318	−0.1043	0.5469	−0.1247
0.81	0.7910	0.2090	0.2874	−0.2328	−0.0988	0.5456	−0.1455
0.82	0.7939	0.2061	0.2850	−0.2337	−0.0934	0.5440	−0.1660
0.83	0.7967	0.2033	0.2827	−0.2346	−0.0880	0.5423	−0.1862
0.84	0.7995	0.2005	0.2803	−0.2355	−0.0825	0.5403	−0.2063
0.85	0.8023	0.1977	0.2780	−0.2363	−0.0771	0.5381	−0.2260
0.86	0.8051	0.1949	0.2756	−0.2370	−0.0718	0.5358	−0.2455
0.87	0.8078	0.1922	0.2732	−0.2377	−0.0664	0.5332	−0.2646
0.88	0.8106	0.1894	0.2709	−0.2384	−0.0611	0.5305	−0.2835
0.89	0.8133	0.1867	0.2685	−0.2389	−0.0558	0.5276	−0.3021
0.90	0.8159	0.1841	0.2661	−0.2395	−0.0506	0.5245	−0.3203
0.91	0.8186	0.1814	0.2637	−0.2400	−0.0453	0.5212	−0.3383
0.92	0.8212	0.1788	0.2613	−0.2404	−0.0401	0.5177	−0.3559
0.93	0.8238	0.1762	0.2589	−0.2408	−0.0350	0.5140	−0.3731

continued

x	$F(x)$	$1-F(x)$	$f(x)$	$f'(x)$	$f''(x)$	$f'''(x)$	$f^{iv}(x)$
0.94	0.8264	0.1736	0.2565	−0.2411	−0.0299	0.5102	−0.3901
0.95	0.8289	0.1711	0.2541	−0.2414	−0.0248	0.5062	−0.4066
0.96	0.8315	0.1685	0.2516	−0.2416	−0.0197	0.5021	−0.4228
0.97	0.8340	0.1660	0.2492	−0.2417	−0.0147	0.4978	−0.4387
0.98	0.8365	0.1635	0.2468	−0.2419	−0.0098	0.4933	−0.4541
0.99	0.8389	0.1611	0.2444	−0.2420	−0.0049	0.4887	−0.4692
1.00	0.8413	0.1587	0.2420	−0.2420	0.0000	0.4839	−0.4839
1.01	0.8438	0.1562	0.2396	−0.2420	0.0048	0.4790	−0.4983
1.02	0.8461	0.1539	0.2371	−0.2419	0.0096	0.4740	−0.5122
1.03	0.8485	0.1515	0.2347	−0.2418	0.0143	0.4688	−0.5257
1.04	0.8508	0.1492	0.2323	−0.2416	0.0190	0.4635	−0.5389
1.05	0.8531	0.1469	0.2299	−0.2414	0.0236	0.4580	−0.5516
1.06	0.8554	0.1446	0.2275	−0.2411	0.0281	0.4524	−0.5639
1.07	0.8577	0.1423	0.2251	−0.2408	0.0326	0.4467	−0.5738
1.08	0.8599	0.1401	0.2227	−0.2405	0.0371	0.4409	−0.5873
1.09	0.8621	0.1379	0.2203	−0.2401	0.0414	0.4350	−0.5984
1.10	0.8643	0.1357	0.2179	−0.2396	0.0458	0.4290	−0.6091
1.11	0.8665	0.1335	0.2155	−0.2392	0.0500	0.4228	−0.6193
1.12	0.8686	0.1314	0.2131	−0.2386	0.0542	0.4166	−0.6292
1.13	0.8708	0.1292	0.2107	−0.2381	0.0583	0.4102	−0.6386
1.14	0.8729	0.1271	0.2083	−0.2375	0.0624	0.4038	−0.6476
1.15	0.8749	0.1251	0.2059	−0.2368	0.0664	0.3973	−0.6561
1.16	0.8770	0.1230	0.2036	−0.2361	0.0704	0.3907	−0.6643
1.17	0.8790	0.1210	0.2012	−0.2354	0.0742	0.3840	−0.6720
1.18	0.8810	0.1190	0.1989	−0.2347	0.0780	0.3772	−0.6792
1.19	0.8830	0.1170	0.1965	−0.2339	0.0818	0.3704	−0.6861
1.20	0.8849	0.1151	0.1942	−0.2330	0.0854	0.3635	−0.6926
1.21	0.8869	0.1131	0.1919	−0.2322	0.0890	0.3566	−0.6986
1.22	0.8888	0.1112	0.1895	−0.2312	0.0926	0.3496	−0.7042
1.23	0.8907	0.1093	0.1872	−0.2303	0.0960	0.3425	−0.7094
1.24	0.8925	0.1075	0.1849	−0.2292	0.0994	0.3354	−0.7141
1.25	0.8944	0.1056	0.1826	−0.2283	0.1027	0.3282	−0.7185
1.26	0.8962	0.1038	0.1804	−0.2273	0.1060	0.3210	−0.7224
1.27	0.8980	0.1020	0.1781	−0.2262	0.1092	0.3138	−0.7259
1.28	0.8997	0.1003	0.1758	−0.2251	0.1123	0.3065	−0.7291
1.29	0.9015	0.0985	0.1736	−0.2240	0.1153	0.2992	−0.7318

x	$F(x)$	$1-F(x)$	$f(x)$	$f'(x)$	$f''(x)$	$f'''(x)$	$f^{iv}(x)$
1.30	0.9032	0.0938	0.1714	−0.2228	0.1882	0.2918	−0.7341
1.31	0.9049	0.0951	0.1691	−0.2216	0.1211	0.2845	−0.7361
1.32	0.9066	0.0934	0.1669	−0.2204	0.1239	0.2771	−0.7376
1.33	0.9082	0.0918	0.1647	−0.2191	0.1267	0.2697	−0.7388
1.34	0.9099	0.0901	0.1626	−0.2178	0.1293	0.2624	−0.7395
1.35	0.9115	0.0885	0.1604	−0.2165	0.1319	0.2550	−0.7399
1.36	0.9131	0.0869	0.1582	−0.2152	0.1344	0.2476	−0.7400
1.37	0.9147	0.0853	0.1561	−0.2138	0.1369	0.2402	−0.7396
1.38	0.9162	0.0838	0.1539	−0.2125	0.1392	0.2328	−0.7389
1.39	0.9177	0.0823	0.1518	−0.2110	0.1415	0.2254	−0.7378
1.40	0.9192	0.0803	0.1497	−0.2096	0.1437	0.2180	−0.7364
1.41	0.9207	0.0793	0.1476	−0.2082	0.1459	0.2107	−0.7347
1.42	0.9222	0.0778	0.1456	−0.2067	0.1480	0.2033	−0.7326
1.43	0.9236	0.0764	0.1435	−0.2052	0.1500	0.1960	−0.7301
1.44	0.9251	0.0749	0.1415	−0.2037	0.1519	0.1887	−0.7274
1.45	0.9265	0.0735	0.1394	−0.2022	0.1537	0.1815	−0.7243
1.46	0.9279	0.0721	0.1374	−0.2006	0.1555	0.1742	−0.7209
1.47	0.9292	0.0708	0.1354	−0.1991	0.1572	0.1670	−0.7172
1.48	0.9306	0.0694	0.1334	−0.1975	0.1588	0.1599	−0.7132
1.49	0.9319	0.0681	0.1315	−0.1959	0.1604	0.1528	−0.7089
1.50	0.9332	0.0668	0.1295	−0.1943	0.1619	0.1457	−0.7043
1.51	0.9345	0.0655	0.1276	−0.1927	0.1633	0.1387	−0.6994
1.52	0.9357	0.0643	0.1257	−0.1910	0.1647	0.1317	−0.6942
1.53	0.9370	0.0630	0.1238	−0.1894	0.1660	0.1248	−0.6888
1.54	0.9382	0.0618	0.1219	−0.1877	0.1672	0.1180	−0.6831
1.55	0.9394	0.0606	0.1200	−0.1860	0.1683	0.1111	−0.6772
1.56	0.9406	0.0594	0.1182	−0.1843	0.1694	0.1044	−0.6710
1.57	0.9418	0.0582	0.1163	−0.1826	0.1704	0.0977	−0.6646
1.58	0.9429	0.0571	0.1145	−0.1809	0.1714	0.0911	−0.6580
1.59	0.9441	0.0559	0.1127	−0.1792	0.1722	0.0846	−0.6511
1.60	0.9452	0.0548	0.1109	−0.1775	0.1730	0.0781	−0.6441
1.61	0.9463	0.0537	0.1092	−0.1757	0.1738	0.0717	−0.6368
1.62	0.9474	0.0526	0.1074	−0.1740	0.1745	0.0654	−0.6293
1.63	0.9484	0.0516	0.1057	−0.1723	0.1751	0.0591	−0.6216
1.64	0.9495	0.0505	0.1040	−0.1705	0.1757	0.0529	−0.6138
1.65	0.9505	0.0495	0.1023	−0.1687	0.1762	0.0468	−0.6057

continued

x	$F(x)$	$1-F(x)$	$f(x)$	$f'(x)$	$f''(x)$	$f'''(x)$	$f^{iv}(x)$
1.66	0.9515	0.0485	0.1006	−0.1670	0.1766	0.0408	−0.5975
1.67	0.9525	0.0475	0.0989	−0.1652	0.1770	0.0349	−0.5891
1.68	0.9535	0.0465	0.0973	−0.1634	0.1773	0.0290	−0.5806
1.69	0.9545	0.0455	0.0957	−0.1617	0.1776	0.0233	−0.5720
1.70	0.9554	0.0446	0.0940	−0.1599	0.1778	0.0176	−0.5632
1.71	0.9564	0.0436	0.0925	−0.1581	0.1779	0.0120	−0.5542
1.72	0.9573	0.0427	0.0909	−0.1563	0.1780	0.0065	−0.5452
1.73	0.9582	0.0418	0.0893	−0.1546	0.1780	0.0011	−0.5360
1.74	0.9591	0.0409	0.0878	−0.1528	0.1780	−0.0042	−0.5267
1.75	0.9599	0.0401	0.0863	−0.1510	0.1780	−0.0094	−0.5173
1.76	0.9608	0.0392	0.0848	−0.1492	0.1778	−0.0146	−0.5079
1.77	0.9616	0.0384	0.0833	−0.1474	0.1777	−0.0196	−0.4983
1.78	0.9625	0.0375	0.0818	−0.1457	0.1774	−0.0245	−0.4887
1.79	0.9633	0.0367	0.0804	−0.1439	0.1772	−0.0294	−0.4789
1.80	0.9641	0.0359	0.0790	−0.1421	0.1769	−0.0341	−0.4692
1.81	0.9649	0.0351	0.0775	−0.1403	0.1765	−0.0388	−0.4593
1.82	0.9656	0.0344	0.0761	−0.1386	0.1761	−0.0433	−0.4494
1.83	0.9664	0.0336	0.0748	−0.1368	0.1756	−0.0477	−0.4395
1.84	0.9671	0.0329	0.0734	−0.1351	0.1751	−0.0521	−0.4295
1.85	0.9678	0.0322	0.0721	−0.1333	0.1746	−0.0563	−0.4195
1.86	0.9686	0.0314	0.0707	−0.1316	0.1740	−0.0605	−0.4095
1.87	0.9693	0.0307	0.0694	−0.1298	0.1734	−0.0645	−0.3995
1.88	0.9699	0.0301	0.0681	−0.1281	0.1727	−0.0685	−0.3894
1.89	0.9706	0.0294	0.0669	−0.1264	0.1720	−0.0723	−0.3793
1.90	0.9713	0.0287	0.0656	−0.1247	0.1713	−0.0761	−0.3693
1.91	0.9719	0.0281	0.0344	−0.1230	0.1705	−0.0797	−0.3592
1.92	0.9726	0.0274	0.0632	−0.1213	0.1697	−0.0832	−0.3492
1.93	0.9732	0.0268	0.0620	−0.1196	0.1688	−0.0867	−0.3392
1.94	0.9738	0.0262	0.0608	−0.1179	0.1679	−0.0900	−0.3292
1.95	0.9744	0.0256	0.0596	−0.1162	0.1670	−0.0933	−0.3192
1.96	0.9750	0.0250	0.0584	−0.1145	0.1661	−0.0964	−0.3093
1.97	0.9756	0.0244	0.0573	−0.1129	0.1651	−0.0994	−0.2994
1.98	0.9761	0.0239	0.0562	−0.1112	0.1641	−0.1024	−0.2895
1.99	0.9767	0.0233	0.0551	−0.1096	0.1630	−0.1052	−0.2797
2.00	0.9772	0.0228	0.0540	−0.1080	0.1620	−0.1080	−0.2700
2.01	0.9778	0.0222	0.0529	−0.1064	0.1609	−0.1106	−0.2603
2.02	0.9783	0.0217	0.0519	−0.1048	0.1598	−0.1132	−0.2506

x	$F(x)$	$1-F(x)$	$f(x)$	$f'(x)$	$f''(x)$	$f'''(x)$	$f^{iv}(x)$
2.03	0.9788	0.0212	0.0508	−0.1032	0.1586	−0.1157	−0.2411
2.04	0.9793	0.0207	0.0498	−0.1016	0.1575	−0.1180	−0.2316
2.05	0.9798	0.0202	0.0488	−0.1000	0.1563	−0.1203	−0.2222
2.06	0.9803	0.0197	0.0478	−0.0985	0.1550	−0.1225	−0.2129
2.07	0.9808	0.0192	0.0468	−0.0969	0.1538	−0.1245	−0.2036
2.08	0.9812	0.0188	0.0459	−0.0954	0.1526	−0.1265	−0.1945
2.09	0.9817	0.0183	0.0449	−0.0939	0.1513	−0.1284	−0.1854
2.10	0.9821	0.0179	0.0440	−0.0924	0.1500	−0.1302	−0.1765
2.11	0.9826	0.0174	0.0431	−0.0909	0.1487	−0.1320	−0.1676
2.12	0.9830	0.0170	0.0422	−0.0894	0.1474	−0.1336	−0.1588
2.13	0.9834	0.0166	0.0413	−0.0879	0.1460	−0.1351	−0.1502
2.14	0.9838	0.0162	0.0404	−0.0865	0.1446	−0.1366	−0.1416
2.15	0.9842	0.0158	0.0396	−0.0850	0.1433	−0.1380	−0.1332
2.16	0.9846	0.0154	0.0387	−0.0836	0.1419	−0.1393	−0.1249
2.17	0.9850	0.0150	0.0379	−0.0822	0.1405	−0.1405	−0.1167
2.18	0.9854	0.0146	0.0371	−0.0808	0.1391	−0.1416	−0.1086
2.19	0.9857	0.0143	0.0363	−0.0794	0.1377	−0.1426	−0.1006
2.20	0.9861	0.0139	0.0355	−0.0780	0.1362	−0.1436	−0.0927
2.21	0.9864	0.0136	0.0347	−0.0767	0.1348	−0.1445	−0.0850
2.22	0.9868	0.0132	0.0339	−0.0754	0.1333	−0.1453	−0.0774
2.23	0.9871	0.0129	0.0332	−0.0740	0.1319	−0.1460	−0.0700
2.24	0.9875	0.0125	0.0325	−0.0727	0.1304	−0.1467	−0.0626
2.25	0.9878	0.0122	0.0317	−0.0714	0.1289	−0.1473	−0.0554
2.26	0.9881	0.0119	0.0310	−0.0701	0.1275	−0.1478	−0.0484
2.27	0.9884	0.0116	0.0303	−0.0689	0.1260	−0.1483	−0.0414
2.28	0.9887	0.0113	0.0297	−0.0676	0.1245	−0.1486	−0.0346
2.29	0.9890	0.0110	0.0290	−0.0664	0.1230	−0.1490	−0.0279
2.30	0.9893	0.0107	0.0283	0.0652	0.1215	−0.1492	−0.0214
2.31	0.9896	0.0104	0.0277	0.0639	0.1200	−0.1494	−0.0150
2.32	0.9898	0.0102	0.0270	0.0628	0.1185	−0.1495	−0.0088
2.33	0.9901	0.0099	0.0264	0.0616	0.1170	−0.1496	−0.0027
2.34	0.9904	0.0096	0.0258	0.0604	0.1155	−0.1496	0.0033
2.35	0.9906	0.0094	0.0252	−0.0593	0.1141	−0.1495	0.0092
2.36	0.9909	0.0091	0.0246	−0.0581	0.1126	−0.1494	0.0149
2.37	0.9911	0.0089	0.0241	−0.0570	0.1111	−0.1492	0.0204
2.38	0.9913	0.0087	0.0235	−0.0559	0.1096	−0.1490	0.0258
2.39	0.9916	0.0084	0.0229	−0.0548	0.1081	−0.1487	0.0311

continued

x	$F(x)$	$1-F(x)$	$f(x)$	$f'(x)$	$f''(x)$	$f'''(x)$	$f^{iv}(x)$
2.40	0.9918	0.0082	0.0224	−0.0538	0.1066	−0.1483	0.0362
2.41	0.9920	0.0080	0.0219	−0.0527	0.1051	−0.1480	0.0412
2.42	0.9922	0.0078	0.0213	−0.0516	0.1036	−0.1475	0.0461
2.43	0.9925	0.0075	0.0208	−0.0506	0.1022	−0.1470	0.0508
2.44	0.9927	0.0073	0.0203	−0.0496	0.1007	−0.1465	0.0554
2.45	0.9929	0.0071	0.0198	−0.0486	0.0992	−0.1459	0.0598
2.46	0.9931	0.0069	0.0194	−0.0476	0.0978	−0.1453	0.0641
2.47	0.9932	0.0068	0.0189	−0.0467	0.0963	−0.1446	0.0683
2.48	0.9934	0.0066	0.0184	−0.0457	0.0949	−0.1439	0.0723
2.49	0.9936	0.0064	0.0180	−0.0448	0.0935	−0.1432	0.0762
2.50	0.9938	0.0062	0.0175	−0.0438	0.0920	−0.1424	0.0800
2.51	0.9940	0.0060	0.0171	−0.0429	0.0906	−0.1416	0.0836
2.52	0.9941	0.0059	0.0167	−0.0420	0.0892	−0.1408	0.0871
2.53	0.9943	0.0057	0.0163	−0.0411	0.0878	−0.1399	0.0905
2.54	0.9945	0.0055	0.0158	−0.0403	0.0864	−0.1389	0.0937
2.55	0.9946	0.0054	0.0155	−0.0394	0.0850	−0.1380	0.0968
2.56	0.9948	0.0052	0.0151	−0.0386	0.0836	−0.1370	0.0998
2.57	0.9949	0.0051	0.0147	−0.0377	0.0823	−0.1360	0.1027
2.58	0.9951	0.0049	0.0143	−0.0369	0.0809	−0.1350	0.1054
2.59	0.9952	0.0048	0.0139	−0.0361	0.0796	−0.1339	0.1080
2.60	0.9953	0.0047	0.0136	−0.0353	0.0782	−0.1328	0.1105
2.61	0.9955	0.0045	0.0132	−0.0345	0.0769	−0.1317	0.1129
2.62	0.9956	0.0044	0.0129	−0.0338	0.0756	−0.1305	0.1152
2.63	0.9957	0.0043	0.0126	−0.0330	0.0743	−0.1294	0.1173
2.64	0.9959	0.0041	0.0122	−0.0323	0.0730	−0.1282	0.1194
2.65	0.9960	0.0040	0.0119	−0.0316	0.0717	−0.1270	0.1213
2.66	0.9961	0.0039	0.016	−0.0309	0.0705	−0.1258	0.1231
2.67	0.9962	0.0038	0.0113	−0.0302	0.0592	−0.1245	0.1248
2.68	0.9963	0.0037	0.0110	−0.0295	0.0680	−0.1233	0.1264
2.69	0.9964	0.0036	0.0107	−0.0288	0.0668	−0.1220	0.1279
2.70	0.9965	0.0035	0.0104	−0.0281	0.0656	−0.1207	0.1293
2.71	0.9966	0.0034	0.0101	−0.0275	0.0644	−0.1194	0.1306
2.72	0.9967	0.0033	0.0099	−0.0269	0.0632	−0.1181	0.1317
2.73	0.9968	0.0032	0.0096	−0.0262	0.0620	−0.1168	0.1328
2.74	0.9969	0.0031	0.0093	−0.0256	0.0608	−0.1154	0.1338
2.75	0.9970	0.0030	0.0091	−0.0250	0.0597	−0.1141	0.1347

x	$F(x)$	$1-F(x)$	$f(x)$	$f'(x)$	$f''(x)$	$f'''(x)$	$f^{iv}(x)$
2.76	0.9971	0.0029	0.0088	−0.0244	0.0585	−0.1127	0.1356
2.77	0.9972	0.0028	0.0086	−0.0238	0.0574	−0.1114	0.1363
2.78	0.9973	0.0027	0.0084	−0.0233	0.0563	−0.1100	0.1369
2.79	0.9974	0.0026	0.0081	−0.0227	0.0552	−0.1087	0.1375
2.80	0.9974	0.0026	0.0079	−0.0222	0.0541	−0.1073	0.1379
2.81	0.9975	0.0025	0.0077	−0.0216	0.0531	−0.1059	0.1383
2.82	0.9976	0.0024	0.0075	−0.0211	0.0520	−0.1045	0.1386
2.83	0.9977	0.0023	0.0073	−0.0208	0.0510	−0.1031	0.1389
2.84	0.9977	0.0023	0.0071	−0.0201	0.0500	−0.1017	0.1390
2.85	0.9978	0.0022	0.0069	−0.0196	0.0490	−0.1003	0.1391
2.86	0.9979	0.0021	0.0067	−0.0191	0.0480	−0.0990	0.1391
2.87	0.9979	0.0021	0.0065	−0.0186	0.0470	−0.0976	0.1391
2.88	0.9980	0.0020	0.0063	−0.0182	0.0460	−0.0962	0.1389
2.89	0.9981	0.0019	0.0061	−0.0177	0.0451	−0.0948	0.1388
2.90	0.9981	0.0019	0.0060	−0.0173	0.0441	−0.0934	0.1385
2.91	0.9982	0.0018	0.0058	−0.0168	0.0432	−0.0920	0.1382
2.92	0.9982	0.0018	0.0056	−0.0164	0.0423	−0.0906	0.1378
2.93	0.9983	0.0017	0.0055	−0.0160	0.0414	−0.0893	0.1374
2.94	0.9984	0.0016	0.0053	−0.0156	0.0405	−0.0879	0.1369
2.95	0.9984	0.0016	0.0051	−0.0152	0.0396	−0.0865	0.1364
2.96	0.9985	0.0015	0.0050	−0.0148	0.0388	−0.0852	0.1358
2.97	0.9985	0.0015	0.0048	−0.0144	0.0379	−0.0838	0.1352
2.98	0.9986	0.0014	0.0047	−0.0140	0.0371	−0.0825	0.1345
2.99	0.9986	0.0014	0.0046	−0.0137	0.0363	−0.0811	0.1337
3.00	0.9987	0.0013	0.0044	−0.0133	0.0355	−0.0798	0.1330
3.01	0.9987	0.0013	0.0043	−0.0130	0.0347	−0.0785	0.1321
3.02	0.9987	0.0013	0.0042	−0.0126	0.0339	−0.0771	0.1313
3.03	0.9988	0.0012	0.0040	0.0123	0.0331	−0.0758	0.1304
3.04	0.9988	0.0012	0.0039	0.0119	0.0324	−0.0754	0.1294
3.05	0.9989	0.0011	0.0038	−0.0116	0.0316	−0.0732	0.1285
3.06	0.9989	0.0011	0.0037	−0.0113	0.0309	−0.0720	0.1275
3.07	0.9989	0.0011	0.0036	−0.0110	0.0302	−0.0707	0.1264
3.08	0.9990	0.0010	0.0035	−0.0107	0.0295	−0.0694	0.1254
3.09	0.9990	0.0010	0.0034	−0.0104	0.0288	−0.0682	0.1243
3.10	0.9990	0.0010	0.0033	−0.0101	0.0281	−0.0669	0.1231
3.11	0.9991	0.0009	0.0032	−0.0099	0.0275	−0.0657	0.1220

continued

x	$F(x)$	$1-F(x)$	$f(x)$	$f'(x)$	$f''(x)$	$f'''(x)$	$f^{iv}(x)$
3.12	0.9991	0.0009	0.0031	−0.0096	0.0268	−0.0645	0.1208
3.13	0.9991	0.0009	0.0030	−0.0093	0.0262	−0.0633	0.1196
3.14	0.9992	0.0008	0.0029	−0.0091	0.0256	−0.0621	0.1184
3.15	0.9992	0.0008	0.0028	−0.0088	0.0249	−0.0609	0.1171
3.16	0.9992	0.0008	0.0027	−0.0086	0.0243	−0.0598	0.1159
3.17	0.9992	0.0008	0.0026	−0.0083	0.0237	−0.0586	0.1146
3.18	0.9993	0.0007	0.0025	−0.0081	0.0232	−0.0575	0.1133
3.19	0.9993	0.0007	0.0025	−0.0079	0.0226	−0.0564	0.1120
3.20	0.9993	0.0007	0.0024	−0.0076	0.0220	−0.0552	0.1107
3.21	0.9993	0.0007	0.0023	−0.0074	0.0215	−0.0541	0.1093
3.22	0.9994	0.0006	0.0022	−0.0072	0.0210	−0.0531	0.1080
3.23	0.9994	0.0006	0.0022	−0.0070	0.0204	−0.0520	0.1066
3.24	0.9994	0.0006	0.0021	−0.0068	0.0199	−0.0509	0.1053
3.25	0.9994	0.0006	0.0020	−0.0066	0.0194	−0.0499	0.1039
3.26	0.9994	0.0006	0.0020	−0.0064	0.0189	−0.0488	0.1025
3.27	0.9995	0.0005	0.0019	−0.0062	0.0184	−0.0478	0.1011
3.28	0.9995	0.0005	0.0018	−0.0060	0.0180	−0.0468	0.0997
3.29	0.9995	0.0005	0.0018	−0.0059	0.0175	−0.0458	0.0983
3.30	0.9995	0.0005	0.0017	−0.0057	0.0170	−0.0449	0.0969
3.31	0.9995	0.0005	0.0017	−0.0055	0.0166	−0.0439	0.0955
3.32	0.9995	0.0005	0.0016	−0.0054	0.0162	−0.0429	0.0941
3.33	0.9996	0.0004	0.0016	−0.0052	0.0157	−0.0420	0.0927
3.34	0.9996	0.0004	0.0015	−0.0050	0.0153	−0.0411	0.0913
3.35	0.9996	0.0004	0.0015	−0.0049	0.0149	−0.0402	0.0899
3.36	0.9996	0.0004	0.0014	−0.0047	0.0145	−0.0393	0.0885
3.37	0.9996	0.0004	0.0014	−0.0046	0.0141	−0.0384	0.0871
3.38	0.9996	0.0004	0.0013	−0.0045	0.0138	−0.0376	0.0857
3.39	0.9997	0.0003	0.0013	−0.0043	0.0134	−0.0367	0.0843
3.40	0.9997	0.0003	0.0012	−0.0042	0.0130	−0.0359	0.0829
3.41	0.9997	0.0003	0.0012	−0.0041	0.0127	−0.0350	0.0815
3.42	0.9997	0.0003	0.0012	−0.00039	0.0123	−0.0342	0.0801
3.43	0.9997	0.0003	0.0011	−0.0038	0.0120	−0.0334	0.0788
3.44	0.9997	0.0003	0.0011	−0.0037	0.0116	−0.0327	0.0774
3.45	0.9997	0.0003	0.0010	−0.0036	0.0113	−0.0319	0.0761
3.46	0.9997	0.0003	0.0010	−0.0035	0.0110	−0.0311	0.0747
3.47	0.9997	0.0003	0.0010	−0.0034	0.0107	−0.0304	0.0734
3.48	0.9997	0.0003	0.0009	−0.0033	0.0104	−0.0297	0.0721

x	$F(x)$	$1-F(x)$	$f(x)$	$f'(x)$	$f''(x)$	$f'''(x)$	$f^{iv}(x)$
3.49	0.9998	0.0002	0.0009	−0.0032	0.0101	−0.0290	0.0707
3.50	0.9998	0.0002	0.0009	−0.0031	0.0098	−0.0283	0.0694
3.51	0.9998	0.0002	0.0008	−0.0030	0.0095	−0.0276	0.0681
3.52	0.9998	0.0002	0.0008	−0.0029	0.0093	−0.0269	0.0669
3.53	0.9998	0.0002	0.0008	−0.0028	0.0090	−0.0262	0.0656
3.54	0.9998	0.0002	0.0008	−0.0027	0.0087	−0.0256	0.0643
3.55	0.9998	0.0002	0.0007	−0.0026	0.0085	−0.0249	0.0631
3.56	0.9998	0.0002	0.0007	−0.0025	0.0082	−0.0243	0.0618
3.57	0.9998	0.0002	0.0007	−0.0024	0.0080	−0.0237	0.0606
3.58	0.9998	0.0002	0.0007	−0.0024	0.0078	−0.0231	0.0594
3.59	0.9998	0.0002	0.0006	−0.0023	0.0075	−0.0225	0.0582
3.60	0.9998	0.0002	0.0006	−0.0022	0.0073	−0.0219	0.0570
3.61	0.9998	0.0002	0.0006	−0.0021	0.0071	−0.0214	0.0559
3.62	0.9999	0.0001	0.0006	−0.0021	0.0069	−0.0208	0.0547
3.63	0.9999	0.0001	0.0005	−0.0020	0.0067	−0.0203	0.0536
3.64	0.9999	0.0001	0.0005	−0.0019	0.0065	−0.0198	0.0524
3.65	0.9999	0.0001	0.0005	−0.0019	0.0063	−0.0192	0.0513
3.66	0.9999	0.0001	0.0005	−0.0018	0.0061	−0.0187	0.0502
3.67	0.9999	0.0001	0.0005	−0.0017	0.0059	−0.0182	0.0492
3.68	0.9999	0.0001	0.0005	−0.0017	0.0057	−0.0177	0.0481
3.69	0.9999	0.0001	0.0004	−0.0016	0.0056	−0.0173	0.0407
3.70	0.9999	0.0001	0.0004	−0.0016	0.0054	−0.0168	0.0460
3.71	0.9999	0.0001	0.0004	−0.0015	0.0052	−0.0164	0.0450
3.72	0.9999	0.0001	0.0004	−0.0015	0.0051	−0.0159	0.0440
3.73	0.9999	0.0001	0.0004	−0.0014	0.0049	−0.0155	0.0430
3.74	0.9999	0.0001	0.0004	−0.0014	0.0048	−0.0150	0.0420
3.75	0.9999	0.0001	0.0004	−0.0013	0.0046	−0.0146	0.0410
3.76	0.9999	0.0001	0.0003	−0.0013	0.0045	−0.0142	0.0401
3.77	0.9999	0.0001	0.0003	−0.0012	0.0043	−0.0138	0.0392
3.78	0.9999	0.0001	0.0003	−0.0012	0.0042	−0.0134	0.0382
3.79	0.9999	0.0001	0.0003	−0.0012	0.0041	−0.0131	0.0373
3.80	0.9999	0.0001	0.0003	−0.0011	0.0039	−0.0127	0.0365
3.81	0.9999	0.0001	0.0003	−0.0011	0.0038	−0.0123	0.0356
3.82	0.9999	0.0001	0.0003	−0.0010	0.0037	−0.0120	0.0347
3.83	0.9999	0.0001	0.0003	−0.0010	0.0036	−0.0116	0.0339
3.84	0.9999	0.0001	0.0003	−0.0010	0.0034	−0.0113	0.0331

continued

x	$F(x)$	$1-F(x)$	$f(x)$	$f'(x)$	$f''(x)$	$f'''(x)$	$f^{iv}(x)$
3.85	0.9999	0.0001	0.0002	−0.0009	0.0033	−0.0110	0.0323
3.86	0.9999	0.0001	0.0002	−0.0009	0.0032	−0.0107	0.0315
3.87	0.9999	0.0001	0.0002	−0.0009	0.0031	−0.0104	0.0307
3.88	0.9999	0.0001	0.0002	−0.0008	0.0030	−0.0100	0.0299
3.89	1.0000	0.0000	0.0002	−0.0008	0.0029	−0.0098	0.0292
3.90	1.0000	0.0000	0.0002	−0.0008	0.0028	−0.0095	0.0284
3.91	1.0000	0.0000	0.0002	−0.0008	0.0027	−0.0092	0.0277
3.92	1.0000	0.0000	0.0002	−0.0007	0.0026	−0.0089	0.0270
3.93	1.0000	0.0000	0.0002	−0.0007	0.0026	−0.0086	0.0263
3.94	1.0000	0.0000	0.0002	−0.0007	0.0025	−0.0084	0.0256
3.95	1.0000	0.0000	0.0002	−0.0006	0.0024	−0.0081	0.0250
3.96	1.0000	0.0000	0.0002	−0.0006	0.0023	−0.0079	0.0243
3.97	1.0000	0.0000	0.0002	−0.0006	0.0022	−0.0076	0.0237
3.98	1.0000	0.0000	0.0001	−0.0006	0.0022	−0.0074	0.0230
3.99	1.0000	0.0000	0.0001	−0.0006	0.0021	−0.0072	0.0224
4.00	1.0000	0.0000	0.0001	−0.0005	0.0020	−0.0070	0.0218
x		1.282	1.645	1.960	2.326	2.576	3.090
$F(x)$		0.90	0.95	0.975	0.99	0.995	0.999
$2[1-F(x)]$		0.20	0.10	0.05	0.02	0.01	0.002

Answers to Selected Exercises and Problems

Chapter 1

Theoretical Exercises

1.20 0.5

1.22 0.25

Applied Problems

1.1 **a.** $k=2: \frac{1}{36}$, $k=8: \frac{5}{36}$,

$k=3: \frac{2}{36}$, $k=9: \frac{4}{36}$,

$k=4: \frac{3}{36}$, $k=10: \frac{3}{36}$,

$k=5: \frac{4}{36}$, $k=11: \frac{2}{36}$,

$k=6: \frac{5}{36}$, $k=12: \frac{1}{36}$,

$k=7: \frac{6}{36}$,

b. $k=2: \frac{1}{36}$, $k=8: \frac{26}{36}$,

$k=3: \frac{3}{36}$, $k=9: \frac{30}{36}$,

$k=4: \frac{6}{36}$, $k=10: \frac{33}{36}$,

$k=5: \frac{10}{36}$, $k=11: \frac{35}{36}$,

$k=6: \frac{15}{36}$, $k=12: \frac{36}{36}$,

$k=7: \frac{21}{36}$,

c. $k=2: \frac{36}{36}$, $k=5: \frac{30}{36}$, $k=8: \frac{15}{36}$, $k=11: \frac{3}{36}$,

$k=3: \frac{35}{36}$, $k=6: \frac{26}{36}$, $k=9: \frac{10}{36}$, $k=12: \frac{1}{36}$,

$k=4: \frac{33}{36}$, $k=7: \frac{21}{36}$, $k=10: \frac{6}{36}$,

1.3 **a.** $k=0: \frac{12}{40}$, **b.** $k=0: 1$, **c.** $k=0: \frac{12}{40}$,

$k=1: \frac{18}{40}$, $k=1: \frac{28}{40}$, $k=1: \frac{30}{40}$,

$k=2: \frac{9}{40}$, $k=2: \frac{10}{40}$, $k=2: \frac{39}{40}$,

$k=3: \frac{1}{40}$ $k=3: \frac{1}{40}$ $k=3: 1$

1.5 **a.** $\frac{25}{108}$ **b.** $\frac{25}{81}$ **c.** 0 **d.** $\frac{25}{189}$

e. $\frac{7}{108}$ **f.** $\frac{7}{27}$ **g.** $\frac{3}{20}$ **h.** $\frac{7}{90}$ **i.** $\frac{1}{6}$

1.7 **a.** $\frac{2}{5}$ **b.** $\frac{2}{5}$ **c.** $\frac{1}{5}$

1.9 $k=1: 0$, $k=5: \frac{10}{126}$, $k=9: \frac{18}{126}$,

$k=2: \frac{1}{126}$, $k=6: \frac{15}{126}$, $k=10: \frac{15}{126}$,

$k=3: \frac{3}{126}$, $k=7: \frac{21}{126}$, $k=11: \frac{11}{126}$,

$k=4: \frac{6}{126}$, $k=8: \frac{20}{126}$, $k=12: \frac{6}{126}$

1.11 3

1.13 **a.** 0.0154 **b.** 0.2315 **c.** 0.1543

1.15 $\frac{244}{495}$

1.17 the former

1.19 a. $\dfrac{1}{2^{n-1}}$ **b.** $\dfrac{31}{32}$ **c.** $\dfrac{2}{3}$

1.21 $\dfrac{w_2(w_1+b_1)+w_1}{(w_1+b_1)(w_2+b_2+1)}$

1.23 $n2^{1-n}$

Chapter 2
Theoretical Exercises

2.1 a. $f(x)=\begin{cases}\dfrac{\binom{13}{x}\binom{39}{7-x}}{\binom{52}{7}}, & x=0,1,2,3,4,5,6,7,\\ 0, & \text{elsewhere}\end{cases}$

b. $F(x)=\begin{cases}1, & x>7,\\ \sum_{k=0}^{[x]}\dfrac{\binom{13}{k}\binom{39}{7-k}}{\binom{52}{7}}, & 1\le x\le 7,\\ 0, & x<0\end{cases}$

2.3 a. $f(x)=\begin{cases}\dfrac{1}{2^x}, & x=1,2,3,\ldots,\\ 0, & \text{elsewhere}\end{cases}$

b. $F(x)=\begin{cases}\sum_{k=1}^{[x]}\dfrac{1}{2^k}, & x\ge 1,\\ 0, & x<1\end{cases}$

2.5 $f(x)=\begin{cases}\dfrac{1}{6}\left(\dfrac{5}{6}\right)^{x-1}, & x=1,2,3,\ldots,\\ 0, & \text{elsewhere}\end{cases}$

$F(x)=\begin{cases}\sum_{k=1}^{[x]}\dfrac{1}{6}\left(\dfrac{5}{6}\right)^{k-1}, & x\ge 1,\\ 0, & x<1\end{cases}$

2.7 $k=\dfrac{60}{77}$

2.9 a. $k=\dfrac{3}{\pi^2}$ **b.** $F(x)=\begin{cases}\dfrac{6}{\pi^2}\sum_{i=1}^{[x]}\dfrac{1}{i^2}, & x\ge 1,\\ 0, & x<1\end{cases}$

Applied Problems

2.1 a. $\dfrac{\binom{13}{5}\binom{39}{2}}{\binom{52}{7}}$ **b.** $\dfrac{\binom{13}{5}\binom{39}{2}}{\binom{52}{7}}-\dfrac{\binom{13}{2}\binom{39}{5}}{\binom{52}{7}}$

2.3 a. $\dfrac{1}{16}$ **b.** $\dfrac{7}{8}$

2.5 a. $\dfrac{22}{77}$ **b.** $\dfrac{22}{77}$

2.7 0.5008

2.9 a. binomial: $\binom{7}{4}\left(\dfrac{1}{6}\right)^4\left(\dfrac{5}{6}\right)^3$

b. geometric: $\left(\dfrac{1}{6}\right)\left(\dfrac{5}{6}\right)^6$

c. negative binomial: $\binom{6}{3}\left(\dfrac{1}{6}\right)^4\left(\dfrac{5}{6}\right)^3$

2.11 a. $\sum_{k=13}^{16}\binom{20}{k}(0.2)^{17}(0.8)^3-\binom{20}{12}(0.2)^{12}(0.8)^8$

b. $\binom{20}{17}(0.2)^{17}(0.8)^3-\binom{20}{12}(0.2)^{12}(0.8)^8$

2.13 a. $\dfrac{\binom{4}{2}\binom{48}{5}}{\binom{52}{7}}$ **b.** $1-\sum_{x=0}^{2}\dfrac{\binom{4}{x}\binom{48}{7-x}}{\binom{52}{7}}$

c. $\sum_{x=0}^{1}\dfrac{\binom{4}{x}\binom{48}{7-x}}{\binom{52}{7}}$

2.15 5

2.17 a. 0.375 **b.** 0.8050

2.19 a. 0.125 **b.** 0.1301 **c.** 0.00005

2.21 a. binomial: 0.455, Poisson: 0.451
b. binomial: 0.9994, Poisson: 0.9999

2.23 Probability she breaks the record: 0.459 (binomial) 0.4512 (Poisson)

2.25 **a.** $\dfrac{\binom{4}{0}\binom{48}{5}}{\binom{52}{5}}$ **b.** $1-\sum_{x=0}^{2}\dfrac{\binom{4}{x}\binom{48}{5-x}}{\binom{52}{5}}$

c. $\sum_{x=0}^{1}\dfrac{\binom{4}{x}\binom{48}{5-x}}{\binom{52}{5}}$

2.27 $\dfrac{11}{42}$

2.29 0.0648

2.31 0.109

2.33 Probability of qualifying: 0.3482, 0.8
Probability of winning: 0.5, 0.9

2.35 0.00099

2.37 **a.** 0.0904 **b.** 0.0054

Chapter 3

Theoretical Exercises

3.1 **a.** $k = \dfrac{1}{2}$

b. $F(x) = \begin{cases} 1, & x \geq \pi, \\ \dfrac{1}{2}(1-\cos x), & 0 < x < \pi, \\ 0, & x \leq 0 \end{cases}$

3.3 **a.** $F(x) = \begin{cases} 1, & x > 1, \\ \dfrac{x^3}{2}+\dfrac{1}{2}, & -1 \leq x \leq 1, \\ 0, & x < -1 \end{cases}$

3.5 **a.** $f(x) = \begin{cases} 1, & 0 < x \leq \dfrac{1}{2}, \\ 2, & \dfrac{3}{4} < x \leq 1, \\ 0, & \text{elsewhere} \end{cases}$

3.9 **a.** $c=3$ **b.** $c=1$ **c.** $c=3$ **d.** $c=\dfrac{1}{2}$

e. $c=\dfrac{1}{\theta}$ **f.** $c=\dfrac{1}{\pi}$ **g.** $c=\dfrac{6}{5}$ **h.** $c=\dfrac{1}{2}$

i. $c=1$ **k.** $c=4$ **l.** $c=\dfrac{1}{6}$ **m.** $c=\dfrac{21}{440}$

n. $c=\dfrac{1-e^{-1}}{1-e}$ **o.** $c=\dfrac{1}{\sqrt{2\pi}}$ **p.** $c=2$

q. $c=1$ **r.** $c=1$ **s.** $c=\dfrac{1}{6}$

Applied Problems

3.1 **a.** $\dfrac{1}{2}$ **b.** 0 **c.** $\dfrac{7}{8}$ **d.** 0 **e.** $\dfrac{7}{8}$

f. $\dfrac{1}{8}$ **g.** $\dfrac{3}{8}$ **h.** $\dfrac{3}{8}$ **i.** 1 **j.** $\dfrac{1}{8}$

3.3 $\left(\dfrac{20}{27}\right)^2$

3.5 **a.** $\dfrac{7}{8}$ **b.** $\dfrac{7}{8}$ **c.** $\dfrac{3}{4}$

3.7 **a.** $\dfrac{2}{59}$ **b.** $\dfrac{91}{118}$

3.9 **a.** $\dfrac{\pi^2+6(3-\sqrt{3})\pi-18}{18(\pi-2)}$

b. $\dfrac{-\sqrt{3}\pi^2+12\pi-72(2-\sqrt{3})}{72(\pi-2)}$

3.11 $\dfrac{3}{2}e^{-\frac{1}{2}}-11e^{-10}$, $\dfrac{17}{e^{16}}$

3.13 0.40

3.15 $250,000

3.17 655

3.19 87.8

3.21 **a.** 0.64 **b.** 0.471196

3.23 0.595

3.25 0.45

3.27 0.274

3.29 0.0276

3.31 0.03213

3.33 0.554, 1, 1.80

3.35 6.7%

3.37 0.77

Chapter 4

Theoretical Exercises

4.1 a. $f(y) = \begin{cases} \frac{1}{12}, & y = -1, \frac{1}{2}, \\ \frac{1}{6}, & y = -\frac{3}{2}, -\frac{1}{2}, \\ \frac{1}{4}, & y = 0, 1, \\ 0, & \text{elsewhere} \end{cases}$

b. $f(z) = \begin{cases} \frac{1}{12}, & z = 1, 10, \\ \frac{1}{4}, & z = 5, 17, \\ \frac{1}{3}, & z = 2, \\ 0, & \text{elsewhere} \end{cases}$

4.3 a. $f(g) = \begin{cases} \dfrac{e^{-\lambda}\lambda^{-y}}{(-y)!}, & y = 0, -1, -2, \ldots, \\ 0, & \text{elsewhere} \end{cases}$

b. $f(z) = \begin{cases} \dfrac{e^{-\lambda}\lambda^{-z^2}}{(z^2)!}, & z = 0, 1, \sqrt{2}, \ldots, \\ 0, & \text{elsewhere} \end{cases}$

c. $f(w) = \begin{cases} \dfrac{e^{-\lambda}\lambda^{(w+1)/2}}{\left(\frac{w+1}{2}\right)!}, & w = -1, 1, 3, 5, \ldots, \\ 0, & \text{elsewhere} \end{cases}$

d. $h(y) = \begin{cases} \dfrac{e^{-\lambda}\lambda^{y^{1/\alpha}}}{(y^{1/\alpha})!}, & y^{1/\alpha} = 0, 1, 2, \ldots \\ & \text{or} \\ & y = 0, 1, 2^\alpha, \ldots, \\ 0, & \text{elsewhere} \end{cases}$

4.5 a. $f(y) = \begin{cases} \binom{n}{\sqrt{y/a}} p^{\sqrt{y/a}} (1-p)^{n-\sqrt{y/a}}, & \sqrt{\dfrac{y}{a}} = 0, 1, \ldots, n \\ & \text{or} \\ & y = 0, a, 4a, \ldots, n^2 a, \\ 0, & \text{elsewhere} \end{cases}$

b. $f(y) = \begin{cases} \binom{n}{\ln y} p^{\ln y} (1-p)^{n-\ln y}, & \ln y = 0, 1, \ldots, n \\ & \text{or} \\ & y = 1, e^1, e^2, \ldots, e^n, \\ 0, & \text{elsewhere} \end{cases}$

4.7 a. $F(y) = \begin{cases} 1 - e^{-(\sqrt{y}-6)}, & y \geq 36, \\ 0, & y < 36 \end{cases}$

$F(x) = \begin{cases} 1 - e^{-(x-6)}, & x \geq 6, \\ 0, & x < 6 \end{cases}$

4.9 a. $f(y) = \begin{cases} \dfrac{y-\beta}{4\alpha^2} e^{-(y-\beta)/2\alpha}, & y > \beta, \\ 0, & \text{elsewhere} \end{cases}$

b. $f(z) = \begin{cases} \dfrac{1}{4} z^{-1} \ln z \, e^{-(\ln z)/2}, & z > 1, \\ 0, & \text{elsewhere} \end{cases}$

c. $f(y) = \begin{cases} \dfrac{1}{4} y^{-3} e^{-1/2y}, & y > 0, \\ 0, & \text{elsewhere} \end{cases}$

d. $f(z) = \begin{cases} \dfrac{1}{2} z^3 e^{z^2/2}, & z > 0, \\ 0, & \text{elsewhere} \end{cases}$

4.11 $f(x) = \dfrac{1}{\pi} \dfrac{\alpha}{\alpha^2 + (x-\mu)^2}, \quad -\infty < x < \infty$

4.13 a. $f(y) = \dfrac{1}{2\sqrt{\pi}} \exp\left[-\dfrac{1}{2}\left(\dfrac{y-1}{2}\right)^2\right], \quad -\infty < y < \infty$

b.
$f(z) = \begin{cases} \dfrac{1}{2\sqrt{2\pi}}\left(\dfrac{z+1}{2}\right)^{-1/2} \exp\left[-\dfrac{1}{2}\left(\dfrac{z+1}{2}\right)^2\right], & z > -1, \\ 0, & \text{elsewhere} \end{cases}$

c. $f(y) = \begin{cases} \dfrac{2\sqrt{2\pi}}{\sqrt{\pi}} y e^{-y^4/2}, & y > 0, \\ 0, & \text{elsewhere} \end{cases}$

4.15 $F(y) = \begin{cases} 1, & y \geq 2, \\ 1 - e^{-3/2}, & -2 \leq y < 2, \\ 0, & y < -2 \end{cases}$

4.17 $g(y) = \begin{cases} 1, & 0 \leq y \leq 1, \\ 0, & \text{elsewhere} \end{cases}$

4.19 $g(y) = \begin{cases} 2\Gamma\left(\dfrac{n+1}{2}\right)\left(1 + \dfrac{y}{n}\right)^{-(n+1)/2}, & y \geq 0, \\ 0, & \text{elsewhere} \end{cases}$

Applied Problems

4.1 **a.** 0.
b. If $4 < a + b$, $\Pr(y < 4) = 0$.
If $\alpha a + b \leq 4 \leq \beta a + b$, $\Pr(y < 4) = \dfrac{4 - \alpha a - b}{a(\beta - \alpha)}$.
If $4 > \beta a + b$, $\Pr(y < 4) = 1$.
c. 0.
d. Here, x must be such that $0 < x < \pi$,
$g(t) = \begin{cases} \dfrac{1}{\pi} \dfrac{1}{\sqrt{1-t^2}}, & -1 < t < 1, \\ 0, & \text{elsewhere.} \end{cases}$
Then, $\Pr\left(0 < T < \dfrac{\pi}{2}\right) = \dfrac{1}{2}$.

4.3 **a.** $1 - 3e^{-2}$ **b.** $7e^{-6}$ **c.** $13e^{-12}$

4.5

a. $h(c) = \dfrac{1}{\sqrt{2\pi}\frac{5}{9}\sigma} \exp\left\{-\dfrac{1}{2}\left[\dfrac{c - \frac{5}{9}(\mu - 32)}{\frac{5}{9}\sigma}\right]^2\right\}$, $-\infty < c < \infty$

4.7 $f(p) = \begin{cases} \dfrac{1}{20}\sqrt{\dfrac{r}{p}}, & \dfrac{100}{r} \leq p \leq \dfrac{400}{r}, \\ 0, & \text{elsewhere} \end{cases}$

Chapter 5

Theoretical Exercises

5.1 $\dfrac{k}{p}$

5.3 **a.** $\dfrac{33}{64}$ **b.** $\dfrac{1723}{20,480}$

5.5 **a.** $c = 2$ **b.** $E(X) = 1$ **c.**

5.9 **a.** $c = \dfrac{1}{2}$ **b.** $E(X) = 0$ **c.** $\text{Var}(X) = 2$

5.11 **a.** $c = \dfrac{1}{95}$ **b.** $\dfrac{79}{19}$ **c.** $\dfrac{1733}{95}$ **d.** $\dfrac{1163}{190}$
e. $\dfrac{1722}{1805}$

5.13 **a.** $c = \dfrac{3}{\pi^2}$ **b.** $E(|X|) = \infty$

5.15 **a.** $\mu = \dfrac{4}{3}$ **b.** $\eta_2 = \dfrac{8}{9}$ **c.** $\gamma = \dfrac{-\sqrt{2}}{2}$
d. $V = \dfrac{\sqrt{2}}{2}$ **e.** $\xi = -\dfrac{3}{2}$

5.17 **a.** $\mu = \dfrac{\beta + \alpha}{2}$ **b.** $\eta_2 = \dfrac{(\beta - \alpha)^2}{12}$ **c.** $\sigma^2 = \dfrac{(\beta - \alpha)^2}{12}$
d. $\gamma = 0$ **e.** $\xi = -\dfrac{6}{5}$

Applied Problems

5.1 $k = 1: E(X) = 3.5$, $k = 2: E(X) = 7$,
$k = 3: E(X) = 10.5$

5.3 **a.** $\dfrac{35}{12}$ **b.** $\dfrac{35}{3}$ **c.** $\dfrac{35}{12}$ **d.** 0

5.5 $\dfrac{13,144}{9075}$

5.7 **a.** σ^2 **b.** $\alpha\sigma^2 + \beta$

5.9 **a.** $E(X) = \dfrac{1}{p}$, $\text{Var}(X) = \dfrac{k(1-p)}{p^2}$
b. $E(X) = \dfrac{k(1-p)}{p}$, $\text{Var}(X) = \dfrac{k(1-p)}{p^2}$

5.11 **a.** 28 **b.** $\dfrac{38}{75}$

5.13 **a.** $E[g(X)] = \beta^k \dfrac{(\alpha + k - 1)!}{(\alpha - 1)!}$ if α is an integer and
$\beta^k(\alpha + k - 1)(\alpha + k - 2)\ldots\alpha$ otherwise

Chapter 6

Theoretical Exercises

6.3 $c = \dfrac{1}{4}$

6.9 $F(x,y) = \begin{cases} 1, & x>1 \text{ and } y>1, \\ y(2x-y), & 0 \le y \le x \le 1, \\ x^2, & 0 \le x \le 1, y > x, \\ y(2-y), & x>1, 0 \le y \le 1, \\ 0, & x<0 \text{ or } y<0 \end{cases}$

6.11 $f(x,y) = \begin{cases} 2, & 1<x<y<2, \\ 0, & \text{elsewhere} \end{cases}$

6.13 $f_1(x) = \begin{cases} \dfrac{1}{3}, & x=1, \\ \dfrac{1}{6}, & x=2,3,4,5, \\ 0, & \text{elsewhere}, \end{cases}$

$f_1(x) = \begin{cases} \dfrac{1}{5}, & y=1,2,\ldots,5, \\ 0, & \text{elsewhere} \end{cases}$

6.15 a. $f_1(x) = \begin{cases} \dfrac{(\lambda p)^x e^{-\lambda p}}{x!}, & x=0,1,2,\ldots, \\ 0, & \text{elsewhere} \end{cases}$

b. $f_2(y) = \begin{cases} \dfrac{\lambda^y e^{-\lambda}}{y!}, & x=0,1,2,\ldots, \\ 0, & \text{elsewhere} \end{cases}$

c. $\Pr(x \le r) = e^{-\lambda p} \sum_{x=0}^{r} \dfrac{(\lambda p)^x}{x!}$

d. $\Pr(y \ge k) = 1 - e^{-\lambda} \sum_{y=0}^{k-1} \dfrac{\lambda^r}{r!}$

6.17 a. $F(x,y) = \begin{cases} (1-e^{-x})(1-e^{-y}), & x,y>0, \\ 1-e^{-x}, & x>y, y=\infty, \\ 1-e^{-y}, & y>0, x=\infty, \\ 0, & x \le 0 \text{ or } y \le 0 \end{cases}$

b. $f_1(x) = \begin{cases} e^{-x}, & x>0, \\ 0, & \text{elsewhere} \end{cases}$

c. $f_2(y) = \begin{cases} e^{-y}, & y>0, \\ 0, & \text{elsewhere} \end{cases}$

d. $F_1(x) = \begin{cases} 1-e^{-x}, & x>0, \\ 0, & x \le 0 \end{cases}$

e. $F_2(y) = \begin{cases} 1-e^{-y}, & y>0, \\ 0, & y \le 0 \end{cases}$

f. $h_1(x|y) = f_1(x)$ **g.** $h_2(y|x) = f_2(y)$

h. $F_X(x|y) = F_1(x)$ **i.** $F_Y(y|x) = F_2(y)$

j. Yes, they are independent

6.21 a. $h_1(x|y) = \begin{cases} \dbinom{y}{x} p^x (1-p)^{y-x}, & x=0,1,\ldots,y, \\ 0, & \text{elsewhere} \end{cases}$

b. $h_2(y|x) = \begin{cases} \dfrac{e^{-\lambda(1-p)}[\lambda(1-p)]^{y-x}}{(y-x)!}, & y=x, x+1, x+2, \ldots, \\ 0, & \text{elsewhere} \end{cases}$

c. X and Y are not independent

6.23 $f_2(y) = \begin{cases} \left(\dfrac{1}{2}\right)^{y+1}, & y \ge 0, \\ 0, & \text{elsewhere} \end{cases}$

Applied Problems

6.1 a. $f(x,y) = \begin{cases} \dfrac{\binom{9}{x}\binom{8}{y}\binom{3}{6-x-y}}{\binom{20}{6}}, & \begin{array}{l} x,y=0,1,2,\ldots,6, \\ \text{such that } 3 \le x+y \le 6, \end{array} \\ 0, & \text{elsewhere} \end{cases}$

b. $\dfrac{\binom{9}{3}\binom{8}{3}\binom{3}{0}}{\binom{20}{6}}$

c. $\dfrac{\binom{9}{1}}{\binom{20}{6}}\left[\binom{11}{5} - \binom{8}{5}\right]$

6.3 a.
$$f(x,y) = \begin{cases} \binom{20}{x,y,20-x-y}\left(\frac{1}{3}\right)^x\left(\frac{1}{2}\right)^y\left(\frac{11}{26}\right)^{20-x-y}, & x,y,=0,1,\ldots,20, \\ 0, & \text{elsewhere} \end{cases}$$

$$f(x,y) = \begin{cases} \dfrac{\binom{13}{x}\binom{26}{y}\binom{44}{20-x-y}}{\binom{52}{40}}, & \begin{matrix}x=0,1,\ldots,13,\\ y=0,1,\ldots,20\\ \text{such that } 7 \le x+y \le 20,\end{matrix} \\ 0, & \text{elsewhere} \end{cases}$$

6.5 a. 0.1 **b.** 0.2124 **c.** 0

6.7 a. $e^{-5/4}$ **b.** $e^{-3}(1-e^{-9})$ **c.** $1-e^{-1/4}$

6.9 a. $f(x) = \begin{cases} \dfrac{\binom{4}{x}\binom{10}{6-x}}{\binom{14}{6}}, & x=0,1,\ldots,4, \\ 0, & \text{elsewhere} \end{cases}$

$f(y) = \begin{cases} \dfrac{\binom{5}{y}\binom{9}{6-y}}{\binom{14}{6}}, & x=0,1,\ldots,5, \\ 0, & \text{elsewhere} \end{cases}$

b. $1 - \dfrac{\binom{10}{2}}{\binom{14}{6}}$

c. $\dfrac{1}{\binom{14}{6}}\sum_{y=2}^{4}\binom{5}{y}\binom{9}{6-y}$

6.11 a.
$$h_1(x|y_0) = \begin{cases} \dfrac{\binom{9}{x}\binom{11}{15-x-y_0}}{\binom{20}{15-y_0}}, & \begin{matrix} x \text{ and } y_0 \text{ integers such that} \\ 0 \le x \le 9, \\ 4-y_0 \le x \le 15-y_0, \\ 0 \le y_0 \le 10, \end{matrix} \\ 0, & \text{elsewhere} \end{cases}$$

b.
$$h_2(y|x_0) = \begin{cases} \dfrac{\binom{10}{y}\binom{11}{15-x_0-y}}{\binom{21}{15-x_0}}, & \begin{matrix} y \text{ and } x_0 \text{ integers such that} \\ 0 \le y \le 10, \\ 4-x_0 \le y \le 15-x_0, \\ 0 \le x_0 \le 9, \end{matrix} \\ 0, & \text{elsewhere} \end{cases}$$

c.
$$h_1(x|5) = \begin{cases} \dfrac{\binom{9}{x}\binom{11}{10-x}}{\binom{20}{10}}, & x \text{ an integer such that } 0 \le x \le 9, \\ 0, & \text{elsewhere} \end{cases}$$

d.
$$h_2(y|5) = \begin{cases} \dfrac{\binom{10}{y}\binom{11}{10-y}}{\binom{21}{10}}, & y \text{ an integer such that } 0 \le y \le 10, \\ 0, & \text{elsewhere} \end{cases}$$

e. X and Y are not independent

6.13 a. $\dfrac{1}{18}$ **b.** $\dfrac{1}{18}$ **c.** $\dfrac{3}{4}$

d. $f(x) = \begin{cases} x, & 0 \le x < 1, \\ 2-x, & 1 \le x < 2, \\ 0, & \text{elsewhere} \end{cases}$

6.15 a. $\dfrac{9}{5}$ **b.** $\dfrac{47}{10}$ **c.** $\dfrac{84}{10}$

6.17 a. Exercise 6.9: $E(X) = \dfrac{2}{3}$

b. Exercise 6.9: $E(Y) = \dfrac{1}{3}$

6.19 a. Exercise 6.2: $E(X)=\dfrac{2n+1}{3}$, $E(X^2)=\dfrac{n(n+1)}{n}$, $\mathrm{Var}(X)=\dfrac{(n+2)(n-1)}{18}$

Exercise 6.6: $E(X)=\Gamma\left(\dfrac{3}{2}\right)$, $E(X^2)=\Gamma(2)$, $\mathrm{Var}(X)=1-\dfrac{\pi}{4}$

b. Exercise 6.2: $\mathrm{Var}(Y)=\mathrm{Var}(X)=\dfrac{(n+2)(n-1)}{18}$

Exercise 6.6: $E(Y)=1$, $E(Y^2)=2$, $\mathrm{Var}(Y)=1$

c. Exercises 6.2 and 6.6: $\mathrm{Cov}(X,Y)=0$

6.23 a. $E(Z_1)=\dfrac{\alpha}{\alpha+\beta}$ **b.** $E(Z_2)-\alpha+\beta$

6.25 a. $E(Z)=-\dfrac{95}{3}$ **b.** $\mathrm{Var}(Z)=\dfrac{241{,}607}{360}$

Chapter 7

Applied Problems

7.1 $f(x,y,z)=\begin{cases}\dfrac{10}{x,y,z,(10-x-y-z)}\left(\dfrac{1}{6}\right)^{x+y+z}\left(\dfrac{1}{2}\right)^{10-x-y-z}, & 0\le x+y+z\le 10,\\ & x,y, \text{ and } z \text{ integers,}\\ 0, & \text{elsewhere}\end{cases}$

7.3 $f(w,x,y,z)=\begin{cases}\dfrac{n}{w,x,y,z,(n-w-x-y-z)}\left(\dfrac{1}{32}\right)^{w}\left(\dfrac{5}{32}\right)^{w}\left(\dfrac{10}{32}\right)^{y}\\ \qquad\left(\dfrac{10}{32}\right)^{z}\left(\dfrac{6}{32}\right)^{n-w-x-y-z}, & 0\le w+x+y\le n,\\ 0, & \text{elsewhere}\end{cases}$

7.5 $f(x,y,z)=\begin{cases}1, & 3\le x\le 4,\ 2\le y\le 3,\ 0\le z\le 1,\\ 0, & \text{elsewhere}\end{cases}$

7.7 $k=\dfrac{32\sqrt{2}}{15}$

7.9 $f(w,x,y,z)=\begin{cases}4, & 0\le w\le 1,\ 1\le x\le \tfrac{3}{2},\ 1\le y\le 2,\ 2\le z\le \tfrac{5}{2},\\ 0, & \text{elsewhere}\end{cases}$

7.11 Problem 7.5: $F(x,y,z) = \begin{cases} 1, & x \geq 4, y \geq 3, z \geq 1, \\ & 3 \leq x < 4, \\ xyz - 3yz - 2xy + 6y, & 2 \leq y < 3, \\ & 0 \leq x < 1, \\ 0, & x < 3, y < 2, z < 0 \end{cases}$

Problem 7.6: $F(x,y,z) = \begin{cases} 1, & x, y, z > 1, \\ 6xyz, & 0 \leq x \leq y \leq z \leq 1, \\ 0, & x < 0 \end{cases}$

7.13 a. Problem 7.1: $f_1(x) = \begin{cases} \binom{10}{x} \left(\frac{1}{6}\right)^x \left(\frac{5}{6}\right)^{10-x}, & 0 \leq x \leq 10, x \text{ an integer,} \\ 0, & \text{elsewhere} \end{cases}$

Problem 7.5: $f_1(x) = \begin{cases} 1, & 3 \leq x \leq 4, \\ 0, & \text{elsewhere} \end{cases}$

Problem 7.7: $f_1(x) = \begin{cases} \dfrac{32\sqrt{2}}{15}\left(2 - 2x^2 + \dfrac{1}{2}x^4\right), & -\sqrt{2} \leq x \leq \sqrt{2}, \\ 0, & \text{elsewhere} \end{cases}$

b. Problem 7.1: $f_2(y) = \begin{cases} \binom{10}{y}\left(\frac{1}{6}\right)^y \left(\frac{5}{6}\right)^{10-y}, & 0 \leq y \leq 10, y \text{ an integer,} \\ 0, & \text{elsewhere} \end{cases}$

Problem 7.5: $f_2(y) = \begin{cases} 1, & 2 \leq y \leq 3, \\ 0, & \text{elsewhere} \end{cases}$

Problem 7.7: $f_2(y) = \begin{cases} \dfrac{64\sqrt{2}}{15}\sqrt{y}(2 - y), & 0 \leq y \leq 2, \\ 0, & \text{elsewhere} \end{cases}$

c. Problem 7.1: $f_{13}(x,z) = \begin{cases} \binom{10}{x, z, (10-x-z)}\left(\frac{1}{6}\right)^{x+z}\left(\frac{2}{3}\right)^{10-x-z}, & 0 \leq x + z \leq 10, x \text{ and } z \text{ integers,} \\ 0, & \text{elsewhere} \end{cases}$

Problem 7.5: $f_{13}(x,z) = \begin{cases} 1, & 3 \leq x \leq 4, 0 \leq z \leq 1 \\ 0, & \text{elsewhere} \end{cases}$

Problem 7.7: $f_{13}(x,z) = \begin{cases} \dfrac{32\sqrt{2}}{15}(z - x^2), & 0 \leq x^2 \leq z \leq 2, \\ 0, & \text{elsewhere} \end{cases}$

d. Problem 7.1: $f_{23}(y,z) = \begin{cases} \binom{10}{y,z,(10-y-z)} \left(\frac{1}{6}\right)^{y+z} \left(\frac{2}{3}\right)^{10-y-z}, & 0 \le y+z \le 10, \ y \text{ and } z \text{ integers}, \\ 0, & \text{elsewhere} \end{cases}$

Problem 7.5: $f_{23}(y,z) = \begin{cases} 1, & 2 \le y \le 5, \ 0 \le z \le 1, \\ 0, & \text{elsewhere} \end{cases}$

Problem 7.7: $f_{23}(y,z) = \begin{cases} \dfrac{64\sqrt{2}}{15}\sqrt{y}, & 0 \le y \le z \le 2, \\ 0, & \text{elsewhere} \end{cases}$

7.15 a. Problem 7.1: $f(x|y) = \begin{cases} \binom{10-y}{x} \left(\frac{1}{5}\right)^x \left(\frac{4}{5}\right)^{10-y}, & x = 0,1,\ldots,(10-y), \\ & \text{where } y = -0,1,\ldots,10, \\ 0, & \text{elsewhere} \end{cases}$

Problem 7.5: $f(x|y) = \begin{cases} 1, & 3 \le x \le 4, \\ 0, & \text{elsewhere} \end{cases}$

Problem 7.7: $f(x|y) = \begin{cases} \dfrac{1}{2\sqrt{y}}, & -\sqrt{y} \le x \le \sqrt{y},\ 0 < y \le 2, \\ 0, & \text{elsewhere} \end{cases}$

b. Problem 7.1: $g(y|z) = \begin{cases} \binom{10-y}{y} \left(\frac{1}{5}\right)^y \left(\frac{4}{5}\right)^{10-z}, & y = 0,1,\ldots,(10-z), \\ & \text{where } z = 0,1,\ldots,10, \\ 0, & \text{elsewhere} \end{cases}$

Problem 7.5: $g(y|z) = \begin{cases} 1, & 2 \le y \le 3, \\ 0, & \text{elsewhere} \end{cases}$

Problem 7.7: $f(y|z) = \begin{cases} \dfrac{3\sqrt{y}}{2z\sqrt{z}}, & 0 \le y \le z \le 2, \\ 0, & \text{elsewhere} \end{cases}$

7.17 a. Problem 7.3: $h(w|x) = \begin{cases} \binom{n-x}{n} \left(\frac{1}{27}\right)^w \left(\frac{26}{27}\right)^{n-x-w}, & w = 0,1,\ldots,(n-x), \\ & \text{where } x = 0,1,\ldots,n \\ 0, & \text{elsewhere.} \end{cases}$

Problem 7.9: $h(w|x) = \begin{cases} 1, & 0 \le w \le 1, \\ 0, & \text{elsewhere.} \end{cases}$

b. Problem 7.3: $f(x|w,y) = \begin{cases} \binom{n-w-y}{x} \left(\frac{5}{21}\right)^x \left(\frac{16}{21}\right)^{n-w-y-x}, & x = 0,1,2,\ldots,(n-w-y), \\ & \text{where } w,y = 0,1,\ldots,n \\ & \text{such that } w+y = 0,1,\ldots,n \\ 0, & \text{elsewhere.} \end{cases}$

Problem 7.9: $f(x|w,y) = \begin{cases} 2, & 1 \le x \le \frac{3}{2}, \\ 0, & \text{elsewhere.} \end{cases}$

Answers to Selected Exercises and Problems **549**

c. Problem 7.3: $g(y|w,x,z) = \begin{cases} \binom{n-w-z}{y}\left(\frac{5}{8}\right)^y\left(\frac{3}{8}\right)^{n-w-x-z-y}, & y=0,1,\ldots,(n-w-x-z), \\ & \text{where } w,x,z=0,1,\ldots,n \\ & \text{such that } w+x+z=0,1,\ldots,n \\ 0, & \text{elsewhere.} \end{cases}$

Problem 7.9: $g(y|w,x,z) = \begin{cases} 1, & 1\leq y\leq 2, \\ 0, & \text{elsewhere.} \end{cases}$

d. Problem 7.3: $l(w|z,y) = \begin{cases} \binom{n-y}{w,z,(n-w-y-z)}\left(\frac{1}{22}\right)^w\left(\frac{10}{22}\right)^z\times\left(\frac{1}{2}\right)^{n-w-y-z}, & w+z=0,1,\ldots,(n-y), \\ & \text{where } y=0,1,\ldots,n \\ 0, & \text{elsewhere.} \end{cases}$

Problem 7.9: $l(w,z|y) = \begin{cases} 2, & 0\leq w\leq 1, 2\leq z\leq \frac{5}{2}, \\ 0, & \text{elsewhere.} \end{cases}$

e. Problem 7.3:

$m(w,x,y|z) = \begin{cases} \binom{n-z}{w,x,y,(n-w-x-y-z)}\left(\frac{1}{22}\right)^w\times\left(\frac{5}{22}\right)^z\left(\frac{10}{22}\right)^y\left(\frac{6}{22}\right)^{n-w-x-y-z}, & w+x+y=0,1,\ldots,(n-z), \\ & \text{where } z=0,1,\ldots,n, \\ 0, & \text{elsewhere.} \end{cases}$

Problem 7.9: $m(w,x,y|z) = \begin{cases} 2, & 0\leq w\leq 1, 1\leq x\leq \frac{3}{2}, 1\leq y\leq 2, \\ 0, & \text{elsewhere.} \end{cases}$

7.19 Problem 7.1: no
Problem 7.2: **a.** no   **b.** no
Problem 7.3: no
Problem 7.5: yes

7.23 **a.** $g(y_1,y_2,\ldots,y_n) = \frac{1}{(2\pi)^{n/2}\sigma^n}\exp\left\{-\frac{1}{2\sigma^2}\left(2\sum_{i=1}^n y_i^2 - 2\sum_{i=2}^n y_i y_{i-1} - y_n^2 - n\mu^2\right)\right\}$, $0<y_1\leq y_2\leq\cdots\leq y_n<\infty$

b. $g(y_1,y_2,\ldots,y_n) = \frac{1}{\beta^{n\alpha}}y_1\prod_{j=2}^n (y_j-y_{j-1})e^{-1/\beta}y_n$, $0<y_1\leq y_2\leq\cdots\leq y_n<\infty$

c. $g(y_1,y_2,\ldots,y_n) = \frac{\lambda^{y_n}e^{-\lambda}}{y_1!\prod_{i=2}^n (y_i-y_{i-1})!}$, $0<y_1\leq y_2\leq\cdots\leq y_n<\infty$

7.27 **a.** $m_Y(t)=\exp(2t+18t^2)$ **b.** $Pr(Y>8)=0.1587$

7.29 **a.** Problem 7.5: $\eta_{1,2,3}=0$
b. Problem 7.5: $\eta_{3,3,3}=0$

Chapter 8

Applied Problems

8.1 $1 - \dfrac{1}{k^2} = \dfrac{7}{10}$

8.3 $n \geq 250$

8.7 $n \geq 23$

8.9 $n \geq 228$

8.13   Yes, by Theorem 8.3.2

8.15 $\dfrac{S_n}{n} \to 0$ as $n \to \infty$ for $c > 1$

8.17   **a.** 0.00002   **b.** $\Phi \approx 0$

8.19   0.00007

8.21   **a.** 0.9979   **b.** 0.00004   **c.** 0.1549

8.25 $n \geq 90$ $n > 109$

8.27   0.9213

Chapter 9

Applied Problems

9.1   **a.**

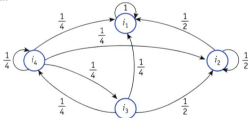

b.

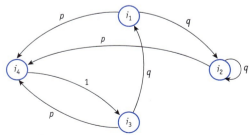

9.3   **a.**, **b.** $\begin{array}{c} \\ c \\ t \\ b \end{array} \begin{bmatrix} \overset{c}{0} & \overset{t}{\tfrac{1}{2}} & \overset{b}{\tfrac{1}{2}} \\ \tfrac{1}{2} & 0 & \tfrac{1}{2} \\ \tfrac{1}{4} & \tfrac{3}{4} & 0 \end{bmatrix}$

c.

9.5 $\begin{array}{c} \\ i_1 \\ i_2 \\ i_3 \\ i_4 \\ i_5 \\ i_6 \\ i_7 \end{array} \begin{bmatrix} 0 & \tfrac{1}{6} & \tfrac{1}{6} & \tfrac{1}{6} & \tfrac{1}{6} & \tfrac{1}{6} & \tfrac{1}{6} \\ \tfrac{1}{3} & 0 & \tfrac{1}{3} & \tfrac{1}{3} & 0 & 0 & 0 \\ \tfrac{1}{3} & \tfrac{1}{3} & 0 & 0 & \tfrac{1}{3} & 0 & 0 \\ \tfrac{1}{3} & \tfrac{1}{3} & 0 & 0 & 0 & \tfrac{1}{3} & 0 \\ \tfrac{1}{3} & 0 & \tfrac{1}{3} & 0 & 0 & 0 & \tfrac{1}{3} \\ \tfrac{1}{3} & 0 & 0 & \tfrac{1}{3} & 0 & 0 & \tfrac{1}{3} \\ \tfrac{1}{3} & 0 & 0 & 0 & \tfrac{1}{3} & \tfrac{1}{3} & 0 \end{bmatrix}$

Answers to Selected Exercises and Problems **551**

9.7 a.

$$\begin{array}{c} i_1\ i_2\ i_3\ i_4\ i_5\ i_6 \\ \begin{array}{c} i_1 \\ i_2 \\ i_3 \\ i_4 \\ i_5 \\ i_6 \end{array} \begin{bmatrix} 0 & 1 & 0 & 0 & 0 & 0 \\ q & 0 & p & 0 & 0 & 0 \\ 0 & q & 0 & p & 0 & 0 \\ 0 & 0 & q & 0 & p & 0 \\ 0 & 0 & 0 & q & 0 & p \\ 0 & 0 & 0 & 0 & 1 & 0 \end{bmatrix} \end{array}$$

eigenvector for $\lambda_1 = 1$: $x_1 \begin{bmatrix} 1 \\ 1 \\ 1 \\ 1 \\ 1 \end{bmatrix}$,

eigenvector for $\lambda_2 = \dfrac{2}{3}$: $x_2 \begin{bmatrix} 0 \\ 2 \\ 2 \\ 2 \\ 3 \end{bmatrix}$,

b.

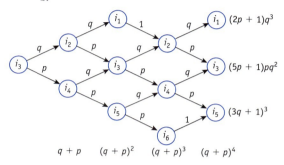

i_1 $(2p+1)q^3$

i_3 $(5p+1)pq^2$

i_5 $(3q+1)^3$

$q+p \quad (q+p)^2 \quad (q+p)^3 \quad (q+p)^4$

eigenvector for $\lambda_1 = 1$: $x_1 \begin{bmatrix} 1 \\ 1 \\ 1 \\ 1 \\ 1 \end{bmatrix}$,

eigenvector for $\lambda_3 = \dfrac{1}{3}$: $x_3 \begin{bmatrix} 0 \\ 0 \\ 0 \\ 0 \\ 1 \end{bmatrix}$,

9.9 a. $(0.6)^3$ **b.** $9(0.6)^4(0.4)^2$

9.13 b. $i_1 : \dfrac{7}{27}$, $i_2 : \dfrac{5}{27}$, $i_3 : \dfrac{2}{27}$, $i_4 : \dfrac{2}{27}$, $i_5 : \dfrac{5}{27}$,

$i_6 : \dfrac{4}{27}$, $i_7 : \dfrac{2}{27}$

eigenvector for $\lambda_1 = 1$: $x_1 \begin{bmatrix} 1 \\ 1 \\ 1 \\ 1 \\ 1 \end{bmatrix}$,

eigenvector for $\lambda_4 = \dfrac{1}{3}$: $x_4 \begin{bmatrix} 0 \\ 0 \\ 0 \\ 4 \\ 1 \end{bmatrix}$,

9.15 $\dfrac{1}{4}, \dfrac{3}{4}, 1$

9.17

$C = \begin{bmatrix} 1 & 0 & 0 & 0 \\ 1 & 2 & 0 & 0 \\ 1 & 2 & 0 & 4 \\ 1 & 3 & 1 & 1 \end{bmatrix}$ and $C^{-1} = \dfrac{1}{4}\begin{bmatrix} 4 & 0 & 0 & 0 \\ -2 & 2 & 0 & 0 \\ 2 & -5 & -1 & 4 \\ 0 & -1 & 1 & 0 \end{bmatrix}$,

$j^k = (A+M)^k = \begin{bmatrix} 1 & 0 & 0 & 0 \\ 0 & \dfrac{2}{3} & 0 & 0 \\ 0 & 0 & \dfrac{1}{3} & 1 \\ 0 & 0 & 0 & 0 \end{bmatrix}^k = \begin{bmatrix} 1 & 0 & 0 & 0 \\ 0 & \left(\dfrac{2}{3}\right)^k & 0 & 0 \\ 0 & 0 & \left(\dfrac{1}{3}\right)^k & k\left(\dfrac{1}{3}\right)^k \\ 0 & 0 & 0 & \left(\dfrac{1}{3}\right)^k \end{bmatrix}$,

$$p^k = CJ^kC^{-1} = C(A+M)^k C^{-1} = \begin{bmatrix} 1 & 0 & 0 & 0 \\ 1-\left(\frac{2}{3}\right)^k & \left(\frac{2}{3}\right)^k & 0 & 0 \\ 1-\left(\frac{2}{3}\right)^k & (2^k-1)\left(\frac{1}{3}\right)^k & \left(\frac{1}{3}\right)^k & 0 \\ 1-\frac{1}{2}(3\cdot 2^k-1)\left(\frac{1}{3}\right)^k & \frac{3}{4}(2^{k+1}-2-k)\left(\frac{1}{3}\right)^k & \frac{3}{4}k\left(\frac{1}{3}\right)^k & \left(\frac{1}{3}\right)^k \end{bmatrix}$$

9.19 $\lambda_1 = 1, \lambda_2 = \lambda_3 = \frac{1}{2}$; $J^k = (A+M)^k = \begin{bmatrix} 1 & 0 & 0 \\ 0 & \left(\frac{1}{2}\right)^k & k\left(\frac{1}{2}\right)^{k-1} \\ 0 & 0 & \left(\frac{1}{2}\right)^k \end{bmatrix}$;

$x_1 = \begin{bmatrix} 1 \\ 1 \\ 1 \end{bmatrix}$, $x_2 = \begin{bmatrix} 0 \\ 1 \\ 0 \end{bmatrix}$, $x_3 = \begin{bmatrix} 2 \\ 1 \\ 0 \end{bmatrix}$; $C = \begin{bmatrix} 1 & 0 & 2 \\ 1 & 1 & 1 \\ 1 & 0 & 0 \end{bmatrix}$; $C^{-1} = \frac{1}{2}\begin{bmatrix} 0 & 0 & 2 \\ -1 & 2 & -1 \\ 1 & 0 & -1 \end{bmatrix}$;

$$P^k = CJ^kC^{-1} = C(A+M)^k C^{-1} = \begin{bmatrix} \left(\frac{1}{2}\right)^k & 0 & 1-\left(\frac{1}{2}\right)^k \\ k\left(\frac{1}{2}\right)^k & \left(\frac{1}{2}\right)^k & 1-(k+1)\left(\frac{1}{2}\right)^k \\ 0 & 0 & 1 \end{bmatrix};$$

$$p^{20} = \begin{bmatrix} \left(\frac{1}{2}\right)^{20} & 0 & 1-\left(\frac{1}{2}\right)^{20} \\ 20\left(\frac{1}{2}\right)^k & \left(\frac{1}{2}\right)^{20} & 1-21\left(\frac{1}{2}\right)^{20} \\ 0 & 0 & 1 \end{bmatrix}$$

Appendix

Review of Preliminary Mathematics

A.0 Introduction

In this appendix, we discuss basic concepts of mathematics. This discussion is likely a review for those who have studied these topics in introductory courses in mathematics, and it fulfills the minimum requirement for those who have had little training in this area. Also, we introduce sine topics of mathematics that are usually not included in introductory courses in calculus—the concept of Jacobians, gamma and beta functions, some useful identities and series, and some matrix theory. You should be familiar with these mathematical concepts to better understand the material covered in the rest of the book. We define these concepts, state some of their important properties, and illustrate with examples a few of their applications.

The material in this appendix provides you with a brief review of some concepts necessary for studying probability theory. If you wish to cover certain areas in greater detail, you are referred to other sources of information.

A.1 Sum and Product Notations

We use the symbol y_i to denote the ith value of the n values $y_1, y_2, ..., y_n$. The sum of these values, $y_1 + y_2 + ... + y_n$, is designated by

$$\sum_{i=1}^{n} y_i$$

or

$$\sum_{i=1}^{n} y_i.$$

The Greek letter Σ (capital sigma) tells you to *sum* elements of a sequence. The letter i is referred to as the *summation index*, and y_i is called the *summand*.

Example A.1.1

a. $\displaystyle\sum_{i=1}^{r} ay_i = ay_1 + ay_2 + \cdots + ay_r$

$= a(y_1 + y_2 + \cdots + y_r)$

$= a \displaystyle\sum_{i=1}^{r} y_i$

b. $\sum_{j=1}^{k} x_j y_j^{j-1} = x_1 + x_2 y_2^1 + \cdots + x_k y_k^{k-1}$

c. $\sum_{j=1}^{n} (x_j - a)^2 = (x_1 - a)^2 + (x_2 - a)^2 + \cdots + (x_n - a)^2$

d. $\sum_{i=2}^{6} (-2)^{i-1} x_{i-1}^i = -2x_1^2 + 4x_2^3 - 8x_3^4 + 16x_4^5 - 32x_5^6$

Therefore, we see that the sum of terms is obtained by letting the summation index i take on those integral values between and including the limits of interest.

If instead of summing the terms y_i, $i = 1, 2, \ldots, n$, we wish to obtain their product, we write

$$y_1 y_2 y_3 \cdots y_n = \prod_{i=1}^{n} y_i,$$

where the Greek letter Π (capital pi) tells you to *multiply* elements of a sequence.

Example A.1.2

a. $\prod_{i=1}^{r} (a + \frac{i}{b})^i = (a + \frac{1}{b})(a + \frac{2}{b})^2 \cdots (a + \frac{r}{b})^r$

b. $\prod_{i=1}^{k} [c + (-\frac{1}{2})^{j-1}] = c(c - \frac{1}{2})(c + \frac{1}{4}) \cdots [c + (-\frac{1}{2})^{k-1}]$

A.2 Set Theory

In this section, we address some basis ideas and concepts of set theory that are essential for a modern introduction to probability.

A *set* is a collection of distinct objects. For example, one of the most familiar, set N, is the set of positive integers $1, 2, 3, \ldots$. Every object belonging to a set is called an *element* of the set. Let A_1 be a given set; if a is an element of A_1, we write

$$a \in A_1.$$

If a is not an element of set A_1, we write

$$a \notin A_1.$$

A set is described either by listing its elements or by stating the properties that characterize the elements of the set. For example, to specify the set A_1 of all positive integers less than 12, we may write

$$A_1 = \begin{cases} \{1, 2, 3, 4, 5, 6, 7, 8, 9, 10, 11, 12\} \\ \{\text{all possible integers less than } 12\}. \\ \{x : x < 12, x \text{ a positive integer}\} \end{cases}$$

A set A_2 is a *subset* of a set A_1 if every element of A_2 is also an element of A_1. We denote this by writing

$$A_2 \subseteq A_1,$$

which is read "A_1 contains A_2" or "A_2 is contained in A_1." For example, if

$$A_2 = \{x : x < 5, x \text{ a positive integer}\},$$

it is obvious that A_2 is a subset of A_1. Also, every set is a subset of itself. Two sets, A_1 and A_3, are *equal*,

$$A_1 = A_3,$$

If and only if $A_1 \subseteq A_3$ and $A_3 \subseteq A_1$. A set A_2 is *proper subset* of the set A_1 if every element of A_2 is an element of A_1 and A_1 contains at least one element that is not an element of A_2. We denote this relationship as

$$A_2 \subset A_1.$$

That is, if

$$A_1 = \{x : x = 1, 2, 3, 4, 5\}$$

and

$$A_2 = \{x : x = 1, 2, 3\}$$

then A_2 is a proper subset of A_1. The set that contains no elements is called the *empty set* or *null set*, and we denote it using the symbol ϕ. The null set is a subset of every set.

A.2.1 Set Operations

Let A_1 and A_2 be arbitrary sets. The union of these sets, denoted as

$$A_1 \cup A_2,$$

is that set containing the elements of A_1, A_2, or both. That is,

$$A_1 \cup A_2 = \{x : x \in A_1 \text{ or } x \in A_2\}$$

The symbol $A_1 \cup A_2$ is read "the union of A_1 and A_2."

Example A.2.1

If $A_1 = \{x : 0 \le x \le 3\}$ and $A_2 = \{x : -2 \le x \le 2\}$, then

$$A_1 \cup A_2 = \{x : -2 \le x \le 3\}.$$

The intersection of two sets A_1 and A_2 is that set containing only those elements common to A_1 and A_2. We denote the intersection of the two sets A_1 and A_2 as

$$A_1 \cap A_2$$

or

$$A_1 A_2;$$

that is,

$$A_1 \cap A_2 = A_1 A_2 = \{x : x \in A_1 \text{ and } x \in A_2\}.$$

The symbol $A_1 \cap A_2$ is read "the intersection of A_1 and A_2."

Example A.2.2

If
$$A_1 = \{x : x = 1, 2, 3, 4, 5, 6, 7\}$$
and
$$A_2 = \{x : x = 4, 5, 6, 7, 8, 9\},$$
then
$$A_1 \cap A_2 = \{x : x = 4, 5, 6, 7\}.$$

If the sets A_1 and A_2 have no elements in common, that is, if $A_1 \cap A_2 = \phi$, then the sets A_1 and A_2 are said to be disjoint or mutually exclusive sets.

The universal set U refers to the totality of elements under consideration—that is, the smallest possible set that is required so that every other set we may consider will be a subset of U. The *complement* of the set $A_1 \subset U$ is the set of all elements that are not in A_1 but are in U. The complement of the set A_1 is denoted as $\overline{A}_1$. The difference of any two sets, A_1 and A_2, is denoted using a $-$ sign; thus, $A_1 - A_2$ is equal to $A_1 \cap \overline{A}_2$ or $A_2 - A_1$ is equal to $A_2 \cap \overline{A}_1$. Therefore,

$$\overline{A}_1 = \{x : x \notin A_1 \text{ but } x \in U\}.$$

Example A.2.3

If $U = \{x : 0 \leq x \leq 25\}$ and $A_1 = \{x : 5 \leq x \leq 10\}$, then

$$\overline{A}_1 = \{x : 0 \leq x < 5 \text{ or } 10 < x \leq 25\}.$$

The usual notation for the complement of a subset A_1 of the universal set U is

$$\overline{A}_1 = U - A_1.$$

Venn diagrams are useful in helping you clarify these set operations. In Figure A.2.1(a), the shaded area represents the operation $A_1 \cup A_2 \cup A_3$ and U is the universal set. Similarly, the shaded area in the Venn diagrams in Figure A.2.1(b), (d), (e), and (f) represent $A_1 \cap A_2$, $\overline{A_1 \cap A_2}$, $A_1 - (A_1 \cap A_2)$, and $\overline{A_1 - (A_1 \cap A_2)}$, respectively, whereas Figure A.2.1(c) shows that $A_2 \subset A_1$.

Appendix 557

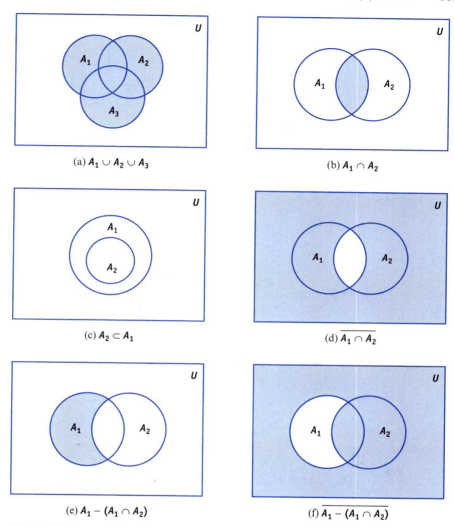

Figure A.2.1

We now state some basic and important properties of the algebra of sets. Let U be the universal set; A_1, A_2, and A_3 be subsets of U; and ϕ be the empty set. The properties of the set operations, that is, union, intersection, and complementation, are given by Theorem A.2.1.

■ **Theorem A.2.1** If A_1, A_2, and A_3 are subsets of the universal set U, then the following laws hold:

1. idempotent law

$$A_1 \cup A_1 = A_1,$$
$$A_1 \cap A_1 = A_1$$

2. commutative law

$$A_1 \cup A_2 = A_2 \cup A_1,$$
$$A_1 \cap A_2 = A_2 \cap A_1$$

3. associative law

$$A_1 \cup A_2 \cup A_3 = A_1 \cup (A_2 \cup A_3),$$
$$A_1 \cap A_2 \cap A_3 = A_1 \cap (A_2 \cap A_3)$$

4. distribution law

$$A_1 \cup (A_2 \cap A_3) = (A_1 \cup A_2) \cap (A_1 \cup A_3),$$
$$A_1 \cap (A_2 \cup A_3) = (A_1 \cap A_2) \cup (A_1 \cap A_3)$$

5. identify law

   a. $A_1 \cup U = U$, $A_1 \cap U = A_1$

   b. $A_1 \cup \phi = A_1$, $A_1 \cap \phi = \phi$

6. complement law

$$A_1 \cup \bar{A}_1 = U,$$
$$A_1 \cap \bar{A}_1 = \phi$$

7. De Morgan's law

$$\overline{A_1 \cup A_2} = \bar{A}_1 \cap \bar{A}_2,$$
$$\overline{A_1 \cap A_2} = \bar{A}_1 \cup \bar{A}_2$$

The two sets A and B are said to be in *one-to-one correspondence* if each element $a \in A$ is paired with one and only one element $b \in B$ in such a manner that each element of B is paired with exactly one element of A.

Sets are classified as finite or infinite. A set is *finite* if it contains exactly n objects, where n is a nonnegative integer. A set is *infinite* if it is not finite.

Example A.2.4

If $A_1 = \{x : x = a_1, a_2, a_3, a_4\}$ and $A_2 = \{x : x = 1, 2, 3, 4\}$, then A_1 and A_2 are equivalent sets; that is, the elements of A_1 are in a one-to-one correspondence with the elements of A_2.

Example A.2.5

If A_1 is a set containing all positive integers less than or equal to 100, that is, if

$$A_1 = \{x : x = 1, 2, 3, ..., 100\},$$

then A_1 is a finite set.

Example A.2.6

If A_2 is a set containing all positive odd integers, that is, if

$$A_2 = \{x : x = 1, 3, 5, ...\},$$

then A_2 is an infinite set.

A set whose elements can be put into a one-to-one correspondence with the set of all positive integers is referred to as being a *countably infinite* set. Also a set is said to be *countable, denumerable,* or *enumerable* if it is finite or countably infinite.

Example A.2.7

The set defined in Example A.2.6, that is,

$$A_2 = \{x : x = 1, 3, 5, ...\},$$

is composed of elements that can be put into a one-to-one correspondence with the set of all positive integers. Thus, this set is countably infinite.

Example A.2.8

If $A_1 = \{x : 0 \leq x \leq 5, x \text{ a real number}\}$, then the set A_1 is not a countable set, because any interval on the real line contains an infinite number of elements. However, a rigorous proof is beyond the scope of this text.

The *product* or *Cartesian product* of the sets A and B is denoted by $A \times B$ and consists of all ordered pairs (a,b), where $a \in A$ and $b \in B$; that is,

$$A \times B = \{(a,b) : a \in A, b \in B\}.$$

Example A.2.9

If
$$A = \{x : x = a_1, a_2, a_3\}$$
and
$$B = \{x : x = 1, 2\},$$
then
$$A \times B = \{(a_1,1),(a_1,2),(a_2,1),(a_2,2),(a_3,1),(a_3,2)\}.$$

The Cartesian product of the sets A, B, and C is given by
$$A \times B \times C = \{(a,b,c) : a \in A, b \in B, c \in C\};$$
that is, the product set $A \times B \times C$ consists of all possible ordered 3-tuples. Hence, the notion of a Cartesian set can be extended to any finite number of sets; that is, $A_1 \times A_2 \times \cdots \times A_n$ is the set of all ordered n-tuples, $(a_1, a_2, ..., a_n)$, where $a_1 \in A_1, a_2 \in A_2, ..., a_n \in A_n$.

These basic concepts of set theory constitute only a brief survey of the area. They are intended to serve only as a helpful review for a better understanding of the formulation of the fundamental concepts of probability theory.

A.3 Permutations, Combinations, and Stirling's Formula

We begin this section by stating two basic rules that are commonly used in combinatorial mathematics: If a particular happening or event can take place in m different ways and a second event can occur in n different ways, then (1) the number of ways that either the first or the second event may occur is $m+n$ and (2) the number of ways that both the first and the second event may occur is $m \cdot n$.

The arrangement of a set of objects in some form or order in a straight line is called a *permutation* of these objects. That is, if we have a set of n distinct elements and we wish to arrange r of them in some order in a straight line, we refer to this arrangement as a permutation of n distinct elements taken r at a time. We denote the total number of such arrangement as P_r^n. Other commonly used symbols are $P(n,r)$, ${}^nP_r$, ${}_nP_r$, P_{nr}, and $(n)_r$.

Example A.3.1

Find the total number of permutations of the elements a_1, a_2, and a_3 taken two at a time, that is, P_2^3.

Solution: Any one of these elements can be put in the first position, and either of the two remaining elements can be put in the second position. Therefore, by rule 2, there are $3 \cdot 2 = 6$ permutations. These are

$$a_1, a_2; \; a_1, a_3; \; a_2, a_3;$$
$$a_2, a_1; \; a_3, a_1; \; a_3, a_2.$$

Example A.3.2

The number of permutations of the four elements a_1, a_2, a_3, and a_4 taken three at a time is similarly $4 \cdot 3 \cdot 2 \cdot 1 = 24$. These are

$a_1, a_2, a_3;$	$a_1, a_2, a_4;$	$a_1, a_3, a_4;$	$a_2, a_3, a_4;$	$a_3, a_1, a_2;$
$a_1, a_3, a_2;$	$a_1, a_4, a_2;$	$a_1, a_4, a_3;$	$a_2, a_4, a_3;$	$a_4, a_1, a_2;$
$a_2, a_1, a_3;$	$a_2, a_1, a_4;$	$a_3, a_1, a_4;$	$a_3, a_2, a_4;$	$a_4, a_1, a_3;$
$a_2, a_3, a_1;$	$a_2, a_4, a_1;$	$a_3, a_4, a_1;$	$a_3, a_4, a_2;$	$a_4, a_2, a_3;$
$a_3, a_2, a_1;$	$a_4, a_2, a_1;$	$a_4, a_3, a_1;$	$a_4, a_3, a_2.$	

By considering similar examples with larger numbers of objects, one arrives at the following formula:

$$P_n^n = n(n-1)(n-2)\cdots 3\cdot 2\cdot 1. \tag{A.3.1}$$

That is, the first position can be filled by any one of the n distinct objects, the next position by any of the remaining $n-1$, and so on. We denote the product of the positive integers, as shown in Equation (A.3.1), as $n!$, which is read n *factorial*. Hence, $5! = 5\cdot 4\cdot 3\cdot 2\cdot 1$ and $(r-1)! = (r-1)(r-2)\cdots 3\cdot 2\cdot 1, 0! = 1(r-1)! = (r-1)(r-2)\ldots 3.2.1$. We defines $0! = 1$, the reason for which is shown in a later section. If r is less than n, we have

$$P_r^n = n(n-1)(n-2)\cdots(n-r+1)$$

$$= \frac{n(n-1)(n-2)\cdots(n-r+1)(n-r)(n-r-1)\cdots 3\cdot 2\cdot 1}{(n-r)(n-r-1)\cdots 3\cdot 2\cdot 1}$$

$$= \frac{n!}{(n-r)!}. \tag{A.3.2}$$

Thus, the number of permutations of 10 objects taken four at a time is

$$P_4^{10} = \frac{10!}{(10-4)!} = \frac{10!}{6!} = 5040.$$

If we are interested in finding the total number of ways in which we can choose r objects out of n, disregarding the order in which they are chosen, we speak of a *combination of n distinct objects taken r at a time;* we denote this using the symbol $\binom{n}{r}$. Other commonly used symbols are $C_{n,r}$, C_r^n, ^nC_r, and $C(n,r)$.

Example A.3.3

The number of combinations of the elements a_1, a_2, and a_3 taken two at time is three, that is, $a_1, a_2; a_1, a_3;$ and a_2, a_3. Each of these combinations gives rise to 2! permutations, as shown in Example A.3.1. Thus, eliminating the order of arrangement, we see that

$$\binom{3}{2} = \frac{P_2^3}{2!} = \frac{3!}{2!(3-2)!} = 3.$$

Example A.3.4

The number of combinations of the elements a_1, a_2, a_3, and a_4 taken three at a time is four, that is, a_1, a_2, a_3; a_1, a_2, a_4; a_1, a_3, a_4; and $a_3, a_2,$ and a_4. Eliminating the order of arrangement in Example A.3.2, we have

$$\binom{4}{3} = \frac{P_3^4}{3!} = \frac{4!}{3!(4-3)!} = 4.$$

Therefore, we can conclude that the total number of ways in which we can choose r objects out of n, $r < n$, disregarding the order of choice, gives

$$\binom{n}{r} = \frac{P_r^n}{r!} = \frac{n!}{r!(n-r)!}.$$

Note that $\binom{n}{n} = \binom{n}{0} = 1$ and $\binom{n}{1} = n$. Also, $\binom{n}{r} = \binom{n}{n-r}$, which implies that if we select n-r objects out of n distinct objects, we select at the same time n-r objects to be left alone.

Example A.3.5

a. $\binom{9}{6} = \dfrac{9!}{6!(9-6)!} = 84$

b. $\binom{a}{a-2} = \dfrac{a!}{(a-2)![a-(a-2)]!} = \dfrac{a!}{2!(a-2)!} = \dfrac{a(a-1)}{2}$

c. $\binom{n-1}{3} = \dfrac{(n-1)!}{3!(n-4)!} = \dfrac{(n-1)(n-2)(n-3)(n-4)!}{3!(n-4)!} = \dfrac{(n-1)(n-2)(n-3)}{3 \cdot 2 \cdot 1}$

The important point in this section: Permutations take order into consideration, whereas combinations do not.

When n, the number of elements in a set, is large, the direct evaluation of $n!$ is impractical. A good approximation for $n!$ is given by *Stirling's formula*:

$$n! \approx n^n e^{-n} \sqrt{2\pi n}.$$

Here, e is the Naperian base, 2.71828....

Using logarithmic tables, it is easy to compute the value of the preceding expression for any number n.

A.4 Binomial and Multinomial Theorems

One of the important theorems in elementary algebra is the binomial theorem. This theorem gives the expansion of the binomial expression $(x+y)^n$, where n is any positive integer, as follows:

$$(x+y)^n = x^n + \frac{n}{1}x^{n-1}y + \frac{n(n-1)}{1 \cdot 2}x^{n-2}y^2 + \frac{n(n-1)(n-2)}{1 \cdot 2 \cdot 3}x^{n-3}y^3 + \cdots$$

$$+ \frac{n(n-1)(n-2)\cdots(n-r+1)}{1 \cdot 2 \cdot 3 \cdots r}x^{n-r}y^r + \cdots + y^n.$$

The proof of this theorem is made by mathematical induction. If the sign in the binomial expression is negative, the expansion begins with a plus and then alternates its sign.

Example A.4.1

Expand and simplify.

a. $(\frac{x}{2}+y)^5 = \frac{1}{32}x^5 + \frac{5}{16}x^4y^2 + \frac{5}{4}x^3y^4 + \frac{5}{2}x^2y^6 + \frac{5}{2}xy^8 + y^{10}$

b. $(2x - \frac{y^2}{2})^6 = 64x^6 - 96x^5y^2 + 60x^4y^4 - 20x^3y^6 + \frac{15}{4}x^2y^8 - \frac{3}{8}xy^{10} + \frac{1}{64}y^{12}$

The binomial theorem can also be expressed in terms of combinations:

$$(x+y)^n = \binom{n}{0}x^n + \binom{n}{1}x^{n-1}y + \binom{n}{2}x^{n-2}y^2 + \cdots + \binom{n}{n-1}x^{n-1}y + \binom{n}{n}y^n$$

This form of theorem is relatively simple to work with in various applications. For example, the coefficient of the sixth term in the expansion of $(x+y)^{20}$ is given by $\binom{20}{5} = \frac{20!}{5!15!}$; in general, the coefficient of the jth term in the expansion of $(x+y)^n$ is

$$\binom{n}{j-1} = \frac{n!}{(j-1)!(n-j+1)!}.$$

A generalization of the binomial theorem is the *multinomial theorem*, which may be expressed as

$$(x_1 + x_2 + \cdots + x_k)^n = \sum \binom{n}{n_1, n_2, \ldots, n_k} x_1^{n_1} x_2^{n_2} \cdots x_k^{n_k}$$

$$= \sum \frac{n!}{n_1! n_2! \ldots n_k!} x_1^{n_1} x_2^{n_2} \cdots x_k^{n_k},$$

where the sum is taken over all permutations of nonnegative integers $n_1, n_2, \ldots, n_k$ such that $n_1 + n_2 + \ldots + n_k = n$.

A physical interpretation of the coefficient of any one of the terms in the multinomial expansion can be thought of as follows: Consider a set A as containing n elements, and let $n_1, n_2, ..., n_k$ be positive integers such that $n_1 + n_2 + ... + n_k = n$. We are interested in the total number of ways in which an unknown number of objects of one kind, of a second kind, and so on, through to an unknown number of objects of a kth kind may be arranged in n cells. This number is given by $\binom{n}{n_1, n_2, ..., n_k}$. In most applications of this theorem, we are interested in obtaining the coefficient of $x_1^2 x_2^4 x_3^0 x_4^3$, which is a term of the generating function or polynomial $(x_1 + x_2 + x_3 + x_4)^9$, that is, $\frac{9!}{2!4!0!3!}$. This number is simply the total number of ways in which two of one kind, four of a second kind, none of a third kind, and three of a fourth kind can be selected and arranged.

Example A.4.2

The coefficient of the term $x_1^2 x_2^3 x_3^4 x_4^0 x_5^1$ in the expansion of the polynomial $(x_1 + x_2 + x_3 + x_4 + x_5)^{10}$ is

$$\binom{10}{2,3,4,0,1} = \frac{10!}{2!3!4!0!1!}$$

We now state some identities and series that are useful in the study of probability theory:

1. If in the binomial theorem, we let $x = p$ and $y = q$ such that $p + q = 1$, then

$$(p+q)^n = \sum_{i=0}^{n}\binom{n}{i} p^{n-i} q^i = 1$$

2. $\dfrac{(1-x^n)}{(1-x)} = \sum_{i=0}^{n-1} x^i$

3. $\dfrac{1}{(1-x^n)} = \sum_{i=0}^{\infty} \binom{n+i-1}{i} x^i$

4. $\dfrac{n(n+1)}{2} = \sum_{i=0}^{n} i$, $\dfrac{n(n+1)(2n+1)}{6} = \sum_{i=1}^{n} i^2$

5. $\binom{n+m}{r} = \sum_{i=0}^{r} \binom{n}{i}\binom{m}{r-i}$

6. $e^x = \exp(x) = \sum_{i=0}^{\infty} \dfrac{x^i}{i!}$

A.5 Jacobians

In many basic mathematical calculations, it is essential to make various transformations, that is, to change from a function of several variables to that of others. When the transformation is from one variable to another variable, the differential is enough to formulate the new function; however, when it is necessary to make transformations that involve more than one variable, we need the concept of *Jacobians*.

In this section, we show how Jacobians are obtained, and we illustrate the procedure with examples.

Suppose that we are given a function $f(x,y)$ and we are interested in obtaining a new function $h(u,v)$, where the relation between the new and the old variables is given by the following mapping:

$$u = h_1(x, y)$$

and

$$v = h_2(x, y).$$

Under the assumption that this mapping is a one-to-one transformation, then its inverse exists:

$$x = g_1(u, v)$$

and

$$y = g_2(u, v).$$

This relationship may be interpreted as mapping a region G_1 of the x-y plane into a region G_2 of the u-v plane. Then, under the condition that both $g_1(u,v)$ and $g_2(u,v)$ have continuous partial derivatives with respect to u and v, we can obtain the new function $h(u,v)$ as follows:

$$h(u,v) = f[g_1(u,v), g_2(u,v)]|J|$$

so that

$$\iint_{G_1} f(x,y) dx dy = \iint_{G_2} f[g_1(u,v), g_2(u,v)] |J| du dv.$$

Here, the J is the so-called Jacobian of the transformation and is defined by the determinant:

$$J = \frac{\partial(x,y)}{\partial(u,v)} = \begin{vmatrix} \frac{\partial x}{\partial u} & \frac{\partial x}{\partial v} \\ \frac{\partial y}{\partial u} & \frac{\partial y}{\partial v} \end{vmatrix}.$$

Similarly, the preceding notion can be extended to a transformation involving three or more variables. Assuming again that we have a continuous partial derivative with respect to

u, v, and w and that they are one to one, we now illustrate how the Jacobian is obtained for the following transformation:

$$x = g_1(u,v,w)$$
$$y = g_2(u,v,w)$$
$$z = g_3(u,v,w)$$

$$J = \frac{\partial(x,y,z)}{\partial(u,v,w)} = \begin{vmatrix} \frac{\partial x}{\partial u} & \frac{\partial x}{\partial v} & \frac{\partial x}{\partial w} \\ \frac{\partial y}{\partial u} & \frac{\partial y}{\partial v} & \frac{\partial y}{\partial w} \\ \frac{\partial u}{\partial u} & \frac{\partial v}{\partial v} & \frac{\partial w}{\partial w} \end{vmatrix}.$$

Example A.5.1

Find the Jacobian of transformation of

$$x = 5u - 3v,$$
$$y = 3u + 2v.$$

Solution:

$$J = \frac{\partial(x,y)}{\partial(u,v)} = \begin{vmatrix} \frac{\partial x}{\partial u} & \frac{\partial x}{\partial v} \\ \frac{\partial y}{\partial u} & \frac{\partial y}{\partial v} \end{vmatrix} = \begin{vmatrix} 5 & -3 \\ 3 & 2 \end{vmatrix} = 1$$

Example A.5.2

Find the Jacobian of transformation of the polar coordinate functions of the form

$$t = r \sin\theta,$$
$$z = r \cos\theta.$$

Solution: Here,

$$J = \frac{\partial(t,z)}{\partial(r,\theta)} = \begin{vmatrix} \frac{\partial t}{\partial r} & \frac{\partial t}{\partial \theta} \\ \frac{\partial z}{\partial r} & \frac{\partial z}{\partial \theta} \end{vmatrix} = \begin{vmatrix} \sin\theta & r\cos\theta \\ \cos\theta & -r\sin\theta \end{vmatrix}$$

$$= -r\sin^2\theta - r\cos^2\theta = -r(\sin^2\theta + \cos^2\theta) = -r.$$

Therefore,

$$|J| = r.$$

A precise theoretical development of this subject can be found in References [5], [8], and [9]

A.6 Gamma Function

The one-parameter integral

$$\Gamma(p) = \int_0^\infty x^{p-1} e^{-x} dx, \, p > 0 \tag{A.6.1}$$

is called the *gamma function*. In Equation (A.6.1), if we replace p with $p+1$ and integrate it by parts, we find that

$$\Gamma(p+1) = \int_0^\infty x^p e^{-x} dx = -x^p e^{-x} \Big|_0^\infty + p \int_0^\infty x^{p-1} e^{-x} dx \tag{A.6.2}$$

$$= p \int_0^\infty x^{p-1} e^{-x} dx$$

$$= p\Gamma(p).$$

If we replace p with $p-1$ in Equation (A.6.2), this becomes

$$\Gamma(p) = (p-1)\Gamma(p-1). \tag{A.6.3}$$

Continuing in this way, we find that

$$\Gamma(p) = (p-1)(p-2)\Gamma(p-2)$$

$$= (p-1)(p-2)(p-3)\Gamma(p-3)$$

$$= (p-1)(p-2)(p-3)(p-4)\cdots\Gamma(1),$$

where

$$\Gamma(1) = \int_0^\infty e^{-x} dx = -e^{-x} \Big|_0^\infty = 1.$$

For example,

$$\Gamma(2) = 1\Gamma(1) = 1,$$

$$\Gamma(3) = 2\Gamma(2) = 2 \cdot 1\Gamma(1) = 2,$$

$$\Gamma(4) = 3\Gamma(3) = 3 \cdot 2\Gamma(2) = 3 \cdot 2 \cdot 1\Gamma(1) = 6,$$

and

$$\Gamma(5) = 4\Gamma(4) = 4 \cdot 3\Gamma(3) = 4 \cdot 3 \cdot 2\Gamma(2) = 4 \cdot 3 \cdot 2 \cdot 1\Gamma(1) = 24.$$

Therefore, for any positive integer n, we have

$$\Gamma(n) = (n-1)(n-2)\cdots 3 \cdot 2 \cdot 1 = (n-1)!.$$

Solving Equation (A.6.3) for $\Gamma(p-1)$ gives

$$\Gamma(p-1) = \frac{\Gamma(p)}{p-1}.$$

This expression may be used to evaluate values of $\Gamma(p)$ when $p < 0$.

In a number of applications of the gamma function, p is either a positive integer or a multiple of $1/2$. Hence, it is important to be able to compute $(1/2)!$, that is, $\Gamma(3/2)$:

$$\Gamma(\tfrac{3}{2}) = \left(\tfrac{1}{2}\right)\Gamma(\tfrac{1}{2}) = \tfrac{1}{2}\int_0^\infty x^{-1/2} e^{-x} dx. \tag{A.6.4}$$

Let $x = t^2/2$ and $dx = t\,dt$ in Equation (A.6.4). We obtain

$$\Gamma(\tfrac{3}{2}) = \left(\tfrac{1}{\sqrt{2}}\right)\int_0^\infty e^{-(1/2)t^2} dt.$$

It can be shown that

$$\left(\tfrac{1}{\sqrt{2\pi}}\right)\int_0^\infty e^{-(1/2)t^2} dt = \tfrac{1}{2}.$$

Therefore,

$$\Gamma(1/2) = \sqrt{\pi}$$

and

$$\Gamma(3/2) = \tfrac{\sqrt{\pi}}{2} = (1/2)!.$$

Example A.6.1

Evaluate the following factorials:

a. $\left(\tfrac{3}{2}\right)! = \left(\tfrac{3}{2}\right)\left(\tfrac{1}{2}\right)! = \tfrac{3\sqrt{\pi}}{4}$

b. $\left(\tfrac{7}{2}\right)! = \left(\tfrac{7}{2}\right)\cdot\left(\tfrac{5}{2}\right)! = \left(\tfrac{7}{2}\right)\cdot\left(\tfrac{5}{2}\right)\cdot\left(\tfrac{3}{2}\right)! = \left(\tfrac{7}{2}\right)\cdot\left(\tfrac{5}{2}\right)\cdot\left(\tfrac{3}{2}\right)\cdot\left(\tfrac{1}{2}\right)!$

$= \left(\tfrac{7}{2}\right)\cdot\left(\tfrac{5}{2}\right)\cdot\left(\tfrac{3}{2}\right)\left(\tfrac{\sqrt{\pi}}{2}\right) = \tfrac{105\sqrt{\pi}}{16}$

Example A.6.2

Evaluate the following integrals:

a. $\int_0^\infty x^{6/7} e^{-x} dx = \Gamma(\tfrac{13}{7})$

b. $\int_0^\infty x^{-1/3} e^{-x} dx = \Gamma(\tfrac{2}{3})$

c. $\int_0^\infty x^3 e^{-3x} dx$; let $t = 3x$, $\tfrac{dt}{3} = dx$; then, $\int_0^\infty x^3 e^{-3x} dx$ becomes $\tfrac{1}{81}\int_0^\infty t^3 e^{-t} dt = \tfrac{1}{81}\Gamma(4) = \tfrac{2}{27}$

For any positive value of p, the gamma function may well be interpreted as the area under the function $f(x) = x^{p-1} e^{-x}$ from $x = 0$ to $x = \infty$, as shown in Figure A.6.1 for various values of p.

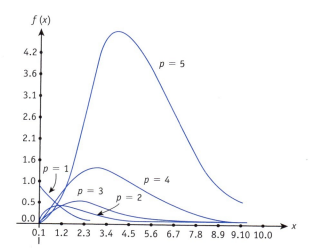

Figure A.6.1

Figure A.6.2 can be used to approximate $\Gamma(p)$ for various positive and negative values of p when the gamma function is difficult to evaluate.

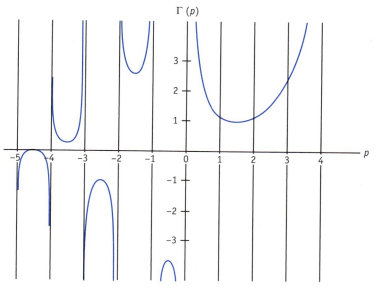

Figure A.6.2

Example A.6.3

From Figure A.6.2, we can approximate the following gamma functions:

a. $\Gamma(1.10) \approx 0.96$

b. $\Gamma(1.30) \approx 0.89$

c. $\Gamma(1.70) \approx 0.91$

d. $\Gamma(1.90) \approx 0.96$

A.7 Beta Functions

The two parameter integral defined by

$$B(p,q) = \int_0^1 y^{p-1}(1-y)^{q-1} dy, \, p > 0, q > 0$$

is called the beat function. This function is related to the gamma function in the following way.

$$B(p,q) = \frac{\Gamma(p)\Gamma(q)}{\Gamma(p+q)}. \qquad (A.7.1)$$

This relationship can be shown as follows:
In the function $\Gamma(p)$, if we apply the transformation $x = y^2$, $dx = 2y\,dy$, we obtain

$$\Gamma(p) = \int_0^\infty x^{p-1} e^{-x} dx = 2\int_0^\infty y^{2p-1} e^{-y^2} dy. \qquad (A.7.2)$$

Similarly,

$$\Gamma(q) = 2\int_0^\infty z^{2p-1} e^{-z^2} dz. \qquad (A.7.3)$$

Multiplying the functions in Equations (A.7.1) and (A.7.3) gives

$$\Gamma(p)\Gamma(q) = 4\int_0^\infty z^{2p-1} e^{-z^2} dz \cdot \int_0^\infty y^{2p-1} e^{-y^2} dy$$

$$= 4\int_0^\infty \int_0^\infty z^{2p-1} y^{2p-1} e^{-(z^2+y^2)} dz\,dy. \qquad (A.7.4)$$

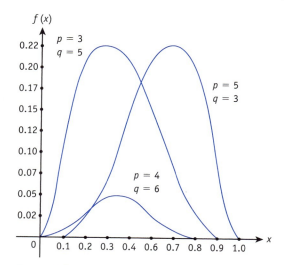

Figure A.7.1

To simplify the double integral in Equation (A.7.4), we must change to polar coordinates. That is, for a positive radius vector r, as shown in Figure A.7.1, we have

$$z = +\sqrt{z^2 + y^2},$$

$$z = r\cos\phi,$$

$$y = r\sin\phi.$$

Substituting these values into Equation (A.7.4); multiplying by $|J|$, the absolute value of the Jacobian that was shown in Example A.5.2 to be equal to r; and restricting ourselves to the first quadrant, that is, ϕ varying from 0 to $\pi/2$ and r from 0 to ∞, we obtain

$$\Gamma(p)\Gamma(q) = 4\int_0^{\frac{\pi}{2}} \{(r\cos\phi)^{2q-1}(r\sin\phi)^{2p-1}e^{-r^2}r\,dr\}d\phi$$

$$= 4\int_0^{\frac{\pi}{2}} \{e^{-r^2}r^{2p+2q-1}\cos^{2q-1}\phi\sin^{2p-1}\phi dr\}d\phi$$

$$= \{2\int_0^{\infty} e^{-r^2}r^{2q+2p-1}dr\}\{2\int_0^{\frac{\pi}{2}}\cos^{2q-1}\phi\sin^{2p-1}\phi d\phi\}. \quad (A.7.5)$$

Replacing q with $p+q$ and z with r in the function in Equation (A.7.3), we have

$$\Gamma(p+q) = 2\int_0^{\infty} r^{2(q+p)-1}e^{-r^2}dr. \quad (A.7.6)$$

Now, if we apply the trigonometric transformation $y = \sin^2\phi$ (which implies $1 - y = \cos^2\phi$), $dy = 2\cos\phi\sin\phi d\phi$ to the beta function, we obtain

$$B(p,q) = \int_0^1 y^{p-1}(1-y)^{q-1}dy$$

$$= 2\int_0^{\frac{\pi}{2}} \sin^{2p-1}\phi\cos^{2q-1}\phi d\phi. \quad (A.7.7)$$

Inspecting Equations (A.7.5) through (A.7.7), it is clear that the relationship

$$B(p,q) = \frac{\Gamma(p)\Gamma(q)}{\Gamma(p+q)}, p,q > 0,$$

is valid. This relationship of the beta function to the gamma function shows that $B(p,q)$ is symmetrical with respect to the parameters p and q; that is,

$$B(p,q) = B(q,p).$$

Example A.7.1

Evaluate the following definite integrals:

a. $\int_0^1 y^4(1-y)^5 dy = \Gamma(5,6) = \frac{\Gamma(5)\Gamma(6)}{B(11)} = \frac{1}{1260}$

b. $\int_0^1 \frac{y^2 dy}{\sqrt{1-y^2}}$; let $u = y^2$, $\frac{du}{2\sqrt{u}} = dy$ so that

$\int_0^1 \frac{y^2 dy}{\sqrt{1-y^2}} = \int_0^1 u^{1/2}(1-u)^{-1/2} du = \frac{1}{2} B(\frac{3}{2}, \frac{1}{2})$

$= \frac{1}{2} \frac{\Gamma(\frac{3}{2})\Gamma(\frac{1}{2})}{\Gamma(2)} = \frac{\Pi}{4}$

c. $\int_0^2 y^3(1-(\frac{y}{2}))^2 dy$, let $u = \frac{1}{2}y$, $2du = dy$ so that

$\int_0^2 y^3(1-(\frac{y}{2}))^2 dy = 16\int_0^1 u^3(1-u)^2 du = 16 B(4,3)$

$= 16 \frac{\Gamma(\frac{3}{2})\Gamma(\frac{1}{2})}{\Gamma(2)} = \frac{4}{15}$

Example A.7.1(c) illustrates an interesting point: the range of the variable y need not be restricted between 0 and 1, because a linear transformation gives us the domain of the beta function.

For various positive values of p and q, the beta function may be interpreted as the area under the function $f(x) = x^{p-1}(1-x)^{q-1}$ from $x=0$ to $x=1$, as shown in the Figure A.7.3. Observe that the parameters p and q determine the shape of the curves.

A.8 Matrices

In this section, we present a brief review of some basic definitions and operations of matrix algebra. This review will be helpful when you begin to study the formulation of the basic concepts of finite Markov chains.

A matrix is a rectangular array of elements of the form

$$A = \begin{bmatrix} a_{11} & a_{12} & a_{13} & \cdots & a_{1n} \\ a_{21} & a_{22} & a_{23} & \cdots & a_{2n} \\ \cdots \\ \cdots \\ a_{m1} & a_{m2} & a_{m3} & \cdots & a_{mn} \end{bmatrix}.$$

The letters a_{ij} denote real numbers or functions, which we refer to as the *elements* of the matrix. The i in the double subscript of a_{ij} indicates the row, and the second subscript letter, j, indicates the column of the array in which the element is located.

We denote the $m \times n$ matrix A as $[a_{ij}]$, where it is understood that $i = 1, 2, ..., m$ and $j = 1, 2, ..., n$. The integers m and n denote the number of rows and columns, respectively, that constitute the array. We usually refer to a matrix by stating its dimension (order); that is, A is an $m \times n$ matrix.

A matrix that has the same number of rows and columns, that is, $m = n$, is called a *square matrix* of the order n or m. In a square matrix, the elements $a_{11}, a_{22}, ..., a_{nn}$ are called the *diagonal elements,* and the sum of these elements is called the trace of the matrix. If $m = 1$, that is, if the matrix consists of a single row, we call it a *row vector*. Analogously if $n = 1$, or the matrix consists of a single column, and we call it a *column vector.* Thus, we can display these two vectors, terming them a and b, as follows:

$$a = [a_1, a_2, a_3, ..., a_n], \quad b = \begin{bmatrix} b_1 \\ b_2 \\ b_3 \\ \cdot \\ \cdot \\ \cdot \\ b_m \end{bmatrix}.$$

Two matrices A and B are *equal,* $A = B$, if and only if they have the same dimensions and each element of A is equal to the corresponding element of B—that is, if and only if

$$a_{ij} = b_{ij}, \quad i = 1, 2, ..., m, \quad j = 1, 2, ..., n.$$

Similarly, we say $A > B$ if $a_{ij} > b_{ij}$ for all i and j. That is, each element of A is greater than the corresponding element of B.

A square matrix with elements in the main diagonal equal to one and all other elements equal to zero is called a *unit matrix,* and we denote it as I. A zero matrix is a matrix in which all elements are zero, and we denote it as 0.

We now define some basic operations and relations of matrices.

1. If A is a matrix with the elements a_{ij} and if c is a scalar, then $cA = [ca_{ij}]$. That is, multiplying matrix A by a constant means multiplying each element of the matrix by its constant.

Example A.8.1

If
$$A = \begin{bmatrix} 3 & -1 \\ 0 & 2 \end{bmatrix},$$

then
$$nA = \begin{bmatrix} 3n & -n \\ 0 & 2n \end{bmatrix}.$$

2. Let $A=[a_{ij}]$ and $B=[b_{ij}]$ be two $m \times n$ matrices. The sum of the matrices, $A+B$, is defined as an $m \times n$ matrix C, where each element of C is the sum of the corresponding elements of A and B; that is,

$$A + B = [a_{ij} + b_{ij}] = [c_{ij}] = C.$$

The difference of the two matrices is defined similarly; that is,

$$A - B = [a_{ij} - b_{ij}] = [d_{ij}] = D.$$

Example A.8.2

If
$$A = \begin{bmatrix} -1 & 2 \\ 3 & 0 \\ 1 & 4 \end{bmatrix}$$

and
$$B = \begin{bmatrix} 2 & 4 \\ 5 & 6 \\ -2 & 1 \end{bmatrix},$$

then
$$A + B = \begin{bmatrix} 1 & 6 \\ 8 & 6 \\ -1 & 5 \end{bmatrix}$$

and
$$A - B = \begin{bmatrix} -3 & -2 \\ -2 & -6 \\ 3 & 3 \end{bmatrix}.$$

Two matrices of different orders cannot be added or subtracted; we speak of such matrices as not being *conformable* for addition or subtraction. Two matrices that can be added or subtracted are conformable.

3. Let A be an $m \times n$ and B an $n \times k$ matrix. The product of these matrices, AB, is an $m \times k$ matrix whose elements are given by $\sum_{t=1}^{n} a_{it} b_{tj}$, $i = 1, 2, ..., m$, $j = 1, 2, ..., k$. That is, the product operation is row by column; each element of the row is multiplied by the corresponding element of the column, and these products are summed.

Example A.8.3

If
$$A = \begin{bmatrix} a_{11} & a_{12} & a_{13} \\ a_{21} & a_{22} & a_{23} \end{bmatrix} \quad (2 \times 3)$$

and
$$B = \begin{bmatrix} b_{11} & b_{12} \\ b_{21} & b_{22} \\ b_{31} & b_{32} \end{bmatrix} \quad (3 \times 2)$$

then
$$AB = \begin{bmatrix} a_{11}b_{11} + a_{12}b_{21} + a_{13}b_{31} & a_{11}b_{12} + a_{12}b_{22} + a_{13}b_{32} \\ a_{21}b_{11} + a_{22}b_{21} + a_{23}b_{31} & a_{21}b_{12} + a_{22}b_{22} + a_{23}b_{32} \end{bmatrix} \quad (2 \times 2).$$

Example A.8.4

If C is an $m \times 1$ matrix and D is a $1 \times n$ matrix, then the product CD in that order is an $m \times n$ matrix. For example, if

$$C = \begin{bmatrix} c_{11} \\ c_{21} \\ \vdots \\ c_{m1} \end{bmatrix}$$

and
$$D = \begin{bmatrix} d_{11} & d_{21} & \cdots & d_{n1} \end{bmatrix} \quad (1 \times n),$$

then
$$CD = \begin{bmatrix} c_{11}d_{11} & c_{11}d_{21} & \cdots & c_{11}d_{n1} \\ c_{21}d_{11} & c_{21}d_{21} & \cdots & c_{21}d_{n1} \\ c_{m1}d_{11} & c_{m1}d_{21} & \cdots & c_{m1}d_{n1} \end{bmatrix} \quad (m \times n).$$

Example A.8.5

If
$$A = \begin{bmatrix} -2 & 1 \\ 3 & 4 \end{bmatrix} \ (2 \times 2)$$

and
$$B = \begin{bmatrix} -1 & 4 & 0 \\ 5 & 3 & -2 \end{bmatrix} \ (2 \times 3),$$

then
$$AB = \begin{bmatrix} 7 & -5 & -2 \\ 17 & 24 & -8 \end{bmatrix} \ (2 \times 3).$$

Example A.8.6

If E is a $1 \times m$ matrix and F is an $m \times 1$ matrix, then EF is a 1×1 matrix. That is, if
$$E = [e_{11}, e_{12}, \ldots, e_{1m}] \ (1 \times m)$$

and
$$F = \begin{bmatrix} f_{11} \\ f_{21} \\ \vdots \\ f_{m1} \end{bmatrix} \ (m \times 1),$$

then
$$EF = e_{11}f_{11} + e_{12}f_{21} + \cdots + e_{1m}f_{m1}$$
$$= \sum_{k=1}^{m} e_{1k} f_{k1}.$$

However, the product FE is an $m \times m$ whose ijth element is $f_{i1}e_{1j}$, $i, j = 1, 2, \ldots, m$.

Hence, for the product of two matrices to exist, the number of columns of the first matrix must be equal to the number of rows of the second matrix. Also, if A and B are $n \times n$ matrices, then the ijth element of the products AB and BA is, respectively,

$$\sum_{k=1}^{n} a_{ik} b_{kj}$$

and

$$\sum_{k=1}^{n} b_{ik} a_{kj}.$$

With respect to the preceding matrix operations, we state the following theorem.

Theorem A.8.1

If the matrices A, B, and C are conformable to the indicated operation, then we have

1. communicative law: $A + B = B + A$
2. associative law:
   a. $[A + B] + C = A + [B + C]$
   b. $[AB]C = A[BC]$
3. scalar law: $cA + cB = c[A + B] = [A + B]c$
4. distributive law:
   a. $AB + AC = A[B + C]$
   b. $AC + BC = [A + B]C$

Let A be a square matrix whose elements are real numbers. By "the determinant of matrix A," we mean a unique real number denoted by $|A|$ or $\det A$ and obtained by

$$|A| = \det A = \sum_r \alpha_{i_1, i_2, \ldots, i_n} a_{1 i_1} a_{2 i_2} \ldots a_{n i_n}.$$

Here, the second subscripts of the product $\alpha_{i_1, i_2, \ldots, i_n} a_{1 i_1} a_{2 i_2} \ldots a_{n i_n}$, when arranged in the sequence $i_1, i_2, \ldots, i_n$, form one of the $n!$ permutations of the integers $1, 2, \ldots n$. In addition, $\alpha_{i_1, i_2, \ldots, i_n}$ is -1 or $+1$ depending on whether the permutation is odd or even, respectively—that is, whether the number of inversions in the subscript of α is odd or even, respectively. Only one element comes from any column. The summation is over all permutations $i_1, i_2, \ldots, i_n$ of the integers $1, 2, \ldots, n$. To determine when $\alpha_{i_1, i_2, \ldots, i_n}$ is -1 and when it is $+1$, we obtain the number of inversion in the permutation of the second subscript of the elements of the product. For example, α_{1234} is $+1$ because the number of inversion of the permutation 1234 is zero, or an even number. But α_{1243} is -1 because the number of inversion of the permutation 1243 is one, or an odd number, and it has one inversion: 4 precedes 3. Finally, α_{1432} is -1 because the number of inversion of the permutation 1432 is three, or an odd number: 4 precedes 3, 4 precedes 2, and 3 precedes 2.

The determinant of $n \times n$ matrix is said to be of the order n. The following examples clarify and illustrate the manner in which we use the preceding definition to obtain the determinant of a square matrix.

Example A.8.7

If

$$A = \begin{bmatrix} a_{11} & a_{12} \\ a_{21} & a_{22} \end{bmatrix},$$

then the determinant A is

$$|A| = \begin{vmatrix} a_{11} & a_{12} \\ a_{21} & a_{22} \end{vmatrix} = \alpha_{12} a_{11} a_{22} + \alpha_{21} a_{12} a_{21}, = a_{11} a_{22} - a_{12} a_{21}.$$

The sign of α_{12} is $+1$ because there is no inversion in the permutation 12, but α_{21} is -1 because the number of inversions of the permutation 21 is one, or an odd number (because 2 precedes 1).

Example A.8.8

If
$$A = \begin{bmatrix} a_{11} & a_{12} & a_{13} \\ a_{21} & a_{22} & a_{23} \\ a_{31} & a_{32} & a_{33} \end{bmatrix},$$

then
$$|A| = \begin{vmatrix} a_{11} & a_{12} & a_{13} \\ a_{21} & a_{22} & a_{23} \\ a_{31} & a_{32} & a_{33} \end{vmatrix} = \alpha_{123}a_{11}a_{22}a_{33} + \alpha_{132}a_{11}a_{23}a_{32} + \alpha_{213}a_{12}a_{21}a_{33}$$
$$+ \alpha_{231}a_{12}a_{23}a_{31} + \alpha_{312}a_{13}a_{21}a_{32} + \alpha_{321}a_{13}a_{22}a_{31}.$$

Note that

a. α_{123} is +1 because there are no inversions in the permutation 123

b. α_{132} is −1 because there is one inversion in the permutation: 3 precedes 2

c. α_{213} is −1 because there is one inversion in the permutation: 2 precedes 1

d. α_{231} is +1 because there are two inversions (even) in the permutation: 2 precedes 1 and 3 precedes 1

e. α_{312} is +1

f. α_{321} is −1

Therefore,
$$|A| = a_{11}a_{22}a_{33} - a_{11}a_{23}a_{32} - a_{12}a_{21}a_{33}, + a_{12}a_{23}a_{31} + a_{13}a_{21}a_{32} - a_{13}a_{22}a_{31}.$$

Example A.8.9

If
$$A = \begin{bmatrix} 3 & 6 & 0 \\ 2 & 4 & -4 \\ -1 & 2 & -3 \end{bmatrix},$$

then
$$|A| = -36 + 12 + 36 + 12 + 0 + 0 = 24.$$

Let A be a square matrix. By "the minor of the element a_{ij} of matrix A," we mean the new matrix obtained when the ith row and ith column are eliminated. It is denoted as m_{ij}. That is, the minor of a_{11} in the 3×3 matrix

$$\begin{bmatrix} a_{11} & a_{12} & a_{13} \\ a_{21} & a_{22} & a_{23} \\ a_{31} & a_{32} & a_{33} \end{bmatrix}$$

is given by

$$m_{11} = \begin{bmatrix} a_{22} & a_{23} \\ a_{32} & a_{33} \end{bmatrix}.$$

By "the cofactor of the element of the ith row and ith column," we mean the determinant of the minor, that is, $|m_{ij}|$ multiplied by $(-1)^{i+j}$. Therefore, the cofactor of a_{23} in the preceding 3×3 matrix is given by

$$(-1)^{2+3} \begin{vmatrix} a_{11} & a_{12} \\ a_{31} & a_{32} \end{vmatrix} = -(a_{11}a_{32} - a_{12}a_{31}).$$

Example A.8.10

If

$$A = \begin{bmatrix} 3 & -1 & 2 \\ 4 & 1 & 7 \\ 0 & 4 & -3 \end{bmatrix},$$

then the cofactor of 4 is given by

$$(-1)^{3+2} \begin{vmatrix} 3 & 2 \\ 4 & 7 \end{vmatrix} = -(21-8) = -13$$

or

$$(-1)^{2+1} \begin{vmatrix} -1 & 2 \\ 4 & -3 \end{vmatrix} = -(3-8) = 5.$$

The determinant of a square matrix can also be defined by the sum of the products formed by multiplying each element of any row or column by its cofactor. That is,

$$|A| = \sum_{j=1}^{n} a_{ij} \{\text{cof } a_{ij}\} = \sum_{i=1}^{n} a_{ij} \{\text{cof } a_{ij}\}.$$

Example A.8.11

If

$$A = \begin{bmatrix} a_{11} & a_{12} & a_{13} \\ a_{21} & a_{22} & a_{23} \\ a_{31} & a_{32} & a_{33} \end{bmatrix},$$

then, choosing the second column, we have

$$|A|=(-1)^{1+2}a_{12}\begin{vmatrix}a_{21} & a_{23}\\ a_{31} & a_{33}\end{vmatrix}+(-1)^{2+2}a_{22}\begin{vmatrix}a_{11} & a_{13}\\ a_{31} & a_{33}\end{vmatrix}+(-1)^{3+2}a_{32}\begin{vmatrix}a_{11} & a_{13}\\ a_{21} & a_{23}\end{vmatrix}$$

$$=-a_{12}(a_{21}a_{33}-a_{23}a_{31})+a_{22}(a_{11}a_{33}-a_{13}a_{31})-a_{32}(a_{11}a_{23}-a_{13}a_{21}).$$

The method of cofactors is perhaps the easiest method for obtaining the determinant of a square matrix.

Example A.8.12

If

$$A=\begin{bmatrix}3 & 2 & -3\\ 0 & 4 & -2\\ 1 & -1 & 0\end{bmatrix},$$

then,

$$|A|=(-1)^{1+1}3\begin{vmatrix}4 & -2\\ -1 & 0\end{vmatrix}+(-1)^{2+1}0\begin{vmatrix}2 & -3\\ -1 & 0\end{vmatrix}+(-1)^{3+1}1\begin{vmatrix}2 & -3\\ 4 & -2\end{vmatrix}$$

$$=3(-2)-0+1(8)=2.$$

We now list some important properties of determinants.

1. Interchanging two rows or columns of a square matrix yields a determinant that is the negative of the original determinant.

2. If two rows or columns of a square matrix are identical, then the determinant is equal to zero.

3. Interchanging the rows and columns of a square matrix does not change the value of the determinant.

4. If any row or column of a square matrix is multiplied by a nonzero constant, the resulting determinant is equal to the constant times the original determinant of the matrix.

5. A common factor of all elements of a row or column of a square matrix may be removed and placed as a multiplier of the resulting determinant.

6. If a multiple of the elements of any row or column of a square matrix is added to or subtracted from the elements of any other row or column, the value of the resulting determinant is the same as that of the original determinant.

7. If A and B are $n \times n$ matrices, then $|AB|=|A||B|$.

The transpose of matrix A, denoted by A^T, is obtained by changing the rows to columns and the columns to rows; that is, the ijth entry of A becomes the jith entry of A^T. A diagonal matrix is one whose only nonzero entries are on the main diagonal. If $n \times n$ matrix A is such that $|A| \neq 0$, then A is of rank n and A is said to be nonsingular.

The inverse of nonsingular square matrix A is denote by A^{-1}, and it is defined by

$$A^{-1} = \left[\frac{\text{cof } a_{ji}}{|A|}\right] = \left[\frac{\text{cof } a_{ij}}{|A|}\right]^T.$$

It can be easily seen that $AA^{-1} = A^{-1}A = I$ and $(A^T)^{-1} = (A^{-1})^T$.

Example A.8.13

Find the inverse of matrix A as defined in Example A.8.12.

Solution: We know that $|A| = 2$, and

$$A^{-1} = \left[\frac{\text{cof } a_{ij}}{|A|}\right]^T$$

$$= \begin{bmatrix} \frac{(-1)^{1+1}\begin{vmatrix}4 & -2\\ -1 & 0\end{vmatrix}}{2} & \frac{(-1)^{1+2}\begin{vmatrix}0 & -2\\ 1 & 0\end{vmatrix}}{2} & \frac{(-1)^{1+3}\begin{vmatrix}0 & 4\\ 1 & -1\end{vmatrix}}{2} \\ \frac{(-1)^{2+1}\begin{vmatrix}3 & -3\\ -1 & 0\end{vmatrix}}{2} & \frac{(-1)^{2+2}\begin{vmatrix}3 & -3\\ 1 & 0\end{vmatrix}}{2} & \frac{(-1)^{2+3}\begin{vmatrix}3 & 2\\ 1 & -1\end{vmatrix}}{2} \\ \frac{(-1)^{3+1}\begin{vmatrix}2 & -3\\ 4 & -2\end{vmatrix}}{2} & \frac{(-1)^{3+2}\begin{vmatrix}3 & -3\\ 0 & -2\end{vmatrix}}{2} & \frac{(-1)^{3+3}\begin{vmatrix}3 & 2\\ 0 & 4\end{vmatrix}}{2} \end{bmatrix}$$

$$= \begin{bmatrix} \frac{1}{2}(-2) & -\frac{1}{2}(2) & \frac{1}{2}(-4) \\ -\frac{1}{2}(-3) & \frac{1}{2}(3) & -\frac{1}{2}(-4) \\ \frac{1}{2}(8) & -\frac{1}{2}(-6) & \frac{1}{2}(12) \end{bmatrix}^T$$

$$= \begin{bmatrix} -1 & -1 & -2 \\ \frac{3}{2} & \frac{3}{2} & \frac{5}{2} \\ 4 & 3 & 6 \end{bmatrix}^T = \begin{bmatrix} -1 & \frac{3}{2} & -2 \\ -1 & \frac{3}{2} & 3 \\ -2 & \frac{5}{2} & 6 \end{bmatrix}.$$

Note that

$$AA^{-1} = \begin{bmatrix} 3 & 2 & -3 \\ 0 & 4 & -2 \\ 1 & -1 & 0 \end{bmatrix}_{(3\times 3)} \cdot \begin{bmatrix} -1 & \frac{3}{2} & 4 \\ -1 & \frac{3}{2} & 3 \\ -2 & \frac{5}{2} & 6 \end{bmatrix}_{(3\times 3)} = \begin{bmatrix} 1 & 0 & 0 \\ 0 & 1 & 0 \\ 0 & 0 & 1 \end{bmatrix}_{(3\times 3)} = I.$$

The basic concepts of matrix algebra constitute only a brief introduction to the subject. They will be a helpful review before your study.

Index

A posteriori probability, 4–5
A priori probability, 3, 25
Absolute probability, 24
Absolutely convergent, 280, 294, 295, 297
Absorbing Markov chain, 451–460
Absorbing state, 449–460
Aitchison, J., 124
Anderson, T. W., 336
Aperiodic state, 451
Arcsine distribution, 128
Aristotle, 1, 89, 141, 221
Associative postulate, 39, 558, 577
Auxiliary variable, 266, 352
Average, 105
Axiomatic definition of probability, 5–9

Banach's problem, 45
Basic concepts of mathematics. *See* Review of preliminary mathematics
Bayes, Thomas, 24
Bayes's theorem, 24–26
Bernoulli, Jakob, 47–48
Bernoulli distribution, 61
Bernoulli's law of large numbers, 372–375
Beta distribution, 112–114
Beta function, 570–572
Bézout, Etienne, 141
Binomial distribution, 61–65, 476–491
Binomial theorem, 563–564
Biographical sketches. *See* Eminent mathematicians
Biological population problem, 418
Bivariate cumulative distribution function, 229–237
Bivariate normal distribution, 302–314
 conditional densities, 306–307
 conditional expectation, 313
 defined, 302
 marginal densities, 304–306
 regression curves, 314
 special cases, 308
Bonaparte, Napoleon, 141, 142
Brown, J. A. C., 124
Buffon's needle problem, 133

Capsule biographies. *See* Eminent mathematicians
Cartesian product, 559, 560
Catenary, 48
Cauchy probability distribution, 115–117
Central limit theorem, 109, 386–389
Central moment of order $k_1 + k_2 + \ldots + k_n$, 355
Central moment of order $k + m$, 282
Chance variable, 50. *See also* Random variable
Characteristic equation, 429
Chebyshev, Pafnuty Lvovich, 370
Chebyshev's inequality, 370–372
Chevalier de Méré's problem, 45
Chi-square distribution, 164, 356
Cicero, 331
Classical probability, 3
Coefficient of correlation, 286–287, 292
Coefficient of skewness, 189, 210–211, 216–217, 218–219
Coefficient of variation, 188, 189, 209, 212–213, 214–215
Cofactor, 579
Column vector, 573
Combination, 561, 562
Combinatorial probability, 30–37
Commutative postulate, 39, 558, 577
Complement, 6, 556
Complement law, 558
Complementation postulate, 39
Compound event, 6
Compound probability, 16
Conan Doyle, Arthur, 331
Conditional cumulative distribution function, 246, 343
Conditional density function, 245
Conditional expectation, 293–302, 357–360
Conditional expected value, 293
Conditional expected value of the variate, 357
Conditional probability, 13–18
Conditional probability density function, 244, 342–345
Conformable matrices, 575
Constant probability, 413

Continuous cumulative probability distribution, 96–99
Continuous probability density function, 93
Continuous probability distributions
 arcsine distribution, 128
 beta distribution, 112–114
 bivariate normal distribution, 302–314
 Cauchy probability distribution, 115–117
 Dirichlet distribution, 352
 exponential distribution, 110
 extreme-value distribution, 127, 128
 gamma probability distribution, 109–110
 Gaussian distribution, 101–109
 Laplace distribution, 117–120
 lognormal probability distribution, 120–124
 Maxwell distribution, 127, 128
 multivariate normal distribution, 336
 normal distribution, 101–109, 526–528
 Pareto distribution, 128
 Rayleigh distribution, 126, 127
 rectangular distribution, 100, 114
 standard normal distribution, 103
 student-t distribution, 409
 tables, 211–220
 triangular distribution, 114
 uniform distribution, 100–101
 unit distribution, 103
 Weibull probability distribution, 124–125, 126
Continuous random variable
 distribution of continuous function, 148–159
 expectation, 179
 n-dimensional, 332
 one-dimensional, 92
 two-dimensional, 225
Continuous sample space, 6
Correction of continuity, 394, 395
Correlation coefficient, 286–287, 292
Countable set, 559
Countably infinite set, 559

583

Covariance, 283, 284, 286
Cumulative distribution function
  conditional, 246, 343
  marginal, 240
  multivariate, 337
  n-dimensional, 337
  one-dimensional
    continuous, 96–99
    discrete, 56–59
    properties, 58, 98, 230, 337
    random variable, 147
  two-dimensional
    continuous, 229–237
    discrete, 229–237

d'Alembert, Jean Le Rond, 141
de Méré, Chevalier, 2
De Moivre, Abraham, 101, 369
De Morgan's laws, 7, 558
Definite (Riemann) integral, 92
Degenerate distribution, 291
Degenerate random variable, 372
DeMoivre-Laplace theorem, 389–398
Density function, 51. *See also* Probability density function
Denumerable set, 559
Dependent event, 27
Dependent variable, 48
Descartes, Rene, 369
Determinant
  matrix, 577
  square matrix, 579
Diagonal elements, 573
Diagonal matrix, 580
Dirichlet distribution, 352
Discrete probability density function, 51–56
Discrete probability distributions, 47–88
  binomial distribution, 61–65, 476–491
  cumulative distribution function, 56–59
  discrete probability density function, 51–56
  geometric distribution, 77–78
  hypergeometric distribution, 73–77, 508–513
  multinomial distribution, 334–335
  negative binomial distribution, 79–81, 514–515
  point binomial distribution, 60–61
  Poisson distribution, 65–73, 492–507
  tables, 208–211
Discrete random variable, 50, 90
  distribution of continuous function, 143–148
  expectation, 168–169
  n-dimensional, 332

  one-dimensional, 51
  two-dimensional, 222
Discrete sample space, 6
Disjoint events, 7
Disquisitiones Arithmeticae (Gauss), 89
Distribution function, 100
Distribution law, 558
Distribution of continuous function of continuous random variable, 148–159
Distribution of continuous function of discrete random variable, 143–148
Distribution of discrete function of continuous variable, 161–162
Distribution of piecewise uniformly continuous function of continuous variate, 160–161
Distributive postulate, 39, 577
Doctrine of Chances, The (De Moivre), 369
Domain, 48
Doubly stochastic, 414

Eigenvalues, 429
Eigenvector, 429, 430
Einstein, Albert, 47
Element
  matrix, 573
  set, 554
Elementary event, 6
Eminent mathematicians
  Aristotle, 221
  Bernoulli, Jakob, 47–48
  De Moivre, Abraham, 369
  Gauss, Carl Friedrich, 89–90
  Kolmogorov, Andrey Nikolayevich, 167
  Laplace, Pierre Simon, 141–142
  Markov, Andrei Andreevich, 411
  Pascal, Blaise, 1–2
  Tukey, John Wilder, 331
Empty set, 6, 555
Enumerable set, 559
Equal events, 6
Equal matrices, 573
Equation of Chapman-Kolmogorov, 426
Ergodic, 449, 450
Essai Philosophique sur les Probabilitiés (Laplace), 142
Euclidean plane, 222, 225
Euclidean space, 332
Event
  compound, 6
  defined, 6
  dependent, 27
  disjoint, 7
  elementary, 6
  equal, 6
  impossible, 6
  independent, 27–30

  mutually exclusive, 7
  simple, 6
Existence postulate, 39
Expectation, 168–182
  binomial distribution, 169
  conditional, 293–302, 357–360
  continuous random variable, 171, 179
  discrete random variable, 168–169
  normal distribution, 171, 179–180
  Poisson distribution, 170
  product of n random variables, 354
  product of two random variables, 271
  product of variates, 273, 354
  properties (one-dimensional), 174–182
  properties (two-dimensional), 274–280
  Rayleigh distribution, 181
  synonyms, 171
  tables, 209, 212–213, 214–215
  uniform distribution, 170
Expected value, 168, 171. *See also* Expectation
Expected value of the product of variates, 273, 354
Expected value of the product of X and Y, 271
Experiment, 5
Exponential distribution, 110
Exponential functions, 516–525
Exponential probability density function, 110
Exposition du Systéme du Monde (Laplace), 142
Extreme-value distribution, 127, 128

Factorial, 561
Factorial moment-generating function, 198
Famous mathematicians. *See* Eminent mathematicians
Fermat, Pierre, 2
Fibonacci sequence of integers, 42
Finite Markov chains, 412. *See also* Markov chains
Finite set, 558
First-passage time (t_k), 464
First-product moment, 286
Formula for posteriori probability, 25
Frequency density function, 51
Frequency function. *See* Probability density function
Function
  beta, 570–572
  defined, 48, 49
  density, 51
  gamma, 567–570
  one-to-one, 143, 157

strictly monotone decreasing, 153
strictly monotone increasing, 148
Functions of a random variable, 141–166
Functions of random variables, 348–353
Functions of two random variables
 auxiliary variable, 266
 one function of two continuous variables, 266
 one function of two discrete variables, 256
 transformations of variables of discrete type, 264
 two functions of two continuous random variables, 258
Fundamental matrix, 452
Fundamental matrix for regular Markov chains, 464

Galton, Francis, 411
Gambler's ruin problem, 417
Gamma function, 567–570
Gamma probability density function, 109, 110
Gamma probability distribution, 109–110
Gauss, Carl Friedrich, 89–90
Gauss, Karl Friedrich, 2
Gaussian distribution, 101–109
Gaussian vs. Weibull probability density, 126
Geiger, Hans, 70
General law of compound probability, 16
General law of total probability, 13
Generalized binomial distribution, 64
Geometric distribution, 77–78
Geometric series, 162
Grundbegriffe der Wahrscheinlichketsrechnung (Kolmogorov), 167

Harris, Bernard, 372
Homogeneous Markov chains, 413
Hypergeometric distribution, 73–77, 508–513

Idempotent law, 558
Identity postulate, 39, 558
Image, 48
Impossible event, 6
Incomplete beta function, 114
Incomplete gamma function, 111
Independent event, 27–30
Independent random variable, 251–255
Independent variable, 48
Infinite set, 558
Initial probability density, 420

Initial probability vector, 420
Intersection, 7, 555
Isochron, 48

Jacobians, 565–566
Joint cumulative distribution function, 337
Joint probability, 19
Joint probability density function, 222–229, 333
Jordan canonical form, 437

Kemeny, J. G., 454
Kitagawa, Toshio, 62, 68, 75
Kolmogorov, Andrey Nikolayevich, 2, 5, 167
Kolmogorov criterion, 385
Kolmogorov's inequality, 372
kth central moment, 184
kth moment, 182
kth moment of $f(x)$, with respect to any point b, 183
kth ordinary moment, 182
Kurtosis, 190, 191

Lagrange, Joseph-Louis, 142
Laplace, Pierre Simon, 2, 101, 141–142
Laplace distribution, 117–120
Law of large numbers
 Bernoulli's law, 372–375
 SLLN, 384–385
 WLLN, 375–384
Leptokurtic, 190
Lieberman, Gerald, 62
Limit, 58
Limit theorems, 369–410
 central limit theorem, 386–389
 Chebyshev's inequality, 370–372
 DeMoivre-Laplace theorem, 389–398
 Kolmogorov criterion, 385
 Kolmogorov's inequality, 372
 normal approximation (binomial distribution), 396
 normal approximation (gamma distribution), 401–403
 normal approximation (Poisson distribution), 399–400
 strong law of large numbers (SLLN), 384–385
 weak law of large numbers (WLLN), 375–384
Lincoln, Abraham, 89
Location parameter, 102
Logarithmic spiral, 48
Lognormal probability distribution, 120–124
Lyapunov, Aleksandr Mikhailovich, 386

Maclaurin series expansion, 194
Malkiel, Burton G., 411
Mapping, 48
Marginal cumulative distribution function, 240
Marginal probability, 18–23
Marginal probability density function, 239, 339–342
Marginal probability distribution, 237–243
Markov, Andrei Andreevich, 411
Markov chains, 411–474
 absorbing, 451–460
 classification of states, 449–467
 classifications, 412
 defined, 412
 ergodic, 449, 450
 homogeneous, 413
 nonrecurrent, 450
 n-step transition probabilities, 424–428
 nontransient, 450
 p^n, 429–448
 regular, 460–467
 stationary transition probabilities, 413
 transient, 450
Markov processes. *See* Markov chains
Mass density function, 51
Mathematical expectation, 171. *See also* Expectation
Matrices, 572–581
 conformable, 575
 determinant, 577–580
 diagonal, 580
 doubly stochastic, 414
 elements, 573
 equal, 573
 first-passage, 464–467
 fundamental, 452
 inverse, 581
 Jordan canonical form, 437
 laws, 577
 n-step transition, 426
 nonsingular, 580
 normal, 452
 positive transition, 449, 460
 regular transition, 460
 steady-state, 463
 stochastic, 414
 sum/difference, 574
 terminology, 573
 transition, 414
 transition probability, 415
 transpose, 580
 unit, 573
 zero, 573
Matrix of the quadratic form, 337
Maxwell distribution, 127, 128
Maxwell-Boltzmann law, 142, 151

Maxwell's demon, 138
Maxwell's velocity distribution, 128
Mean, 105
Mean density function. *See* Probability density function
Mean first-passage matrix, 464–466
Mean value, 171. *See also* Expectation
Mécanique Céleste (Laplace), 142
Mesokurtic, 190
MGF. *See* Moment-generating function (MGF)
Modal value, 62
Mode, 62
Molina, Edward, 62, 68
Moment, 182–191
 central moment of order $k_1 + k_2 + ... + k_n$, 355
 central moment of order $k + m$, 282
 defined, 182
 gamma distribution, 183
 kth central, 184
 kth ordinary, 182
 kth ordinary with respect to a point, 183
 order $k_1 + k_2 + ... + k_n$, 355
 order $k + m$, 280
Moment of order $k_1 + k_2 + ... + k_n$, 355
Moment-generating function (MGF), 191–200
 binomial distribution, 192, 196
 defined, 191
 factorial MGF, 198
 gamma distribution, 193
 normal distribution, 193
 Poisson distribution, 192
 properties, 194–195
 tables, 210–211, 216–217, 218–219
 uniform distribution, 192
Multinomial distribution, 334–335
Multinomial theorem, 563, 564
Multivariate cumulative distribution function, 337
Multivariate normal distribution, 336
Multivariate probability density function, 333
Mutually exclusive events, 7
Mutually independent events, 28
Mutually independent random variable, 251

$n!$, 561
Natural (Naperian) antilogarithms, 516
n-dimensional continuous random variable, 332
n-dimensional discrete random variable, 332
n-dimensional Euclidean space, 332
Negative asymmetry, 189
Negative binomial distribution, 79–81, 514–515

Nonincreasing sequence, 58
Nonindependent event, 27
Nonnull Markov chain, 449
Nonperiodic Markov chain, 449
Nonrecurrent state, 450
Nonsingular matrix, 580
Nontransient state, 450
Normal approximation
 binomial distribution, 396
 gamma distribution, 401–403
 Poisson distribution, 399–400
Normal curve of error, 102
Normal distribution, 101–109, 526–528
Normal matrix, 452
Normal probability paper, 122
n-step transition matrix, 426
n-step transition probabilities, 426–428
Null set, 555

One-dimensional continuous random variable, 92
One-dimensional discrete random variable, 51
One-to-one correspondence, 558
One-to-one function, 143, 157
Owen, Donald, 62

Parameter, 60
Pareto distribution, 128
Parzen, E., 412
Pascal, Blaise, 1–2
Pascal, Etienne, 2
Path probabilities, 421–423
Peakedness of a distribution, 190
Pearson product-moment coefficient of correlation, 287
Perfect correlation, 292
Pericles, 221
Periodic state, 451
Permutation, 560–562
Persistent state, 451
Plato, 167
Playkurtic, 190
p^n, 429–448
Poincaré, H., 13
Point binomial, 60
Point binomial distribution, 60–61
Poisson, Siméon, 65
Poisson distribution, 65–73, 492–507
Positive (skewness) asymmetry, 189
Positive definite form, 337
Positive definite symmetrical matrix, 337
Positive transition matrix, 449, 460
Preliminary mathematics. *See* Review of preliminary mathematics
Probability
 a posteriori, 4–5
 a priori, 3
 absolute, 24

 axiomatic definition, 5–9
 classical, 3
 combinatorial, 30–37
 compound, 16
 conditional, 13–18
 constant, 413
 joint, 19
 marginal, 18–23
 measure of belief, as, 3
 n-step transition, 424–428
 path, 421–423
 relative frequency, 4–5
 stationary, 413
 transition, 413
Probability density, 37
Probability density function
 bivariate
  conditional, 244
  continuous, 225
  discrete, 222
  marginal, 239
 continuous, 93
 discrete, 51
 joint, 222–229
 multivariate, 333
 n-dimensional
  conditional, 342–345
  marginal, 339–342
 symmetric, 189
Probability distribution vector, 422
Probability distributions, 100. *See also* Continuous probability distributions; Discrete probability distributions
Probability histogram, 56
Probability mass function. *See* Probability density function
Probability sample space, 19
Probability space, 8
Probability tables. *See* Tables
Probability vector, 414, 422
Product (set), 559, 560
Product notations, 553–554
Provinciales and Pensées (Pascal), 2
Pushkin, Aleksandr Sergeyevich, 167
Pythagorean formula, 126

Quadratic form, 337

Random process, 411
Random sample, 30
Random variable, 37
 classes, 50
 continuous. *See* Continuous random variable
 defined, 48, 49, 142
 degenerate, 372
 denotation of, 50
 discrete. *See* Discrete random variable
 independent, 251–255

Index

standardized, 188
synonyms, 50
Random-walk problem, 416
Range, 48
Rayleigh distribution, 126, 127
Rectangular distribution, 100, 114
Recurrent state, 451
Refraction seismology, 136
Regression curve, 298–302
Regular Markov chain, 460–467
Regular transition matrix, 460
Relative-frequency probability, 4–5
Review of preliminary mathematics, 553–581
    beta function, 570–572
    binomial and multinational theorems, 563–564
    gamma function, 567–570
    Jacobians, 565–566
    matrices, 572–581. *See also* Matrices
    permutations and combinations, 560–562
    set theory, 554–560
    Stirling's formula, 562
    sum and product notations, 553–554
Riemann integral, 92
Row vector, 573
Rutherford, Ernest, 70
Rutherford and Geiger experiment, 70

Sample mean, 357
Sample point, 5
Sample space, 5
    continuous, 6
    discrete, 6
Sample variance, 357
Sampling
    random, 30
    schemes, 30–32
Sampling with replacement, 30–32
Sampling without replacement, 30–32
Scalar law, 577
Sequence of independent random variables, 345–347
Set
    complement, 6, 556
    countably infinite, 559
    defined, 554
    elements, 554
    empty, 6, 555
    equal, 6, 555
    finite/infinite, 558, 559
    intersection, 7, 555
    product (Cartesian product), 559, 560
    properties, 557, 558
    union, 7, 555
Set theory, 2, 3, 554–560
Shape parameter, 102
Simple event, 6

Skewness, 189, 210–211, 216–217, 218–219
SLLN. *See* Strong law of large numbers (SLLN)
Snell, J. L., 454
Square matrix, 573
Standard deviation, 105, 186
Standard normal distribution, 103
Standardized random variable, 188
State
    absorbing, 449–460
    aperiodic, 451
    defined, 412
    nonrecurrent, 451
    nontransient, 450
    periodic, 451
    persistent, 451
    recurrent, 451
    transient, 450, 451
State diagram, 414, 415
State space, 412
Stationary probability, 413
Statistic, 60
Statistical tables. *See* Tables
Steady-state matrix, 463
Stirling's approximation, 390
Stirling's formula, 77, 399, 562
Stochastic matrix, 414
Stochastic process, 411, 412
Stochastic variable, 50. *See also* Random variable
Strictly monotone decreasing function, 153
Strictly monotone increasing function, 148
Strong law of large numbers (SLLN), 384–385
Student-t distribution, 409
Subset, 6, 554, 555
Sum and product notations, 553–554
Summand, 553
Summation index, 553
Symmetric probability density function, 189

Tables, 208–219, 475–538
    binomial distribution, 476–491
    coefficient of skewness, 210–211, 216–217, 218–219
    coefficient of variation, 209, 212–213, 214–215
    continuous distributions, 211–219
    discrete distributions, 208–211
    expected value, 209, 212–213, 214–215
    exponential functions, 516–525
    hypergeometric distribution, 508–513
    moment-generating function, 210–211, 216–217, 218–219

    negative binomial distribution, 514–515
    normal distribution, 526–538
    Poisson distribution, 492–507
    variance, 209, 212–213, 214–215
Terminology, 100
Theorem of absolute probability, 24
Théorie Analytique des Probabilitiés (Laplace), 142
Time-term method of refraction seismology, 136
Traite Du Triangle Arithmetique (Pascal), 1
Transient state, 450, 451
Transition diagram, 414
Transition matrix, 414, 422
Transition probability, 413
    n-step, 424–428
Transition probability matrix, 415
Transpose (matrix), 580
Tree diagram, 420, 421
Triangular distribution, 114
Tsokos, Chris, 2
Tukey, John Wilder, 331
Two-dimensional continuous random variable, 225
Two-dimensional cumulative distribution, 229–237
Two-dimensional discrete random variable, 222
Two-dimensional random variable, 222

Uniform distribution, 100–101
Union, 7, 555
Unit distribution, 103
Unit matrix, 573
Universal set, 6

Variance, 105, 186, 187
    sample, 357
    tables, 209, 212–213, 214–215
Variance first-passage matrix, 466–467
Variate, 50. *See also* Random variable
von Leibniz, Gottfried Wilhelm, 48
von Mises, R., 4

Weak law of large numbers (WLLN), 375–384
Weber, Wilhelm, 89
Weibull, Waloddi, 124
Weibull probability distribution, 124–125, 126
WLLN. *See* Weak law of large numbers (WLLN)

Zeicon, 86
Zero matrix, 573